Advances in Science, Technology & Innovation

IEREK Interdisciplinary Series for Sustainable Development

Advances in Science, Technology & Innovation (ASTI) is a series of peer-reviewed books based on important emerging research that redefines the current disciplinary boundaries in science, technology and innovation (STI) in order to develop integrated concepts for sustainable development. It not only discusses the progress made towards securing more resources, allocating smarter solutions, and rebalancing the relationship between nature and people, but also provides in-depth insights from comprehensive research that addresses the **17 sustainable development goals (SDGs)** as set out by the UN for 2030.

The series draws on the best research papers from various IEREK and other international conferences to promote the creation and development of viable solutions for a **sustainable future and a positive societal** transformation with the help of integrated and innovative science-based approaches. Including interdisciplinary contributions, it presents innovative approaches and highlights how they can best support both economic and sustainable development, through better use of data, more effective institutions, and global, local and individual action, for the welfare of all societies.

The series particularly features conceptual and empirical contributions from various interrelated fields of science, technology and innovation, with an emphasis on digital transformation, that focus on providing practical solutions to **ensure food, water and energy security to achieve the SDGs.** It also presents new case studies offering concrete examples of how to resolve sustainable urbanization and environmental issues in different regions of the world.

The series is intended for professionals in research and teaching, consultancies and industry, and government and international organizations. Published in collaboration with IEREK, the Springer ASTI series will acquaint readers with essential new studies in STI for sustainable development.

ASTI series has now been accepted for Scopus (September 2020). All content published in this series will start appearing on the Scopus site in early 2021.

Haroun Chenchouni · Zhihua Zhang ·
Deepak Singh Bisht · Matteo Gentilucci ·
Mingjie Chen · Helder I. Chaminé ·
Maurizio Barbieri · Mahesh Kumar Jat ·
Jesús Rodrigo-Comino · Dionysia Panagoulia ·
Amjad Kallel · Arkoprovo Biswas · Veysel Turan ·
Jasper Knight · Attila Çiner · Carla Candeias ·
Zeynal Abiddin Ergüler
Editors

Recent Advancements from Aquifers to Skies in Hydrogeology, Geoecology, and Atmospheric Sciences

Proceedings of the 2nd MedGU, Marrakesh 2022 (Volume 1)

Springer

Editors
Haroun Chenchouni
Higher National School of Forests
Khenchela, Algeria

Deepak Singh Bisht
Western Himalayan Regional Centre
National Institute of Hydrology
Jammu, India

Mingjie Chen
Sultan Qaboos University
Muscat, Oman

Maurizio Barbieri
Department of Chemical Engineering Materials
Environment
Sapienza University of Rome
Rome, Italy

Jesús Rodrigo-Comino
University of Granada
Granada, Spain

Amjad Kallel
ENIS
University of Sfax
Sfax, Tunisia

Veysel Turan
Bingol University
Bingöl, Türkiye

Attila Çiner
Istanbul Technical University
Istanbul, Türkiye

Zeynal Abiddin Ergüler
Kütahya Dumlupınar University
Kütahya, Türkiye

Zhihua Zhang
Shandong University
Jinan, China

Matteo Gentilucci
Geology Division, School of Science
and Technology
University of Camerino
Camerino, Italy

Helder I. Chaminé
School of Engineering (ISEP)
Polytechnic of Porto
Porto, Portugal

Mahesh Kumar Jat
Malaviya National Institute of Technology Jaipur
Jaipur, Rajasthan, India

Dionysia Panagoulia
National Technical University of Athens
Athens, Greece

Arkoprovo Biswas
Department of Geology, Institute of Science
Banaras Hindu University
Varanasi, Uttar Pradesh, India

Jasper Knight
University of the Witwatersrand
Johannesburg, South Africa

Carla Candeias
GeoBioTec
University of Aveiro
Aveiro, Portugal

ISSN 2522-8714 ISSN 2522-8722 (electronic)
Advances in Science, Technology & Innovation
ISBN 978-3-031-47078-3 ISBN 978-3-031-47079-0 (eBook)
https://doi.org/10.1007/978-3-031-47079-0

This Springer imprint is published by the registered company Springer Nature Switzerland AG
The registered company address is: Gewerbestrasse 11, 6330 Cham, Switzerland

About the Conference

About MedGU

Steps toward the creation of a Mediterranean Geosciences Union (MedGU)

Mediterranean Geosciences Union (MedGU) aims to create a unique federation that brings together and represents the Mediterranean geoscience community specializing in the areas of Earth, planetary, and space sciences.

MedGU will be structured along the lines of American Geophysical Union (AGU) and European Geosciences Union (EGU).

The plan is to establish a large organization for the Mediterranean region that is more influential than any one local geoscience society with the objective of fostering fundamental geoscience research, as well as applied research that addresses key societal and environmental challenges.

MedGU's overarching vision is to contribute to the realization of a sustainable future for humanity and for the planet.

The creation of this union will give the Earth sciences more influence in policy-making and in the implementation of solutions to preserve the natural environment and to create more sustainable societies for the people living in the Mediterranean region. It is hoped that the union will also provide opportunities to Mediterranean geoscientists to undertake interdisciplinary collaborative research. MedGU plans to recognize the work of the most active geoscientists with a number of awards and medals.

Although MedGU has not yet been officially inaugurated, its first annual meeting was organized in November 2021 in Istanbul (MedGU-21). This has provided a forum to achieve a consensus for the formation of this non-profit international union of geoscientists. Membership will be open to individuals who have a professional engagement with the Earth, planetary, and space sciences and related studies, including students and retired seniors.

Nabil Khélifi (MedGU Founder, Germany) and Attila Çiner (MedGU Interim President, Turkey) in collaboration with Abdelaziz Mridekh (MedGU-22 Local Chair, Morocco) have taken an ambitious approach to the launch of the second MedGU Annual Meeting 2022 and hope to develop it in the near future into the largest international geoscience event in the Mediterranean and the broader MENA region. Its mission is to support geoscientists based in this region by establishing a Global Geoscience Congress.

It is expected that hundreds of participants from all over the world will attend this second MedGU Annual Meeting 2022, making it one of the largest and most prominent geosciences events in the region. So far, over 1300 abstracts have been submitted from 95 countries. The meeting's sessions will cover a wide range of topics with more details available on the conference tracks.

This second 2022 Annual Meeting will have a "hybrid" format, with both in-person and virtual participation. Springer, its official partner, will publish the proceedings in a book series (indexed in Scopus) as well as a number of special issues in diverse scientific journals (for more details, see Publications). The official journal of MedGU is Mediterranean Geoscience Reviews (Springer).

Conference Tracks

The scientific committee of the MedGU invites research papers on all cross-cutting themes of Earth sciences, with a main focus on the following 18 conference tracks:

- Track 1. Atmospheric Sciences, Meteorology, Climatology, Oceanography
- Track 2. Biogeochemistry, Geobiology, Geoecology, Geoagronomy
- Track 3. Earthquake Seismology and Geodesy
- Track 4. Environmental Earth Sciences
- Track 5. Applied and Theoretical Geophysics
- Track 6. Geo-Informatics and Remote Sensing
- Track 7. Geochemistry, Mineralogy, Petrology, Volcanology
- Track 8. Geological Engineering, Geotechnical Engineering
- Track 9. Geomorphology, Geography, Soil Science, Glaciology, Geoarchaeology, Geoheritage
- Track 10. Hydrology, Hydrogeology, Hydrochemistry
- Track 11. Marine Geosciences, Historical Geology, Paleoceanography, Paleoclimatology
- Track 12. Numerical and Analytical Methods in Mining Sciences and Geomechanics
- Track 13. Petroleum and Energy Sciences and Engineering
- Track 14. Sedimentology, Stratigraphy, Paleontology, Geochronology
- Track 15. Structural Geology, Tectonics and Geodynamics, Petroleum Geology
- Track 16. Special Session on Astrogeology, Impact Craters and Meteorites
- Track 17. Special Session on climate and sea-level change during the Cenomanian-Turonian Anoxic Event: Synthesis of sedimentological, micropaleontological and geochemical records
- Track 18. Special Session on hydrogeological and climatic risks, their management and the effect of climate change on groundwater quality

About the Conference Steering Committee

General Committee

Honorary Chair

A. M. Celâl Sengör
Associate Editor, *Mediterranean Geosciences Reviews* (Springer)
Eurasia Institute of Earth Sciences,
Istanbul Technical University Istanbul, Turkey

Conference Supervisor

Nabil Khélifi
Senior Publishing Editor
MedGU-21 Supervisor

Program Chair

Mustapha Meghraoui
Editorial Board Member, *Mediterranean Geosciences Reviews* (Springer)
Editor of *Arabian Journal of Geosciences* (Springer)
IPG Strasbourg, France

Publications Chair

Attila Çiner
MedGU (Interim) President
Founding Editor-in-Chief, *Mediterranean Geosciences Reviews* (Springer)
Chief Editor—Tracks 11 and 14, *Arabian Journal of Geosciences* (Springer)
Eurasia Institute of Earth Sciences, Istanbul Technical University, Turkey

Conference Manager

Mohamed Sahbi Moalla
Performer—The Leading Conference Organiser, Tunisia
Journal Coordinator, *Euro-Mediterranean Journal for Environmental Integration* (Springer)
ISET, University of Sfax, Tunisia

Conference Support

Mourad Amer
Founder and CEO of IEREK
Editor of ASTI Series (Springer/IEREK)
IEREK, Alexandria, Egypt

Local Committee

Chair

Abdelaziz Mridekh
Associate Editor, *Arabian Journal of Geosciences* (Springer)
University Ibn Tofail, Kenitra, Morocco

Coordinators

El Hassane Chellai
University Cadi Ayyad, Marrakech, Morocco

Bouabid El Mansouri
University Ibn Tofail, Kenitra, Morocco

Noureddine Laftouhi
University Cadi Ayyad, Marrakech, Morocco

Azzouz Kchikach
University Cadi Ayyad, Marrakech, Morocco

Jean-Louis Bodinier
Program Lead Geology and Sustainable Mining
Mohammed VI Polytechnic University

Members

Abdallah El Hmaidi, Moulay Ismail University, Meknes, Morocco
Abdelhadi El Ouali, Moulay Ismail University, Meknes, Morocco
Aicha Benmohammadi, Ibn Tofail University, Kenitra, Morocco
Ali Essahlaoui, Moulay Ismail University, Meknes, Morocco
Annis Moumen, National Applied School, Kenitra, Morocco
Atika Fahmi, Ibn Tofail University, Kenitra, Morocco
Driss El Azzab, Sidi Mohamed Ben Abdellah University, Fes, Morocco
Hassan Echarfaoui, Ibn Tofail University, Kenitra, Morocco
Jamal Chao, Ibn Tofail University, Kenitra, Morocco
Jaouad Dabounou, Hassan 1 University, Settat, Morocco
Malika Kili, Ibn Tofail University, Kenitra, Morocco
Mohamed Bouhaddioui, National School of Mine, Rabat, Morocco
Mohamed Jaffal, Cadi Ayyad University, Marrakech, Morocco
Mohammed Ouhcine, Ibn Tofail University, Kenitra, Morocco
Moulay Hachem Awragh, Moulay Ismail University, Meknes, Morocco
Mustapha Boualoul, Moulay Ismail University, Meknes, Morocco
Mustapha Boujemaoui, Moulay Ismail University, Meknes, Morocco
Nabil El Mouçaid, Central School of Engineering, Casablanca, Morocco
Souad Haida, Ibn Tofail University, Kenitra, Morocco
Abdullah Sukkar, Department of Geomatics Engineering, Istanbul Technical University, Istanbul, Turkey
Melek Rebai, Performer—The Leading Conference Organiser, Tunisia
M. Bassem Abdelhedi, Performer—The Leading Conference Organiser, Tunisia
Oumayma Abidi, Performer—The Leading Conference Organiser, Tunisia
Toka M. Amer, IEREK—International Experts for Research Enrichment and Knowledge Exchange, Egypt

Advisory Committee

Hans Thybo
President of International Lithosphere Program (ILP)
Editor-in-Chief of *Earth and Planetary Science Letters* (EPSL)
Professor at:
• Eurasia Institute of Earth Sciences, Istanbul Technical University, Turkey
• Center for Earth Evolution and Dynamics, University of Oslo, Norway

A. M. Celâl Sengör
Associate Editor, *Mediterranean Geosciences Reviews* (Springer)
Eurasia Institute of Earth Sciences,
Istanbul Technical University Istanbul, Turkey

François Roure
Chief Editor—Track 15
Arabian Journal of Geosciences (Springer)
IFP—Energies Nouvelles, France

Giovanni Bertotti
Associate Editor, *Mediterranean Geosciences Reviews* (Springer)
Geoscience and Engineering, Delft University of Technology, The Netherlands

Abdullah Al-Amri
Founder and Editor-in-Chief
Arabian Journal of Geosciences (Springer)
King Saud University, Saudi Arabia

Akiça Bahri
Director for Africa at the International Water Management Institute (IWMI), Ghana (2005–2010)
Coordinator of the African Water Facility (AWF)
at the African Development Bank (2010–2015)
Director of Research at the National Research Institute for Agricultural Engineering, Water, and Forestry (INRGREF), Tunisia (since 2016)
Professor at the National Agricultural Institute of Tunisia (INAT), Tunisia (since 2017)
Awardee of the International Water Association (IWA) Women in Water Prize (2018)
Associate Editor, *Euro-Mediterranean Journal for Environmental Integration* (Springer) (since 2019)
Minister of Agriculture, Water Resources and Fisheries in Tunisia (2019–2020)

Program Committee

Chair

Mustapha Meghraoui
Editorial Board Member, *Mediterranean Geosciences Reviews* (Springer)
Editor of *Arabian Journal of Geosciences* (Springer)
IPG Strasbourg, France

Members

Attila Çiner
MedGU (Interim) President
Founding Editor-in-Chief, *Mediterranean Geosciences Reviews* (Springer)
Chief Editor—Tracks 11 and 14, *Arabian Journal of Geosciences* (Springer)
Eurasia Institute of Earth Sciences, Istanbul Technical University, Turkey

Luc Bulot
North Africa Research Group
The University of Manchester, UK

Sami Khomsi
Georesources Lab, CERTE, University of Carthage, Tunis, Tunisia

Hasnaa Chennaoui Aoudjehane
Meteoritical Society Fellow
Laureate, "Prix Paul Doistau–Émile Blutet" from the French Academy of Sciences
Editor of *Arabian Journal of Geosciences* (Springer)
Professor, Hassan II University of Casablanca, Morocco

Catherine Kuzucuoglu
Associate Editor, *Mediterranean Geosciences Reviews* (Springer)
Research Director Emeritus
CNRS, Laboratoire de Géographie Physique
UMR 8591, Meudon, France

Elena Xoplaki
Chief Editor, *Euro-Mediterranean Journal for Environmental Integration* (Springer)
Justus-Liebig-University Giessen, Germany

Publications Committee

Chair

Attila Çiner
MedGU (Interim) President
Founding Editor-in-Chief, *Mediterranean Geosciences Reviews* (Springer)
Chief Editor—Tracks 11 and 14, *Arabian Journal of Geosciences* (Springer)
Eurasia Institute of Earth Sciences, Istanbul Technical University, Turkey

Members

Zeynal Abiddin Erguler
Chief Editor—Track 8
Arabian Journal of Geosciences (Springer)
Dumlupinar University, Kutahya, Turkey

Amjad Kallel
Chief Editor—Track 4
Arabian Journal of Geosciences (Springer)
Managing and Development Editor, *Euro-Mediterranean Journal for Environmental Integration* (Springer)
ENIS, University of Sfax, Tunisia

Mourad Bezzeghoud
School of Sciences and Technology (ECT)
Insititut of Earth Sciences (IIFA)
University of Évora, Portugal

Hesham El-Askary
Professor of Remote Sensing and Earth Systems Science
Editor of *Arabian Journal of Geosciences* (Springer)
Director Computational and Data Sciences
Graduate Programs
Center of Excellence in Earth Systems Modeling
and Observations
Schmid College of Science and Technology,
Chapman University, USA

Zakaria Hamimi
President of ArabGU
IAGETH VP for Africa and IAGETH National Chapter
for Egypt
Editor of *Arabian Journal of Geosciences* (Springer)
Professor, Benha University, Benha, Egypt

Syed E. Hasan, Ph.D.; RG; FGS
Chair, Environmental Characterization and Remediation
Technical Working Group, AEG
Department of Earth and Environmental Sciences
University of Missouri-Kansas City, USA

François Roure
Chief Editor—Track 15
Arabian Journal of Geosciences (Springer)
IFP—Energies Nouvelles, France

Anastasia Kiratzi
Professor of Seismology
Faculty of Sciences, Aristotle University of Thessaloniki, Greece

Broder Merkel
Chief Editor—Track 10
Arabian Journal of Geosciences (Springer)
Associate Editor of *Environmental Earth Science* (Springer)
Publisher of Freiberg Online Geoscience (FOG)
Institute of Geology, Technische Universität Bergakademie Freiberg, Germany

Preface

The proceedings of the 2nd Mediterranean Geosciences Union (MedGU-2) held in Marrakesh onsite and online in November 2022 consisted of four volumes titled:

Proceedings Volume 1: Recent Advancements from Aquifers to Skies in Hydrogeology, Geoecology, and Atmospheric Sciences

Proceedings Volume 2: Recent Research on Sedimentology, Stratigraphy, Paleontology, Geochemistry, Volcanology, Tectonics, and Petroleum Geology

Proceedings Volume 3: Recent Research on Geotechnical Engineering, Remote Sensing, Geophysics, and Earthquake Seismology

Proceedings Volume 4: Recent Research on Environmental Earth Sciences, Geomorphology, Soil Science, and Paleoenvironments

These volumes are based on the accepted conference papers for either oral/poster presentations or selected for online publication during the MedGU-2.

This first volume contains 68 papers that discuss the latest advances in hydrogeosciences from diverse backgrounds, including global change, geoecology, biogeochemistry, water resources management, and environmental monitoring and assessment. It shares insights of experienced scientists from, but not limited to, research institutes worldwide, focused on the Mediterranean and Middle East regions on how the understanding of ecological, climatological, oceanic, and hydrological processes is the key to improving practices in environmental management, including the eco-responsibility, scientific integrity, social, and ethical dimensions.

The proceedings of the MedGU-2 are of interest to all researchers, experts, and students in all fields of geosciences.

Khenchela, Algeria — Haroun Chenchouni
Jinan, China — Zhihua Zhang
Jammu, India — Deepak Singh Bisht
Camerino, Italy — Matteo Gentilucci
Muscat, Oman — Mingjie Chen
Porto, Portugal — Helder I. Chaminé
Rome, Italy — Maurizio Barbieri
Jaipur, India — Mahesh Kumar Jat
Granada, Spain — Jesús Rodrigo-Comino
Athens, Greece — Dionysia Panagoulia
Sfax, Tunisia — Amjad Kallel
Varanasi, India — Arkoprovo Biswas
Istanbul, Türkiye — Attila Çiner
Bingöl, Türkiye — Veysel Turan
Johannesburg, South Africa — Jasper Knight
Aveiro, Portugal — Carla Candeias
Kütahya, Türkiye — Zeynal Abiddin Ergüler
June 2023

Contents

Biogeochemistry, Geobiology and Geoecology

Atmospheric and Oceanographic Sciences

About the Editors

Haroun Chenchouni Higher National School of Forests, Khenchela, Algeria
Dr. Haroun Chenchouni is an associate professor and research scientist (Ecologist) at the Higher National School of Forests (Khenchela, Algeria). He is a former associate professor at the University of Tebessa (Algeria). He holds a doctorate degree in Ecology and Environment from the University of Batna 2 (Algeria) and a M.Sc. (Magister) in Dryland Ecology from the University of Ouargla (Algeria). He graduated as an engineer in Plant Ecology and Forest Ecosystems from the Department of Biological Sciences (University of Batna, Algeria). His research interests are fairly broad; he uses statistical modeling approaches to understand how natural environments, mainly climatic and edaphic factors, and anthropogenic perturbations influence biological interactions, shape trends in population dynamics, and influence community diversity. He uses various biological models to investigate biological interactions and community ecology of arid and semiarid ecosystems of North Africa. At various universities in Algeria, he teaches forest ecology, biostatistics, and ecological modeling. He has published more than 100 peer-reviewed publications and internationally recognized research papers. He is also involved in national and international research projects. In 2017, he joined the *Arabian Journal of Geosciences* (AJGS) as an associate editor. Then in 2019, he was assigned as a chief editor of Topic 2 (biogeochemistry, geobiology, geoecology, and geoagronomy) to handle submissions dealing with various fields of biogeosciences, geoecology, climate change, plant and soil science, agricultural and forest environment, and environmental sciences.

Zhihua Zhang Shandong University, China, China
Zhihua Zhang is Taishan Distinguished Professor at Shandong University (China) and is leading an interdisciplinary big data mining research group. His long-standing researches focus on Big Earth Data, climate change mechanisms, ocean dynamics, environmental evolution, and sustainability. Recently, Prof Zhang's researches have highlighted many times by New Scientist (UK), China Science Daily, and China Social Science Daily, and his four monographs published in Elsevier and Springer have been used widely in 60 countries around the world. Currently, Prof. Zhang is serving as Editor-in-Chief of *International Journal of Big Data Mining for Global Warming*

(World Scientific); Chief Editor of *Arabian Journal of Geosciences* (Springer); Associate Editor of *Environment, Development and Sustainability* (Springer), *EURASIP Journal on Advances in Signal Processing* (Springer), and *International Journal of Climate Change Strategies and Management* (Emerald); Topical Editor of *Big Earth Data* (Taylor); and Editorial Board Member of *Earth Science Informatics* (Springer), *PLoS ONE*, *Open Geosciences* (DeGruyter), and *International Journal of Global Warming* (Indersci).

Deepak Singh Bisht National Institute of Hydrology, Roorkee, India
Dr. Deepak Singh Bisht is working as Scientist 'B' in the National Institute of Hydrology, an autonomous society under the Ministry of Jal Shakti, River Development and Ganga Rejuvenation, Government of India. He did his Ph.D. and Masters from Indian Institute of Technology Kharagpur in the domain of hydrology and climate change. His research interests primarily include, but not limited to, hydrological modeling, urban drainage, flood modeling, climate change impact analysis, remote sensing applications, and springshed management.

Matteo Gentilucci University of Camerino, School of Science and Technology, Geology division, Camerino, Italy
Matteo Gentilucci had his B.Sc. degree in "Geological Sciences" from the University of Camerino, Italy, in 2009 and then earned M.Sc. degree in 2012 from the University of Camerino, Italy, in "Geoenvironmental Resources and Risks." He earned a first-level master's degree in GIS for the governance of the territory in 2012 and a Ph.D. in "Sciences and Technology: Earth Science" from the University of Camerino in 2017. From February 2017 to October 2018, he was a researcher at the Experimental Geophysical Observatory of Macerata, and in 2018, he became an honorary fellow in Advanced GIS. In 2018 and 2019, he was a contract professor at the University of Camerino. Since 2022, he has been a researcher at the University of Camerino and holds courses in meteorology, climatology, and GIS. He is a member of editorial board of the *Journal of Experimental Sciences, of Journal Sustainability*, and of the *Arabian Journal of Geosciences* and of the International Research Conference Committee. He is the author of numerous articles in international journals. Dr. Matteo Gentilucci is involved in numerous research projects related to climate. The main topics of interest and expertise are: geographic information systems applied to the environment, relationship between climate and phenology, climate change, global warming, and extreme events. https://www.scopus.com/authid/detail.uri?authorId=57201441740

Mingjie Chen Sultan Qaboos University, Muscat, Oman
Dr. Mingjie Chen holds a BE in Environmental Engineering (1997) from Tsinghua University (China), a M.Sc. in Environmental Sciences (2000) from Peking University (China), and a Ph.D. degree in Environmental Sciences (2005) from the University of California, Santa Barbara (USA). After over 10 years of research experiences in prestigious institutions (Los Alamos National Laboratory, Tufts University, and Lawrence Livermore National Laboratory) in the USA, Dr. Chen joined Water Research Center, Sultan Qaboos University (Oman), in 2014 as a senior hydrogeologist. His research focuses on using field data, laboratory experiment, and numerical models to study fluid flow and contaminant transport in subsurface area. He has conducted more than 20 research projects and published over 50 peer-reviewed journal papers on underground environment remediation, hydrocarbon reservoirs, CO_2 utilization and sequestration, geothermal reservoir, and groundwater modeling and management. Dr. Chen served or is serving as the associate editor for *Hydrogeology Journal*, *Journal of Hydrology*, and *Arabian Journal of Geosciences*.

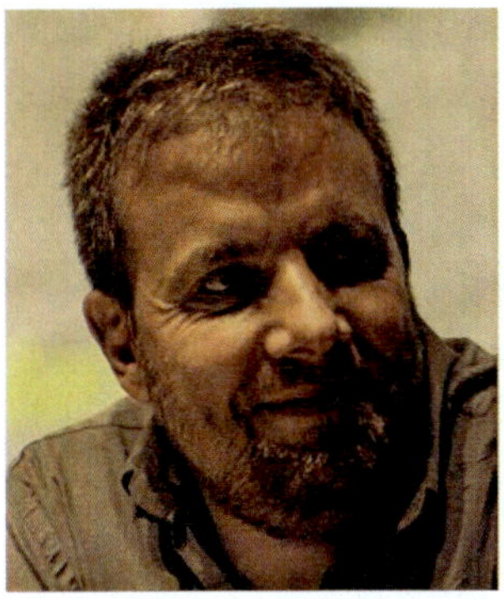

Helder I. Chaminé School of Engineering of Porto (ISEP), Portugal
Helder I. Chaminé is a skilled geologist (B.Sc., Ph.D., D.Sc.) and a professor of engineering geosciences at the School of Engineering (ISEP) of the Polytechnic of Porto, Portugal, with over 33 years of experience in multidisciplinary geosciences research and practice. Before joining the academy in 2001, he worked for over 13 years on international and national projects for mining hydrogeomechanics and geology, structural geology mapping, applied mineralogy, rock engineering, exploration geology, and groundwater. His major research interests are GIS mapping techniques for applied geology, geotechnics and natural hazards, engineering geosciences and hydrogeomechanics, hard-rock hydrogeology, urban groundwater, water resources, and thermal waters management. He has interests in geomining heritage, geoethics, history of cartography, military geosciences, higher education dissemination, and geoprofessional core values. Presently, he is Head of the Laboratory of Cartography and Applied Geology (LABCARGA|ISEP), Senior Researcher at Centre GeoBioTec|U.Aveiro and Centre IDL|U.Lisbon, as well as belongs to the executive board of the M.Sc. Geotechnical and Geoenvironmental Engineering program (OE + EUR-ACE Label) and the Department of Geotechnical Engineering (ISEP). Furthermore, he was a consultant and responsible for over 70 projects of applied geology, hydrogeomechanics, slope geotechnics, mining geology, exploration hydrogeology, hard-rock hydrogeology, water resources, urban groundwater, and applied mapping (Mozambique, Portugal, and Spain). He has co-authored over 220 publications in indexed journals, conference proceedings/full papers, book chapters, and technical and professional papers. He co-edited over 16 special volumes (journals and Springer Series Books) and is presently editing topical collections for several international journals. He serves as Associate Editor

(*SN Applied Sciences*, *Discover Water*, *Arabian Journal of Geosciences*, *Euro-Mediterranean Journal for Environmental Integration*) and on the editorial or advisory boards (*Mediterranean Geoscience Reviews*, *Geotechnical Research ICE*, *Journal of Geoethics and Social Geosciences*, *Cadernos do Laboratório Xeolóxico de Laxe*, *Revista Geotecnia SPG*, *Geología Aplicada a la Ingeniería y al Ambiente ASAGAI*). He also served as Associate Editor in the *Hydrogeology Journal*. In addition, he has a broad activity as a reviewer for several international journals. In 2021, the Journal *SN Applied Sciences* awarded him an outstanding guest editor and editorial board member. In addition, he integrates as a moderator or session chair in several conferences, workshops, and meetings. He is the vice-president of the Portuguese Chapter of the International Association of Hydrogeologists (IAH) and deputy secretary of the Technical Commission of Environmental Geotechnics from the Portuguese Geotechnical Society (SPG). Furthermore, he served as invited Expert Evaluator of the Bologna Geoscience program for DGES (Portugal) and Scientific Projects Evaluation for NCST, 2017–2019 (Kazakhstan) and NRF|RISA, 2019 (South Africa), as well as Coordinator of "Geology on Summer/Ciência Viva" program at ISEP (2005–2019) for geoscience dissemination. He has also been active in teaching and supervising many Ph.D., M.Sc., and undergraduate students.

Maurizio Barbieri University of Rome La Sapienza, Italy
Maurizio Barbieri is a skilled geologist (B.Sc., Ph.D.) with over 27 years of experience and a full professor of Geochemistry at the Department of Chemical, Materials, and Environmental Engineering (DICMA) of the Sapienza University of Rome, Italy. Presently, he is Fellow of the Association of Applied Geochemists and European Chair (Italy) of the Society for Environmental Geochemistry and Health (https://segh.net/about-us). He has been on the editorial board of the *Arabian Journal of Geosciences*, *Chemie der Erde*, *Environmental Geochemistry and Health*, *Euro-Mediterranean Journal for Environmental Integration*, *Water*, *Geofluids*, and *Discover Water*. In addition, he is an editor-in-chief (hydrogeology section) for *Geosciences MDPI*. The main R&D fields are: environmental geochemistry; environmental hydrogeology; hydrogeochemistry; water-rock interaction; water resources management; and geoenvironment and geohazards.

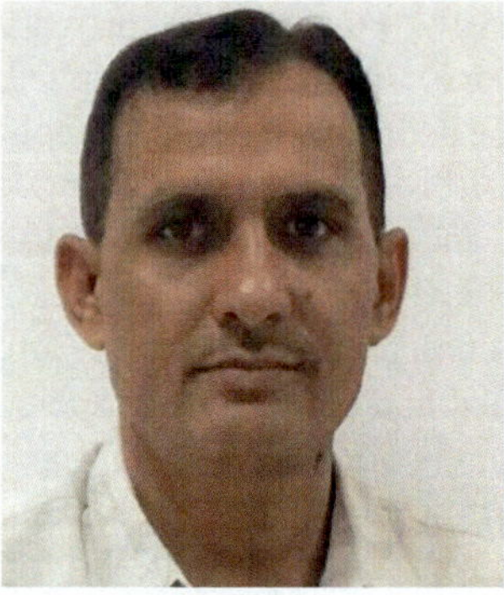

Mahesh Kumar Jat Malaviya National Institute of Technology Jaipur, India
Dr. Mahesh Kumar Jat is a professor of Water Resources at Malaviya National Institute of Technology (MNIT) Jaipur (India). He is teaching and doing research in the area of hydrological modeling, integrated water management, geospatial technologies, and land use land cover change modeling. He has published more than 100 research articles in ISI/Scopus Journals and Conferences (total citations: 2330; h-index—21; i-index 30). He has completed 12 research projects amounting to more than 65 million Indian rupees. He is a member of

many national and international professional bodies such as the European Geosciences Union (EGU), ASCE, IWRS, IAH, IRS. He sits as a member of the board of studies of many Institutes in India. He is Associate Editor and Editorial Member in 2 ISI journals. He has delivered 40 keynote and invited talks at international and national conferences. He has edited and reviewed many articles. He has supervised 11 Ph.D. students and 36 M.Tech. students (with a thesis). Professor Jat has widely traveled abroad visiting many countries to present his research findings.

Jesús Rodrigo-Comino University of Granada, Spain
Jesús Rodrigo-Comino (Ph.D. D.Eng.) is a member of the Department of Regional and Physical Geography at the University of Granada (Spain). He completed his first Ph.D. in Geography at the University of Málaga (Spain) and Trier (Germany) in 2018 and the second one in the engineering of Geomatics and Topography at the Polytechnic University of Valencia (2023). His current research interests include soil geography, regional geography, and land degradation. He coordinates the Terra Lab 2 EGEMAP (Environmental Geography and Mapping) www.egemap.eu.

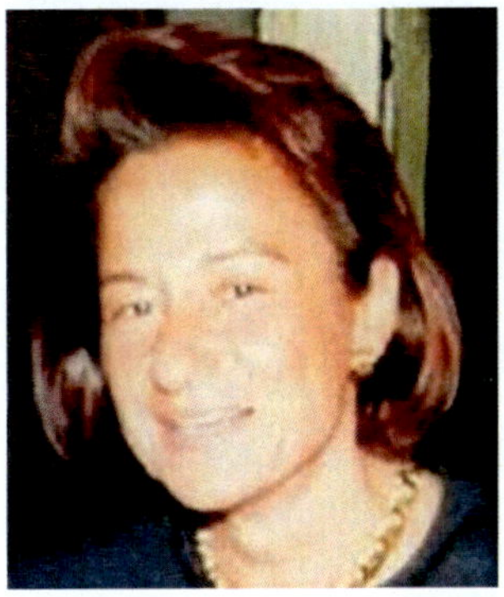

Dionysia Panagoulia National Technical University of Athens, Greece
Dr. Dionysia G. Panagoulia is an associate professor at the National Technical University of Athens, Greece, with expertise in hydrology, hydroclimatology, and water systems. She is the author of more than 145 published research works, including being the co-editor of the book *River Basin Management—Under a Changing Climate* and lately has extended her research work to water economics theory and complex time/dark matter approaches. She has over 30 years of research experience in floods and their risk and hazard, extreme events, precipitation, global climate contribution to local climate, climate change, low flows, droughts, maximum/minimum temperatures, sediment transport, groundwater/streamflow interaction, ANN, WEF Nexus, and water economics and management. She has cooperated in joint research with the University of Stuttgart, Germany; CNRS Laboratory of Oceanology and Geosciences Wimereux, France; Wageningen University, Netherlands; and McGill University, Canada. She was/is a member of twelve scientific societies, a reviewer and guest editor for thirty-two international journals, and a recipient of the 2018 Outstanding Reviewer Award from Water MDPI.

Amjad Kallel Sfax National School of Engineering, University of Sfax, Sfax, Tunisia
Dr. Amjad Kallel is currently Associate Professor of Environmental Geology at the Sfax National School of Engineers at the University of Sfax, Tunisia. He holds a B.Eng. in Georesources and Environment (1998) from the University of Sfax (Tunisia) and an M.Sc. degree and a Ph.D. degree in Georesources and Environment (2004) from Hokkaido University (Japan). He joined Venture Business Laboratory (VBL) at

Akita University, Japan (2005–2006), as a researcher focusing on refining and recycling technologies for the recovery of rare elements from natural and secondary sources. On his return to Tunisia, he worked at the University of Gabes from 2006 to 2011, where he contributed to the elaboration of teaching programs at the Higher Institute of Water Sciences and Technologies of Gabes. Since 2011, he has joined the Sfax National School of Engineers. There, he has also been involved in various research projects related to environmental geology and environmental geotechnics. Dr. Kallel has co-organized many prestigious workshops, seminars, and international conferences. In 2016, Dr. Kallel joined the *Arabian Journal of Geosciences* (Springer) and the *Euro-Mediterranean Journal for Environmental Integration* (Springer) as Chief Editor and Managing Editor, respectively.

Arkoprovo Biswas Banaras Hindu University, India
Dr. Arkoprovo Biswas is Assistant Professor at the Department of Geology, Institute of Science, Banaras Hindu University (BHU), Varanasi. He received his B.Sc. (2002) in Geology from Presidency College, University of Calcutta, M.Sc. (2004) in Geological Science, M.Tech. (2006) in Earth and Environmental Science from IIT Kharagpur, and PG Diploma (2009) in Petroleum Exploration from Annamalai University. He joined Geostar Surveys India Pvt. Ltd. as Geophysicist in 2006 and later joined WesternGeco Electromagnetics, Schlumberger, as On-Board Data Processing Field Engineer/Geophysicist in 2007 and served there till 2008. In 2013, he received his Ph.D. in Exploration Geophysics from IIT Kharagpur. Later, he joined the Department of Earth and Environmental Sciences, Indian Institute of Science Education and Research Bhopal as Visiting Faculty in 2014 and completed his tenure in 2015. He again joined the Wadia Institute of Himalayan Geology (WIHG), Dehradun, in 2016 as Research Associate, and later, he joined BHU in October 2017. He is an experienced geophysicist with research interests in near surface geophysics, integrated electrical and electromagnetic methods, geophysical inversion, mineral, and groundwater exploration and subsurface contamination. He has published more than 50 papers on theoretical modeling, inversion, and application in practical geoscience problems in peer-reviewed international and national journals and 6 book chapters. He also published two edited books with Springer on *Advances in Modeling and Interpretation in Near Surface Geophysics* and *Self-Potential Method: Theoretical Modeling and Applications in Geosciences*. Dr. Biswas received the Prestigious M. S. Krishnan Medal of the Indian Geophysical Union (IGU), Hyderabad, for the year 2019 and the B. C. Patnaik Memorial Gold Medal from the Society of Geoscientists and Allied Technologists (SGAT), Bhubaneshwar, Odisha, India, in 2019. He is also Life Fellow of the Geological Society of India and International Science Congress Association, Fellow of the Society of Earth Scientists, Life Member of the Indian Geophysical Union, and Active Member of the Society of Exploration Geophysicist, USA. He is Editor of the Journal *Results in Earth Sciences* as well as

Associate Editor of the *International Journal of Geophysics*, *Results in Geophysical Sciences*, *Contributions to Geophysics and Geodesy*, *Journal of Earth System Sciences*, *Arabian Journal of Geosciences*, and *Spatial Information Research*. Dr. Biswas is also a member of the International Editorial Advisory Board of *Natural Resources Research*, and *Bitlis Eren University Journal of Science and Technology* Journals.

Veysel Turan Bingol University, Turkey
Veysel Turan is currently an associate professor at the Faculty of Agriculture, Bingöl, Türkiye. His research focuses on the interactions among soil science, environmental remediation, food chemistry and climate change. His Google Scholar H-index is 38, Web of Science ESI top papers (13 as "highly cited papers" and 4 as "hot paper"). He was awarded an Excellence in Reviewing for the *Journal of Environmental Management, Resources, Conservation and Recycling* and Best Reviewer Award for *Critical Reviews in Environmental Science and Technology* in 2022.

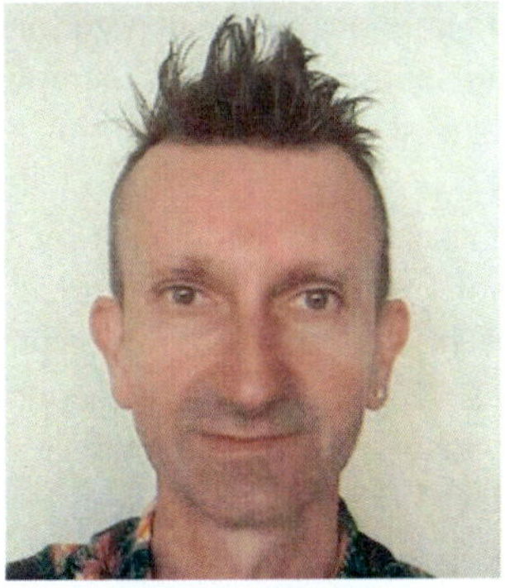

Jasper Knight University of the Witwatersrand, South Africa
I am a geoscientist with research interests in the spatial and temporal variability in morphosedimentary system responses to rapid hemispheric-scale climatic and environmental changes during the late Pleistocene and Holocene. I focus thematically on glaciers, rivers, coasts, and mountains. I focus geographically on Africa, Ireland, northwest USA, Australia, the European Alps, and various places in Asia and South America.

Attila Çiner Istanbul Technical University, Turkey
Attila Çiner is a sedimentology and quaternary geology professor at the Eurasia Institute of Earth Sciences at Istanbul Technical University, Turkey. After graduating from the Middle East Technical University in Ankara (1985), he obtained his M.Sc. degree at the University of Toledo, USA (1988), and his Ph.D. at the University of Strasbourg, France (1992). He works on the tectono-sedimentary evolution of basins and quaternary depositional systems such as moraines, fluvial terraces, alluvial fans, and deltas. He uses cosmogenic nuclides to date these deposits. He primarily focuses on the glacial deposits and landscapes and tries to understand paleoclimatic and paleoenvironmental changes since the Last Glacial Maximum. Lastly, he was part of the Turkish Antarctic Expedition. He spent two months working on the site recognition and decision of the future Turkish scientific research station to be implemented on the continent. He is the founding editor-in-chief of *Mediterranean Geoscience Reviews* and chief editor of *Arabian Journal of Geosciences*, both published by Springer. He received the Humboldt Foundation Georg Forster lifetime achievement award in 2022. He has published more than 100 peer-reviewed articles and book chapters.

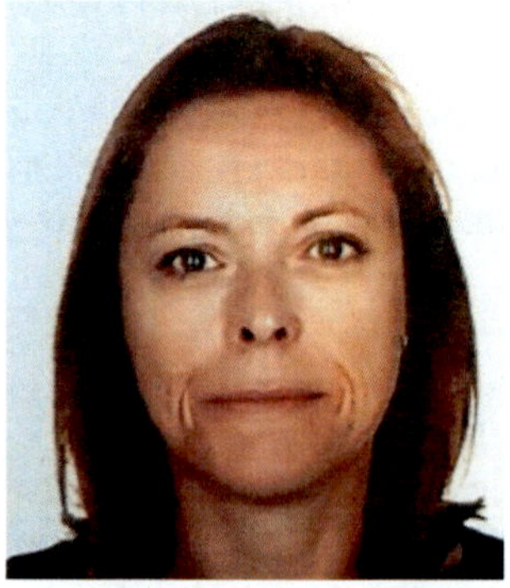

Carla Candeias GeoBioTec, Geosciences Department, University of Aveiro, Portugal

Carla Candeias is Researcher at GeoBioTec Research Centre, Geosciences Department, University of Aveiro, 3810-193 Aveiro, Portugal; candeias@ua.pt, Researcher ID: A-2521-2014/Scopus: 37062910200/Ciencia ID: D912-6FC4-79CC/ORCID: 0000-0001-6664-8545 Keywords: Medical geology, water and sediment quality, soil quality, air quality, indoor and outdoor dust, pelotherapy, environmental geochemistry, human health risk assessment, epidemiology, geophagy students' supervisor. Carla Candeias is Member of international and national projects, including as PI. She is Member of scientific and organizing committees of international conferences. Knowledge dissemination in the scientific and social communities. She is Editor of books and special issues. She published papers indexed on Science Citation Index, e.g.: (1) Carla Candeias et al. (2022). Potentially toxic elements dynamics in the soil rhizospheric-plant system in the active volcano of Fogo (Cape Verde) and interactions with human health. Catena 209(1), 105843, https://doi.org/10.1016/j.catena.2021.105843. (2) Lara Almeida et al. (2022). Geochemical and mineralogical characterization of Ria de Aveiro (Portugal) saltpan sediments for pelotherapy application. Environmental Geochemistry and Health, https://doi.org/10.1007/s10653-022-01407-5. (3) Archi Mishra et al. (2023). Chemical fractionation of particulate-bound metal(loid)s to evaluate its bioavailable fractions, sources and assessment of associated cancer risk. Science of the Total Environment, https://doi.org/10.1016/j.scitotenv.2022.159516. (4) Retshepile Evelyn Malepe et al. (2023). Geophagy and its potential human health implications—A review of South African cases. *Journal of African Earth Sciences*, https://doi.org/10.1016/j.jafrearsci.2023.104848. (5) Bernardino Bernardo et al. (2022). Soil properties and environmental risk assessment of soils in the surrounding area of Hulene-B waste dump, Maputo (Mozambique). *Journal of Environmental Earth Sciences* 81:542, https://doi.org/10.1007/s12665-022-10672-7. Projects, e.g., (1) European Commission/Directorate-General for Research and Innovation Grant Agreement with European Plate Observing System—European Research Infrastructure Consortium (EPOS ERIC) and other beneficiaries (number 871121—EPOS SP—H2020-INFRADEV-2018-2020/H2020-INFRADEV-2019-2). Key Researcher, responsible for the Environmental Geology and Medical Geology studies. (2) FIRE—Fogo Island volcano: multidisciplinary Research on 2014 Eruption. FCT (PTDC/GEO-GEO/1123/2014). Task 9 Coordinator. (3) COST Action IS1408, Industrially Contaminated Sites and Health Network (ICSHNet). Project Member. (4) Scientific Cooperation Agreement Portugal and Poland, Waters Geochemical evolution in abandoned mining areas in Portugal and Poland. Funding FCT 31027/2014. Project Member. (5) Wireless sensors network as a base solution for environmental water quality assessment and monitoring. Funding OHM/CNRS (France), competitive call. Project Coordinator.

Zeynal Abiddin Ergüler Kutahya Dumlupinar University, Geological Engineering Department, Kutahya, Turkey

Prof. Dr. Zeynal Abiddin Ergüler is a full professor at the Geological Engineering Department at Kutahya Dumlupinar University (Turkey). Dr. Ergüler holds a B.Sc. (1998), an M.Sc. (2001), and a Ph.D. degree (2007) in Geological Engineering from Hacettepe University (Turkey). His research interests mainly focus on rock mechanics, engineering geology, environmental geology, and soil mechanics. His current investigation is to understand and model the thermo-hydro-mechanical behavior of shale rocks in the area of shale gas production. In addition to performing many types of research and industry-funded projects, he has also taught and supervised undergraduate and graduate students. In 2017, Dr. Erguler joined the *Arabian Journal of Geosciences* (AJGS) as an editor responsible for evaluating submissions in the fields of rock mechanics, engineering geology, environmental geology, and soil mechanics.

Hydrology, Groundwater and Water Resource Management

Eigenvector Method for Demarcation of Groundwater Potential Zones on a Gridded Domain

A. Keerthana and Archana Nair

Abstract

Groundwater being the most abundant source of freshwater for the survival of flora and fauna is expected to meet the demands of agriculture, industry, and drinking water requirements. Due to climatic changes and over-exploitation, the total groundwater reserve has decreased. The only way out of this quagmire is to use the resource judiciously and recharge frequently. Not all regions are suitable for recharge due to differences in hydrological settings. As a result, determining the ability of an area to retain water is pertinent. This study proposes a demarcation of groundwater potential zones using eigenvectors on a grid within the Periyar River Basin (PRB) in Kerala, India. The PRB is Kerala's second largest river basin, serving the domestic, agricultural, and industrial needs of the metropolitan Kochi. Over PRB, the grid of approximately 11.1 km-by-11.1 km in length and points within 1.11 km of each other is considered. The variables used for the analysis are Normalized Difference Vegetation Index, Soil Moisture, Lineament Density, Sand content, Silt content, Clay content, Drainage Density, Slope, Rainfall and Temperature. The eigenvectors are computed for the matrix deformed using the grid point variable values. Satty's Scale and Analytical Hierarchical Process are used to calculate the weights. The basic method employs several weighting approaches to the layer subclasses, which increases the uncertainty in the final output. This volatility has been morphed using the developed method. On a seasonal scale, the dominant eigenvector that explains the most variability is used to form the groundwater potential zones (GWPZ), namely low, moderate and high during the months of January, April, August and November of 2020. Validation is based on actual groundwater levels (GWL) in the 9 observation wells managed by the Central Groundwater Board (CGWB). Based on the results, seven wells in January and six in November reached the expected potential zone class, while one of the wells encountered over-exploitation at the considered PRB grid point. Overall, the agreement between the groundwater levels and the potential zone delineation is 72.2%. This developed method can be applied in different locations to delineate GWPZs. An accurate potential map aids in source monitoring, assessment and protection.

Keywords

Analytical hierarchical process · Eigenvector · Groundwater level · Groundwater potential zones · Periyar River Basin

1 Introduction

Groundwater, the primary source of usable freshwater has started to deplete and degrade. The groundwater resources are being monitored and safeguarded by India's Central Groundwater Board (CGWB). The depth to groundwater levels is measured in meters below ground level (mbgl) four times a year in January, April, August and November. A greater depth in groundwater levels can be due to a lack of water storage capacity or over-exploitation. The hydro-geological setting can explain the capacity beneath the earth's surface. Groundwater is exploited if a region has a high depth even with enough capacity. The delineation of water holding capacity or potentiality aids in determining the exact groundwater scenario (Negm et al., 2020). The potential zone helps to determine if a location is suitable for the storage of future groundwater and thus indirectly ensuring recharge. Although discharge and land cover influence the

A. Keerthana · A. Nair (✉)
Department of Mathematics, Amrita Vishwa Vidyapeetham, Kochi Campus, Ernakulam, Kerala 682024, India
e-mail: archananair@kh.amrita.edu

H. Chenchouni et al. (eds.), *Recent Advancements from Aquifers to Skies in Hydrogeology, Geoecology, and Atmospheric Sciences*, Advances in Science, Technology & Innovation, https://doi.org/10.1007/978-3-031-47079-0_1

recharge rates, the focus is on locating groundwater potential zones (GWPZ).

The commonly used methods in delineation are Analytical Hierarchical Process (AHP), Weighted Overlay Analysis (WOA), Geographic Information System (GIS) and Multi-Criterion Decision Making (MCDM) (Doke et al., 2021). The usual AHP reclassifies each variable layer and applies ranks. For example, an NDVI of 0.41 will be in class 1 with a rank of 5, while 0.42 will be in class 2 with a rank of 10. This would increase ambiguity in the outcome. Hence, we have normalized each variable that acts as a rank to each value rather than a sub-class. To the best of our knowledge, no study has included the mathematical derivation of eigenvectors to delineate GWPZ. This study introduces a new dimension of potential zone analysis using AHP, GIS and eigenvector technique. A grid from a tropical river basin of Kerala state, India, is chosen for the analysis considering January, April, August and November (JAAN) months of 2020.

2 Materials and Methods

The state of Kerala is in India's southwestern region. The Periyar River is the longest of the 44 rivers flowing through the state during various seasons, originating from the Sivagiri Hills and discharging into the Arabian Sea. The Periyar River Basin (PRB) is Kerala's second largest river basin, home to a diverse range of plants and animals. It is located between latitudes 9° 15′ N to 10° 20° N and longitudes 76° 10′ E to 77° 30′ E. Over PRB, a 0.1° × 0.1° grid was selected and 121 points at 0.01° interval were measured (Fig. 1). The grid variable values are acquired and inferred. Groundwater levels in 2020 were downloaded from India Water Resources Information System (https://indiawris.gov.in). The Normalized Difference Vegetation Index (NDVI) is assimilated from Copernicus Global Land Services (https://land.copernicus.eu/global/index.html). Monthly soil moisture (SM) (in mm) provided by the Climate Prediction Center of resolution 0.5° × 0.5° was acquired from IRI Digital Library (http://iridl.ldeo.columbia.edu/). The Cartosat-1 Digital Elevation Model (DEM) of 30 m resolution was obtained from Bhuvan (https://bhuvan.nrsc.gov.in/home/index.php). Using this DEM, Lineament Density (LD), Drainage Density (DD) and Slope were created in the GIS environment. The sand, silt and clay content were obtained from International Soil Reference and Information Centre—World Soil Information (https://www.isric.org/). The daily rainfall (RF) (in mm) with 0.25° × 0.25° resolution (Pai et al., 2021), maximum and minimum temperatures (TP) (in °C) with 1° × 1° resolution (Srivastava et al., 2009) is downloaded from Indian Meteorological Department. These variable values of 121 grid points were extracted and normalized on a scale from −1 to 1. The layer weights were assigned using Saaty's scale of 1–9 (Saaty, 1987). AHP entails creating Pairwise Comparison

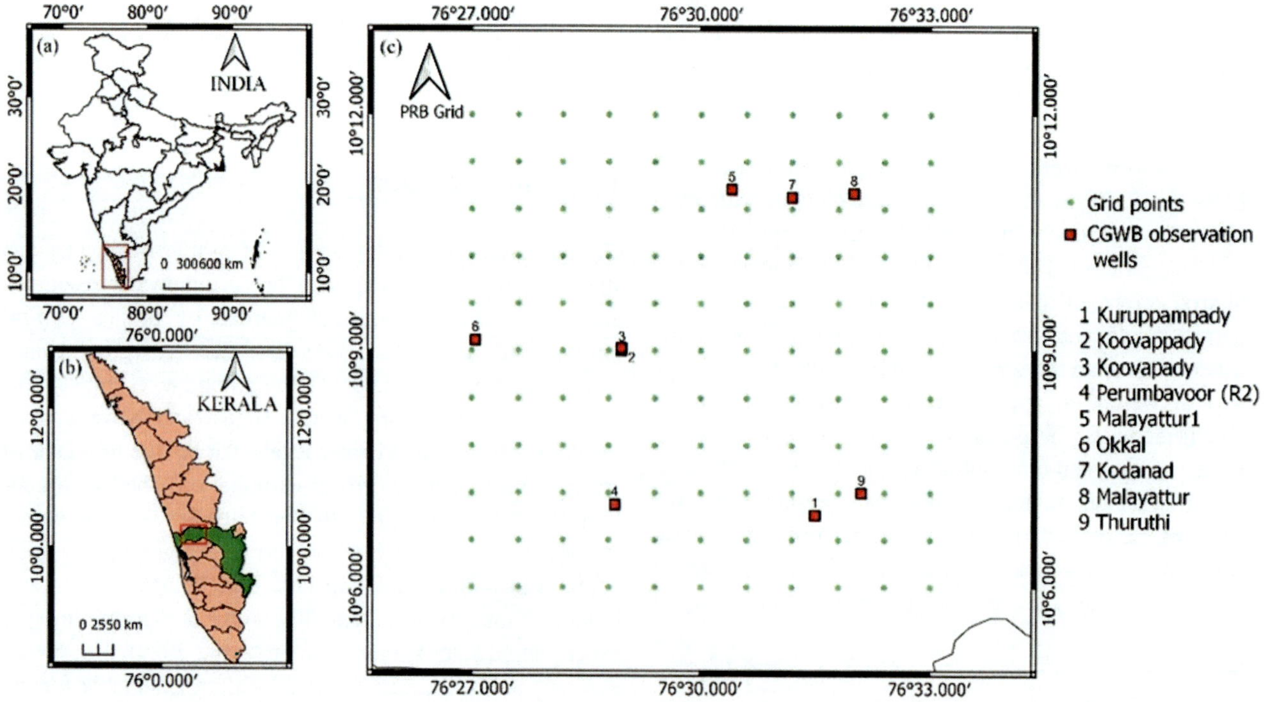

Fig. 1 Study area map of **a** India showing Kerala state, **b** Kerala state showing Periyar River Basin, and **c** the studied grid points with CGWB observation well locations

Matrix (PCM) and Normalized PCM (NPCM) and evaluating the consistency using Principal Eigenvalue (λ_{max}), Consistency Vector Matrix (CVM), Consistency Index (CI) and Consistency Ratio (CR).

$$\text{CI} = \frac{\lambda_{max} - n}{n - 1} \text{and CR} = \frac{\text{CI}}{\text{RI}} \quad (1)$$

where n is the number of thematic layers and RI is the random consistency index of reciprocal matrices. PCM is considered consistent if CR is less than or equal to 0.1. The normalized weights of the variables are then signed depending on their relationship to water holding capacity. The matrix of normalized variable was multiplied by the signed weights to generate matrix Y and converted into a square matrix M by multiplication and transposition. Eigenvector of matrix M produces a diagonal matrix D and an eigenvector matrix V, related by Eq. 2.

$$M \times V = V \times D \quad (2)$$

3 Results

The matrices obtained while performing AHP are given in Table 1. The average of CVM, λ_{max} is 10, and the RI for $n = 10$ is 1.49. Substituting these values in Eq. 1 gives CI and CR values to be approximately 0. The weightage (W) is signed according to the relation to water holding capacity as 14, −13, 11, 11, −11, −11, −9, −9, 7 and −4 for NDVI, SM, LD, sand, silt, clay, DD, slope, RF and TP. The normalized variables are multiplied by the signed weights and arranged to form matrix Y of order 121 × 10. Multiply Y with its transpose to obtain square matrix M of order 121 × 121. The eigenvalues and eigenvectors of M are calculated using Eq. 2. The most contributing six eigenvectors are extracted, leading to V (121 × 6) and D (6 × 6). The eigenvector of the greatest eigenvalue is rescaled into −1 to 1 and interpolated using the Inverse Distance Weighted method to obtain the potential map. The ranges are reclassified into low (red), moderate (green) and high (blue) using natural breaks (Jenks) (Fig. 2). The classification ranges for the month of January are −1 to 0.145 as low, 0.146 to 0.655 as moderate and 0.656 to 1 as high potential zones. That of April are −1 to −0.678 for low, −0.677 to −0.137 for moderate and −0.136 to 1 as high. The ranges for the month of August are −1 to 0.161 as low, 0.162 to 0.678 as moderate and 0.679 to 1 as high. Finally, the November month ranges are −1 to −0.122 for low, −0.121 to 0.373 for moderate and 0.374 to 1 for high potential zones. The regions of high potential zones are found enough during January and August, while the lowest is during April. In November, all zones are distinguishable. It was found that 7, 97, 7 and 30 grid points lie in low potential zones, 23, 17, 8 and 30 in moderate zones and 91, 7, 106 and 61 grid points lie in high potential zones, respectively, in monthly order.

4 Discussion

The potential zones are validated using the CGWB's actual groundwater levels for the months of January and November in nine observational locations. Out of the nine wells, seven during January and six during November predicted the same potential zone class. Among the non-matching wells, over-exploitation occurred in two wells in January and one in November. In these areas, the water withdrawal rate is high, which makes the water level drop below the actual limit. Six wells in April and one well in November were in the low potential zone. The moderate zone has two wells in August and three wells in each of the other months. Six wells in January, seven in August and five in November lie in the high potential zone. In April and November, the Malayattur well lies in a low potential

Table 1 Resulting matrices using AHP on the ten variable layers

Layer	Assigned weight	NPCM row average (W)	Weightage $= W * 100$	Product matrix $P = \text{PCM} * W$	CVM $= P./W$
NDVI	8	0.145	14	1.454	10
SM	7	0.127	13	1.272	10
LD	6	0.109	11	1.091	10
Sand	6	0.109	11	1.091	10
Silt	6	0.109	11	1.091	10
Clay	6	0.109	11	1.091	10
DD	5	0.091	9	0.909	10
Slope	5	0.091	9	0.909	10
RF	4	0.073	7	0.727	10
TP	2	0.036	4	0.364	10

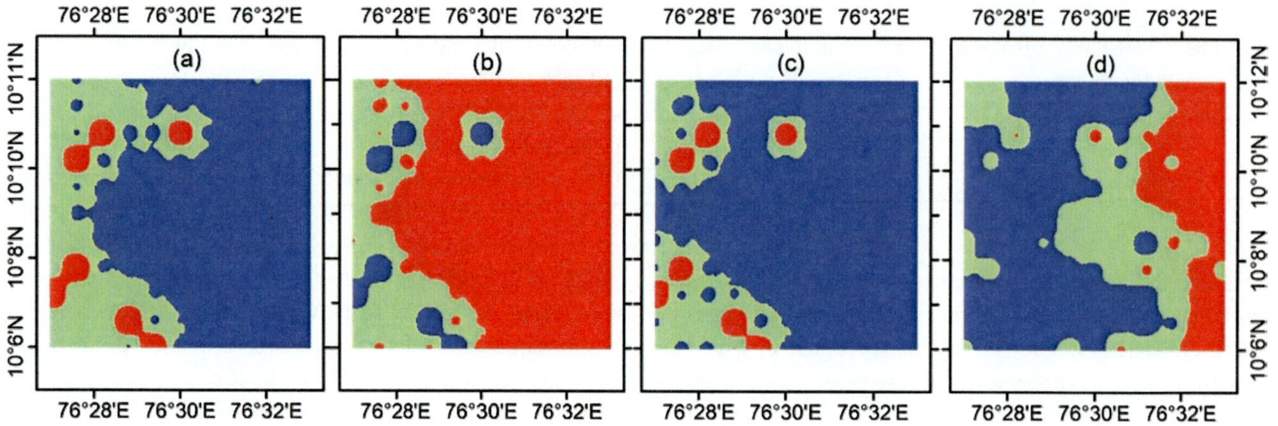

Fig. 2 GWPZ for the year 2020 in **a** January, **b** April, **c** August and **d** November

zone, whereas the Malayattur1 well location is over-exploited during January and November. Identifying these regions helps farmers, businesspeople and hydrologists in decision-making.

5 Conclusions

The zone-wise area percentages are 5.64, 71.45, 4.74 and 20.36% for low, 27.71, 23.29, 18.94 and 30.36% for moderate, and 66.65, 5.26, 76.32 and 49.28% for high potential zone, for JAAN months, respectively, with the low potential zones on the edge of drying off. The eigenvector delineation of potential zones in a gridded domain of PRB resulted in 77.78 and 66.67% agreement with the actual groundwater levels during January and November 2020. The accuracy averages 72.2%, indicating that the proposed methodology can be efficiently applied in demarcating groundwater potential zone across any domain with differential characteristics.

References

Doke, A. B., Zolekar, R. B., Patel, H., & Das, S. (2021). Geospatial mapping of groundwater potential zones using multi-criteria decision-making AHP approach in a hardrock basaltic terrain in India. *Ecological Indicators, 127*, 107685. https://doi.org/10.1016/j.ecolind.2021.107685

Negm, A., Bouderbala, A., Chenchouni, H., & Barcelo, D. (2020). *Water resources in algeria-Part I: assessment of surface and groundwater*. Springer, Cham. https://doi.org/10.1007/978-3-030-57895-4

Pai, D. S., Rajeevan, M., Sreejith, O. P., Mukhopadhyay, B., & Satbha, N. S. (2021). Development of a new high spatial resolution (0.25° × 0.25°) long period (1901–2010) daily gridded rainfall data set over India and its comparison with existing data sets over the region. *MAUSAM, 65*, 1–18. https://doi.org/10.54302/mausam.v65i1.851

Saaty, R. W. (1987). The analytic hierarchy process-what it is and how it is used. *Mathematical Modelling, 9*, 161–176. https://doi.org/10.1016/0270-0255(87)90473-8

Srivastava, A. K., Rajeevan, M., & Kshirsagar, S. R. (2009). Development of a high resolution daily gridded temperature data set (1969–2005) for the Indian region. *Atmospheric Science Letters, 10*, 249–254. https://doi.org/10.1002/asl.232

3D Geological Modeling of Saïss Basin (Northern Morocco)

Latifa Bouib, Fouad Amraoui and Youssef Arjdal

Abstract

The Moroccan septentrional region includes the Bassin du Saïss. Due to this bassin's enormous surface area and substantial depth, our current understanding of it is still limited and fragmentary. To improve understanding of the aquifer, a 3D geological model was developed using data from over two hundred boreholes, eight geological maps, and a digital terrain model. The data was processed using a Geographic Information System (GIS) and a geological modeling software, with consideration given to five lithostratigraphic units: Paleozoic, Triassic, Jurassic, Miocene, and Plio-Quaternary. The developed model provides a precise visualization of the geometry and stratigraphy of the basin, as well as a better understanding of the deep Lias aquifer and its hydrogeology. The study results provide a valuable conceptual model that can be used to reinterpret existing geophysical data and assess the extent of the Lias aquifer. Furthermore, the results can contribute to the development of a hydrogeological model that simulates the hydrodynamic functioning of the aquifer under variable climate and exploitation conditions.

Keywords

Hydrogeological Modeling · Stratigraphy · Geometry · Saïss Basin · Morocco

L. Bouib (✉) · F. Amraoui
Laboratory of Applied Geosciences to Development Engineering, Faculty of Sciences Ain Chock, Hassan II University of Casablanca, B.P 5366 Mâarif, Casablanca, Morocco
e-mail: latifa.bouib@gmail.com

F. Amraoui
e-mail: amraoui_f@hotmail.com

Y. Arjdal
Natural Resources & Sustainable Development Laboratory, Earth Sciences Department, Faculty of Sciences, Ibn Tofail University, Kenitra 14000, Morocco
e-mail: youssef.arjdal@uit.ac.ma

1 Introduction

Groundwater resource management is a critical aspect of sustainable development worldwide, with geological factors playing a pivotal role in shaping groundwater dynamics. Understanding the distribution and characteristics of aquifers is essential for effective water resource planning and utilization in various regions globally (Chenchouni et al., 2022, 2023). The Saïss sedimentary basin, located in northern Morocco, covers an area of about 2100 km^2 and is centered on the Fez-Meknes region (Fig. 1). The basin comprises various lithologies of varying thicknesses, including sands, marls, sandstones, clays, limestones, and dolomitic limestones (Ait Brahim, 1991; Amraoui, 2005; Essahlaoui, 2000). The Paleozoic basement is mainly formed by pelites with intercalation of quartzites. Four lithostratigraphic units have been discovered above this bedrock, ranging from Jurassic to Plio-Quaternary, containing two main regional reservoirs: the Plio-Quaternary and the Lias dolomitic limestone.

Several authors have used geological and geometric modeling to study and characterize the sediments in different basins of Morocco (Ben Moussa et al., 2019; Bouazza et al., 2013; Chamrar et al., 2019; Hamidi et al., 2021). Although there are numerous issues and concerns, no regional 3D geological modeling of the Saïss basin has been performed yet. To address these issues and improve the understanding of the geological and hydrogeological systems, a 3D geological model of the basin is being implemented. The objectives of this modeling are to verify the 3D geological mapping of the study area, to assess the shape and size of the aquifers, to define the potential connections between the Jurassic limestones and the Plio-Quaternary aquifers, and to create a conceptual model that can be used for future 3D hydrogeological modeling.

H. Chenchouni et al. (eds.), *Recent Advancements from Aquifers to Skies in Hydrogeology, Geoecology, and Atmospheric Sciences*, Advances in Science, Technology & Innovation, https://doi.org/10.1007/978-3-031-47079-0_2

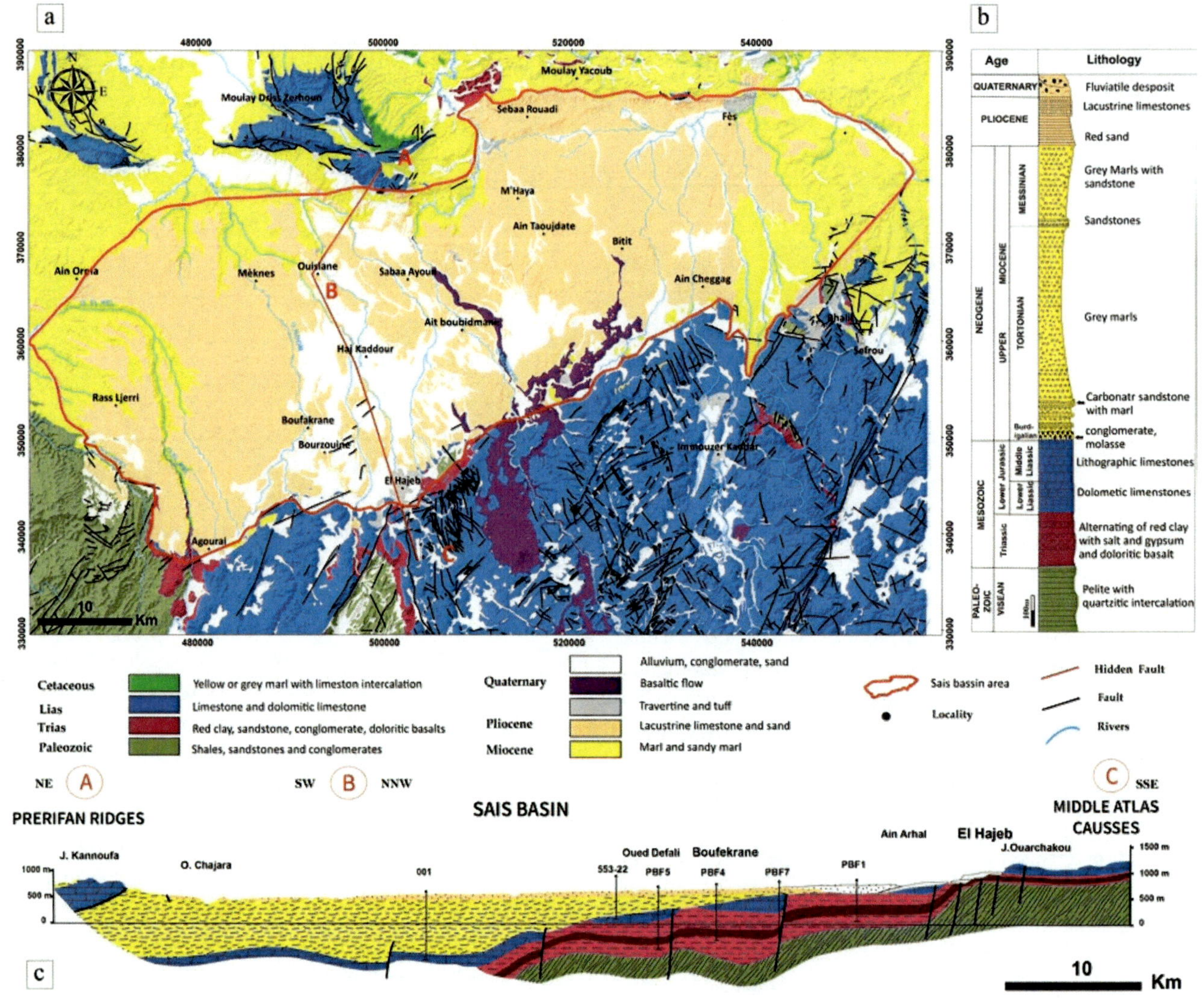

Fig. 1 **a** Geological map of the Saïss sedimentary basin and the Middle Atlas Causses, **b** lithostratigraphic column, **c** cross section A-B-C

2 Materials and Methods

To construct a 3D model of the geology of the Saïss Basin, we employed a GIS platform to process our database, which included geological data sourced from maps. A digital terrain model was used to represent the topographic surface, while we analyzed 245 borehole diagraphs gathered by the Sebou River Basin Agency (ABHS) (Fig. 2). We then compiled, analyzed, encoded, and integrated the borehole data into our Geological Modeling software.

Furthermore, we carried out a geostatistical study and incorporated specific prediction maps into our geoscientific information system. To choose the appropriate semivariogram and cross-validate our results, we applied normality tests and pattern analysis to each measurement.

3 Results and Discussion

An elaborate three-dimensional model (Fig. 3a) based on the available geological data made it possible to visualize the architecture of the Saïss Basin and to realize the geological sections (Fig. 3b) in all directions and in almost all parts in the frame of the model.

The resulting model shows five geological strata from Paleozoic to Plio-Quaternary, from bottom to top.

The Paleozoic substratum consists of schists, sandstones, and quartzites. It outcrops mainly in the southwestern border of the basin, and is also in the southern borders. In addition, the Paleozoic terrain can be observed in the slopes of the Causses, northwest of El Hajeb. The Triassic sediments are composed of gypsiferous and saliferous clays (evapotic

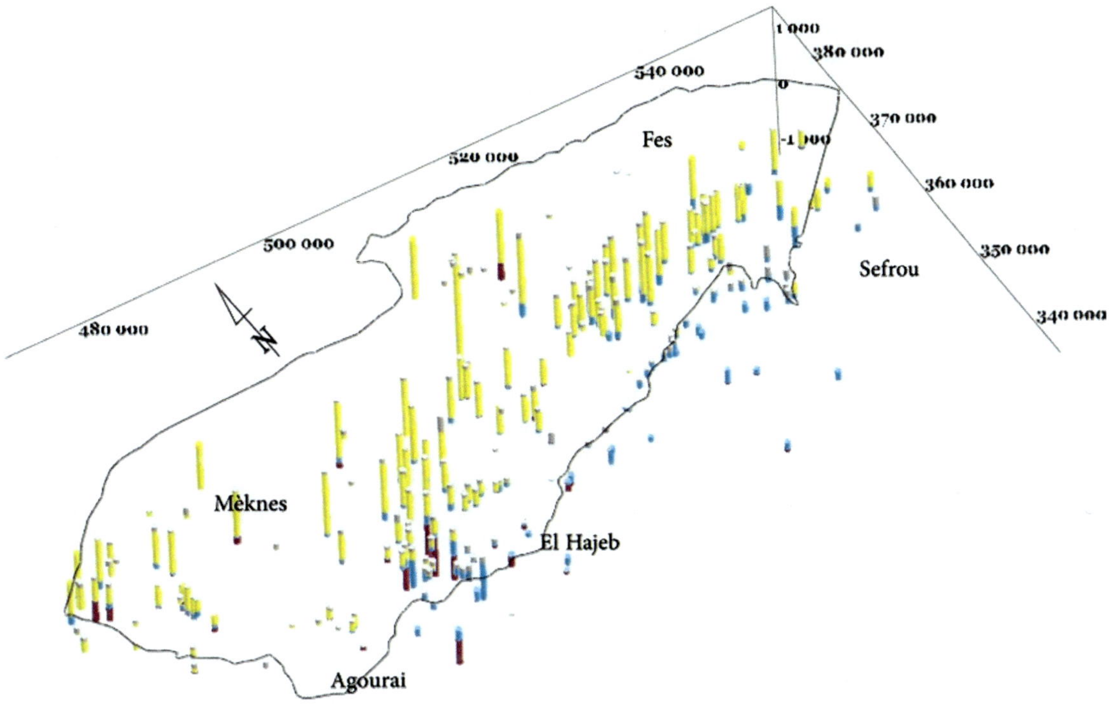

Fig. 2 Borehole distribution in the study area

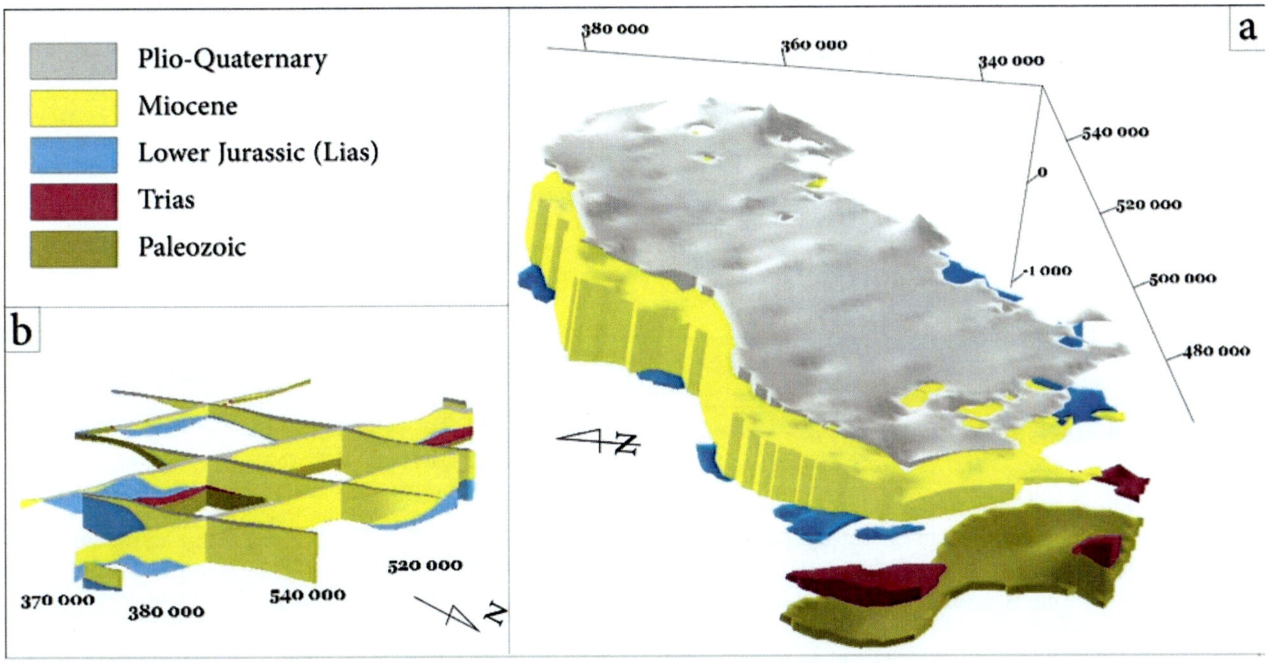

Fig. 3. 3D lithostratigraphical model of Saïss Basin in **a** solid, **b** N-S, E-W cross sections

deposits), as well as intercalations of doleritic basalts. This formation, with variable thickness, reaches a maximum depth of 2161 m. It outcrops north of Sefrou, near Bhalil (Figs. 1 and 3), and at El Hajeb, where one can observe an elongated dome extending in the SW-NE direction, where the Triassic rises to the surface. It is also found in the Agouraï region.

Next, predominantly dolomitic carbonate formation constitutes the main groundwater reservoir of the basin. This formation belongs to the Lower and Upper Lias. It outcrops in the south of the basin, on the Middle Atlas Causse, where the Lias aquifer is unconfined. In the basin, the thickness of the Lias is either thinner or reinforced. At the level of the plain of Fez, the thickness of the Lias exceeds 60 m and even reaches

368 m (borehole N°IRE 2625/15) in border of the road of Sefrou at approximately 16 km SE of the city of Fez. On the other hand, the presence of limestones in certain zones is uncertain, which would correspond to zones of erosion of the liasic formations (absence or thinning of the Lias).

The fourth formation is composed of yellow-gray-blue marls with intercalations of sandstone or sandy layers and dates to the Miocene. This formation constitutes the impermeable cover of the deep aquifer and the substrate of the superficial aquifer. The topography of this marl layer varies irregularly from south to north. The thickness varies from a few meters in the south to over 1000 m in the north near the pre-Rif ridges (Fig. 1). The marls are exposed in several parts of the basin, notably in the eastern part, from the east to the north of Sefrou, as well as in the northern borders of the basin (Figs. 1 and 3). Finally, the upper layer covering most of the basin is composed of sandy-clay or limestone beds, lacustrine limestone, and sandy marls. Its thickness exceeds 100 m. Additionally, the thickness of the Plio-Quaternary is considerable along the edges of the Causses.

The construction of a 3D model provides significant advantages for visualizing the structure of the aquifer and underlying geological formations in the Saïss basin. Cross sections provided by this model allow for comparison of the thickness and extent of the various formations that make up the aquifer, providing a better understanding of the internal geometry of the aquifer. The interpretive geological sections show the horst and graben or half-graben structures, as well as areas of thinning or disappearance of Liassic formations in the basin's substrate. The sub-meridian sections also confirm the structure observed in the schematic geological sections and previous geophysical surveys and presented in Amraoui (2005).

4 Conclusions

In conclusion, this study has greatly enhanced our understanding of the structure of the Saïss basin. Through the use of 245 mechanical boreholes and geological maps, a 3D geological model of the basin was created, which provides valuable insights into the basin's deep aquifer and internal structure. Although faults were not included in the model, it still allows for a visualization of the basin's general architecture. The study revealed that the basin's internal structure is characterized by horsts and grabens, controlled by sets of normal faults that were highlighted by geophysical data. Overall, the findings of this study have allowed for the development of a conceptual model that is useful in both the reinterpretation of existing geophysical data and the establishment of a hydrogeological model.

References

Ait Brahim, L. (1991). Tectonique cassante et états de contraintes récents au Nord du Maroc. Contribution à l'étude du risque sismotectonique, Thèse de doctorat d'Etat de l'Université de Rabat, Maroc. Es Sci, N° 233p.

Amraoui, F. (2005). Contribution à la connaissance des aquifères karstiques: cas du lias de la plaine du Saïss et du causse Moyen atlasique tabulaire (Maroc). Thèse de doctorat d'Etat de l'Université Hassan II, Ain Chock, Casablanca, N° 249p.

Ben Moussa, A., Mridekh, A., El Mansouri, B., & Mazini, I. A. L. (2019). The mio-plio-quaternary Volubilis basin evolution (prerif ridges, NW Morocco), contribution of geophysical imagery. *Journal of African Earth Sciences*, 103601.

Bouazza, M., Khattach, D., Houari, M. R., & Kaufmann, O. (2013). Apport du modèle géologique 3D à l'étude de la structure de l'aquifère profond d'Aïn Béni Mathar, Maroc Oriental. *Bull. Inst. Sci. (Rabat), 35*, 53–61.

Chamrar, A., Oujidi, M., El Mandour, A., & Jilali, A. (2019). 3D geological modeling of Gareb-Bouareg basin in northeast Morocco. *Journal of African Earth Sciences, 154*, 172–180.

Chenchouni, H., Chaminé, H. I., Khan, M. F., Merkel, B. J., Zhang, Z., Li, P., Kallel, A., Khélifi, N. (2022). *New Prospects in Environmental Geosciences and Hydrogeosciences*. Springer, Cham.

Chenchouni, H., Chaminé, H. I., Zhang, Z., Khelifi, N., Ciner, A., Ali, I., & Chen, M. (2023). Recent research on hydrogeology, geoecology and atmospheric sciences. Springer, Cham.

Essahlaoui, A. (2000). Contribution à la reconnaissance des formations aquifères dans le bassin de Meknès-Fès (Maroc) prospection géoélectriques, étude hydrogéologique et inventaire des Ressources en eau. Doct. Sc. App. Ecole Mohammadia d'Ingénieurs, Rabat Maroc, N° 258p.

EL Hamidi, M. J., Larabi, A., & Faouzi, M. (2021) Numerical modeling of saltwater intrusion in the Rmel-OuladOgbane coastal aquifer (Larache, Morocco) in the climate change and sea-level rise context (2040). *Water, 13*(16), 2167.

Deep-Circulating Salty-Fluids in Cambrils del Pirineu (Pyrenees, NE Iberian)

Elisabet Playà, Jordi Pujadas, Juan Diego Martín-Martín, Irene Cantarero, Vinyet Baqués, Marta Martín, Sergi Casals, Eloi Carola and Anna Travé

Abstract

Salt springs in Cambrils del Pirineu (Pyrenees, NE Iberian Peninsula) are related to diapirism of Triassic (Keuper) evaporite-bearing deposits. Such brines, which are close to NaCl saturation, have been used at least since s. XVIII to produce non-marine table salts. The present research aims to study the hydrogeological system in a high salinity spring water from mountain source in the southern Pyrenees using geochemical study of water and its association with local geology. The research includes total conductivity, major anion and element analyses by means of HPLC and ICP-OES and water flow measurements for one-year period. Sampled waters include a salty spring and nine closer freshwater springs. Conductivity of the salt water is close to 44 mS/cm, chloride is about 17% and Na is close to 10%, with minor amounts of sulphate, K, Ca, Mg and Sr. Water flow is about 4 L/min (July 2021 to June 2022). Minor increase in the flow rate is related to superficial inflows during last metres of brine circulations before outcropping. Freshwater springs have HCO_3-Ca compositions but show a slight Mg-SO_4 enrichment due to nearby Triassic evaporites. Eh of brine at source is only of 12 mV, thus evidencing its reductive behaviour. Local geology suggests large brine circulation along Triassic-Cretaceous contact until surface discharge. This model differs from previous interpretations, which pointed to more superficial recharge and fluid circulation.

Keywords

Salty mountain source · Deep-circulating fluids · Triassic evaporites dissolution · Pyrenees

1 Introduction

Understanding the dynamics of deep-circulating salty-fluids in various geological settings is crucial for comprehending their role in natural processes and human activities, such as salt production and potential environmental impacts. Investigating the hydrogeological systems associated with salt springs provides valuable insights into fluid circulation patterns and the interactions between surface and subsurface environments. Salt springs in Cambrils del Pirineu (Pyrenees, NE Iberian Peninsula; Fig. 1) are related to diapirism of Triassic (Keuper) evaporite-bearing deposits (Mata-Perelló et al., 2011). Such brines, which are close to NaCl saturation, have been profited at least since s. XVIII to produce non-marine table salts. Hydrogeological system has not been previously studied, having preliminary ideas about surficial-fluid circulation after meteoric water recharge and infiltration through evaporites (Salines de Cambrils Museum, Pers. Com., 2019).

2 Materials and Methods

We have studied four water samples from a salty spring (Cambrils saltwork source, sampled from November 2019 to July 2021) and nine samples from nearby freshwater sources (September 2020). The study includes total conductivity, major anion and element analyses by means of HPLC with Waters IC-Pak anions column (chloride, sulphate; HPLC equipped with conductivity Waters 432 and UV/V Kontron 332 detectors), potentiometric methods (bicarbonate; Metrohm Tritino 702 SM) and ICP-OES (Na, K, Mg, Ca; Optima 8300 Perkin Elmer). Water flow was

E. Playà · J. D. Martín-Martín · I. Cantarero · V. Baqués · E. Carola · A. Travé (✉)
Departament de Mineralogia, Petrologia i Geología Aplicada, Facultat de Ciències de la Terra, Universitat de Barcelona (UB), c/Martí i Franquès s/n., 08028 Barcelona, Spain
e-mail: atrave@ub.edu

J. Pujadas
JPF Consultores, SL. Montblanc, Tarragona, Spain

M. Martín · S. Casals
Salí de Cambrils, Cambrils (Odèn), 25283 Lleida, Spain

H. Chenchouni et al. (eds.), *Recent Advancements from Aquifers to Skies in Hydrogeology, Geoecology, and Atmospheric Sciences*, Advances in Science, Technology & Innovation, https://doi.org/10.1007/978-3-031-47079-0_3

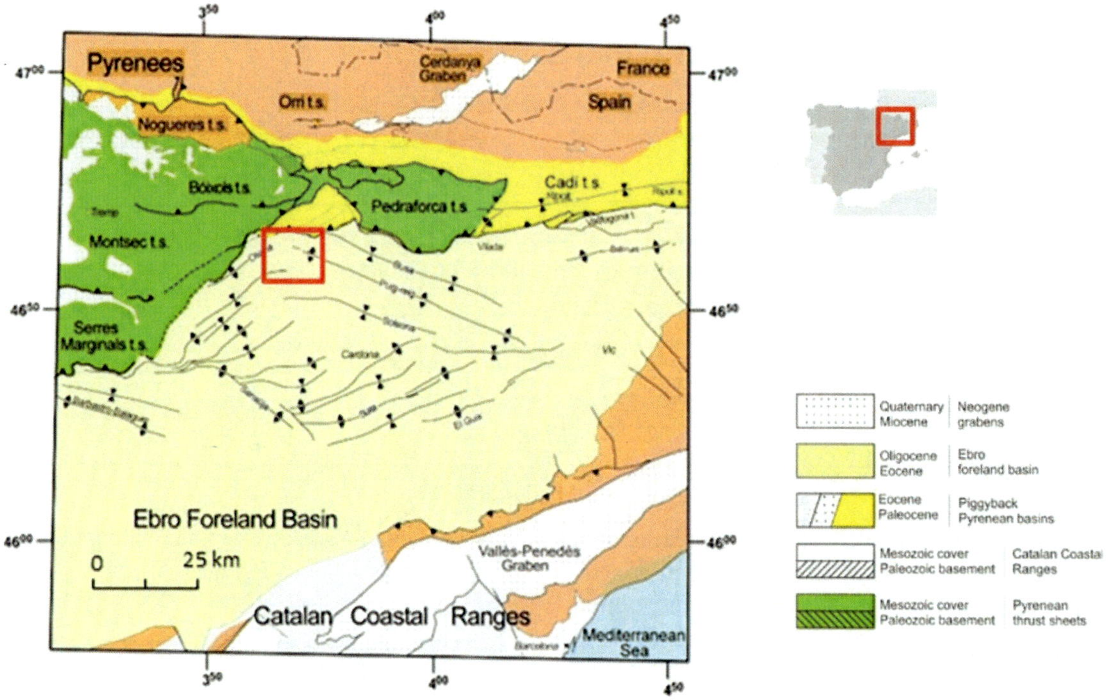

Fig. 1 Geological location (Cambrils del Pirineu, Pyrenees, NE Spain). Modified from Vergés (1993)

hand-measured monthly for one-year period (measuring water volume collected during 10 s).

3 Results

Conductivity of the salty water is close to 44 mS/cm. Chloride is about 17% and Na is close to 10%, with minor amounts of sulphate, K, Ca and Mg (Fig. 2). Eh of brine at source is 12 mV, thus evidencing its reductive behaviour.

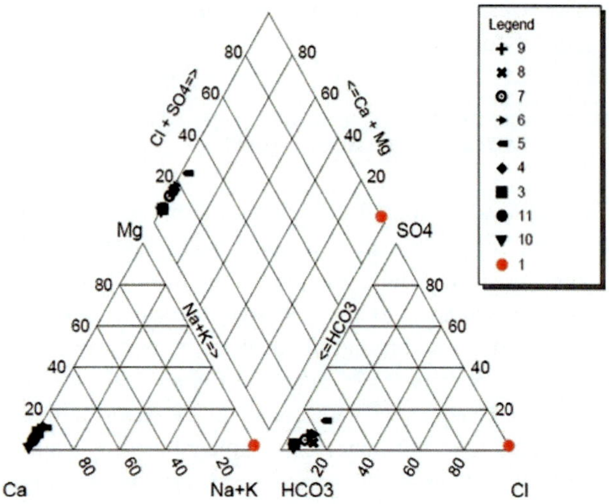

Fig. 2 Piper plot of freshwater sources (samples 3–11) and Cambrils salty spring (1, representative of four samples)

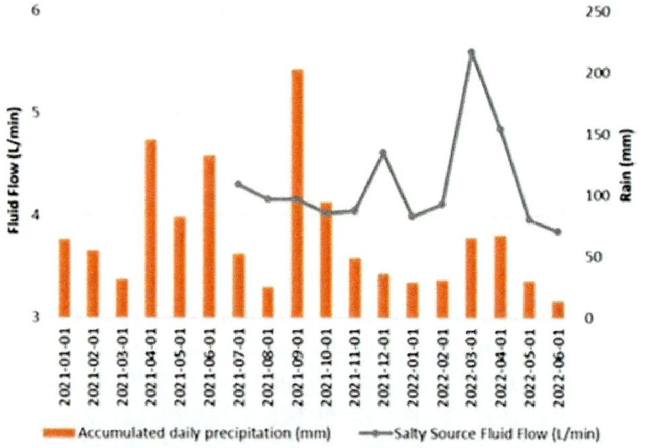

Fig. 3 Mean monthly fluid flows from studied salty source and local precipitation

Freshwater springs display HCO_3-Ca compositions but show a light tendency to Mg-SO_4 (Fig. 2).

Mean flow rate of the salty spring is 4.3 L/min, oscillating between 3.8 and 5.6. Figure 3 displays mean-monthly fluid flows of 49 measures, compared to the local accumulated rain (*Port del Compte* meteorological station) (Fig. 3).

4 Discussion

Brine composition clearly indicates water lixiviating evaporites, including halite (Na, Cl and K-Mg) and minor amounts of sulphates and carbonates (SO_4, HCO_3, Ca, Mg).

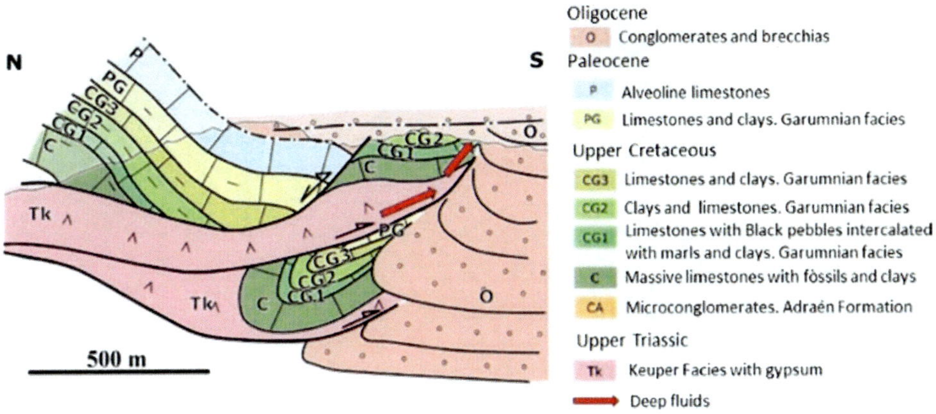

Fig. 4 Model of fluid circulation based on the cross section in Casas et al. (2022)

Freshwater springs display typical HCO_3-Ca compositions but show a light tendency to Mg-SO_4 enrichment due to the closer Triassic evaporites. Moreover, the hydrogeological system of salty-fluids is mainly disconnected from the freshwater ones.

Fluid flow in the salty spring is relatively stable, close to 4 L/min, showing two spikes not directly related to the local precipitation curve, suggesting deep circulation. Minor flow raises are probably related to very superficial inflows during last metres of brine circulations before outcropping.

Local cross section suggests large brine circulation along Triassic-Cretaceous mechanic contact until final outcropping (Fig. 4). Cretaceous limestones are highly fractured, thus promoting water infiltration through Triassic evaporites. Brines outcrop after deep water ascension. Eh of brine at source is only of 12 mV, evidencing its reductive behaviour and reinforcing deep origin. This model differs from previous preliminary interpretations, which pointed to more superficial recharge and fluid circulation.

5 Conclusion

Deep-circulating salty-fluids outcrop in Cambrils del Pirineu (Pyrenees, NE Iberian Peninsula), after dissolving Triassic halite-sulphate-carbonate rocks. Reductive brines are close to the NaCl saturation. Therefore, continuous salty water supply seems to be ensured and continuity of salt industry is guaranteed.

Acknowledgements This research has been financed with the projects DGICYT PID2021-122467NB-C22, PGC2018-093903-B-C22, PID2020-118999GB-I00 and 2021 SGR 00349.

References

Casas, M., Guinau, M., Travé, A., Saura, E., & Garcia, D. (2022). Dynamics, conditioning factors and possible causes of the formation of the Tartera de Cambrils (Solsonès, Lleida). *Revista de la Sociedad Geológica de España, 35*(1), 3–14.

Mata-Perelló, J. M., Restrepo, C., Montané, P., & Alet, A. (2011). La restauración del Salí de Cambrils. Su papel en la ruta de la sal y en el parque Geológico y Minero del Solsonès – Alt Urgell (Pirineos, Catalunya) (pp. 259–270). In *I Congreso Internacional sobre Patrimonio Geominero, Geología y Minería Ambiental de Bolivia y de los Andes*.

Vergés, J. (1993). *Estudi Geològic Del Vessant Sud Del Pirineu Oriental I Central. Evolució Cinemàtica En 3 D* [PhD thesis, Universitat de Barcelona, Barcelona, Spain].

Artificial Intelligence-Based Decision Support System for Groundwater Management Under Climate Change: Application to Mornag Plain in Tunisia

Youssef Tfifha, Manel Ennahedh and Nehla Debbabi

Abstract

This research aims to investigate the influence of climate change on the groundwater level (GWL) in Mornag plain in Tunisia. Due to the spatiotemporal variability of rainfall (RF) and temperature, aquifers all over the world have seen significant water level decline in recent decades. Therefore, it is crucial to analyze and estimate the GWL variability for reliable groundwater (GW) management in the context of climate change. In this study, we focus on the plain of Mornag, located in the southeast of Tunisia, since it contributes 33% of the national agricultural production. From this plain, we have collected historical piezometric and RF data covering the period 2005–2015. Knowing the RF data, our goal is to forecast the GWL. This issue has already been investigated using numerical GW modeling tools such as Modflow and Feflow. Unfortunately, these techniques are data and time-consuming. To overcome all these drawbacks, we propose to use an Artificial Intelligence (AI) approach that has shown great performance in the literature for recurrent data modeling and forecasting. This approach corresponds to the Long Short-Term Memory (LSTM) Neural Network. Compared to Modflow, LSTM showed a notable improvement in terms of minimizing the mean square error, confirming its suitability for GWL forecasts. Using the proposed AI prediction model, the impact of climate change on Mornag GWL has been studied under two Representative Concentration Pathway (RCP) scenarios; RCP 4.5 and RCP 8.5 for three future periods: 2015–2040 (short term), 2041–2065 (medium term), and 2066–2100 (long term). As expected, the results reveal a future decline for Mornag GWL. The performed study of future Mornag GWL behavior using LSTM could classify this AI approach as a good decision support system that could be used to optimize the management of our limited water resources to satisfy the population needs for drinking water and agricultural production, as well as to avert upcoming drought.

Keywords

Artificial intelligence · Climate change · Decision support system · Forecasting · Groundwater level · Representative concentration pathway

1 Introduction

Groundwater (GW) is one of the most significant and important water sources in the world, it has an impact on numerous aspects of human life, including industrial development, agricultural productivity, the availability of drinking water, etc. (Vaux, 2011). Unfortunately, over recent decades, aquifers all over the world have experienced significant Groundwater Level (GWL) decline that makes water resource management challenging (Anand et al., 2020). Two key factors are behind this decline: the increase in water demand and climate change. Indeed, due to population, industrial, and agricultural growth, the consumption of water resources has skyrocketed. Climate change, on the other hand, has played a significant role in GWL variability, owing primarily to the spatiotemporal unpredictable changes in rainfall (RF) and temperature (Jeppesen et al., 2015; Negm et al., 2020). As a matter of fact, according to UNESCO, GW is the sole way for 2.5 billion people throughout the world to meet their daily water demands (UNESCO, 2012). However, climate change constantly endangers these resources. With full knowledge of the facts, a complete study of historical, current, and future

Y. Tfifha · N. Debbabi (✉)
ESPRIT School of Engineering, 2083 Tunis, Tunisia
e-mail: nehla.debbabi@esprit.tn

M. Ennahedh
National Agronomic Institute of Tunisia, 1082 Tunis, Tunisia

H. Chenchouni et al. (eds.), *Recent Advancements from Aquifers to Skies in Hydrogeology, Geoecology, and Atmospheric Sciences*, Advances in Science, Technology & Innovation, https://doi.org/10.1007/978-3-031-47079-0_4

GWL variability is required for policymakers and practitioners to develop water resource planning and management strategies, as well as avert drought for the upcoming years. Both research and public interest in the projected climatic consequences of GW have intensified in recent years. In fact, GWL has long been anticipated using a variety of numerically based conceptual models, such as Feflow and Modflow (Alloisio et al., 2004). They are widely considered a worldwide benchmark for simulating and forecasting GW. The complexity of the water systems and their response to climatic conditions make the use of numerical models challenging. In addition, these models are mostly data and time-consuming, hard to set up and maintain, and therefore expensive. In recent years, the limitations of traditional numerical models have been widely addressed through the application of Artificial Intelligence (AI) models since they offer high accuracy with relatively less parametrization. Among these models, we find the Long Short-Term Memory (LSTM) Neural Network model (Zhang et al., 2021) that has shown great performance in the literature when handling recurrent data, for instance, for sign language translation (Guo et al., 2018), video prediction (Wang et al., 2018), or weather forecasting (Karevan & Suykens, 2020), to name a few. Since LSTM is well adapted to deal with sequential data, such as time series, we propose in this paper to use it to model the GWL of the Mornag plain in Tunisia. Corresponding to our region of interest, the Mornag plain is one of the most important plains in Tunisia. Given its significance in the bandwidth of GW dynamic flow in the country, modeling the Mornag GWL is critical for better national water resource management. Indeed, available data from Mornag plain corresponds to RF and GWL observations measured over the period 2005–2015. By incorporating new features that improve the description of RF dependencies based on the standard climate change index, such as the Standard Precipitation Index, we propose using an LSTM model to study the direct impact of the climate on the GWL variability based on chronological RF data (SPI). In fact, the goal of this research is to develop a model that accurately predicts future Mornag GWL behavior using only RF data in order to study the effects of climate change on GWL using two Representative Concentration Pathway (RCP) scenarios, RCP 4.5 and RCP 8.5, over the period of 2015–2100, which are global reference data for short-, medium-, and long-term RF values. This AI approach, on the other hand, will be regarded as a decision support system that will be used to optimize the management of our limited water resources in order to meet the needs of the population in terms of drinking water and agricultural production, as well as prevent an upcoming drought. The remainder of this paper is structured as follows: In Sect. 2, we describe the study area as well as the used dataset. The proposed model for GWL forecasting is detailed in Sect. 3. Findings and outcomes are discussed in Sect. 4. Finally, conclusions are drawn in Sect. 5.

2 Materials and Methods

2.1 Region of Interest: The Mornag Plain

This study focuses on the Mornag plain, which is in northern Tunisia, 20 km southeast of the capital Tunis. The study area has a moderate climate that ranges from arid to semiarid. The annual RF is approximately 526 mm (Zaafouri, 2020). As illustrated in Fig. 1, this plain is drained by two major rivers (Meliane and El Hma). The surface area of the Mornag aquifer is about 200 km^2. It stretches over 14 km from Tunis Gulf (the Mediterranean Sea) in the North to the Khledia hills in the South. It is limited to the west by the Rades hills and their surroundings, and to the east by the J. Rourouf mountain and its surroundings. Its hydraulic system is made up of an unconfined (GW lodged in recent Quaternary series) and a confined aquifer (a deep aquifer that groups a series of 4 systems that occurred in ancient

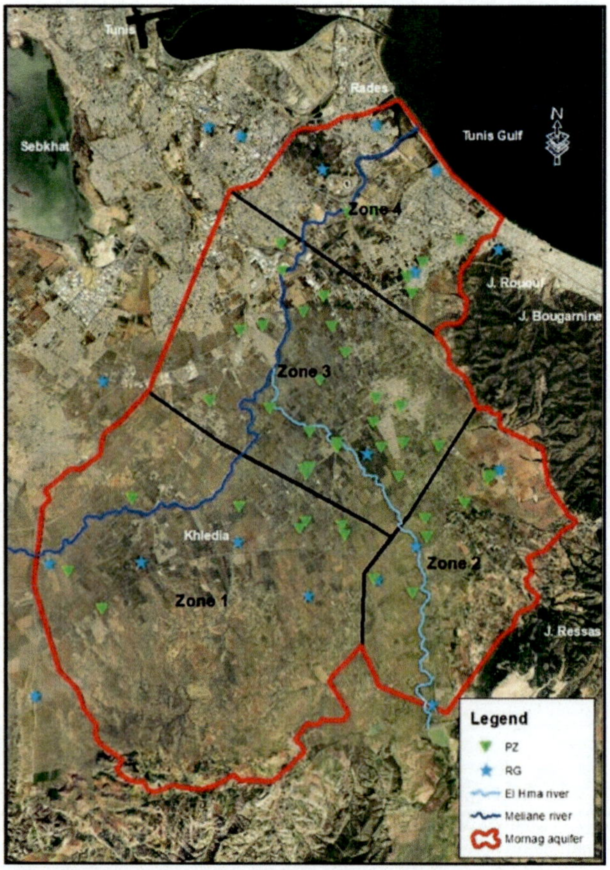

Fig. 1 General location map for the Mornag plain

Quaternary, Oligocene, Miocene, and Eocene sediments). The aquifer system of the Mornag plain is characterized by the presence of the densest observation system of GWL in Tunisia (Horriche, 2004). In this study, we focus on the unconfined aquifer. Thus, it is more and more exploited: in 2015, the exploitation rate reached 195% with a deficit equal to −6.62 mm^3/year (Menani, 2015). This massive exploitation caused a significant piezometric depression and, as a result, an increase in the salinity of the groundwater studied (Zaafouri, 2020). This aquifer is monitored by 44 piezometric stations and 18 pluviometric observation points, which record the GWL and RF, respectively, and allow hydrologists and researchers to better understand and investigate the Mornag aquifer system.

2.2 Data Understanding and Preprocessing

The Mornag plain dataset and the additional features that will be used to improve the GWL forecast are both described in this section. The initial version of our dataset has 285,480 rows and 4 features: daily RF (RFd), Rain Gauge (RG), GWL, and Piezometer (PZ). In fact, GWL data has been collected twice annually using the 44 PZs since 1971, and RF has been measured every day using RGs since 2005. We used mean interpolation to handle the missing values in the collected dataset (8.8% for RFd and 4% for GWL). It will be possible to monitor both spatial changes and temporal changes in the GWL of the Mornag aquifer (on a daily, seasonal, inter-annual, and longer-time scale). As we tracked the GWL over time, we identified 4 distinct delimited zones (Fig. 1); each one groups PZs with comparable variability. As a result, the first new feature, called Zone, designates the piezometric location. GWLs can be affected by nearby surface water bodies, such as rivers, even though seasonal climatic changes are the primary cause of GWL variations. Due to the significant correlation between GWL and both long-term and short-term RF (Ahmadi & Sedghamiz, 2007), we have on the one hand included data on seasonal influences such as monthly RF (RFm), trimestral RF (RFt), semestral RF (RFm), and yearly RF (RFy). On the other hand, we have included the wetness and dryness category described by the feature SPI-C, which will be better understood by studying the Standardized Precipitation Index (SPI) (Cancelliere et al., 2007). Finally, we take into account the month, year, as well as two features due to the chronological dependencies. Summary of the 13 features that make up the new dataset: RG, PZ, SPI-C, and Zone are four categorical features, and RFd, RFm, RFt, RFs, RFy, SPI, Month, Year, and GWL are nine continuous features.

3 Artificial Intelligence for Groundwater Level Modeling

Long Short-Term Memory (LSTM) networks are an improved version of the traditional Recurrent Neural Networks (RNNs) (Bianchi et al., 2017) that are frequently used to handle sequential data, such as time series. As stated in the literature, RNNs suffer from the "vanishing gradient" problem during back-propagation, where the gradient gets smaller with every layer during network training, until it is too small to reach the deepest levels. This drawback makes basic RNN memory unable to learn past information (Zhang et al., 2021). Alternatively, because they were specifically designed to address this issue, LSTMs came to recall long-term dependencies. With LSTM, training errors retain their values, which overcomes the vanishing gradient problem and allows learning from sequences hundreds of time steps long (Zhang et al., 2021). The initial LSTM model consists of a single hidden LSTM layer that could be composed of several Memory Blocks (MBs). MB is shown in Fig. 2b, which is composed of three sigmoids (σ) separate work layers and one hyperbolic tangent (Tanh) corresponds to the key contribution of the LSTM neural network since the decision to consider or throw away information is taken inside. Indeed, the LSTM MB has three gates to govern the information flow; for a given time t, coupling the input x_t and the output of the previously hidden state h_{t-1}, the forget gate f_t regulates which and how much cell information is forgotten, the input gate i_t controls which inputs are used to update the old cell state (conveying important information) c_{t-1}, into the new cell state c_t and as for the output gate o_t it defines which cell memory elements are used to update the hidden state h_t of the LSTM cell (Bianchi et al., 2017).

Using a unique LSTM model as initially developed, the past information stored inside a sequence is barely captured (Zhang et al., 2021). Alternatively, one could use the stacked LSTM, which is a model expansion that involves several hidden LSTM layers, with many MBs for each of them, enhancing the ability to capture more complicated associations in the dataset.

Based on experimentation, in this paper, we propose the stacked LSTM architecture illustrated in Fig. 2a, for which two LSTM layers are used, each one consisting of 50 MBs. As input, our stacked LSTM model uses 30 time steps of the encoded sequence X_1 to X_{12} corresponding to our 12 features. The stacked LSTM is followed by a dropout layer to reduce overfitting while improving model performance. Finally, we add a Fully Connected (FC) layer, giving rise to the output layer, denoted by Y and corresponding to the GWL in our study.

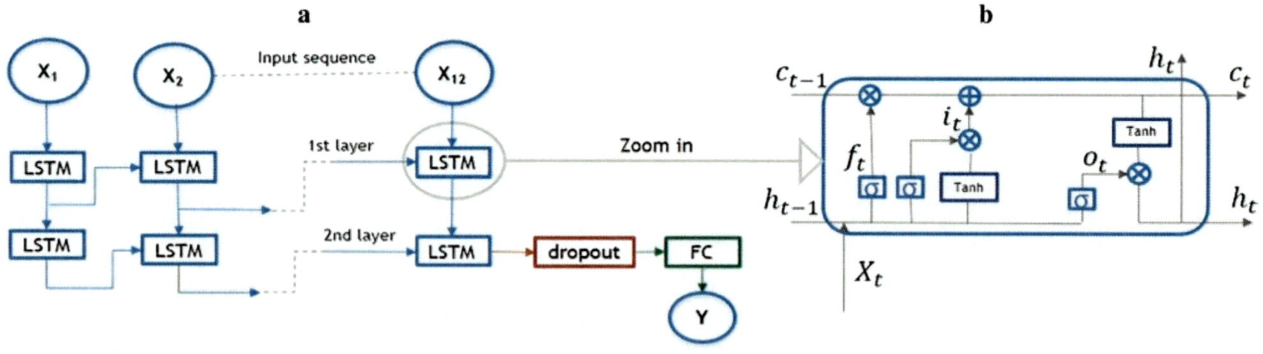

Fig. 2 Illustration of the proposed LSTM Neural Network architecture for GWL forecasting

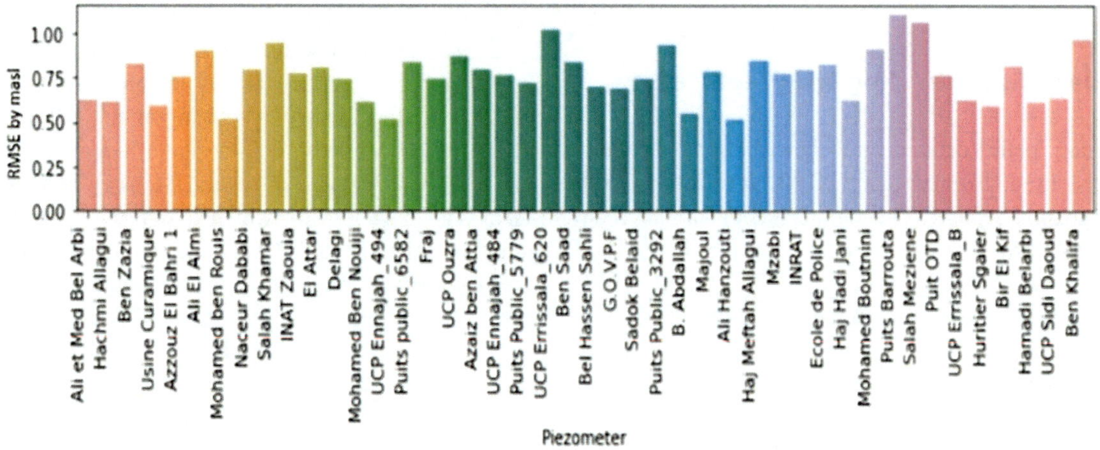

Fig. 3 RMSE of all piezometric stations using data from 2013 to 2015

4 Results and Discussion

To train the proposed stacked LSTM model, we used 70% of our dataset, which covers the years 2004 through 2011. We kept 30% of the data for testing purposes, covering the years 2011 through 2015. In order to model hyperparameters, we have used the "Adam" optimizer, a stochastic gradient descent technique that has demonstrated excellent efficacy and resilience. The corresponding batch size has been set to 512, and 8 epochs have been taken into account. We have taken into account a rate of 0.2 for the dropout layer and have used the RMSE as the loss function. After being trained, we used test data to predict GWL using the suggested stacked LSTM model. We use RMSE to represent training and testing losses (on a log scale) in Fig. 4. A good fit is demonstrated by the fact that both loss curves are descending and reaching a low point with a small gap attending less than 10^{-2}. Additionally, we display the obtained RMSEs for each PZ in Fig. 3; the results show very respectable values. Using the same conditions, data, and overall PZs, we also compared the forecasting outcomes using the proposed stacked LSTM and the Modflow

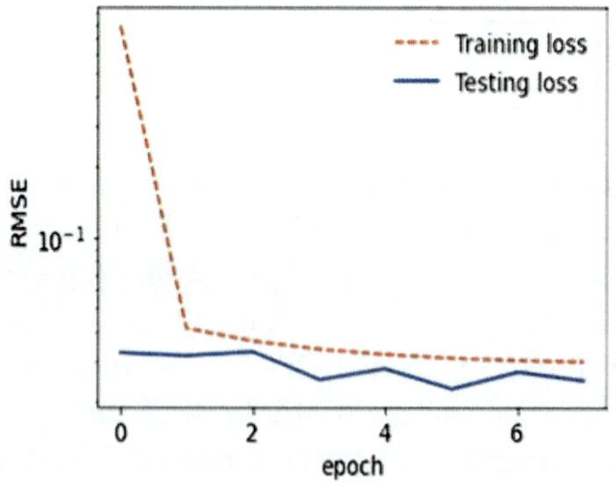

Fig. 4 Validation and train losses

one; LSTM outperforms the Modflow, which has an RMSE of 6.9 m (Ennahedh et al., 2020), with an RMSE of 0.85 m. The study's findings demonstrate how AI techniques are used in Mornag GWL forecasting, which makes it a useful decision support system for managing water resources.

The forecasting will be performed using both the Representative Concentration Pathway RCP 4.5 and 8.5, which are two of the Intergovernmental Panel on Climate

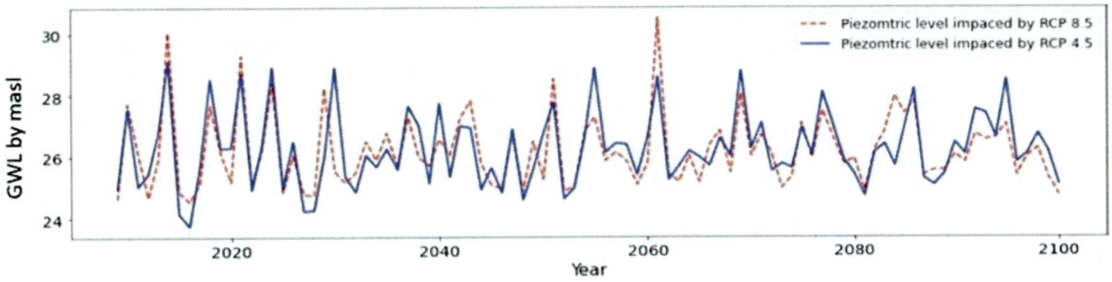

Fig. 5 Sample of the forecasting results under RCP 4.5 and 8.5 of the "Ben Saad" piezometric station

Change (IPCC) scenarios of radiative forcing trajectories to the year 2100. Because of constraints such as the selection of future virtual situations, an imprecise physical understanding of various self-connections, and computational capabilities, these scenarios are unexpected. As a result, they were readily rectified by utilizing the mean values of the various models constituting these scenarios. The predicted results of the piezometric station "Ben Saad" are presented in Fig. 5.

For our simulated results up to 2100, GW resources in the Mornag aquifer were affected by climate change due to a decline in natural recharge from reduced precipitation (the mean will be 19.42% less at the end of the century for RCP 4.5; and 44.86% less for RCP 8.5). As Fig. 5 shows, the absolute variations in GWL under RCP 4.5 and RCP 8.5 may seem small (between 1 and 5 masl), but the fact that we are investigating a shallow aquifer (thickness between 30 and 50 masl) reinforces the importance of the results in terms of water availability for vegetation and agriculture. A drop of tens of centimeters (depending on aquifer thickness) can be critical for plants during hot and dry periods if GW is no longer accessible.

Due to the lack of GW, some farms in the study area would have to be abandoned, which raises serious concerns about the future of irrigated agriculture there. However, there is a pressing need for adaptation strategies that consider the effects of climate change on GW resources, particularly to increase agricultural productivity. Examples of such strategies include a widespread switch from gravity irrigation to drip irrigation and the growth of crops that are both water-saving and climate change-resistant.

5 Conclusions

The Mornag plain has already experienced water scarcity because of anthropogenic activities such as over-pumping and climate change conditions; an increasing number of drought years has been observed over the last few decades. Therefore, the present study focuses on the assessment of the pressure that climate change will impose on GWL in the future. However, an Artificial Intelligence Decision Support System was conducted, while SPI was adopted for the first time in Mornag Plain, to identify climate change impacts on groundwater resources and, at the same time, explore the use of SPI as an indicator to predict GW responses to climate conditions and the use of AI modeling techniques in designing and planning GW management. The developed intelligent model LSTM produced significant results in GWL prediction, with an RMSE of less than 1 m. Furthermore, the main finding of the climate change impact assessment indicates that the predicted hydrological drought events will affect water table fluctuation in the medium and long term, with a drawdown of up to 5 m. Thus, these results are of great importance as key information for decision-makers regarding the future of the sustainable exploitation of groundwater resources in the aquifer.

References

Ahmadi, S. H., & Sedghamiz, A. (2007). Geostatistical analysis of spatial and temporal variations of groundwater level. *Environmental Monitoring and Assessment, 129*(1), 277–294.

Alloisio, S., et al. (2004). *Groundwater modeling for large-scale mine dewatering in Chile: MODFLOW or FEFLOW*. Water Management Consultants.

Anand, B., Karunanidhi, D., Subramani, T. et al. (2020). Long-term trend detection and spatiotemporal analysis of groundwater levels using GIS techniques in Lower Bhavani River Basin, Tamil Nadu, India. *Environment, Development, and Sustainability, 22*(4), 2779–2800. https://doi.org/10.1007/s10668-019-00318-3

Bianchi, F. M., Maiorino, E., Kampffmeyer, M. C., Rizzi, A., & Jenssen, R. (2017). Recurrent neural networks for short-term load forecasting: An overview and comparative analysis.

Cancelliere, A., Mauro, G. D., Bonaccorso, B., & Rossi, G. (2007). Drought forecasting using the standardized precipitation index. *Water Resources Management, 21*(5), 801–819.

Ennahedh, M., Hariga-Tlatli, N., Tarhouni, J. (2020). Hydrogeological modeling for the aquifer system of the Mornag plain (Tunisia) for future real-time management. In *3rd Conference of the Arabian Journal of Geosciences*.

Guo, D., Zhou, W., Li, H., & Wang, M. (2018). Hierarchical LSTM for sign language translation. In *Proceedings of the AAAI Conference on Artificial Intelligence* (Vol. 32(1)).

Horriche, F. (2004). *Contribution à l'analyse et à la rationalisation des réseaux piézométriques* [PhD thesis]. Tunis El Manar University, ENIT.

Jeppesen, E., Brucet, S., Naselli-Flores, L. et al. (2015). Ecological impacts of global warming and water abstraction on lakes and reservoirs due to changes in water level and related changes in salinity. *Hydrobiologia, 750*(1), 201–227. https://doi.org/10.1007/s10750-014-2169-x

Karevan, Z., & Suykens, J. A. (2020). Transductive LSTM for time-series prediction: An application to weather forecasting. *Neural Networks, 125*, 1–9.

Menani, M. (2015). Evaluation du risque de conflit autour des eaux transfrontalières du système aquifère du Sahara septentrional (SASS). *Larhyss Journal*, 22, 59–59.

Negm, A., Bouderbala, A., et al. (2020). *Water Resources in Algeria-Part I: Assessment of Surface and Groundwater*. Springer, Cham.

UNESCO. (2012). *World's groundwater resources are suffering from poor governance*. UNESCO Natural Sciences Sector News.

Van Vuuren, D. P., Edmonds, J., Kainuma, M., et al. (2011). The representative concentration pathways: An overview. *Climatic Change, 109*(1), 5–31. https://doi.org/10.1007/s10584-011-0148-z

Vaux, H. (2011). Groundwater under stress: The importance of management. *Environmental Earth Sciences, 62*(1), 19–23.

Wang, Y., Jiang, L., Yang, M. H., Li, L. J., Long, M., & Fei-Fei, L. (2018). Eidetic 3D LSTM: A model for video prediction and beyond. In *International Conference on Learning Representations*. https://openreview.net/forum?id=B1lKS2AqtX

Zaafouri, A. (2020). *Impact de l'urbanisation sur la zone humide de sebkha Ariana et dynamique sédimentaire du littoral (Golfe de Tunis)* [Doctoral dissertation]. University of Sfax. Sfax National Engineering School: ENIS.

Zhang, J., Zeng, Y., & Starly, B. (2021). Recurrent neural networks with long-term temporal dependencies in machine tool wear diagnosis and prognosis. *SN Applied Sciences, 3*(4), 442. https://doi.org/10.1007/s42452-021-04427-5

Groundwater Pollution Risk Mapping Using Index Methods (North-East Tunisia)

Omeyma Gasmi, Mourad Louati, Ammar Mlayah and Juan José Gomez Alday

Abstract

The aim of this study is to assess, within a GIS, the pollution risk of the Mornag groundwater using the DRASTI-LU and SI index methods, and to validate the resulting maps with geochemical data. To achieve this objective, four risk maps were created by combining hydrogeological and geochemical data. 41 samples were analyzed to validate the risk maps developed. Each risk map was calcified using four reclassification methods so that a total of 16 risk maps were produced. The spatial distribution of groundwater contamination risk allowed results to be compared by classifying them into five risk classes: very low, low, medium, high, and very high. All the resulting risk maps show that the downstream part of the aquifer is the riskiest to pollution compared to the other areas. In fact, the results of validation of the risk maps show that relatively the best map with the lowest correlation index is the one made by the DRASTI-LU method obtained by the additive combination with a weight of 5 for the land use parameter. The coincidence between the nitrate concentration distribution and the spatial distribution of risk degree confirms that the study area is affected by anthropogenic pollution more precisely agricultural pollution by nitrates.

Keywords

Mornag groundwater · Pollution · Risk · GIS · Index methods · Geochemistry

1 Introduction

Groundwater is one of the most important natural resources in the world. In recent decades, pollution has been one of the most global problems threatening these water resources (Chenchouni et al., 2022). Due to its location in North Africa, between the Mediterranean and the Sahara, Tunisia is subject to a predominantly semi-arid to arid climate and therefore has low water resources (Elloumi, 2016). The country's groundwater resources have shown the most acute signs of overexploitation and anthropogenic contamination manifested by a decrease in piezometric level, the invasion of salinization and contamination of shallow groundwater (Lachaal et al., 2018). Since 1960, the Mornag Basin (NE Tunisia) has experienced a major agricultural and industrial development (Yousfi et al., 2019). Indeed, the infiltration of pesticides and fertilizers in cultivated areas causes contamination and degradation of groundwater quality in this region. In this context, a pollution risk assessment of the Mornag groundwater was carried out within a Geographic Information System (GIS).

O. Gasmi (✉) · A. Mlayah
Georesources Laboratory, Centre de Recherches Et Des Technologies Des Eaux CERTE, Technopole Borj Cedria, BP 273, 8020 Soliman, Tunisia
e-mail: Omayma.gasmi@gmail.com

O. Gasmi
Faculty of Sciences of Bizerte, University of Carthage, 7021 Bizerte Jarzouna, Tunisia

M. Louati
Higher Institute of Environmental Technologies, Urban Planning and Building (ISTEUB), University of Carthage, Carthage, Tunisia

J. J. G. Alday
Group of Hydrogeology, Biotechnology and Natural Resources Laboratory, Institute for Regional Development, University of Castilla-La Mancha, Albacete, Spain

H. Chenchouni et al. (eds.), *Recent Advancements from Aquifers to Skies in Hydrogeology, Geoecology, and Atmospheric Sciences*, Advances in Science, Technology & Innovation, https://doi.org/10.1007/978-3-031-47079-0_5

2 Materials and Methods

A delineation of the zones of different pollution risk in the Mornag aquifer was carried out using the DRASTI-LU and SI models. Six major hydrogeological factors such as water body depth, net recharge, aquifer media, soil properties, topography, unsaturated zone impact and also land use were considered as variables in these two models. Four risk maps were developed. Each map was classified according to 4 reclassification methods, so a total of 16 risk maps were produced. To verify the resulting water pollution risk maps of the Mornag aquifer, hydrochemical data were used (Gasmi et al., 2022). Ten hydrochemical parameters (pH, TDS, $Ca^{2+}, Mg^{2+}, Na^{+}, K^{+}, HCO_3^{-}, SO_4^{2-}, Cl^{-}, NO_3^{-}$) were used for the calculation of the water quality index (WQI) to validate the resulting risk maps.

3 Results and Discussion

The spatial distribution of the risk of groundwater pollution allowed to comparate the results by classifying them in five risk classes: very low, low, medium, high, and very high. The different specific vulnerability maps produced by additive combination according to the different classification methods show that the risk index varies from one place to another and the classification of degrees of risk varies from one map to another when we change the classification method. These maps prove that the swallow part is the riskiest part with a very high degree of risk.

The risk maps drawn up according to the DRASTI-LU method carried out by additive combination with a weight of LU equal to 20 show that the risk index varies between 64 and 312. Thus, the downstream part is the common part which is the riskiest to pollution for all the maps drawn up according to the different reclassification methods, this is the part which presents a strong anthropic activity.

The risk maps produced according to the multiplicative combination and according to the different reclassification methods show that the contamination risk index varies between 0 and 23,760 and that the northern part is the most at risk of pollution. The risk maps developed using the SI method show that the SI index varies from 24.08 to 74.25.

Validation by nitrates

See Fig. 1 and Table 1.

Validation by the WQI quality index

The results of the validation of the different risk maps elaborated using several methods of reclassification by nitrates and by the quality index show that the map realized by the DRASTI-LU method realized by the additive combination with a weight equal to 5 for the land use parameter is the best map with the lowest correlation index (Figs. 1 and 2). Important differences appear between the results of the risk assessment methods. The results show that as soon as we change the method of combination, classification and the weight of the LU parameter, the risk map changes. Indeed, all the risk maps confirm that the downstream part of the aquifer is the riskiest to pollution. This area is the one occupied by urban areas and cultivated lands and also with high and very high vulnerability because of its low slope, its permeability of the aquifer medium and its shallow water body. (Table 2).

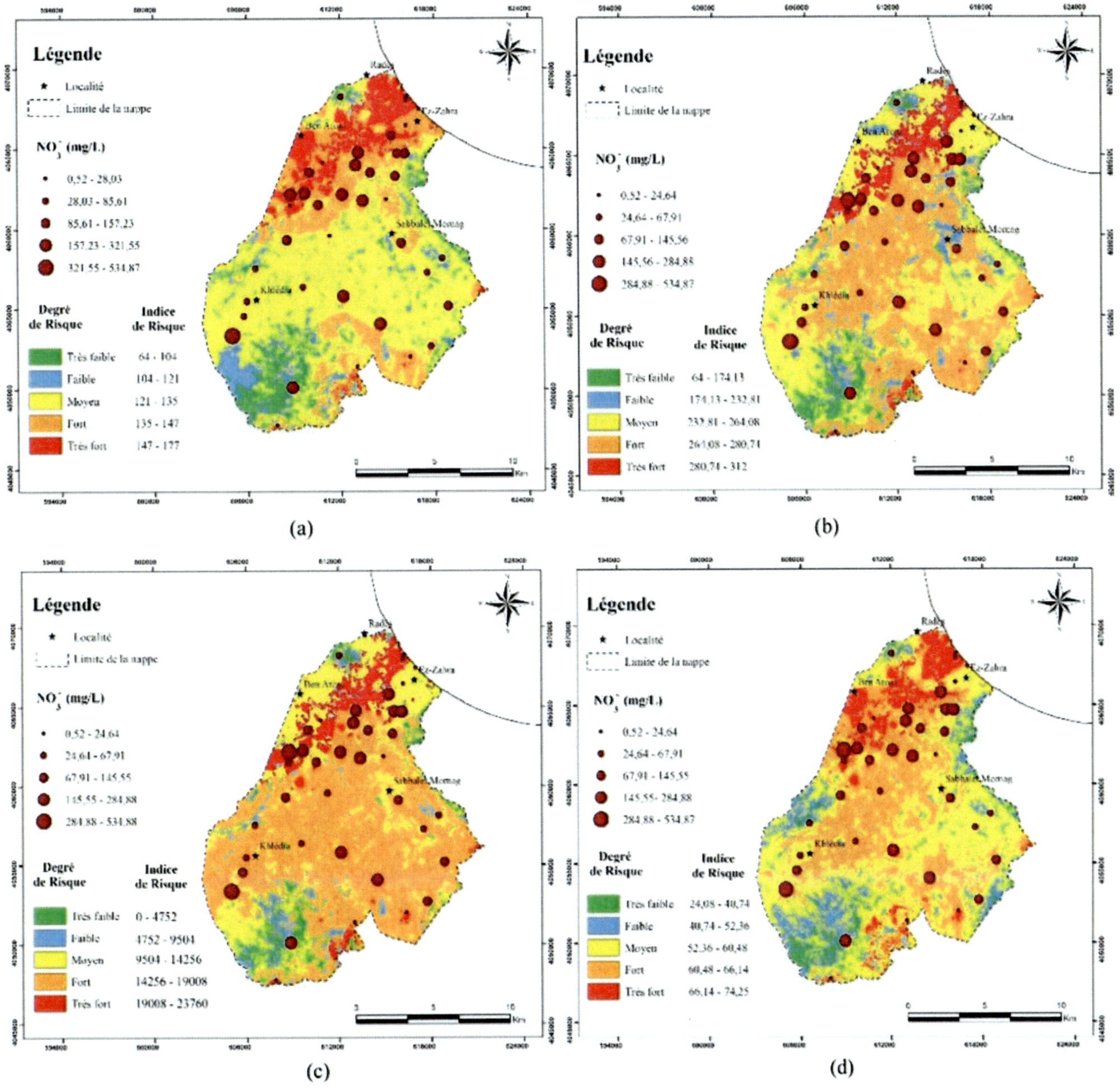

Fig. 1 Distribution of nitrates in the risk maps

Table 1 Correlation index corresponds to the different risk maps and nitrate concentrations

	(a) DRASTI-LU by additive combination with a weight equal to 5	(b) DRASTI-LU by additive combination with a weight of 20	(c) DRASTI-LU by multiplicative combination	(d) SI Specific Vulnerability Map
NO_3-	52	59	58	55

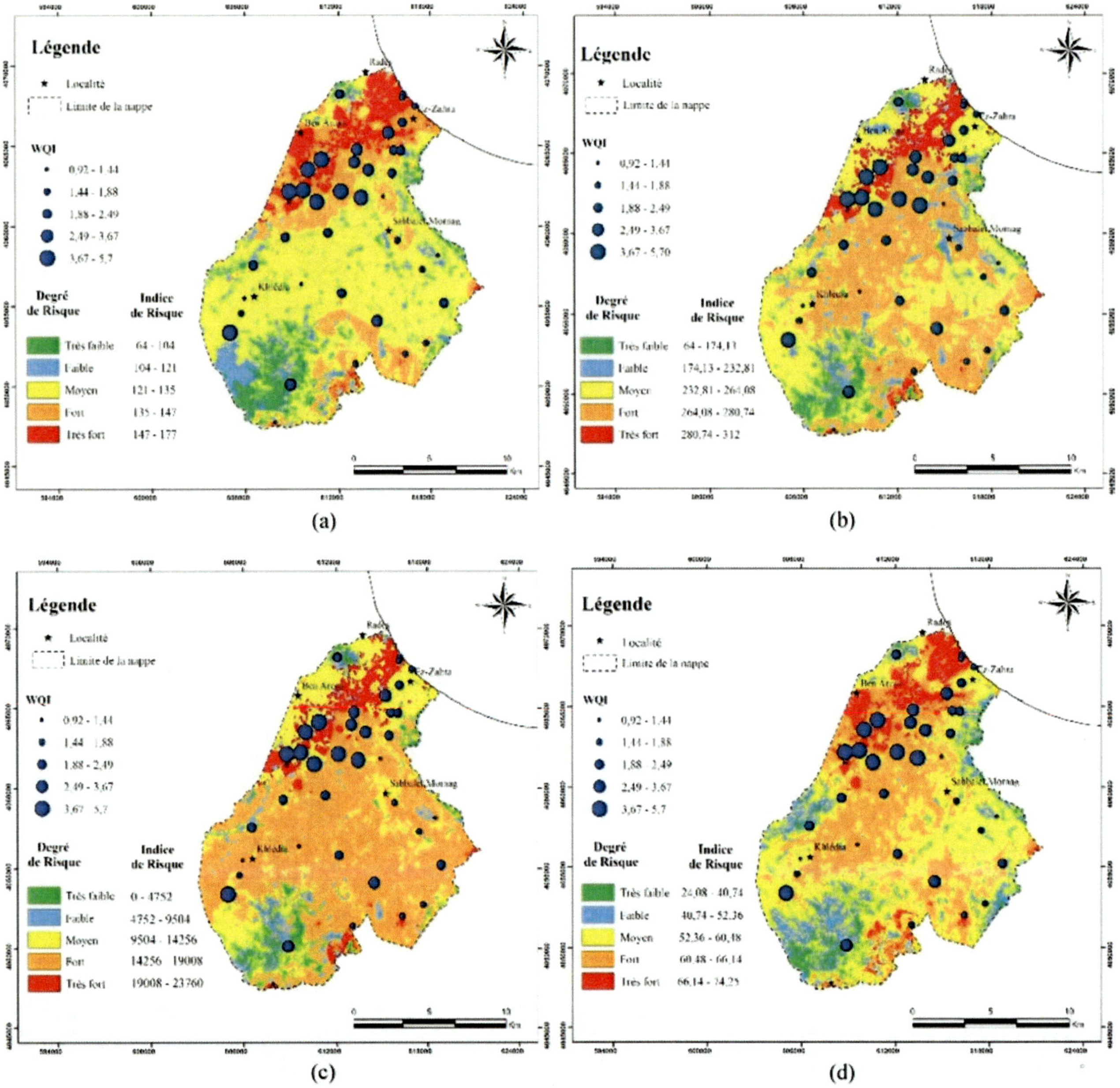

Fig. 2 Distribution of the WQI in the risk maps

Table 2 Correlation index corresponds to the different risk maps and the WQI

	(a) DRASTI-LU by additive combination with a weight equal to 5	(b) DRASTI-LU by additive combination with a weight of 20	(c) DRASTI-LU by multiplicative combination	(d) SI Specific vulnerability map
WQI	48	53	54	48

4 Conclusion

The application of DRASTI-LU and SI models of water risk assessment to pollution in the study area shows that they can provide a reliable and better vision of the threats affecting the groundwater resources of the Mornag plain which are mainly human activities. Also, this highlights the reliability of each model. Therefore, the Mornag water table must be protected by developing a monitoring strategy to resist water degradation and ensure its restoration to its natural state. Overall, the results obtained from this study have demonstrated the important influence that humans have on the sustainable use of groundwater in the Mornag Plain. These results should attract the attention of decision makers and stimulate their reaction to stop any efflux of contamination.

References

Elloumi, M. (2016). *La gouvernance des eaux souterraines en Tunisie*. International Water Management Institute project (IWMI). https://gw-mena.iwmi.org/wp-content/uploads/sites/3/2017/04/Rep.7-Groundwater-governance-in-Tunisia_final_cover.pdf

Chenchouni, H., Chaminé, H. I., Khan, M. F., Merkel, B. J., Zhang, Z., Li, P., Kallel, A., & Khélifi, N. (2022). *New Prospects in Environmental Geosciences and Hydrogeosciences*. Cham, Springer. https://doi.org/10.1007/978-3-030-72543-3

Gasmi, O., Louati, M., Chekirbane, A., Menchen, A., Twihri, A., Alday, J. J. G., & Mlayah, A. (2022). Assessment of groundwater quality and pesticide distribution in Mornag aquifer using GIS-based technique (Northeast Tunisia). *Arabian Journal of Geosciences, 15*(1042). https://doi.org/10.1007/s12517-022-10210-6

Lachaal, F., Ben Messaoud, R., Jellalia, D., et al. (2018). Impact of water resources management on groundwater hydrochemical changes: Case of Grombalia shallow aquifer, NE of Tunisia. *Arabian Journal of Geosciences, 11*, 304. https://doi.org/10.1007/s12517-018-3656-6

Yousfi, R., El Ouear, Z., Dahri, N., Ouddane, B., & Rigane, H. (2019). Evaluating the heavy metals-associated ecological risks in soil and sediments of a decommissioned Tunisian Mine. *Polish Journal of Environmental Studies, 28*(4), 2981–2993. https://doi.org/10.15244/pjoes/86198

Groundwater Dynamics in the Haouz Plain: Analysis of the Interactions Between Vegetation, Water and Climate Data

Imane El Bouazzaoui, Yassine Ait Brahim, El Mahdi El Khalki, Adam Najmi, Adelhakim Amazirh and Blaid Bougadir

Abstract

The Haouz aquifer faces multiple environmental and socioeconomic challenges largely related to climatic aridity and to the growth of the agriculture sector. By combining several methods such as cascade analysis, cross-correlation and Principal Component Analysis (PCA), this study conducts a comprehensive analysis of hydroclimatic data in order to comprehend the interactions between water management components. The findings indicate that there are three different groundwater functioning systems. The first system is noticeable in areas with extensive irrigation from groundwater. The water cycle's natural balance is disturbed, and surface water's contribution to groundwater recharge is negligible. The second and the third systems are manifested in areas where pumping is low. The hydrological cycle balance is preserved, which exhibits a cascading effect in all of its studied elements and a noticeable contribution from surface water to groundwater recharge. To promote effective groundwater management, this study has made data and graphics available to reveal the challenges related to groundwater vulnerability to climatic aridity and overexploitation in the Haouz region.

Keywords

Haouz · Irrigation · Aquifer · Pumping · Water cycle · Cascade

1 Introduction

Drought is a serious problem that hinders the development of countries (Chenchouni et al., 2023). Understanding the impact of drought on groundwater dynamics is essential for regions facing water scarcity, as it directly affects agricultural sustainability and socioeconomic development. By analyzing these interactions, valuable insights can be gained into effective groundwater management strategies that are adaptable and cost-efficient, applicable across diverse aquifer types and terrains globally (Kallel et al., 2023). In Morocco, semi-arid regions are among the most impacted by water scarcity and socioeconomic development (Ouhamdouch et al., 2019). This situation is becoming more alarming as it constitutes a serious threat to surface water resources, and consequently, to the natural recharge of groundwater, widely used for irrigation (Montginoul & Molle, 2019). This could endanger the sustainability of the agricultural sector, considered the principal activity in the Haouz region. This study seeks to analyze the impact of drought propagation on groundwater dynamics in the Haouz plain (Fig. 1), which might help gain insights into groundwater management, using simple and economical methods, adapted to different types of aquifers and terrains.

2 Methods

The current study intends to analyze the impact of many factors, such as rainfall, runoff, the Normalized Difference Vegetation Index (NDVI), the Leaf Area Index (LAI), on groundwater dynamics in the Haouz plain. It also seeks to identify the effects of drought on said components. This

I. El Bouazzaoui (✉) · B. Bougadir
Laboratory of Sciences Applied to the Environment and Sustainable Development, Cadi Ayyad University, 20000 Essaouira, Morocco
e-mail: imane.elbouazzaoui@ced.uca.ma

Y. Ait Brahim · E. El Khalki
International Water Research Institute, Mohammed VI Polytechnic University, 43150 Ben Guerir, Morocco

A. Najmi
Laboratory of Georesources, Geo-Environments, and Civil Engineering, Cadi Ayyad University, 40000 Marrakech, Morocco

A. Amazirh
Center for Remote Sensing Applications, Mohammed VI Polytechnic University, 43150 Ben Guerir, Morocco

H. Chenchouni et al. (eds.), *Recent Advancements from Aquifers to Skies in Hydrogeology, Geoecology, and Atmospheric Sciences*, Advances in Science, Technology & Innovation, https://doi.org/10.1007/978-3-031-47079-0_6

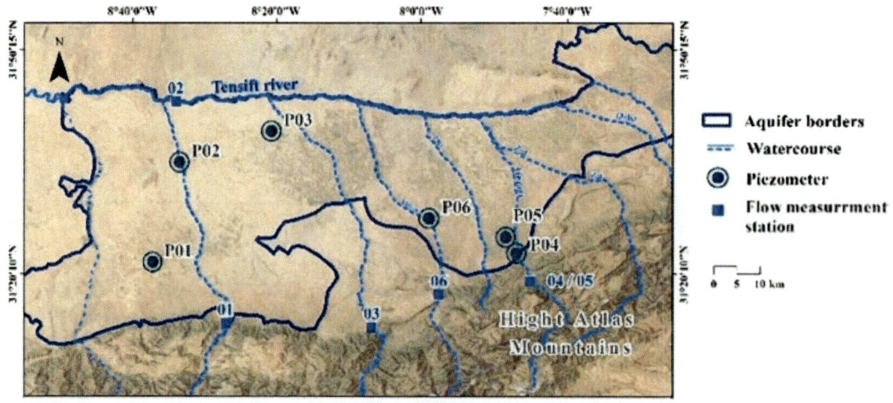

Fig. 1 Map of the study area, showing the location of the studied piezometers and runoff measurement stations

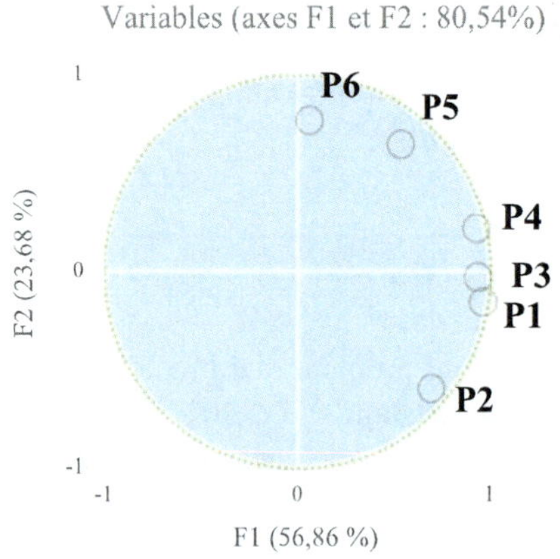

Fig. 2 Principle component analysis performed on the six studied piezometers

was done first through the PCA (Zitko, 1994) of the six piezometers, then by cross-correlation in order to measure the similarities between the datasets associated with each factor. This method is already in use in many regions of the world to analyze the natural recharge of groundwater and its interactions with other components of the water cycle (Fiorillo & Doglioni, 2010; Ha et al., 2006; Howell et al., 2019; Kim et al., 2016; Lee et al., 2018; Mondal & Singh, 2004). The following part consists in analyzing the temporal co-evolution of the factors in order to identify whether there is a cascading effect between them, which was applied in California for the same purpose (Massoud et al., 2020).

3 Results

The PCA (Fig. 2) and the cross-correlation (Table 1) show the presence of three distinct groundwater operating systems. Piezometers P4, P3 and P1 constitute the first system, the second is indicated by piezometers P6 and P5, while P2 fluctuations show a unique behavior different from the two other systems. For the first system, the correlation between surface water and the PL is insignificant (Table 1). This system is also defined by a significant and negative correlation between the vegetation and the PL, except for P4, which correlates positively with the local NDVI. For the second system, the PL has a positive correlation with rainfall, runoff and vegetation (Table 1). For the average LAI, the correlation with the PL is significant, while for the local NDVI, it is significant at P6 and less at P5. As for P2, it has an instantaneous and a positive response to rainfall and runoff. It correlates significantly with the average LAI and insignificantly with the local NDVI. For the second and the third systems, the cascading effect is more noticeable. The effects of drought spread from rain to runoff and PL, but not as much to vegetation because of the use of surface water for irrigation in these areas.

Table 1 Cross-correlation between PL, rainfall, runoff and vegetation indices

		Rainfall	Runoff	Average LAI	Local NDVI
1st system	P1			−0.86 (lag = −2)	−0.67 (lag = −4)
	P3			−0.96 (lag = −3)	−0.79 (lag = −5)
	P4		0.4 (lag = 1)	−0.88 (lag = −1)	0.98 (lag = −4)
2nd system	P5	0.4 (lag = −4)	0.4 (lag = −4)	0.9 (lag = 0)	0.4 (lag = 0)
	P6	1 (lag = −3)	0.82 (lag = 2)	1 (lag = 0)	1 (lag = 0)
3rd system	P2	0.65 (lag = 0)	0.92 (lag = 0)	0.5 (lag = 0)	

4 Discussion

The first groundwater operating system is characterized by a disturbed balance of the water cycle, reflected by a low correlation between surface water and the PL, due to the piezometer's location in areas dominated by groundwater irrigation (Bouazzaoui et al., 2022). Moreover, rainfall recovery periods are insufficient for groundwater recharge, resulting in a very pronounced and prolonged PL deficit, noticeable at P1 and P3 (Fig. 3). For P4, the positive correlation between the PL and the local NDVI can be explained by the fact that this region requires little pumping, because of its proximity to the Rocade Canal, which enables water to be transferred from the Oum Erbia basin to the Haouz plain (Bouazzaoui et al., 2022). Watercourses are experiencing drought throughout the driest year of the study period (2011–2012), while the PL is also at its lowest level for P4, resulting in a complete cascade. Regarding the second system, the water cycle is preserved due to the piezometer's location in low-pumping areas close to the Rocade canal (Bouazzaoui et al., 2022) and to the contribution of surface water to groundwater recharge, especially at P6. This system starts to experience water table stress from 2011 to 2012 (Fig. 3) given the minimum recharge episodes occurring in that year (Hajhouji, 2018). Therefore, a complete cascade occurred at P5 succeeded by another one the following year, while P6 showed an incomplete cascade that did not reach the PL. As for the third system, surface water contributes significantly to groundwater recharge. The low correlation between the local NDVI and groundwater at P2 suggests that groundwater is substantially less used for irrigation at this location, where rain-fed crops are cultivated. P2 also shows a complete cascade during 2011–2012 (Fig. 3). Moreover, the proximity of watercourses to piezometers that show a correlation between runoff and the PL suggests that watercourses can be considered a potential recharge site for groundwater. This finding was supported by the study made by Abourida on groundwater recharge in the Houz plain (Abourida, 2007).

5 Conclusion

This study was conducted in order to understand the challenges associated with groundwater overexploitation under semi-arid climate conditions in the Haouz region. This alarming state of the groundwater requires pro-active attention in the setting up of public policies with long-term goals, and an efficient management based on sustainable alternatives that are less water intensive and more profitable for farmers, such as the use of crop varieties adapted to water stress or even the use of rationalized irrigation techniques like the model-based control.

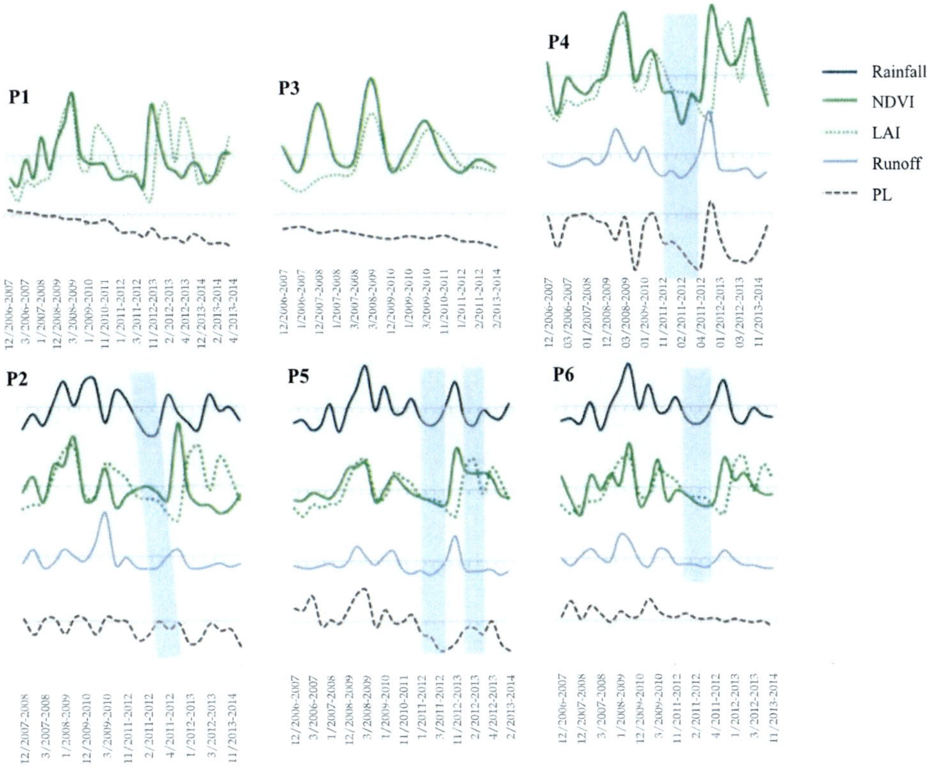

Fig. 3 Rainfall, vegetation, runoff and PL temporal dynamics of the three systems. The blue area shows the transfer of drought from rainfall toward vegetation, runoff and groundwater resulting in complete or incomplete cascades

References

Abourida, A. (2007). *Approche hydrogéologique de La nappe du Haouz (Maroc) par télédétection, isotopie, SIG et modélisation.* Université Cadi Ayyad: Faculté des sciences Semlalia, Marrakech.

El Bouazzaoui, I., Ait-Brahim, Y., El Khalki, E. M., Najmi, A., & Bougadir, B. (2022). A summary analysis of groundwater vulnerability to climate variability and anthropic activities in the Haouz Region, Morocco. Sustainability, *14.* https://doi.org/10.3390/su142214865

Chenchouni, H., Chaminé, H. I., Zhang, Z., Khelifi, N., Ciner, A., Ali, I., & Chen, M. (2023). Recent Research on Hydrogeology, Geoecology and Atmospheric Sciences. Springer, Cham. https://doi.org/10.1007/978-3-031-43169-2

Fiorillo, F., & Doglioni, A. (2010). The relation between karst spring discharge and rainfall by the cross-correlation analysis. *Hydrogeology Journal, 18,* 1881–1895. https://doi.org/10.1007/s10040-010-0666-1

Ha, K., Ko, K.-S., Koh, D.-C., Yum, B.-W., & Lee, G. (2006). Time series analysis of the responses of the groundwater levels at multi-depth wells according to the river stage fluctuations. *Economic and Environmental Geology, 39,* 269–284.

Hajhouji, Y. (2018). *Modélisation Hydrologique Du Bassin Versant de l'oued Rheraya et Sa Contribution à La Recharge de La Nappe Du Haouz (Bassin Du Tensift, Maroc).* Université Paul Sabatier-Toulouse III.

Howell, B., Fryar, A., Benaabidate, L., Bouchaou, L., & Farhaoui, M. (2019). Variable responses of karst springs to recharge in the middle atlas region of Morocco. *Hydrogeology Journal, 27,* 1–18. https://doi.org/10.1007/s10040-019-01945-w

Kallel, A., Barbieri, M., Rodrigo-Comino, J. (2023). *Selected Studies in Environmental Geosciences and Hydrogeosciences.* Springer, Cham. https://doi.org/10.1007/978-3-031-43803-5

Kim, I., Park, D., Kyung, D., Kim, G., Kim, S., & Lee, J. (2016). Comparative influences of precipitation and river stage on groundwater levels in near-river areas. *Sustainability, 8,* 1. https://doi.org/10.3390/su8010001

Lee, J. M., Park, J. H., Chung, E., & Woo, N. C. (2018). Assessment of groundwater drought in the Mangyeong River Basin, Korea. *Sustainability, 10,* 831. https://doi.org/10.3390/su10030831

Massoud, E., Turmon, M., Reager, J., Hobbs, J., Liu, Z., & David, C. H. (2020). Cascading dynamics of the hydrologic cycle in California explored through observations and model simulations. *Geosciences, 10(2),* 71. https://doi.org/10.3390/geosciences10020071

Mondal, N., & Singh, V. (2004). A new approach to delineate the groundwater recharge zone in hard rock terrain. *Current Science, 87,* 658–662. https://www.jstor.org/stable/24109057

Montginoul, M., & Molle, F. (2019). Modernisation Des Périmètres Irrigués Au Maroc: Une Solution Aux Effets Contrastés Pour Les Agriculteurs et La Ressource En Eau (p. 47). Le Cas d'un Périmètre Du N'Fis - Haouz; IRD-Institut de Recherche pour le Developpement; UMR G-Eau.

Ouhamdouch, S., Bahir, M., Ouazar, D., Carreira, P., & Zouari, K. (2019). Evaluation of climate change impact on groundwater from semi-arid environment (Essaouira Basin, Morocco) using integrated approaches. *Environmental Earth Sciences, 78.* https://doi.org/10.1007/s12665-019-8470-2

Zitko, V. (1994). Principal component analysis in the evaluation of environmental data. *Marine Pollution Bulletin, 28,* 718–722. https://doi.org/10.1016/0025-326X(94)90329-8

Modeling CO_2 Geological Storage and CO_2-Circulated Geothermal Harvest in a Heterogeneous Reservoir in North Oman

Mingjie Chen, Ali Al-Maktoumi, Azizallah Izady and Sulaiman Al-Hashmi

Abstract

CO_2 geological storage has been investigated and carried out as a strategic approach to mitigate climate change caused by drastic increase of CO_2 emission to the atmosphere in the past 20 years. To offset the cost associated with sequestration, partial stored CO_2 can be circulated to harvest geothermal energy from the deep reservoir, which has been demonstrated more efficient than conventional water-based method. This study aims to model CO_2 sequestration and subsequent geothermal production in a heterogeneous reservoir in North Oman. CO_2 injected, net storage, spatiotemporal distribution of reservoir pressure, temperature and CO_2 plume, and recovered geothermal energy are assessed based on the model simulation. The findings provide a preliminary feasibility evaluation of operating CO_2 sequestration and CO_2-circulated geothermal recovery in similar reservoirs in North Oman and worldwide.

Keywords

Geothermal energy · CO_2 injection · CO_2 circulation · Geological heterogeneity · Horizontal well

M. Chen (✉) · A. Al-Maktoumi · A. Izady
Water Research Center, Sultan Qaboos University, Muscat, Oman
e-mail: mingjie@squ.edu.om

A. Al-Maktoumi
Department of Soils, Water and Agricultural Engineering, Sultan Qaboos University, Muscat, Oman

S. Al-Hashmi
Environmental Studies and Research Center, Sultan Qaboos University, Muscat, Oman

1 Introduction

Renewable energy is becoming an attractive alternative to meet energy demands due to increasing concerns of CO_2 emission and the depletion of fossil energy reserves. Exploring potential renewable energy resources is strategically significant to diversify Oman's petroleum-dominant economy, besides the demonstrated environmental benefits over fossil fuels. Compared with solar and wind energy sources, geothermal energy has many advantages, including being unaffected by weather changes, providing stable base load power, and high thermal efficiency (Axelsson, 2012). There are considerable geothermal resources in North Oman Mountains due to the tectonic movements, where some oil fields are located (Al-Lamki & Terken, 1996). Recently, an emerging technique called CO_2 Plume Geothermal (CPG) production has been considered to be an effective approach to harvest low-enthalpy geothermal energy from saline aquifers owing to CO_2's favorable thermophysical properties (Adams et al., 2014). It can be integrated with CO_2 geological storage, which is believed an effective mitigation approach to reduce atmospheric CO_2. This study aims to numerically model and assess CO_2-circulated geothermal production from a thin fault block reservoir, where 3 megatonnes (Mt) of CO_2 is injected.

2 Methodology and Modeling

As shown in Figs. 1 and 2, the model domain is simplified from a reservoir block in Daleel oil field in North Oman (Zhang et al., 2007). The domain is 2° inclined clockwise around Y-axis, extended 2000, 3000, and 20 m in XYZ, and discretized into 50 $m^3 \times 50$ $m^3 \times 1$ m^3 cells. The porosity field is generated using an exponential variogram model (Cirpka, 2003). The field data shows that the porosity mean and variance are 0.25 and 0.002, respectively. The correlation lengths are 1000, 1000, and 1 m in X, Y, and Z

H. Chenchouni et al. (eds.), *Recent Advancements from Aquifers to Skies in Hydrogeology, Geoecology, and Atmospheric Sciences*, Advances in Science, Technology & Innovation, https://doi.org/10.1007/978-3-031-47079-0_7

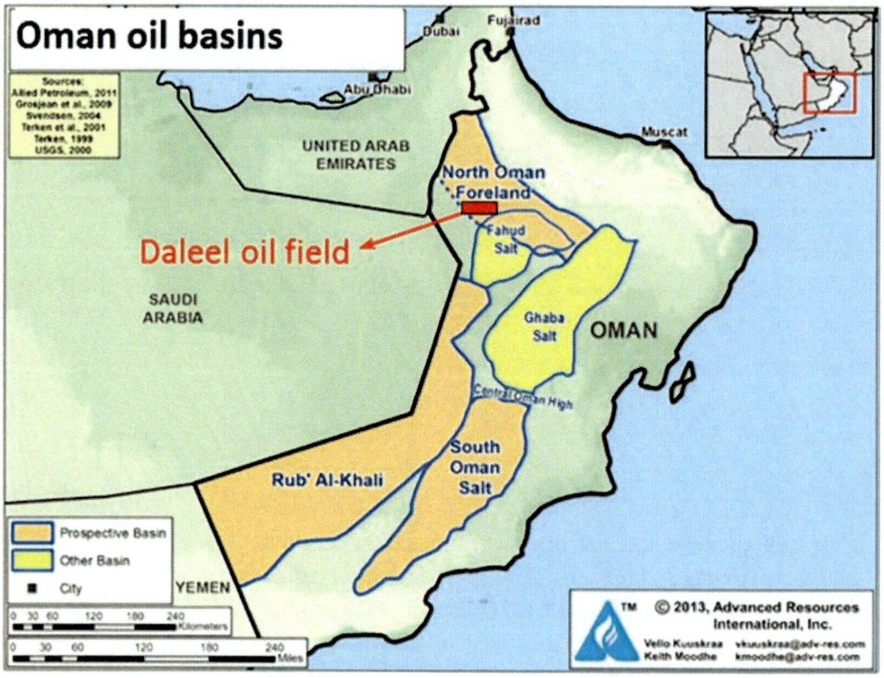

Fig. 1 Location of study area: Daleel oil field

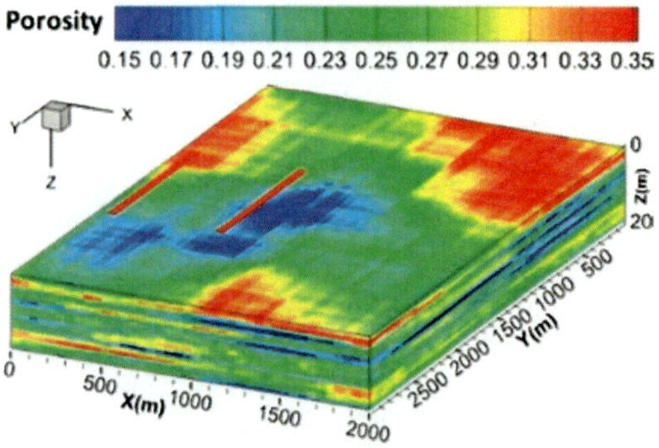

Fig. 2 3D model domain and heterogeneous porosity field. Z-scale is exaggerated by 20

directions. Permeability field is estimated from porosity using the relationship $k = 192\phi^3/(1 - \phi^2)$ regressed from Daleel field data (Abbaszadeh et al., 2000), where k and ϕ represent permeability (in mD) and porosity, respectively. The other hydrothermal parameters are treated homogeneous: rock density, compressibility, specific heat, and thermal conductivity are 2650 kg/m^3, 4.5×10^{-10} Pa^{-1}, 1000 J/(kg °C), and 2.1 W/(m °C), respectively. Residual CO_2 and water saturations are set as 0.05 and 0.3. van Genuchten α and n parameters are 5.1×10^{-5} Pa^{-1} and 0.46. Brine salinity is 180 ppt based on field measurement.

The left top is at 1500 m deep, and initial pressure is assigned to each model cell according to the depth and hydrostatics. Initial temperatures are assumed uniform 90 °C for the entire reservoir. All the model boundaries are closed except the right side. A constant 0.05 W/m^2 geothermal heat source applies to the bottom. A doublet horizontal well placement, that is, one 1-km injection well (IW) and one parallel production well (PW) 600 m away, are horizontally placed between $Y = 1000$–2000 m on the top layer to circulate CO_2. The cells representing the two wells are assigned a high uniform porosity of 0.4, and shown as two parallel red lines in Fig. 2. A 15 MPa overpressure is specified in IW in both CO_2 injection (1st) and circulation (2nd) stages. The PW starts to produce fluids at a constant rate of 5 kg/s after injection of 3 Mt CO_2 in the 1st stage, to circulate CO_2 for geothermal production. Both overpressure in IW and production rate in PW are linearly increased in the 1st year to the specified values and remain thereafter. The numerical models are developed using Non-Isothermal Unsaturated–saturated Flow and Transport (NUFT) software (Nitao, 1998) to simulate 21 years of CO_2 injection and 50 years of CO_2 circulation. NUFT code is integrated an equation of state for CO_2 thermodynamics and has been applied in multiple CO_2 geothermal projects (Chen et al., 2022, 2023; Rajabi et al., 2021).

3 Results and Discussion

Figure 3 shows the time series of CO_2 injection, production, and resulting storage rates, as well as pressure, temperature, CO_2 cut, and heat recovery rate in PW during the 50 years of CO_2 circulation (2nd stage). The injection/storage rate in the end of the 1st stage (onset of the 2nd stage) is 3.47 kg/s. CO_2 injection rate increases rapidly in the first 3 years of the 2nd stage and slowly to over 8 kg/s in 50th year (Fig. 3a). To maintain 15 MPa overpressure in IW, the increase of injection rate is induced by the production in PW, which is

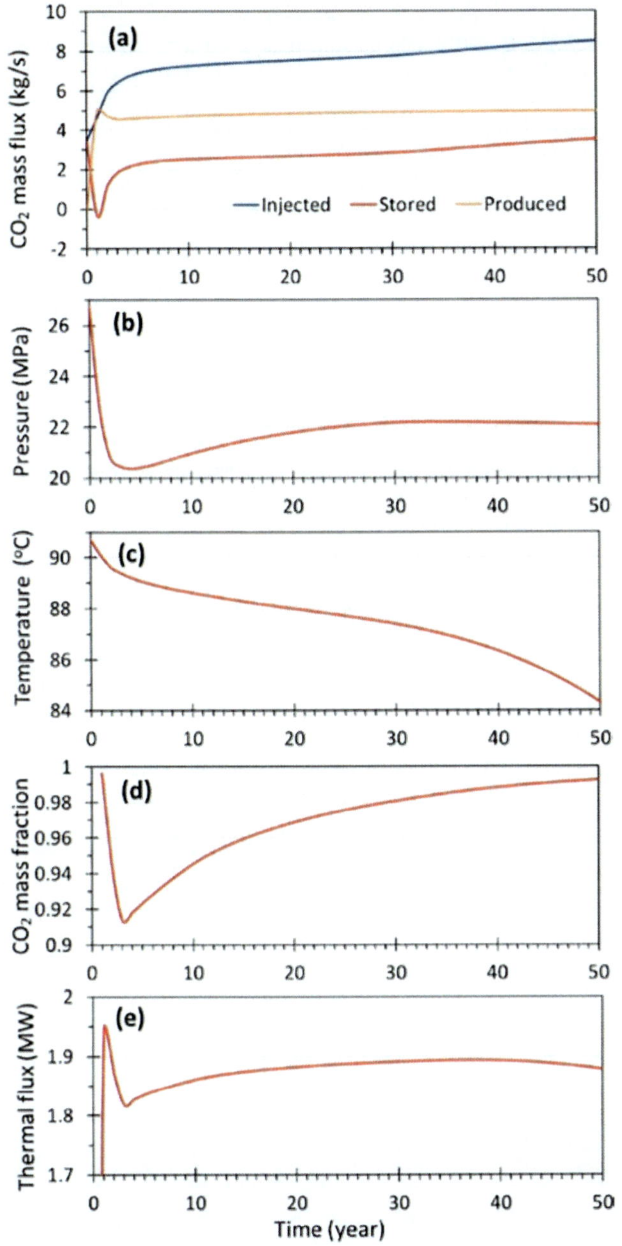

Fig. 3 CO_2 mass flux, and produced fluid pressure, temperature, CO_2 cut, and thermal flux during 50 years of CO_2 circulation

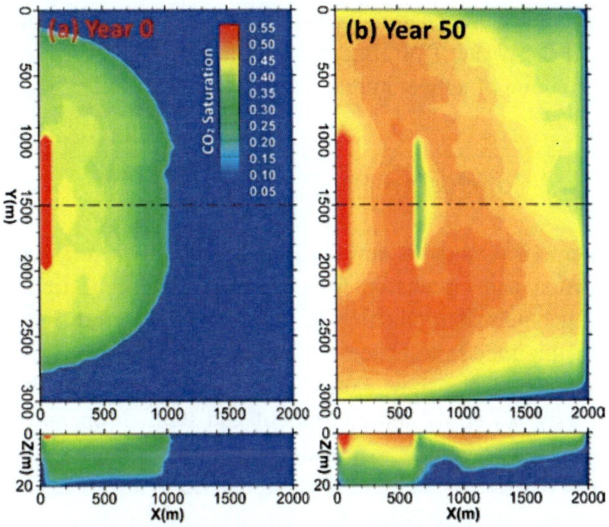

Fig. 4 CO_2 saturation distribution on top XY and middle XZ section at the times of onset and 50th year of CO_2 circulation

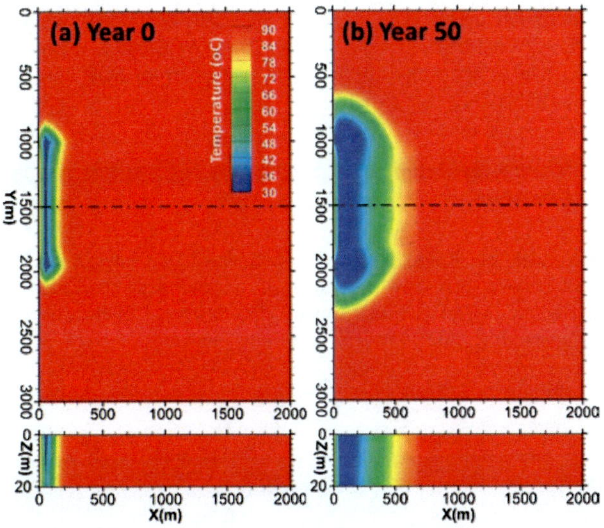

Fig. 5 Temperature distribution on top XY and middle XZ section at the times of onset and 50th year of CO_2 circulation

increased linearly from 0 to 5 kg/s in the 1st year, and then remains constant. The storage rate is the difference between injection and production rate. It decease sharply in the 1st year and then increase smoothly. Brine is co-produced with CO_2, as demonstrated by CO_2 cut curve shown in Fig. 3d. The pressure in PW declines drastically from 27 to almost 20 MPa in the first 2 years, and then increases smoothly to 22 MPa (Fig. 3b). The maximal decline of pressure is about 7 MPa, less than half of the allowed overpressure of the reservoir. It won't impact geomechanical stability of the reservoir. The temperature in PW drops 6 °C smoothly in 50 years of CO_2 circulation (Fig. 3c). Produced thermal flux is a product of mass flux and specific enthalpy determined by pressure and temperature. It increases sharply from 0 to 1.95 MW in the 1st year, in response to the CO_2 production, and then changes relatively slowly with the dynamics of CO_2 mass flux, pressure and temperature (Fig. 3e). The averaged thermal recovery is about 1.86 MW.

CO_2 plume is formed within 1 km in *X* after 3 Mt CO_2 is injected (Fig. 4a). As PW starts to produce fluids, cold CO_2 continues to migrate from IW and some bypasses the PW toward the right boundary (Fig. 4b). CO_2 in lower layer of the right side of PW, however, is abstracted to PW. Heterogeneous effects on CO_2 plume distribution are apparent in both XY and XZ sections. As shown in Fig. 5, temperature propagation is restricted in the sweeping region between IW and PW. Heterogeneity affects only horizontal distribution of temperature.

4 Conclusion

A numerical model is developed to simulate sequential CO_2 injection and circulation for integrated CO_2 sequestration and geothermal production in a heterogeneous thin reservoir simplified from Daleel oil field block B in North Oman. Simulated results show 3.0 and 4.35 Mt CO_2 are stored during 21 years of 1st and 50 years of 2nd stage of operation, respectively. Average geothermal heat recovery rate is about 1.9 MW. Heterogeneity of formation affects CO_2 migration more significantly during CO_2 circulation (2nd stage) than during CO_2 injection only stage (1st stage). Impact of heterogeneity is moderate on temperature propagation horizontally, but negligible vertically in this thin reservoir.

Acknowledgements This work was financially supported by grants #IG/DVC/WRC/22/02 and #RC/RG-DVC/WRC/21/02. Authors would like to appreciate the technical support from SQU research Group DR/RG/17.

References

Abbaszadeh, M., Koide, N., & Murahashi, Y. (2000). Integrated characterization and flow modeling of a heterogeneous carbonate reservoir in Daleel Field, Oman. *SPE Reservoir Evaluation and Engineering, 3*(02), 150–159.

Adams, B. M., Kuehn, T. H., Bielicki, J. M., Randolph, J. B., & Saar, M. O. (2014). On the importance of the thermosiphon effect in CPG (CO_2 plume geothermal) power systems. *Energy, 69*, 409–418.

Al-Lamki, M., & Terken, J. M. (1996). The role of hydrogeology in Petroleum Development Oman. *GeoArabia, 1*(4), 495–510.

Axelsson, G. (2012). The physics of geothermal energy. In C. A. Sayigh (Eds.), *Comprehensive renewable energy* (Vol. 7, Chap. 7.02, pp. 3–50). Elsevier.

Chen, M., Nikoo, M., Al-Maktoumi, A., Izady, A., Cai, J., & Dong, Y. (2022). Use closed reservoirs for CO_2 storage and heat recovery: A two-stage brine-extraction and CO_2-circulation scheme. *Sustainable Energy Technologies and Assessments, 52*, 102346.

Chen, M., Nikoo, M., Al-Maktoumi, A., Izady, A., & Rajabi, M. (2023). The impact of geological heterogeneity on coupled CO_2 storage and geothermal extraction in inclined reservoirs. *Journal of Hydrology, 617*, 128950.

Cirpka, O. A. (2003). Generation of random, autocorrelated, periodic fields. http://m2matlabdb.ma.tum.de/download.jsp?MC_ID=5&MP_ID=31

Nitao, J. J. (1998). Reference manual for the NUFT flow and transport code, version 2.0. Technical report UCRL-MA-130651, Lawrence Livermore National Laboratory, Livermore, CA.

Rajabi, M., Chen, M., Bozorgpour, A., Al-Maktoumi, A., & Izady, A. (2021). Stochastic techno-economic analysis of CO_2-circulated geothermal energy production in a closed reservoir system. *Geothermics, 96*, 102202.

Zhang, L., Zhou, L., Al-Mugheiry, M. A., & Xu, K. (2007). Horizontal Waterflooding in Shuaiba carbonate reservoir of Daleel field in Oman: From pilots performance to development era. Society of Petroleum Engineers.

Hydrogeological and Hydrogeochemical Characterization of the Aquifer System of Regueb (Central Tunisia)

Mouez Gouasmia, Ferid Dhahri, Abdelkader Mhamdi and Mohamed Soussi

Abstract

In the region of Regueb, central Tunisia, agriculture represents the first socio-economic activity and has experienced a considerable extension of irrigated areas which has been accompanied by intensive exploitation of deep waters. The management of these resources is becoming increasingly difficult, and hydrogeological studies of groundwater resources are furthermore required. The multi-layered aquifer system of Regueb is yielded in sandy and sandy-clay formations intercalated by impermeable to semi-permeable Mio-Plio-Quaternary levels. It includes a superficial aquifer level exploited by surface wells and a semi-deep to deep level drilled by deep wells, especially in the south-eastern and north-eastern parts of the basin. The examination of the piezometric data from available wells allowed us to distinguish two areas with different piezometries apparently separated by an E–W edge. The northern area is restricted to the north of Regueb city and exhibits an outflow towards Sebkha Mchiguig, and the southern one is relatively wide and extends overall the basin between Regueb and Mezzouna cities. The groundwater salinity ranges between 2 and 8.5 g/L and increases to the north of the basin, however, nitrates concentrations are relatively increased in the eastern part of the basin in correlation with agricultural activity.

Keywords

Groundwater quality · Hydrochemistry · Hydrogeology · Aquifer system · Regueb · Central Tunisia

1 Introduction

The quantity of fresh water available on earth is limited and, in recent years, its quality is under pressure. This alteration has been increased by the intensification of human activities including the overexploitation of groundwater and its pollution by the intensive use of chemical fertilizers in agriculture (Chenchouni et al., 2022). In arid and semi-arid regions, groundwater is the most important water resource (Negm et al., 2020). In the Regueb region, located in central Tunisia (Fig. 1), the availability of surface water resources is exceptionally linked to flooding events (Dhahri & Boukadi, 2017). Ground waters are used to satisfy the demands of irrigated agriculture in the region which engender intensive exploitation of deep groundwater. The management of these resources is becoming increasingly difficult and further hydrogeological studies of groundwater resources in the region are welcome.

2 Site Description

The Regueb basin, which is part of the Tunisian Eastern Atlas, is located at the eastern limit of the North–South axis (Fig. 1). This basin is filled by Tertiary-Quaternary deposits however, the lithostratigraphic outcrops at its northern,

M. Gouasmia (✉) · F. Dhahri · A. Mhamdi
Faculty of Sciences of Gafsa, University of Gafsa, Sidi Ahmed Zarroug, 2112 Gafsa, Tunisia
e-mail: mouez.gouasmia@yahoo.fr

M. Gouasmia · A. Mhamdi · M. Soussi
Faculty of Sciences of Tunis, LB18ES07: Sedimentary Basins and Petroleum Geology, University of Tunis El Manar, El Manar II, 2092 Tunis, Tunisia

F. Dhahri
Faculty of Sciences of Tunis, Laboratory LR18ES37: Geodynamics, Geomaterials and Geo-digital, University of Tunis El Manar, El Manar II, 2092 Tunis, Tunisia

H. Chenchouni et al. (eds.), *Recent Advancements from Aquifers to Skies in Hydrogeology, Geoecology, and Atmospheric Sciences*, Advances in Science, Technology & Innovation, https://doi.org/10.1007/978-3-031-47079-0_8

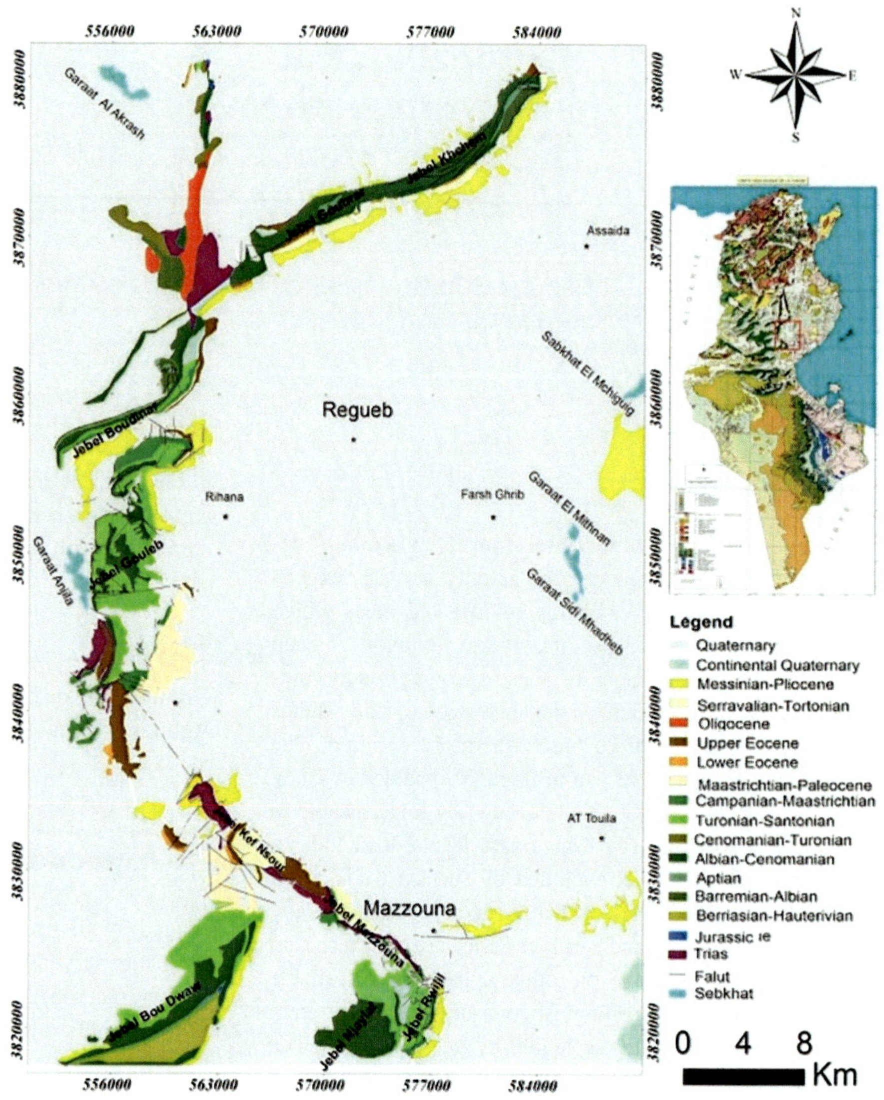

Fig. 1 Geological map of the study area

western and southern mountainous borders are from Triassic to Tertiary ages (Fig. 1) (Dhahri & Boukadi, 2017). The Mio-Pliocene deposits within this basin are made of fine to coarse detrital sediments that rest unconformably over the previous formations. The Quaternary outcrops are widespread over large areas and exhibit heterogenous sedimentation from the detrital, fluvial and eolian origin both of high permeability. Many faults cross the Regueb basin and should control the geometry of the hydrogeological units and the circulation of groundwater. These waters, not suitable for drinking, sometimes exceed the standards for irrigation (> 7.7 g/L).

The aquifer system of Regueb is of multilayer type contained in sandy and sandy-clay formations intercalated by impermeable to semi-permeable Mio-Plio-Quaternary levels. This system includes a superficial level exploited by surface wells and a semi-deep to deep level drilled particularly in the south-eastern and north-eastern parts of the basin (Smida et al., 2005).

3 Results and Discussion

3.1 Hydrodynamics of the Mio-Plio-Quaternary Aquifer

The piezometric map established in 2020 (Fig. 2) allowed us to distinguish two piezometric areas with a general hydraulic gradient of 1.5%. The first one is in the

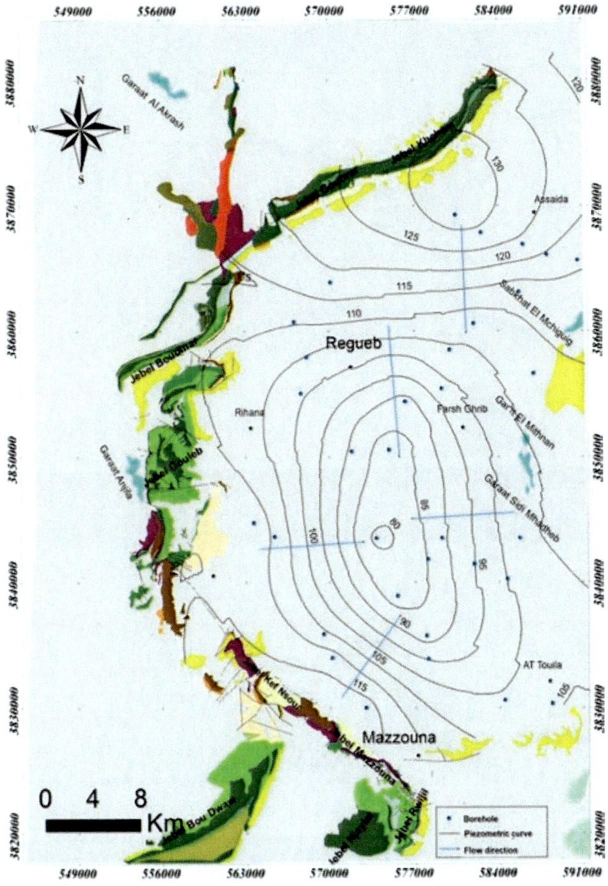

Fig. 2 Piezometric contour map (m) and groundwater flow (blue arrows) (2020)

north-eastern part of the basin. It exhibits a groundwater flow towards the NW–SE. The piezometric curves are well-spaced with concavity oriented towards Sabkhat El Mchiguig which constitutes the natural outlet of the aquifer. This is also the case of the surface flow characterized by endorheic conditions in this region. The second area is relatively widest and extends the overall rest of the basin from Regueb to Mezzouna city in the south (Fig. 1). In this area, the piezometric curves are concentric and flow arrows converge towards the centre. The piezometric depressions indicate significant exploitation of the aquifer due to water abstraction for irrigation. The groundwater balance in this region recorded a deficit of 12.5 Mm^3/year in the last decade and an overexploitation rate of 97%. The water table recharge is ensured mainly by the surface flow from wadis during wet periods and by direct infiltration of meteoric water at the piedmont of the reliefs adjoining the basin.

3.2 Hydrochemistry

The total mineralization map (Fig. 3a) shows that salinity values generally oscillate between 2 and 8.5 g/L with spatial marked zonation. A zone with low salinity (2.5 g/L) characterizes the areas at the piedmonts of the reliefs (recharge zone). Another zone with a salinity of about 4 g/L is located in the southern part, to the north of Mezzouna. This salinity increase is presumably due to the dissolution of Triassic and Upper Eocene evaporite rocks that outcrops along the Meknassy-Mazzouna corridor (Dhahri & Boukadi, 2017). The third zone type is characterized relatively by high salinity ranging from 4 to 8.5 g/L and matching with endorheic depressions such Sabkhat El Mchiguig and Garaat Sidi Mhadhab (Fig. 1). They represent the natural outflows of the surface drainage in which salts accumulate and percolate by infiltration downward to contaminate the aquifer system.

The spatial distribution of nitrates in groundwater is generally related to agricultural activity and especially irrigated cultures often characterized by a large use of pesticides and chemical fertilizers (Missaoui et al., 2022). In the study area, nitrate concentration was measured in 47 water samples and it shows that nitrate levels vary from 40 to 180 mg/L with high concentrations found around Sabkhat El Mchiguig and Garaat El Mithnan due to contamination by the septic fuss and the leaching of the chemical products. The lowest concentrations characterize the border areas, where the water quality seems to be acceptable (Fig. 3b).

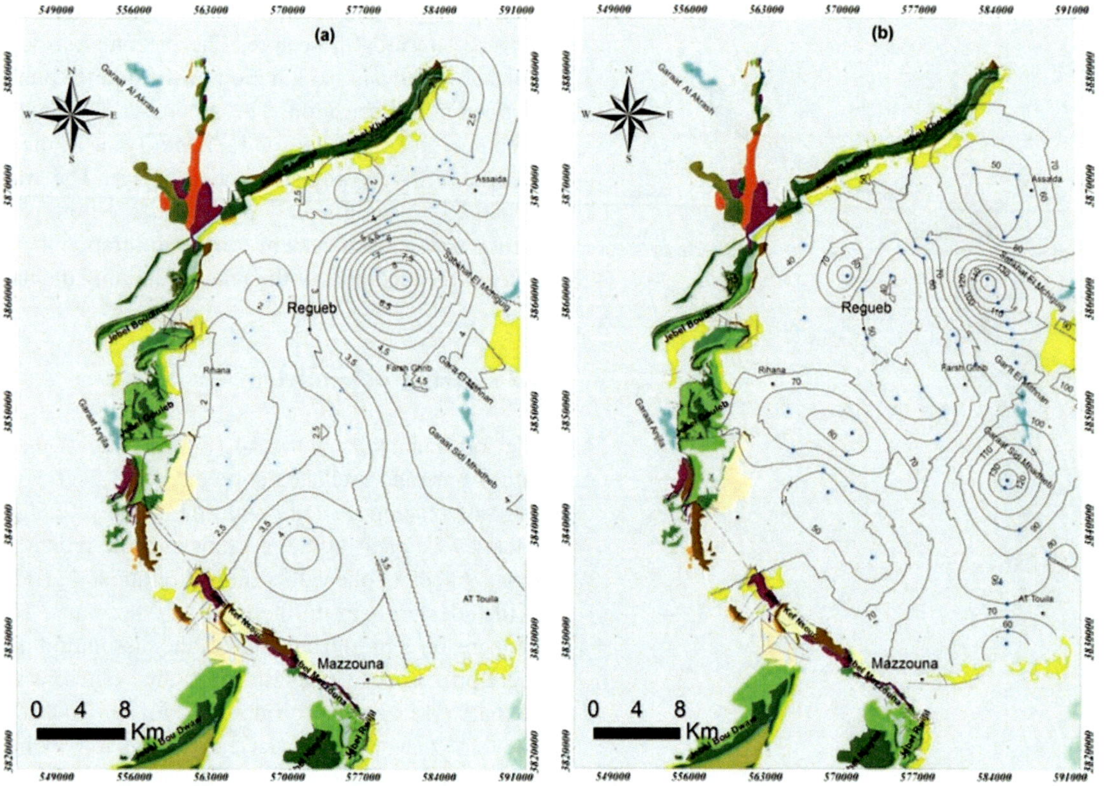

Fig. 3 **a** Spatial variation of salinity (in g/L) and **b** nitrate concentrations (mg/L) in the Regueb basin

4 Conclusions

Groundwater in the region of Regueb, central Tunisia, is yielded in a multi-layered aquifer system made of Mio-Plio-Quaternary detrital units. The hydrogeological potentialities of this aquifer system are respectable however they faced over-exploitation during the last decades to satisfy the increased demands of the emerging agricultural activity in the region. This overexploitation together with the recharge limitation engendered an alarming deficit in the groundwater balance of about 12.5 Mm^3/year and an overexploitation rate of 97%. In addition, the intensive use of chemical fertilizers contributed to the nitrates increase and the degradation of groundwater quality in the basin, especially around the cultivated areas.

References

Chenchouni, H., Chaminé, H. I., Khan, M. F., et al. (2022). *New Prospects in Environmental Geosciences and Hydrogeosciences*. Cham, Springer.

Dhahri, F., & Boukadi, N. (2017). Triassic salt sheets of Mezzouna, Central Tunisia: New comments on Late Cretaceous halokinesis and geodynamic evolution of the northern African margin. *Journal of African Earth Sciences, 129*, 318–329. https://doi.org/10.1016/j.jafrearsci.2017.01.016

Missaoui, R., Abdelkarim, B., Ncibi, K., Hamed, Y., Choura, A., & Essalami, L. (2022). Assessment of groundwater vulnerability to nitrate contamination using an improved model in the Regueb Basin, Central Tunisia. *Water, Air, and Soil Pollution, 233*(8), 1–16.

Negm, A., Bouderbala, A., Chenchouni, H., & Barcelo, D. (2020). *Water Resources in Algeria-Part I: Assessment of Surface and Groundwater*. Springer, Cham.

Smida, H., Zairi M., Trabelsi, R., & Ben Dhia, H. (2005). Etude et gestion des Ressources en eau dans une région aride par le SIG: Cas de la région de Regueb - Sidi Bouzid – Tunisie (p. 20). In *Conférence francophone ESRI*, Issy Les Moulineaux, Paris France.

Chemical and Isotopes Indicators of Mixing Between Multilayered Aquifer Systems of Tadla Plain, Morocco

Mohammed Hssaisoune, Lhoussaine Bouchaou, Mustapha Namous, Mohamed Beraaouz and Tarik Tagma

Abstract

The Tadla plain located in the Oum Er-Rbia River Basin constitutes one of the principal agricultural production areas in Morocco. The groundwater resources are derived from the karst aquifer of the Atlas Mountains and the multilayered system (superficial and deep aquifers) of the Tadla plain. The Turonian constitutes the main productive aquifer in the area. The isotopic composition and concentration of strontium in 43 groundwater samples, combined with solute concentration data, provide important details regarding groundwater geochemical evolution, flow pathways and mixing processes in the Tadla multilayered aquifers. Shallow aquifers are characterized by significantly higher salinity, particularly in irrigated perimeter areas (Beni Amir and Beni Moussa). The relationship between chloride and EC shows a similar correlation for all water groups, except for some Atlas springs with conspicuously lower chloride/EC ratios. Stable isotopes suggest that the waters originated from meteoric water that was infiltrated into the different aquifers through permeable formations without secondary surface evaporation. The strontium isotope ratios of low saline water from the Turonian aquifer and the High Atlas spring waters overlap and show mixing (0.7078–0.7092) between Sr derived from the Turonian age limestone aquifer (0.703) and a higher $^{87}Sr/^{86}Sr$ source. Two types of chloride saline water were found in all aquifers: (i) low $^{87}Sr/^{86}Sr$ (0.7078) and Br/Cl (1.7×10^{-4}) source that affects groundwater in the Turonian and Senonian aquifers. This type of water is originated from the dissolution of gypsum and halite deposits. (ii) High $^{87}Sr/^{86}Sr$ (0.7110) and marine Br/Cl (~ 1.5×10^{-3}) source that occurs mostly in the Eocene aquifer.

Keywords

Strontium isotope · Mixing process · Multilayered aquifer · Groundwater · Tadla plain

M. Hssaisoune (✉) · L. Bouchaou
Laboratory of Applied Geology and Geo-Environment, Ibn Zohr University, 80035 Agadir, Morocco
e-mail: m.hssaisoune@uiz.ac.ma

M. Hssaisoune · M. Beraaouz
Faculty of Applied Sciences, Ibn Zohr University, 86153 Ait Melloul, Morocco

International Water Research Institute (IWRI), Mohammed VI Polytechnic University (UM6P), 43150 Ben Guerir, Morocco

M. Namous
Data4Earth Laboratory, Sultan Moulay Slimane University, Beni Mellal, Morocco

T. Tagma
MRI Laboratory, Polydisciplinary Faculty of Khouribga, Sultan Moulay Slimane University, Beni Mellal, Morocco

1 Introduction

The multilayered aquifers system of Tadla plain, lowland area of Oum Er-Rbia River Basin, localized in northern-central part of Morocco are important groundwater sources—besides karst aquifers of Atlas Mountains—for the many cities residents and their surrounding areas. An increasing demand for water for industrial (mining and agro-food activities) and agriculture uses has led to the overexploitation of aquifers of relatively shallow depths (Hssaisoune et al., 2017, 2020). Therefore, a large number of active boreholes (more than 30,000) have been drilled in the Tadla area with depths varying from a few meters to several hundred meters. Most of them penetrate the Mio-Plio-Quaternary fluvio-lacustrine sediments of the multilayered aquifers. Although some of the wells are used for industrial or drinking water, most of water up to 94% is used for irrigation (ABHOER, 2017).

H. Chenchouni et al. (eds.), *Recent Advancements from Aquifers to Skies in Hydrogeology, Geoecology, and Atmospheric Sciences*, Advances in Science, Technology & Innovation, https://doi.org/10.1007/978-3-031-47079-0_9

The groundwater resources in the area are derived from two sources—the karst aquifer of the Atlas Mountains and the multilayered system of the Tadla plain, which consists of four aquifers: Mio-Plio-Quaternary, Eocene, Senonian, and Turonian. The main productive aquifer in the Tadla plain is the Turonian aquifer, which is separated by fractured formations that allow for significant hydraulic intercommunications. The faulted structure of the various formations also facilitates hydrodynamic relations between the different aquifers in the region. These findings are based on studies conducted by Bouchaou et al. (1997, 2009).

The groundwater resources are derived from the karst aquifer of the Atlas Mountains.

A multi-tracer approach was employed with strontium isotopes ratios, radium isotopes and stable isotopes composition of groundwater system used to evaluate groundwater mixing. In addition, to identify water–rock interaction processes leading to the hydrochemical evolution of groundwater in the catchment (Warner et al., 2013; Vinson et al., 2013). It is anticipated that these findings can be applied in other semi-arid regions worldwide to help stakeholders to manage groundwater resources.

2 Material and Methods

2.1 Environmental Settings

The study area is located in the Oum Er-Rbia River Basin, central part of Morocco (Fig. 1). It is one of the largest Moroccan basins occupying an area of nearly 35,000 km^2 and contain one of the principle agricultural production areas in the country. It is demarcated in the southeast by the Atlas Mountains (Middle and High Atlas), that constitute the main source of groundwater recharge for different aquifers in the region, limited in the north by the Phosphates plateau and in the west by the Atlantic Ocean (Fig. 1).

The Tadla plain, as a vast depression limited by Atlas Mountains in the South and by phosphate plateau in the North, is composed of Mio-Plio-Quaternary sediments (sands, gravels and lacustrine limestone), which covers a Mesozoic syncline in the north of the basin and a Paleozoic schistose basement in the south. The Plio-Quaternary strata in the plain are locally heterogeneous both in vertical and lateral directions (Fig. 2).

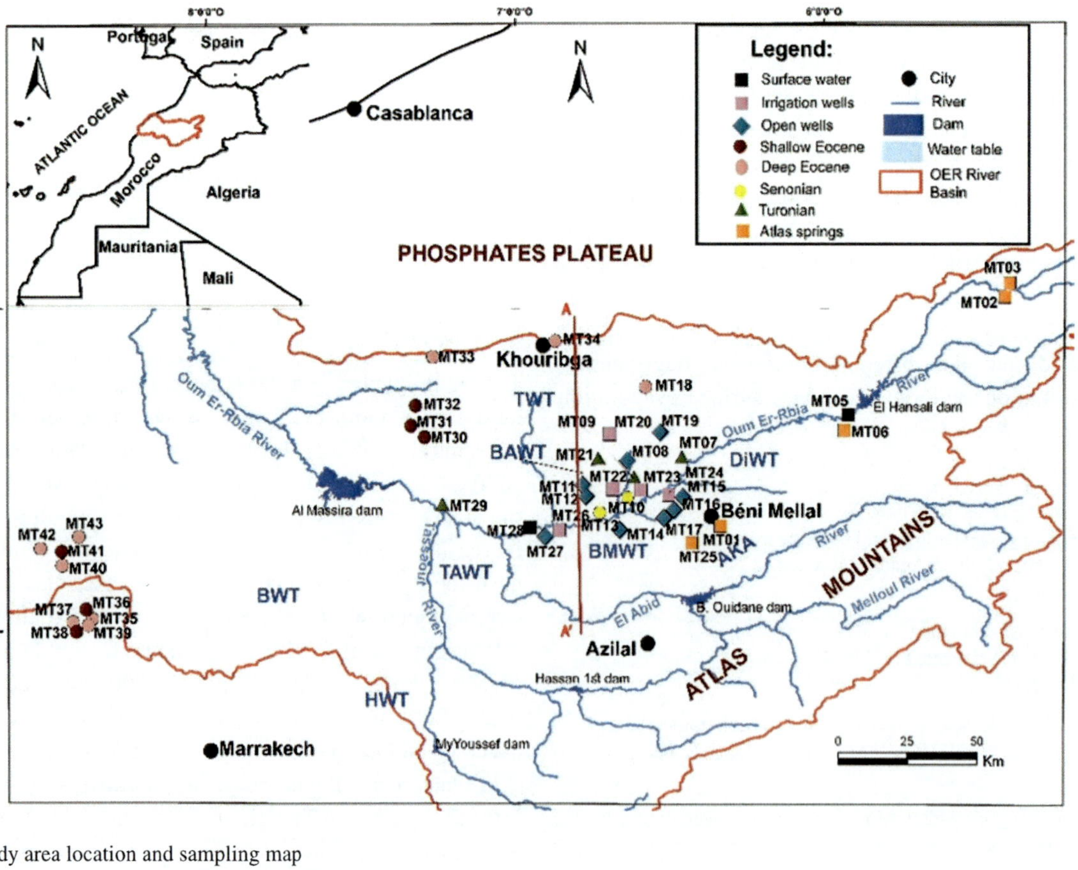

Fig. 1 Study area location and sampling map

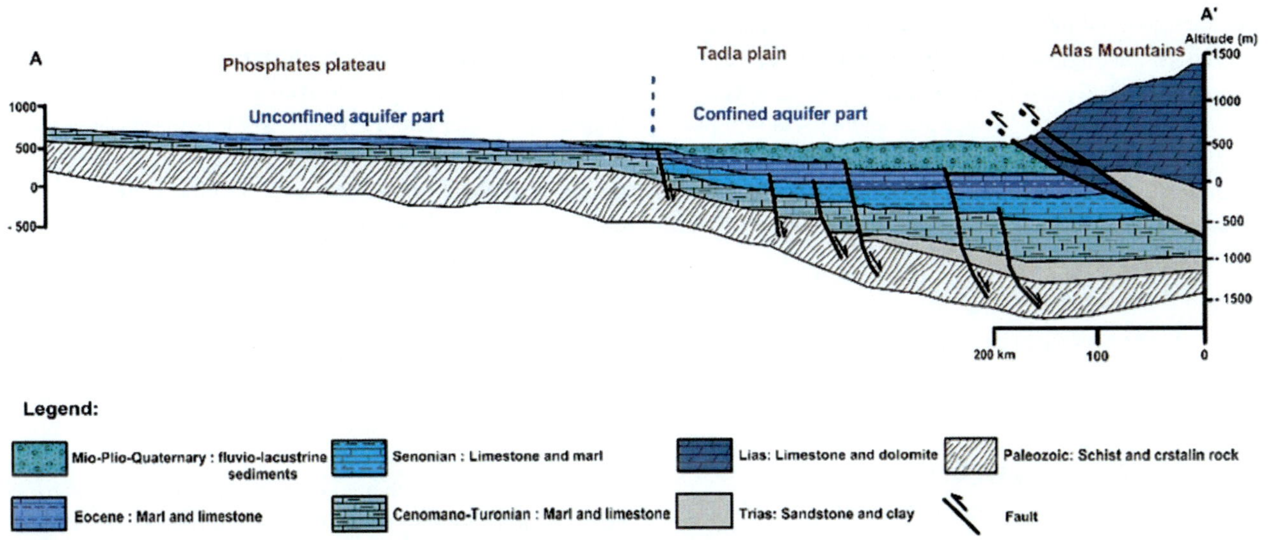

Fig. 2 Geological cross section (line A–A′ in Fig. 1) through the Tadla Plain

2.2 Methods

Forty-three surface water, spring and wells were collected from throughout the Tadla Basin (Fig. 1). Samples were analyzed in the field for pH, temperature, electrical conductivity, and dissolved oxygen. Samples were field filtered and delivered to the laboratory facilities at Duke University for analysis of major ions and trace metals. On a sub-set of samples strontium, isotope ratios were measured on a ThermoFisher Thermal ionization mass spectrometer (TIMS), at Duke University. 87Sr/86Sr values are normalized to 86Sr/88Sr of 0.1194, the long-term average of the NBS 987 standard measured at Duke University is 0.710247 ± 0.00001 SD. Oxygen and hydrogen isotopes were analyzed on a cavity ring-down Picarro L2120-i at University of Ibn Zohr.

3 Results

3.1 Physico-Chemical Characterization

The salinity of confined and unconfined groundwater from the Eocene, Senonian, Turonian and karstic aquifers is usually low to moderate, mostly under the limits allowed by Moroccan standards for drinking water supplies (< 2700 µS/cm and < 750 mg/L of chlorides). However, shallow superficial aquifers are characterized by significantly higher salinity, particularly in irrigated perimeter areas. The relationship between chloride and EC shows a similar correlation for all water groups, except for some Atlas springs with conspicuously lower chloride/EC ratio (Fig. 3a).

Figure 3b shows the Cl/Br ratio versus chloride concentration plot of different water samples from Tadla region.

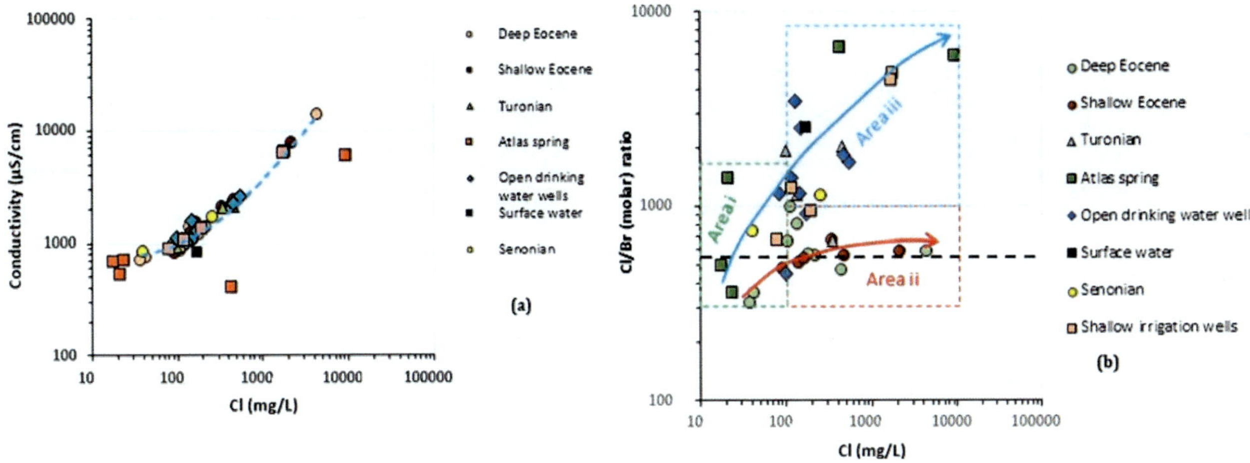

Fig. 3 Relationship between chlorides and electrical conductivity of sampled aquifers

Three areas of salinity in Tadla multi-layered aquifer system can be distinguished: (i) different types of recharge water (High altitude and Inland areas); (ii) water affected by anthropogenic pollution (wastewater and agricultural pollution); (iii) water including leaching or dissolution of natural evaporates (halite). The blue arrow indicates the freshwater to salt water and the red one shows the evolution of freshwater to contaminated water.

3.2 Isotopes Characterization

The plot shows δ^2H versus δ^{18}O values as compared to the Global Meteoric Water Line (GMWL) and the Local Meteoric Water Line (LMWL).

The similarity to the water meteoric water lines suggest that the different waters originated from recharge of meteoric water and water–rock interaction has not modified the stable isotope composition (Fig. 4a).

The ^{87}Sr/^{86}Sr ratios ranged from 0.707884 (Atlas saline springs) to 0.712986 (Shallow and deep Eocene) with an average value of 0.70909. The 87Sr/86Sr of Eocene aquifers samples was higher than that of the other aquifers (Fig. 4b).

4 Discussion

According to studies in the similar context (Hssaisoune et al., 2019, 2022; Bouchaou et al., 2008, 2009), the stable isotopes δ^{18}O and δ^2H indicate that aquifer recharge is due to direct infiltration of oceanic precipitation without significant evaporation.

The strontium isotope ratios of low saline water from the Turonian aquifer and High Atlas spring waters suggest mixing between Sr derived from the Turonian age limestone aquifer and a higher ^{87}Sr/^{86}Sr source. This mixing is evidenced by overlapping ratios of 0.7078–0.7092. Saline waters rich in sulfate and chloride were found in all aquifers. Two types of water were distinguished based on their isotopic ratios: (1) low 87Sr/86Sr (0.7078) and Br/Cl (1.7×10^{-4}) source affecting groundwater in the Turonian and Senonian aquifers, originating from the dissolution of Cretaceous gypsum and halite deposits; and (2) high 87Sr/86Sr (0.7110) and marine Br/Cl ($\sim 1.5 \times 10^{-3}$) source occurring mostly in the Eocene aquifer. The measured 87Sr/86Sr ratios in the groundwater are significantly higher than the expected carbonate marine composition, ranging from 0.703 in Turonian to 0.7078 in Eocene.

5 Conclusions

This study aimed to evaluate mixing processes between different aquifers in the Tadla plain:

- Sulfate and chloride rich saline waters were found in all aquifers;
- Low ^{87}Sr/^{86}Sr and Br/Cl source that affects groundwater in the Turonian and Senonian aquifers and originated from dissolution of gypsum and halite deposits;
- High ^{87}Sr/^{86}Sr and marine Br/Cl source that occurs mostly in the Eocene aquifer.

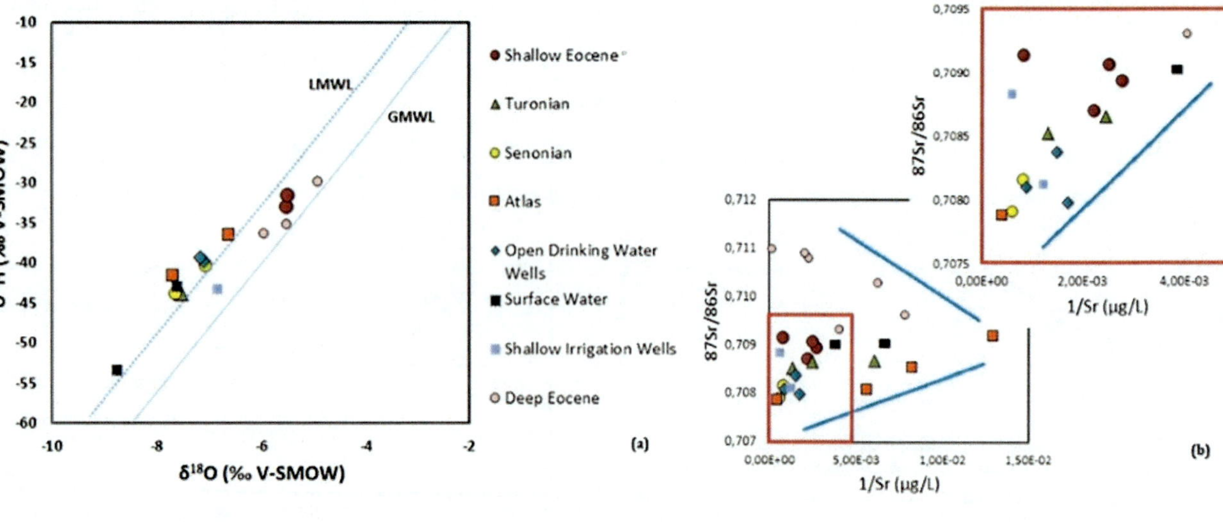

Fig. 4 **a** δ^2H plotted versus δ^{18}O and **b** ^{87}Sr/^{86}Sr plotted versus 1/Sr concentration

Acknowledgements We are grateful for the help of all laboratories where analyses were made, NATO, Academy Hassan II for Sciences and Techniques (CHARISMA), Isotrace and GEANTech projects financed by OCP fundation.

References

ABHOER (Agence du Bassin Hydraulique de l'Oum Er-Rbia). (2017). Assistance technique pour l'intégration et l'évaluation des risques climatiques dans la planification et le développement des ressources en eau du bassin de l'Oum Er-Rbia - Maroc. Rabat, Maroc.

Bouchaou, L., Chauve, P., Mudry, J., Mania, J., & Hsissou, Y. (1997). Structure et fonctionnement d'un hydrosystème karstique de montagne sous climat semi-aride: Cas de l'Atlas de Beni-Mellal (Maroc). *Journal of African Earth Sciences, 25*, 225–236. https://doi.org/10.1016/s0899-5362(97)00100-0

Bouchaou, L., Michelot, J. L., Qurtobi, M., Zine, N., Gaye, C. B., Aggarwal, P. K., Marah, H., Zerouali, A., Taleb, H., & Vengosh, A. (2009). Origin and residence time of groundwater in the Tadla basin (Morocco) using multiple isotopic and geochemical tools. *Journal of Hydrology, 379*, 323–338. https://doi.org/10.1016/j.jhydrol.2009.10.019

Bouchaou, L., Michelot, J. L., Vengosh, A., Hsissou, Y., Qurtobi, M., Gaye, C. B., Bullen, T. D., & Zuppi, G. M. (2008). Application of multiple isotopic and geochemical tracers for investigation of recharge, salinization, and residence time of water in the Souss-Massa aquifer, southwest of Morocco. *Journal of Hydrology, 352*, 267–287. https://doi.org/10.1016/j.jhydrol.2008.01.022

Hssaisoune, M., Bouchaou, L., N'da, B., Malki, M., Abahous, H., & Fryar, A. E. (2017). Isotopes to assess sustainability of over-exploited groundwater in the Souss-Massa system (Morocco). *Isotopes in Environmental and Health Studies*. https://doi.org/10.1080/10256016.2016.1254208

Hssaisoune, M., Bouchaou, L., Matsumoto, T., Araguas, L., Kraml, M., & Aggarwal, P. (2019). New evidences on groundwater dynamics from the Souss-Massa system (Morocco): Insights gained from dissolved noble gases. *Applied Geochemistry*, 104395. https://doi.org/10.1016/j.apgeochem.2019.104395

Hssaisoune, M., Bouchaou, L., Qurtobi, M., Marah, H., Beraaouz, M., & Stitou El Messari, J. (2022). Isotopic and chemical tracing for residence time and recharge mechanisms of groundwater under semi-arid climate: Case from Rif Mountains (Northern Morocco). *Geosciences, 12*, 74. https://doi.org/10.3390/geosciences12020074

Hssaisoune, M., Bouchaou, L., Sifeddine, A., Bouimetarhan, I., & Chehbouni, A. (2020). Moroccan groundwater resources and evolution with global climate changes. *Geosciences, 10*, 81. https://doi.org/10.3390/geosciences10020081

Vinson, D. S., Tagma, T., Bouchaou, L., Dwyer, G. S., Warner, N. R., & Vengosh, A. (2013). Occurrence and mobilization of radium in fresh to saline coastal groundwater inferred from geochemical and isotopic tracers (Sr, S, O, H, Ra, Rn). *Applied Geochemistry, 38*, 161–175.

Warner, N. R., Lgourna, Z., Bouchaou, L., Boutaleb, S., Tagma, T., Hsaissoune, M., & Vengosh, A. (2013). Integration of geochemical and isotopic tracers for elucidating water sources and salinization of shallow aquifers in the sub-Saharan Draa Basin, Morocco. *Applied Geochemistry, 34*, 140–151. https://doi.org/10.1016/j.apgeochem.2013.03.005

Mixing Processes in Wells Tapping Confined Aquifers: Quality and Risks Assessments for Public Drinking Water Supply

Chiara Zanotti, Agnese Redaelli, Alice Palazzi, Letizia Fumagalli, Mariachiara Caschetto, Camilla Stano, Tullia Bonomi and Marco Rotiroti

Abstract

In the last decades, scientific attention has grown toward vulnerability assessment of drinking water quality. Groundwater is a crucial resource for drinking water needs, and its vulnerability to contaminations is associated with natural and anthropic processes. The vulnerability of wells tapping confined aquifers is frequently associated with natural contamination by reduced species (e.g., As), while anthropic contamination is considered less relevant. Nevertheless, mixing processes induced by water pumping and related to the well structure can determine the presence of a shallower water fraction in the extracted water, which increases wells vulnerability to anthropic contamination. In this study, redox zonation is applied to water quality data of 16 drinking water wells tapping confined aquifers to assess the main processes affecting water quality. All the considered wells resulted under reducing conditions at different redox states. Three wells showed evidence of mixing processes between shallower and deeper waters. In these cases, mixing processes lower the concentration of As, but increase the vulnerability toward anthropic contamination.

Keywords

Drinking water · Water quality · Arsenic · Risk assessment · Redox zonation

C. Zanotti (✉) · A. Redaelli · A. Palazzi · L. Fumagalli · M. Caschetto · T. Bonomi · M. Rotiroti
University of Milano-Bicocca, Milan, Italy
e-mail: chiara.zanotti@unimib.com

C. Stano
A2A Ciclo Idrico S.P.A, Brescia, Italy

1 Introduction

Groundwater is a vital resource for fulfilling drinking water needs. Confined aquifers often constitute a crucial source of drinking water, since they are in most of cases physically unconnected with the surface, thus protected by local anthropic contamination. However, confined aquifers are susceptible to high risks of natural pollution, especially when groundwater is under reducing conditions, as in the case of arsenic which threatens many areas worldwide. Furthermore, drinking water wells often tap several aquifers through different screens, inducing mixing processes by drawing water from aquifers at different depths, including shallower aquifers. This implies that, when assessing drinking water vulnerability to natural and anthropic contaminants, a detailed assessment of the processes governing the quality of the water extracted by each well is crucial (Eberts, 2014; Negm et al., 2020a, b; Zanotti et al., 2022). Here, quality data of 16 wells are used to perform a redox zonation of groundwater and successively to untangle the occurrence of potential mixing processes controlled by well-structures which can influence the quality and vulnerability of the extracted groundwater.

2 Materials and Methods

The study area covers five municipalities in the Po Plain, N Italy, hosting a complex layered structure of confined aquifers system characterized by natural reduced hydrofacies (Zanotti et al., 2022). The area is served by 16 drinking wells with variable depths, number of screens and screen depths.

The initial dataset consists of the analyses of 973 samples, collected between January 2009 and May 2020. The available data include major ions (Ca, Mg, Na, K, HCO_3, SO_4, Cl), metals and semimetals (As, Fe, Mn), alkalinity expressed as $CaCO_3$ and Total Organic Carbon (TOC). Not all parameters listed above were measured for each sample.

H. Chenchouni et al. (eds.), *Recent Advancements from Aquifers to Skies in Hydrogeology, Geoecology, and Atmospheric Sciences*, Advances in Science, Technology & Innovation, https://doi.org/10.1007/978-3-031-47079-0_10

The adopted methodology consists in: (i) data preprocessing, (ii) exploratory data analysis through box-plot and Piper diagrams and (iii) data analysis through redox zonation and correlation analysis of major ions and redox-sensitive species with respect to different well screens and gravel-pack structures.

Data preprocessing was aimed at managing the dataset and purging the data from sampling and analytical errors. Data preprocessing consisted in (a) substitution of non-detected data with the LOD/2 value, (b) the Chemical Balance Error (CBE) calculation for each sample and the exclusion of samples exceeding the $\pm 5\%$ CBE threshold (Müller et al., 2006) and (c) an outlier analysis through the Interquartile Range Method (IQR) (Wang et al., 2018).

The exploratory data analysis was performed through graphical visualization methods such as Piper diagrams. Successively, a redox zonation has been performed for each available sample, using threshold concentrations of the redox-sensitive species O_2, NO_3, Mn, Fe and SO_4.The threshold values proposed by the USGS (Jurgens et al., 2009) were used, with some changes (sulfate reduction with $SO_4 < 1.5$ mg/L) and improvements (distinction between initial and advanced sulfate reduction with SO_4 respectively higher and lower than 10 mg/L) according with Rotiroti et al. (2021). Finally, each well's redox characteristics were further investigated by elaborating scatterplots of redox-sensitive species.

3 Results

Results of the redox zonation are shown in Fig. 1a. It emerges that wells were classified with decreasing redox states moving downstream with respect to the groundwater flow and toward deeper aquifers: (a) V1 is the only well under denitrification, it has shallow screens and it is upstream of the groundwater flow and therefore closer to the recharge area; (b) V2–V4 and M1–M4, M6 and BB1 were classified under Fe-oxide reduction and/or early-stage sulfate reduction, (c) M5 and M7 and SG1 were classified under Fe-oxide reduction and/or advanced sulfate reduction and (d) P1 and P2 in the southern part of the study tap groundwater under methanogenesis.

Figure 1b shows scatterplots of different redox-sensitive species, color-coded by wells. It emerges a strong direct correlation of As with NH_4 and TOC. The wells which have on average higher concentrations of NH_4 and TOC (SG1, BB1, P1 and P2) are also the richest in dissolved As. On the contrary, M1–M7 and V1–V4 have lower As, NH_4 and TOC.

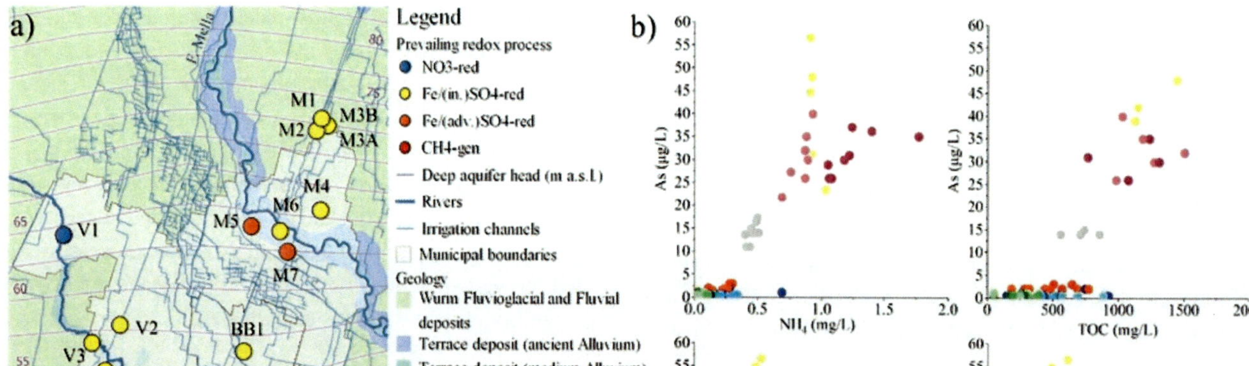

Fig. 1 a Study area and analyzed wells, color-coded by results of the redox zonation. b Scatterplot of redox-sensitive species, color-coded by well; the dashed blue line indicates a possible mixing line

4 Discussion

Arsenic concentration drops when the reduction of Fe-oxides and SO_4 occurs simultaneously, due to As sequestration via co-precipitation in FeS or through direct precipitation of As-sulfide (Rotiroti et al., 2021). Therefore, highest arsenic concentrations are expected during Fe-reduction (with no SO_4 reduction) and later during methanogenesis, while low As in groundwater is found during sulfate reduction. Based on this, SO_4 and As behave as mutually exclusive species. However, three peculiar situations emerged from Fig. 1b, concerning P1, SG1 and BB1 which show the co-presence of As and SO_4. Their simultaneous presence is interpreted as the likely result of mixing of water at different redox stages.

As regards P1, it emerges that its extracted water is the result of mixing between deep and shallower groundwater, plausibly in contact with the neighboring irrigation channel (Fig. 1a) due to the presence of a shallower screen. While the deeper water is more prone to the presence of reduced species and As, shallower water, especially if in contact with surface water, is more vulnerable to local anthropic contamination. Therefore, the vulnerability profile of this kind of well happens to be a combination of the vulnerabilities typical of deeper and shallower water. SG1 and BB1 show similar conditions, exacerbated in BB1. In Fig. 1b, the position of SG1 samples is consistent with a mixing between water with characteristics similar to those of the P1–P2 wells (under methanogenesis), and waters similar to those of the M and V groups (Fe-oxide and sulfate reduction). Similarly, although BB1 shows evidence of sulfate reduction, with high SO_4 concentrations, it has the highest As concentrations in the dataset. In this case, it is possible to assume a mixing between shallower water (similar to water extracted by the M or V wells under Fe/SO_4 reduction), and a deeper water which does not figure among the examined water samples and should have As concentrations even higher than the P1–P2 wells. In the SG1 and BB1 cases the mixing induced by the presence of shallower screens lowers the As concentrations (if only deep water was extracted, the As concentration would have been higher) but, also in this case, it increases the vulnerability profile concerning the risk of anthropic contamination from shallower water.

5 Conclusion

The results of this study highlight how a detailed assessment of the processes governing the quality of the extracted water is crucial when assessing drinking water vulnerability to contaminations. In this case, 16 wells tapping confined aquifers were analyzed, and three of them showed evidence of mixing processes between deeper and shallower groundwater. While the deeper groundwater increases the risk of undesired reduced species, the fraction of shallower groundwater increases the vulnerability toward local anthropic contaminations.

References

Eberts, S. M. (2014). If groundwater is contaminated, will water from the well be contaminated? *Groundwater, 52*, 3–7. https://doi.org/10.1111/gwat.12260

Jurgens, B. C., McMahon, P. B., Chapelle, F. H., & Eberts, S. M. (2009). An Excel® workbook for identifying redox processes in ground water. In *U.S. geological survey open-file report no. 2009-1004* (p. 8).

Müller, D., Blum, A., Hart, A., Hookey, J., Kunkel, R., Scheidleder, A., et al. (2006). Final proposal for a methodology to set up groundwater threshold values in Europe. In *Deliverable D18, BRIDGE project* (Vol. 63).

Negm, A., Bouderbala, A., Chenchouni, H., & Barcelo, D. (2020a). *Water Resources in Algeria-Part I: Assessment of Surfaceand Groundwater*. Springer, Cham. https://doi.org/10.1007/978-3-030-57895-4

Negm, A., Bouderbala, A., Chenchouni, H., & Barcelo, D. (2020b). *Water Resources in Algeria-Part II: Water Quality, Treatment, Protection and Development*. Springer, Cham. https://doi.org/10.1007/978-3-030-57887-9

Rotiroti, M., et al. (2021). Overlapping redox zones control arsenic pollution in Pleistocene multi-layer aquifers, the Po Plain (Italy). *Science of the Total Environment, 758*, 143646. https://doi.org/10.1016/j.scitotenv.2020.143646

Wang, C., Caja, J., & Gómez, E. (2018). Comparison of methods for outlier identification in surface characterization. *Measurement, 117*, 312–325. https://doi.org/10.1016/j.measurement.2017.12.015

Zanotti, C., et al. (2022). A cost-effective method for assessing groundwater well vulnerability to anthropogenic and natural pollution in the framework of water safety plans. *Journal of Hydrology, 613*, 128473. https://doi.org/10.1016/j.jhydrol.2022.128473

The Geothermal Systems of the Vallès Fault (NE of Spain): Fracture Network Characterization and Weathering Patterns

Agathe Jullien-Sicre, Anna Travé, Damien Guinoiseau, Vinyet Baqués, Bertrand Saint-Bezar, Frank Despinois, Irene Cantarero, Elisabet Playà and Antonio Benedicto

Abstract

The ongoing energetic transition needs critical raw materials such as lithium or other metals and renewable energies, such as low to mid-enthalpy geothermal energy. Geothermal systems, mostly located near plate boundaries, undergo strong structural controls essentially through crustal faults, also known to drain fluids. Furthermore, dissolved metals in geothermal fluids can be exploited as a byproduct in geothermal plants. In the present, study we define spatial variations in metal content and the structural control along the Vallès Fault, as a field analogue of a passive margin domain. The Vallès Fault is a crustal fault that juxtaposes the Miocene detrital basin infill to a syn- and late-Hercynian granitic basement. With a crustal thermal anomaly, a geothermal gradient > 35 °C/km, several geothermal systems of low enthalpy (< 130 °C at 1 km depth) are closely associated to the Fault. These geothermal systems are revealed by several springs with temperatures ranging from 20 to 75 °C. The present work combines hydrogeochemical, structural and petrological approaches to characterize the architecture and the relative chronology of the fracture networks of the sites of hot springs; and the metal content of thermal waters to reconstruct the spatial evolution of the metal content of the geothermal fluids. From the structural and petrological point of view, three main results are highlighted: (1) the footwall of the Vallès Fault is characterized by a network of polyphase fractures affecting the granite basement and constituting an important network of permeability by fracturing; (2) a thick fault zone with cataclastic to ultracataclastic fault rocks constitutes, a priori, a barrier for fluids circulation; and (3) an arkosic Miocene basin infill, affected by secondary faults, constitutes a very good reservoir. The metal content of hot waters is low but significant to establish differences according to their structural position. Tectonic structures and surrounding lithologies seem to be decisive in water compositions.

Keywords

Geothermal systems · Vallès Fault · Metals · Aqueous geochemistry

1 Introduction

Critical raw materials such as lithium or other metals as well as renewable energies, such as low to mid-enthalpy geothermal energy, are key to the ongoing energetic transition. Mostly located near plate boundaries (Moeck, 2014), geothermal systems undergo strong structural controls essentially through crustal faults, also known to drain fluids (Duwiquet et al., 2021). Often seen as a nuisance in geothermal plants, dissolved metals in geothermal fluids can be exploited as a byproduct. This study aims at defining spatial and temporal variations in metal content and the structural control along the Vallès crustal fault, as a field analogue of a passive margin domain.

2 Materials and Methods

An analysis of the fracturing network at the regional scale has been carried out to reconstructing the kinematic evolution. A relative chronology of each family of fractures has been established to deconvolute episodes of fracturing.

A. Jullien-Sicre · D. Guinoiseau · B. Saint-Bezar · A. Benedicto
UMR CNRS 8148 GEOPS, Université Paris-Saclay, Orsay, France

A. Travé (✉) · V. Baqués · I. Cantarero · E. Playà
c/Martí i Franquès s/n., Universitat de Barcelona (UB), 08028 Barcelona, Spain
e-mail: atrave@ub.edu

F. Despinois
TotalEnergies—OneTech R&D Sustainability Line, Pau, France

H. Chenchouni et al. (eds.), *Recent Advancements from Aquifers to Skies in Hydrogeology, Geoecology, and Atmospheric Sciences*, Advances in Science, Technology & Innovation, https://doi.org/10.1007/978-3-031-47079-0_11

Fracturing network has been described from sampled rocks in the fault zone which were carefully oriented in the field and sliced perpendicular to the orientation of the fault. The fault rocks were classified according to the petrological criteria.

Hot water springs (36–55 °C) and surficial cold waters were collected in the vicinity of the fault zone. Physical parameters (temperature and electrical conductivity) were measured in situ. Dissolved major and trace concentrations were determined using ion chromatography and quadrupole ICP-MS, respectively.

3 Results and Discussion

3.1 Fracturing in the Vallès Fault Zone

Ductile and brittle deformation have been observed in the field (Fig. 1). Six families of fractures have been identified and measured (Fig. 2). The six families have been grouped according to orientation, dip, affected units and style of deformation when petroglyphs are visible.

The samples have been chosen because they are representative of the organization of the fault zone, starting from a little deformed rock where the protolith is recognizable, and going to a very fine gouge, in which the material is crushed, and the elements of the protolith difficult to identify. The recorded ductile deformation is associated with a Hercynian compressive episode with an NE–SW axis (stage 1). The Mesozoic extension is then expressed by the presence of NE–SW faults associated with a NW–SE stretch (stage 2). The E–W faults are formed during the Pyrenean compressional episode of N–S axis (stage 3). The Neogene NW–SE extension (stage 4) at the origin of the formation of the semi-graben of the Vallès-Penedès is complex because it is interspersed with compressive episodes along the NW–SE axis (stage 5). Neogene deformation even affects the filling of the basin (stage 6). The intense fracturing affecting granites of the lower block makes this compartment a good reservoir. The upper block, made up of fractured arkosic sediments, can also be a good reservoir. To understand the system, it is very important to take a closer look at the fault zone. The petrological study of the Vallès crustal fault shows that it consists of a thick fault core with cataclastic to ultracataclastic fault rocks (gouges) which constitutes, a priori, a screen/barrier to fluid circulation. This impermeable gouge is also observed further south characterizing the fault (Aldega et al., 2019).

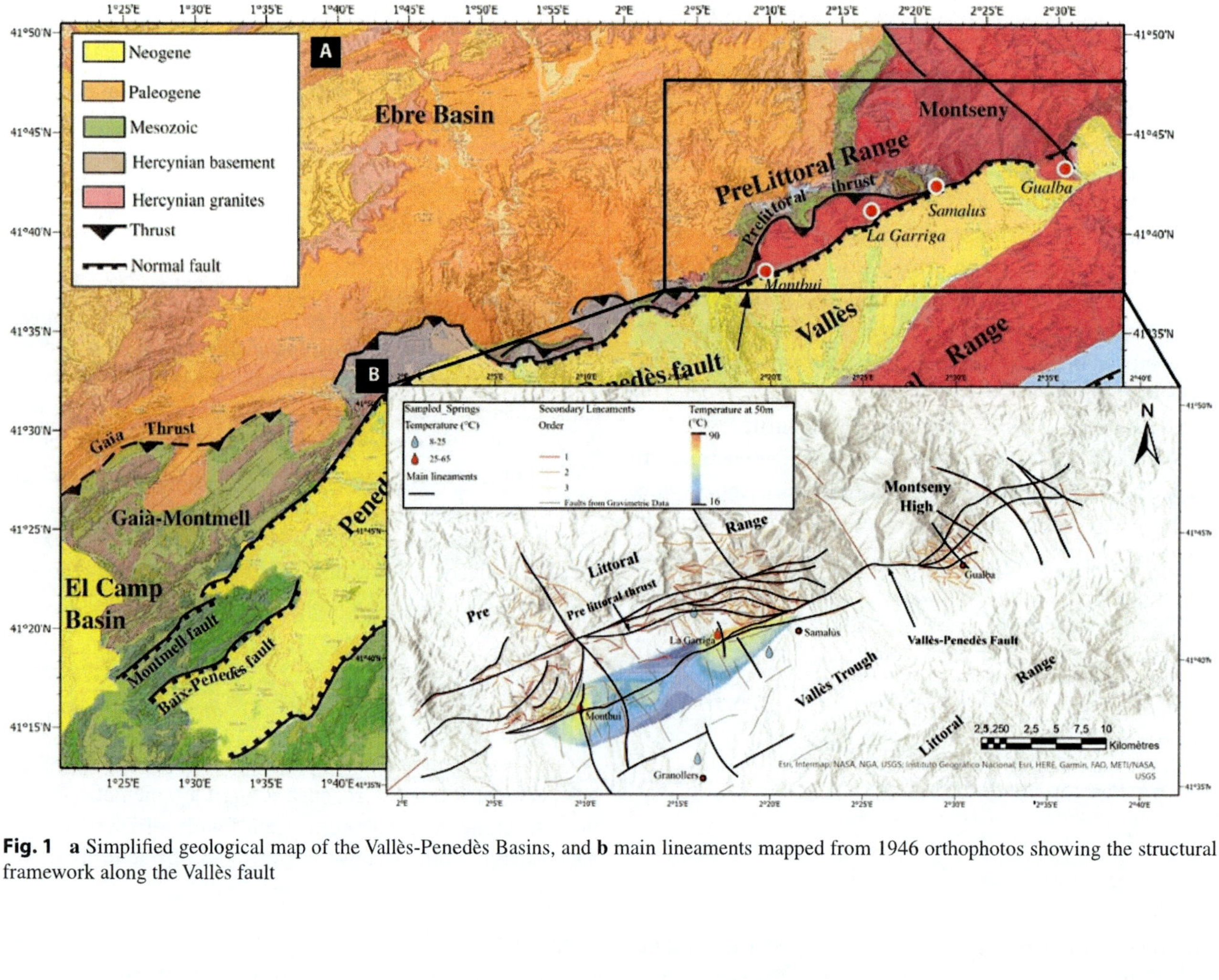

Fig. 1 **a** Simplified geological map of the Vallès-Penedès Basins, and **b** main lineaments mapped from 1946 orthophotos showing the structural framework along the Vallès fault

3.2 Geochemical Signature of Vallès Surficial and Thermal Waters

Larger total dissolved solids as well as Li concentrations were measured in hot springs and in cold waters. Hot springs are located at the back of the Vallès normal fault. Two distinct geochemical signatures are highlighted using element ratio diagram in hot springs and cold waters (Fig. 3). Surficial cold waters are reflecting the chemical weathering of silicate and carbonate rocks. Hot springs are typically the result of the drainage of evaporitic structures, especially of halite (NaCl).

Strong differences of mineralization are observed between the two hot springs (La Garriga and Montbui, Fig. 4) sampled in the Vallès fault area, evidencing distinct water flow paths and/or fluid-rock interactions in the fractured zone at a scale of 5–10 km. Contrary to other geothermal systems [e.g., Upper Rhine Graben, (Sanjuan

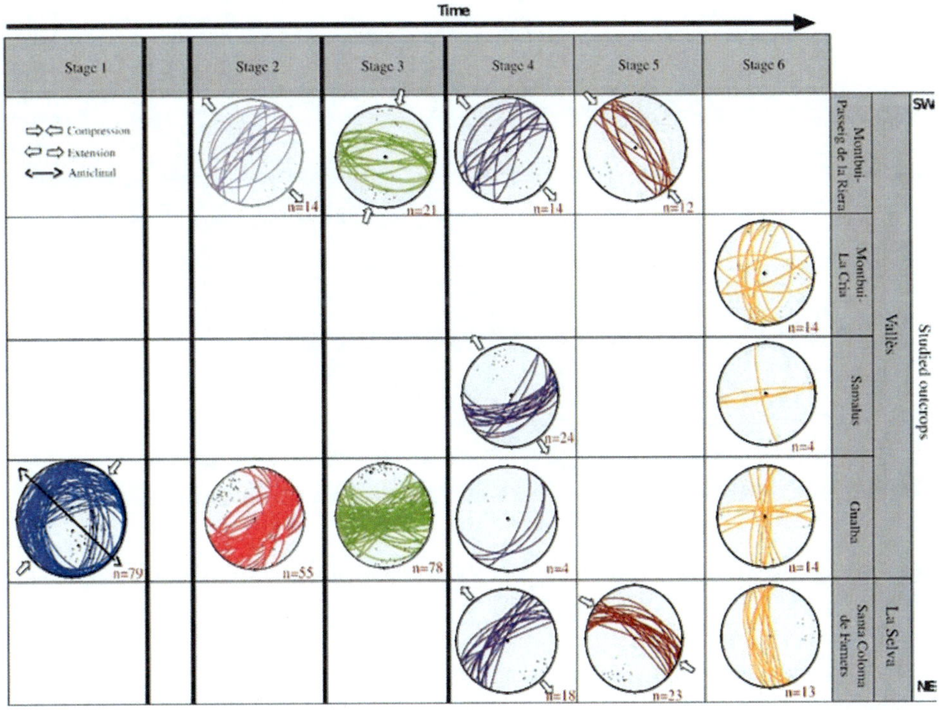

Fig. 2 Stereo Plots of structural data measured in the studied outcrops and projected onto a Wulf canvas. They are sorted according to the different stages of deformation

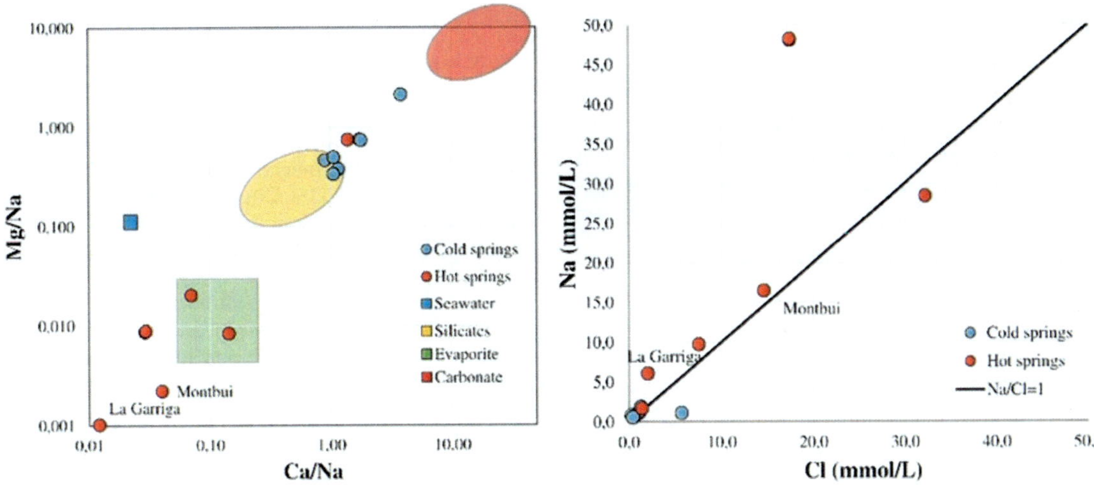

Fig. 3 Identification of weathering patterns in cold and hot springs using element ratio. Silicate, carbonate, and evaporitic end-members are from Gaillardet et al. (1999)

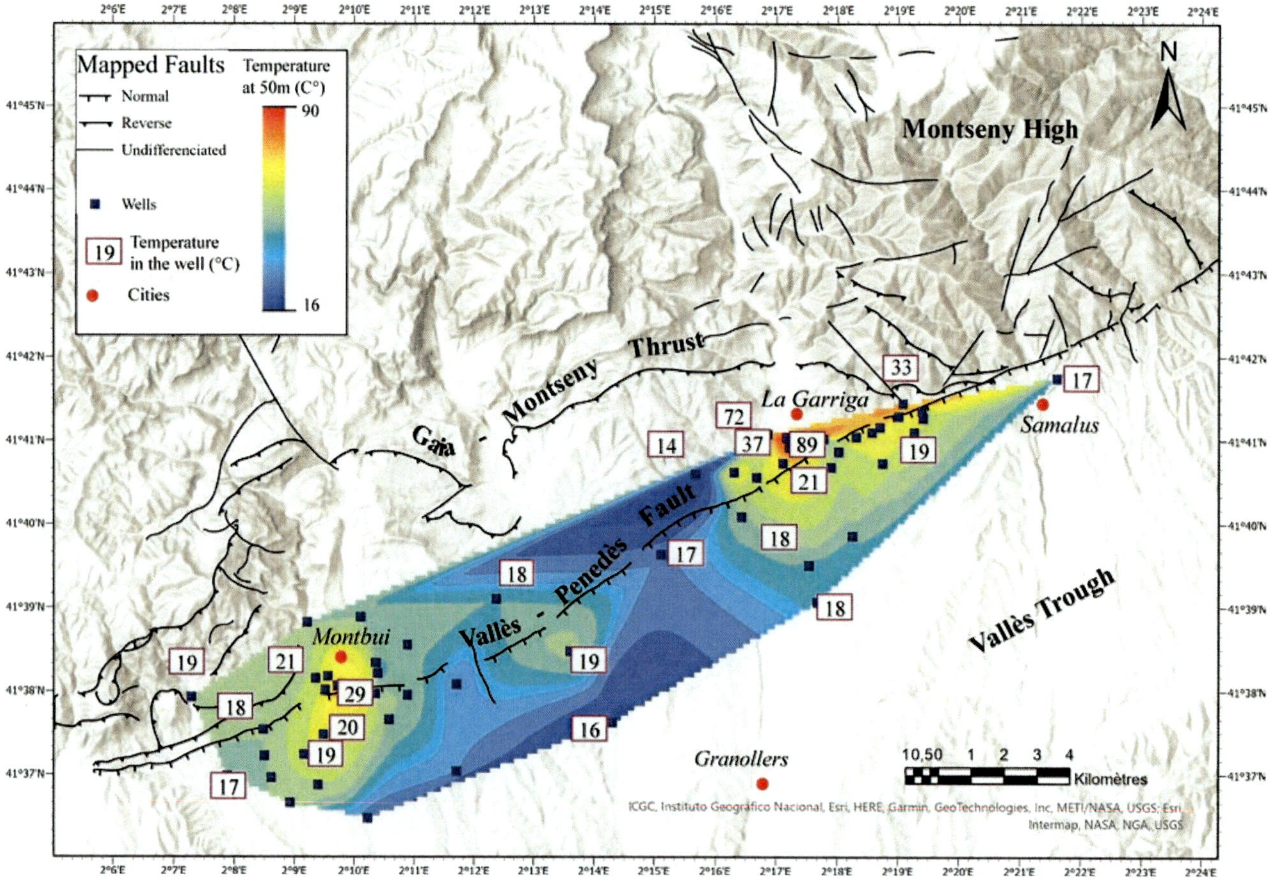

Fig. 4 Location of the Vallès geothermal along the Vallès fault. Map of the local geothermic gradient between the Montbui and La Garriga hot-springs, obtained by nearest neighbour interpolation of 56 wells data

et al., 2022)], the hot spring temperature and the limited Li enrichment preclude strong interactions between fluids and granites, rather a dissolution of evaporitic layers moderately rich in Li.

4 Conclusion

Six different stages of deformation have been characterized by analyzing the fracture network. The Vallès system is associated with two good reservoirs on either side of the main fault. Structural map suggests that hot springs are located at the intersection of the NE–SW Vallès fault and NW–SE transverse faults. Nevertheless, the impermeable gouge which constitutes the core of the Vallès fault zone screens the circulation of fluid and confines the geothermal system in the lower block. Contrary to other geothermal systems, the hot spring temperature and the limited Li enrichment preclude strong interactions between fluids and granites, rather than a dissolution of evaporitic layers moderately rich in Li.

Acknowledgements We thank TotalEnergies, DGICYT PID2021-122467NB-C22 and PGC2018-093903-B-C22 (MICIU/AEI/https://doi.org/10.13039/501100011033/FEDER) and Grup Consolidat de Recerca "Geologia Sedimentària" (2021 SGR-CAT 00349) for financial support.

References

Aldega, L., Viola, G., Casas-Sainz, A., Marcén, M., Román-Berdiel, T., & van der Lelij, R. (2019). Unraveling multiple thermotectonic events accommodated by crustal-scale faults in Northern Iberia, Spain: Insights from K–Ar dating of clay gouges. *Tectonics, 38*, 3629–3651.

Duwiquet, H., Guillou-Frottier, L., Arbaret, L., Bellanger, M., Guillon, T., & Heap, M. J. (2021). Crustal fault zones (CFZ) as geothermal power systems: A preliminary 3D THM model constrained by a multidisciplinary approach. *Geofluids, 2021*, 1–24.

Gaillardet, J., Dupré, B., Louvat, P., & Allègre, C. J. (1999). Global silicate weathering and CO_2 consumption rates deduced from the chemistry of large rivers. *Chemical Geology, 159*, 3–30.

Moeck, I. S. (2014). Catalog of geothermal play types based on geologic controls. *Renewable and Sustainable Energy Reviews, 37*, 867–882.

Sanjuan, B., Gourcerol, B., Millot, R., Rettenmaier, D., Jeandel, E., & Rombaut, A. (2022). Lithium-rich geothermal brines in Europe: An up-date about geochemical characteristics and implications for potential Li resources. *Geothermics, 101*, 102385.

Preliminary Evaluation of the Geothermal Influence on the Hydrogeochemistry of Rhodopes' Coastal Aquifer (NE Greece)

Ekaterini Sachsamanoglou, Evangelos Tziritis and Paschalis Dalampakis

Abstract

The coastal region of Rhodope (NE Greece) is characterized by increased salinization risk leading to groundwater quality degradation and water scarcity. In the context of the MEDSAL Project (www.medsal.net), a comprehensive assessment has been carried out concerning the hydrogeochemical characteristics of the area while evaluating the potential impact of geothermal fluids circulation and their ability to move upward and contaminate more surficial water bodies. According to previously conducted surveys, three geothermal wells have been identified in the study area, two of which are representative of the geothermal reservoir, which is hosted within Tertiary molassic sediments. To get an insight into the current hydrogeochemical conditions, a sampling campaign was performed, during which 46 groundwater samples were collected and analyzed for a wide set of parameters. Based on the analytical results, elevated concentrations of certain trace elements presumably associated with geothermal activity (indicatively referred to as B, Li, Sr, and U) have been detected in groundwater wells close to the geothermal one, indicating a possible connection. The relatively elevated groundwater temperature, which reaches up to 25.1 °C, as well as the electrical conductivity values, additionally supports the aforementioned connection. In order to better understand the potential connection of the geothermal influence, the analytical results have been further processed using different tools. The outcomes of the spatial distribution of Li, U, and Ce denoted that the groundwater wells, which are located close to the reported tapped geothermal reservoir and the dominant tectonic structures that prevail in the area, seem to follow a similar hydrogeochemical pattern.

Keywords

Groundwater salinization · Geothermal energy · Hydrogeochemistry · Coastal aquifer · Spatial interpolation

1 Introduction

The study area (RHO) is located in the Rhodope (NE Greece) region and constitutes a coastal aquifer heavily affected by groundwater salinization. The area is predominately covered by agricultural land resulting in the overexploitation of groundwater reserves to meet the increasing water demands. Two main aquifers with no direct hydraulic connection have been identified; the upper, semi-confined one is situated in the recent alluvial deposits (sand, gravel, etc.) and the lower confined aquifer is located in the Miocene–Pliocene formations (mainly consisting of marls and sandstones) (Kallioras, 2008). The tectonic features of the RHO pilot site influence the groundwater flow and define the boundaries of the geothermal field and, thus, its potential impact on groundwater resources. The main tectonic features are two: a normal NE–SW direction fault, lying parallel to Aspropotamos River and a system of faults with the same direction, forming a horst/graben feature, lying SW and SE of Pagouria village and Ismarida Lake. In the wider area, high heat flow regime has been observed while the water temperatures vary between 30 and 45 °C at depths of 400–500 m (Dotsika et al., 2021). Furthermore, based on a detailed survey conducted by the former Greek Institute of Geological and Mineralogical Research (IGME) (Kolios, 2000) the RHO study area is impacted by the presence of a geothermal field (Komotini basin). This study aimed to get a preliminary insight into the potential influence of the geothermal reservoir on the groundwater system by using chemical analysis data.

E. Sachsamanoglou (✉) · E. Tziritis · P. Dalampakis
ELGO-DIMITRA, Soil and Water Resources Institute (SWRI), 57400 Thessaloníki, Greece
e-mail: k.sachsamanoglou@swri.gr

H. Chenchouni et al. (eds.), *Recent Advancements from Aquifers to Skies in Hydrogeology, Geoecology, and Atmospheric Sciences*, Advances in Science, Technology & Innovation, https://doi.org/10.1007/978-3-031-47079-0_12

2 Methodology

In the context of MEDSAL project (www.medsal.net), a sampling campaign was performed in July 2020, corresponding to the beginning of the irrigation period.

A total of forty-six (46) samples were collected from groundwater (GW) wells that are mainly used for irrigation purposes, as seen in Fig. 1, and analyzed for a wide set of parameters, including B, Li, Sr, and U, which are often associated with geothermal activity. According to the survey, three exploration geothermal wells have been identified in the broader area, two of which can be characterized as the most representative of the geothermal reservoir. The charge balance error (CBE) was calculated using Geochemist Workbench® (GWB) software, expressed as the deviation from the ideal electroneutral condition to evaluate the accuracy and validity of the analytical results of the water sample. The following equation was used (Appelo & Postma, 2005).

$$\mathrm{CBE} = \frac{\sum \mathrm{cations} + \sum \mathrm{anions}}{\sum \mathrm{cations} - \sum \mathrm{anions}} \times 100 \quad (1)$$

The groundwater samples (42 out of 46) and the respective samples from the geothermal wells with CBE < 10% were included in the data processing.

3 Results

The results of the hydrochemical analysis (focusing on the B, Li, Sr, and U, fingerprints, which are often associated with geothermal activity) revealed some noteworthy anomalies, located on the east side of the RHO study area, along the horst-graben feature and close to the identified geothermal well (W250). Boron (B) has a median value of 31.6 μg/L. Some of the highest concentrations, however, occur close to the main tectonic structures and near the referenced geothermal well (ranging from 89.8 to 138.5 μg/L) denoting a possible hydrothermal influence. Lithium (Li) concentrations, indicative of geothermal activity, varied from 80 to 244 μg/L, with a median value of 154 μg/L. Elevated concentrations (>132 μg/L) are detected in wells close to the horst-graben feature and to the geothermal well, suggesting a possible connection. Strontium (Sr) exhibits

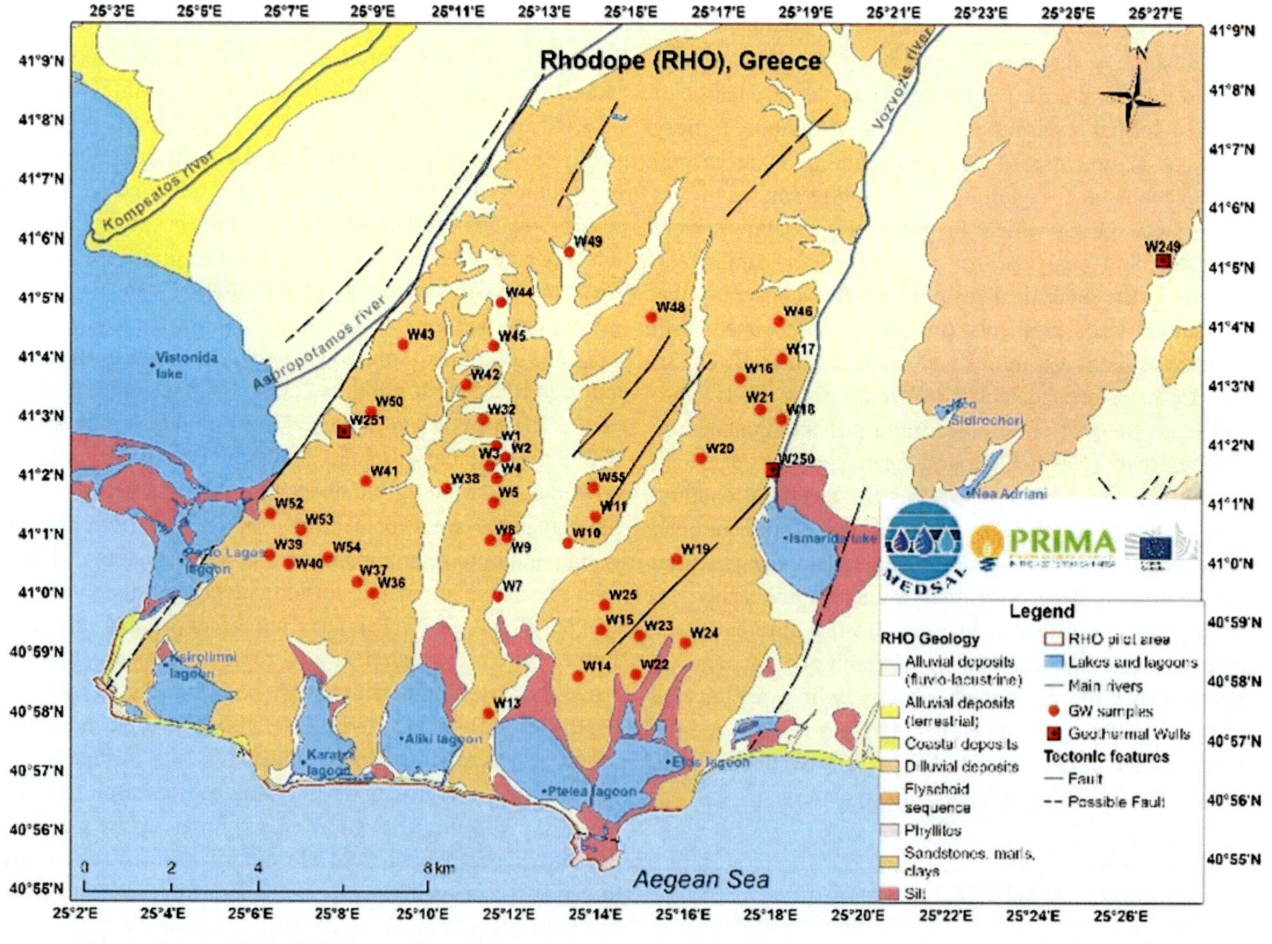

Fig. 1 Geological map of the location of the RHO sampling sites, the GW samples, and the geothermal wells

values ranging from 223 to 10,450 μg/L, with a median value of 1135 μg/L, denoting significant water–rock interaction. However, the highest values are spread in the study area and oriented parallel to the coastline, indicating the impact of seawater intrusion. Uranium (U) concentrations were all below the WHO recommended limit of 30 μg/L. The concentrations recorded ranged from 0.4 reaching up to 15.4 μg/L and the median value was 4.25 μg/L. Interestingly, the highest U concentrations (>15 μg/L) were detected in W11 close to the horst-graben and in W21, which is close to the geothermal well.

4 Discussion

According to the outcomes, there are some indications of the impact of the geothermal activity observed in the hydrochemistry of certain site-specific wells. The substrate of the study area consists of volcano-metamorphic formations (gneisses, orthogneisses, granites, etc.), in the deepest layers of which lies the geothermal reservoir. Furthermore, the geothermal reservoir is hydraulically connected through preferential flow paths with the aquifer system (hosted in Quaternary deposits). The spatial distribution exhibited local enrichment of Li and U in the east side of the study area, close to the referenced geothermal well and along the horst/graben tectonic feature, potentially attributed to the upward circulation of the geothermal fluids. A similar pattern has been observed with cerium (Ce), the anomaly of which is also oriented along the NE–SW direction system of normal faults on the easternmost side of the area (Tziritis et al., 2021). The RHO study area is heavily impacted by the elevated salinity, mostly caused by seawater intrusion. Based on the spatial distribution of the electrical conductivity, the areas close to the horst-graben feature and to the referenced geothermal well have medium to low concentrations. Therefore, the local enrichment of Li and U, denoting their potential common source, is not affected by the elevated salinity and is probably caused by the weathering of the surrounding geological formations favored by the high permeability through preferential flows.

5 Conclusion

The RHO study area is characterized by a complex hydrogeochemical regime impacted by seawater intrusion, water–rock interactions, and the impact of the nearby geothermal reservoir. Reviewing the potential origin and connection of the B, Li, Sr, and U concentrations, can further support the conceptualization of the possible interaction of the geothermal fluids and the deep groundwater circulation. Thus, allowing the delineation of the range of influence of the geothermal reservoir.

References

Appelo, C., & Postma, D. (2005). *Geochemistry, groundwater and pollution* (2nd ed.). Balkema.

Dotsika, E., Dalampakis, P., Spyridonos, E., et al. (2021). Chemical and isotopic characterization of the thermal fluids emerging along the North-Northeastern Greece. *Science and Reports, 11*, 16291.

Kallioras, A. (2008). *Groundwater resources management of aquifers subjected to seawater intrusion regime. The case of western coastal plain of the Prefecture of Rhodope* [Ph.D. Thesis]. Democritus University of Thrace (in Greek).

Kolios, N. (2000). *Geothermal Survey of eastern plain in Xanthi—Komotini (area of Lake Mitrikou)*. IGME (in Greek).

Tziritis, E., Aschonitis, V., Dalampakis, P., Doulgeris, C., Kinigopoulou V., Panagopoulos, A., Pisinaras, V., Sachsamanoglou, K., & Vrouhakis, I. (2021). Identification of cerium (Ce) anomalies in the groundwater system of coastal Rhodope (NE Greece). In 48th *international association of hydrogeologists IAH Congress, 6–10 September, Brussels*.

Lithium Contents in Non-marine Salty Springs

Elisabet Playà, Juan Diego Martín-Martín, Irene Cantarero, Vinyet Baqués, Eloi Carola and Anna Travé

Abstract

Salt springs in the Pyrenees (NE Iberian Peninsula) are common due to diapirism and dissolution of Triassic evaporite-bearing deposits. Various salty sources from Eastern Pyrenees have been analysed in order to elucidate their accurate chemical and isotopic composition. The study includes trace analyses by means of ICP-OES and ICP-MS, and, sulphate (S–O) and water molecule (H–O) isotope analyses by mass spectrometry. Outcropping Na–Cl rich brines contain B, Li (from 0.17 to 2.43 ppm), Rb, Zn and Mn as main trace elements. The origin of the salty composition is due to fluids circulating through evaporite Keuper facies (Triassic; halite-sulphates, calcium carbonates and clays), as clearly revealed by brine S–O isotope signature. Sulphate molecule has not suffered sulphate reduction, thus evidencing simple rock dissolution. Noticeable lithium content in seawater brines is well known due to the conservative behaviour of such element, which increases during evaporation. Studied salty springs are non-marine in origin, and not clearly related to hydrothermalism or volcanic fluids, which will enrich brines in lithium. The negative H–O and C isotope signatures of water also support this evidence. Thus, Li origin is mainly due to dissolution of Keuper evaporites, both considering Li contained in salt rocks and/or in fluid inclusions. Study of such brines will contribute to a better understanding of Li circulation and storage in non-marine salty springs.

Keywords

Salty Mountain Source · Lithium concentration · Triassic evaporites dissolution · Pyrenees

1 Introduction

Salt springs in the Pyrenees (NE Iberian Peninsula) are common due to diapirism and dissolution of Triassic evaporite-bearing deposits. Lithium (Li) contents and detailed isotope compositions in such brines are unknown, although many of such brines have been historically exploited in non-marine saltworks to produce high-quality Table Salt.

Increasing interest in Li is due to its strategic metal use in batteries for electric vehicles, mobile phones, etc. or for its properties or indications on mood and for improving the effectiveness of other trace elements in the human body (Thomas et al., 2020). Study of such brines will contribute to better understanding of Li circulation in non-marine salty springs.

2 Materials and Methods

We have studied several salty sources from Eastern Pyrenees (NE Spain), including Gerri de la Sal, Vilanova de la Sal and Cambrils del Pirineu brines (Fig. 1 and Table 1). Minor and trace element analyses have been performed by means of ICP-OES (Sr) and ICP-MS (Al, As, B, Ba, Cu, Fe, La, Li, Mn, Pb, Rb, Ti, Y, Zn). Dissolved sulphate (S–O) and water molecule (H–O) isotope analyses have also been measured using mass spectrometry.

E. Playà · J. D. Martín-Martín · I. Cantarero · V. Baqués · E. Carola · A. Travé (✉)
Departament de Mineralogia, Petrologia i Geología Aplicada, Facultat de Ciències de la Terra, Universitat de Barcelona (UB), c/Martí i Franquès s/n., 08028 Barcelona, Spain
e-mail: atrave@ub.edu

H. Chenchouni et al. (eds.), *Recent Advancements from Aquifers to Skies in Hydrogeology, Geoecology, and Atmospheric Sciences*, Advances in Science, Technology & Innovation, https://doi.org/10.1007/978-3-031-47079-0_13

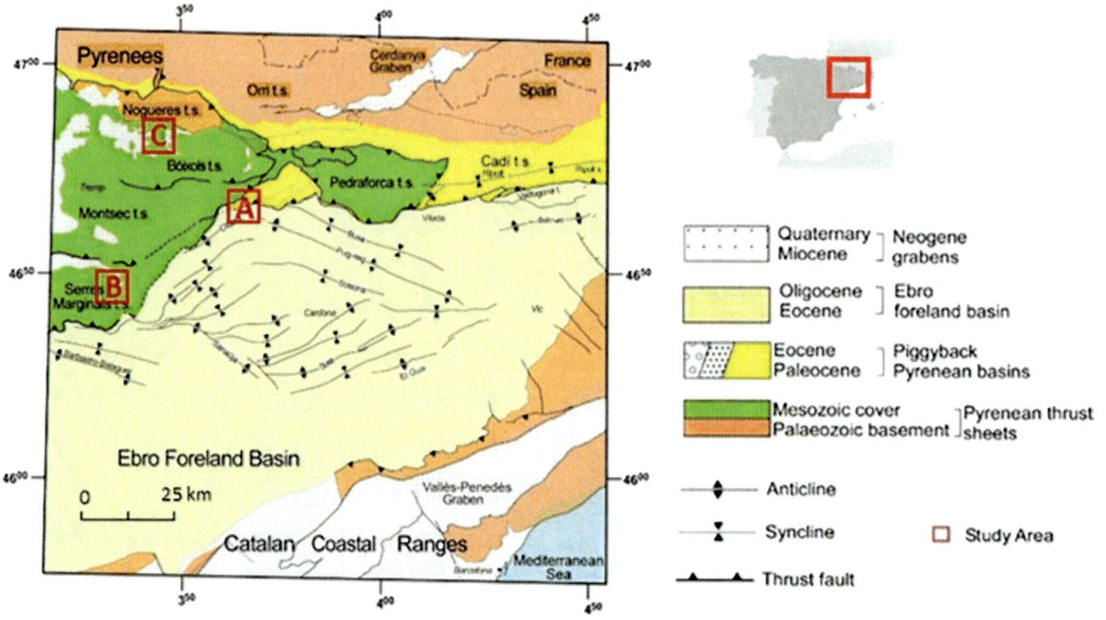

Fig. 1 Geological location. A: Cambrils, B: Vilanova de la Sal, C: Gerri de la Sal

Table 1 Collected saline waters

Sample	Sampling data	Location	Salinity (g NaCl/kg)
CAM-3	2019, Nov	Cambrils saltwork source	253
VIL-1	2019, Dec	Vilanova de la Salt saltwork source	78
GE-1	2020, Jan	Gerri de la Sal saltwork source	58
FS-1	2020, Sept	Cambrils saltwork source	NA

NA Not analysed

3 Results

Outcropping Na–Cl rich brines contain Li (0.17–2.43 ppm), B, Rb, Zn and Mn as main trace elements, followed by minor amounts of Ba and Cu. The rest of the elements are mostly below their detection limits (Al, As, Fe, Pb, Ti, Y; Table 2). Isotopic signatures of salty fluids are shown in Table 3.

4 Discussion

Noticeable Li content in seawater brines is well known due to the conservative behaviour of such element, which increases during evaporation as it is not hosted in most evaporite minerals lattices. From seawater (Li concentration 0.18–0.20 mg/L (Berner & Berner, 1996)), Li in seawater brines will only be extracted during precipitation of last Mg-rich evaporites (Li replaces Mg), retained in interstitial brines and/or in fluid inclusions in chlorides. Therefore, dissolution of fossil seawater evaporites is considered one of the plausible origins of Li in deep fluids. However, the highest Li contents, up to 6400 mg/L (López Steinmetz et al., 2018), are found in non-marine "salars" (arid to semi-arid salt pans and ephemeral salt lakes in tectonically active sedimentary basins). Li contents of the studied salty brines (< 3 ppm) are not contemptible, and are similar to some geothermal fluids from NE Spain (Jullien-Sicre, 2022), including the widely consumed bottled drinking waters in Spain, with 1.2 ppm Li (Perea Horno & Catalan, 1988).

Studied salty springs are non-marine in origin, and not clearly related to hydrothermalism or volcanic fluids, which would enrich brines in lithium. Their negative H–O and C isotope water signatures also support this evidence. Thus, Li origin is mainly due to dissolution of Keuper evaporites, both considering Li contained in salt rocks and/or in fluid inclusions. The S–O isotope signature in the Cambrils area (CAM and FS samples) is clearly in accordance with isotope values of Keuper units (Ortí et al., 2022) and with their Na/Cl ratio close to 1 (Playà et al., this volume), indicating simple rock dissolution. Moreover, higher *S* values in the rest of the springs may suggest bacterial sulphate reduction processes in the sulphate molecule or other enrichment processes, such as reservoir effect by basin closure during sulphate precipitation. The rest of the detected trace elements

Table 2 Geochemical composition of salty sources (in ppb)

Samples	Ba	Li	As	Pb	B	Al	Cu	Mn	Rb	Ti	Zn	Y	Fe
CAM-3	*bdl*	2434	*bdl*	*bdl*	7396	*bdl*	*bdl*	659	971	*bdl*	610	10	*bdl*
VIL-1	112	172	148	13	1431	310	65	47	75	22	483	*bdl*	*bdl*
GE-1	64	616	*bdl*	*bdl*	2275	*bdl*	30	128	480	*bdl*	309	*bdl*	*bdl*
FS-1	27	2755	*bdl*	*bdl*	7721	*bdl*	16	793	1143	*bdl*	217	*bdl*	267

bdl Below detection limit

Table 3 Sulphate and water isotopic composition of salty sources

Samples	$\delta^{18}O_{VSMOW}$ (‰)	$\delta^{34}S_{VCDT}$ (‰)	$^{87}Sr/^{86}Sr$	$\delta^{2}H_{VSMOW}$ (‰)	$\delta^{18}O_{VSMOW}$ (‰)
GE-1	+13.5	+16.9	0.70780	− 54.3	− 8.4
VIL-1	+13.4	+15.4	0.70806	− 57.5	− 6.6
CAM-3	+9.7	+13.5	0.70792	− 47.1	− 8.3

(B, Rb, Zn, Mn) are also related to halite dissolution and accompanying minerals dissolution-interaction.

Main reported Mg-minerals in Keuper units are dolomite, although not predominant, and Mg-chlorides are not reported (Korte et al., 2003). Thus, dissolved salts in the studied brines are due to fluids circulating through the evaporite Keuper facies (Triassic), mainly related to halite dissolution and opening of their fluid inclusions (interstitial brines are not considered as they have been expulsed during burial). Sr isotopy reveals similar compositions among the samples and indicates marine evaporite deposition (Ortí et al., 2017).

5 Conclusion

Some salty springs of NE Spain display amounts of Li between 0.17 and 2.43 ppm, which are noticeable values compared to seawater and similar to some hydrothermal springs from NE Spain. Higher Li values are related to the most saline brines. Na–Cl brines outcrop after deep circulation and dissolution through Upper Triassic marine evaporites (Keuper), evidenced by their S–O, Sr, and H–O isotope compositions.

Acknowledgements The research is under the frame of the DGICYT PID2021-122467NB-C22, PGC2018-093903-B-C22 (MICINN/https://doi.org/10.13039/501100011033/FEDER), PID2020-118999GB-I00, and Grup de Recerca Reconegut "Sedimentary Geology" (2021 SGR-Cat 00349).

References

Berner, E., & Berner, R. (1996). *Global environment, water and air geochemical cycles* (p. 311). Prentice Hall.

Jullien-Sicre, A. (2022). *Étude d'un analogue de terrain de faille crustale associant une exploitation géothermique à fluides caloporteurs enrichis en lithium: Faille du Vallès- Penedès et système géothermique du fossé du Vallès (Espagne)* [Master Thesis]. Université Paris-Saclay.

Korte, C., Kozur, H. W., Bruckschein, P., & Veizer, J. (2003). Strontium isotope evolution of Late Permian and Triassic seawater. *Geochimica Et Cosmochimica Acta, 67*(1), 47–62.

López Steinmetz, R. L., Salvi, S., García, M. G., Peralta Arnold, Y., Béziat, D., Franco, G., Constantini, O., Córdoba, F. E., & Caffe, P. J. (2018). Northern Puna Plateau-scale survey of Li brine-type deposits in the Andes of NW Argentina. *Journal of Geochemical Exploration, 190*, 26–38.

Ortí, F., Pérez-López, A., Pérez-Valera, F., & Benedicto, C. (2022). Isotope composition ($\delta^{34}S$, $\delta^{18}O$) of the Middle Triassic-Early Jurassic sulfates in eastern Iberia. *Sedimentary Geology, 431*, 106104.

Ortí, F., Pérez-López, A., & Salvany, J. M. (2017). Triassic evaporites of Iberia: Sedimentological and palaeogeographical implications for the western Neotethys evolution during the Middle Triassic-Earliest Jurassic. *Palaeogeography, Palaeoclimatology, Palaeoecology, 471*, 157–180.

Perea Horno, M. A. (1988). Vichy Catalan. *Boletín Sociedad Española Hidrología Médica, III*(3), 149–151.

Playà, E., Pujadas, J., Martín-Martín, J. D, Cantarero, I., Baqués, V., Martín, M., Casals, S., & Travé, A. (this volume). *Deep-fluids circulation in high Mountain Salty Springs*. MedGU Conference.

Thomas, B., Magnan, J. F., Rupp, A., & Laroche, N. (2020). From mine to mind and mobiles: society's increasing dependence on lithium. *Elements, 16*(4), 265–270.

Contribution to the Hydrogeochemical and Bacteriological Knowledge of the Aquifer Systems of the EL Oued Area (Northern Rif, Morocco)

Redouan Alilouch, Karim Elmorabiti, Abdelaziz Elmrihi, Nicolas Rollo and Bachaer Ayed

Abstract

The study is the first in the El Oued region. It was based on a descriptive statistical approach to analyze and process hydrogeochemical and bacteriological data of groundwater draining detrital and carbonate sedimentary geological formations to develop water quality classes for the aquifers in the study area and to valorize them. The study area is in northern Morocco, in the northern Rif. The latter is a part of the Flyschs aquifer and a part of the Calcareous-Dorsal of the Rif chain southeast of the Tetouan city. The principal Calcareous-Dolomitic and detrital geomorphic unit formations provide an important reservoir in the study area. Recharge of these aquifers is deeply linked to the direct precipitation. The atomic spectrometry and the titrimetric methods analyzed fifteen groundwater samples from springs, boreholes, and hand-dug wells. They were the subject of a physical–chemical (EC, temperature, pH) measurement in situ, a chemical-bacteriological analysis (Ca^{2+}, Na^{+}, K^{+}, Mg^{2+}, TAC, TDS, Cl−, HCO_3^-, SO_4^{2}−, NO_3^-, fecal coliforms) in the laboratory, and a hydrochemical study. The aquifers in the study area are generally of two types: fissured aquifers and porous media aquifers. The physicochemical and bacteriological data show that the water in the aquifer system of the study area was generally clear of Nitrate and of excellent to very good quality. These results were confirmed by the Water Quality Index (WQI $\leq$ 41.46) and the Principal Component Analysis (PCA). Three geochemical water facies were distinguished: a calcium chloride dominated water (Numidian Flysch), a calcium bicarbonate dominated water and a calcium magnesium bicarbonate dominated water (Beni Ider Flysch, Limestone and Dolostone). Based on the Electric Conductivity (EC) and the sodium percentage (%Na) classification, 93.34% of the water samples were considered excellent and 6.66% were considered as good for irrigation purposes. While, based on the EC and the SAR classification, 26.7% and 73.3% of groundwater samples indicate excellent and a good to somewhat good water quality, respectively.

Keywords

Aquifer System · Groundwater · Water Quality Index · Descriptive statistics · Irrigation

R. Alilouch (✉)
Laboratory of Geosciences and Environment Techniques, Department of the Geology, Faculty of Sciences, Chouaib Doukkali University, El Jadida, Morocco
e-mail: alilouch.r@ucd.ac.ma

K. Elmorabiti · A. Elmrihi
LGO, Department of Geology, Faculty of Science-Tetouan, Abdelmalek Essaadi University, Tetouan, Morocco

N. Rollo
LETG-Nantes Géolittomer, University of Nantes, UMR 6554, BP 81 227, Nantes, France

B. Ayed
Laboratory of Water, Energy and Environment (LR3E), BPW 3038, Sfax, Tunisia

1 Introduction

Water is a precious and complex resource for humanity, whose access to this vital resource on earth has become a strategic and sovereignty issue. The demand for drinking water is set to increase significantly over the coming decades, mainly due to demographic pressure and irrigated agriculture. This leads to increasing depletion and contamination of groundwater resources putting water security at risk (Blanco & Donoso, 2020).

H. Chenchouni et al. (eds.), *Recent Advancements from Aquifers to Skies in Hydrogeology, Geoecology, and Atmospheric Sciences*, Advances in Science, Technology & Innovation, https://doi.org/10.1007/978-3-031-47079-0_14

Groundwater enjoys a very central role in the domestic water delivery system in the Northern Rif of Morocco. It is particularly the main water supply for the rural population, which forms about 1.3% of the total population of the Tangier Tetouan Al Houceima region (R.G.P.H., 2014). The study area is a part of the internal domain (Calcareous-Dorsal), the external domain and the flysch domain of the Rifan chain southeast of the Tetouan city. It is a rural municipality with a concentrated population of about 11,288 inhabitants (R.G.P.H., 2014). Groundwater is often used for domestic purposes as well as for the irrigation of some agricultural products but is always threatened by septic tanks.

The focus of water quality in humanitarian programs is primarily in terms of drinking water and irrigation (Lamastra et al., 2016; Bouaroudj et al., 2019). The Water Quality Index (WQI) is a very useful method for detecting groundwater quality and suitability for drinking (Akhtar et al., 2021; Allaoua et al., 2024). At this location in Talassemtane National Park, we conducted research based on classifications of the chemical and bacterial quality of the water to give a comprehensive groundwater evaluation.

2 Materials and Methods

The groundwater assessment was achieved according to whole methodology. So, a research methodology was followed based on field and laboratory work. It is to highlight that systematic sampling, measurements, and analyses were conducted during a first dry period (2019–2020) and a second wet period (2019–2020). To attempt our objectives, the first step consists of the fieldwork where fifteen water points were inventoried. The second step was groundwater sampling and some of physical parameter's measurements in situ. The third step was the bacteriological and chemical analyses in the Laboratory of ABHL and Water Mineral chefchaouen. The final step was the results of the anions and cations analyses used to calculate the WQI (Sunitha & Reddy, 2022), from which the drinking water quality was derived.

Water samples taken for chemical analysis are kept in 1-l plastic bottles previously rinsed three times with the water to be collected, while samples for bacteriological analysis required clean and sterile plastic flasks, and cold-stored. Atomic absorption spectrometry was used to determine Ca^{2+} concentrations while a UV spectrophotometer was used to determine the NO_3^- concentration. The Na^+ and the K^+ concentrations were determined using flame analysis. Titrimetric and colorimetric methods were used in determining the Cl−, the HCO_3^- and the SO_4^{2}– concentrations respectively. While for bacteriological analysis, the membrane filtration techniques were used to measure fecal coliforms 'FC' (Rodier et al., 2009).

3 Results and Discussion

3.1 Geochemical Facies and Parentage

Based on the chemical analyses (15 samples) (Cotruvo, 2017), the mean and median values of the chemical elements concentrations of the studied aquifer systems (Table 1) and the graphical representations of three groundwater types were distinguished: (i) water dominated by calcium chloride (Sw6), (ii) water dominated by calcium bicarbonate (13 samples) and (iii) water dominated by calcium and magnesium bicarbonate (Sw11). Therefore, the chemical features of the groundwater were conditioned by the geological formation nature in the study area such us Flysch (Numidian and Beni Ider), Limestone and Dolostone (Fig. 1).

The statistical method PCA allows us to determine three groundwater groups (Fig. 2): (i) Cluster (A) includes 5 water points (Sw4, Sw5, Sw1, Sw2, Sw3), plotted on the negative side of the F1 axis, it includes the least natural mineralized samples in the region. (ii) Cluster (B) consists of only 2 groundwater points (Sw6, Sw12); these two samples were characterized by a mixture between the less mineralized water of cluster (A) and the moderately mineralized water of cluster (C). And (iii) Cluster (C) which was made up of 8 groundwater points and was positively correlated with the F1 factor. It represents the samples whose lithological formation is related to the flysch aquifer systems of Beni Ider and to the aquifer systems of the Calcareous-Dolomitic Dorsal. It gathers the most mineralized samples of the area.

3.2 Water Quality

Based on the WQI results for individual samples, these statistical values, and the basic classification of the WQI (Şener et al., 2017), it was evident that the studied aquifer groundwater was of an excellent quality for human drinking. Therefore, the studied aquifers' groundwater chemical quality is evolving from West (less mineralized) to East (well mineralized). In another words, from the Numidian flysch aquifers of the calcareous-dolomitic dorsal aquifers through the Quaternary and Beni Ider Flysch aquifers. The Riverside diagram shows two classes: the first class was characterized by an excellent water quality

Table 1 Synthesis of the statistical description of the physical–chemical and bacterial parameters

Parameter	Unit	(WHO standards) (Rodier et al., 2009)	Min	Max	Mean	Standard deviation
Na^+	mg/L	200	3	35.5	17.36	10.7
K^+	mg/L	12	0.4	5.3	1.63	1.29
T°	°C	24	10	14.5	11.72	1.64
pH		6, 5–8, 5	5.6	7.6	6.9	0.61
EC	µS/cm	1500	113	1033	541.27	240.61
TDS	mg/L	500	72	561	339.55	141.09
Ca^{+2}	mg/L	75	8.2	176.35	76.42	45.45
Mg^{+2}	mg/L	30	4.86	37.68	12.25	9.06
Cl^-	mg/L	250	9.23	132.86	55.83	41.63
HCO_3-	mg/L	500	24.5	375.32	213.16	95.37
SO_4^{2-}	mg/L	200	11.75	46.5	22.15	9.9
NO_3-	mg/L	45	0	0.5	0.22	0.18
TH	°F	300	4.7	34.1	19.63	8.51
TAC	°F	200	2.01	30.75	17.46	7.81
FC	UFC/100 mL	20	0	178	20.47	45.77
WQI			13.19	41.46	26.93	7.66

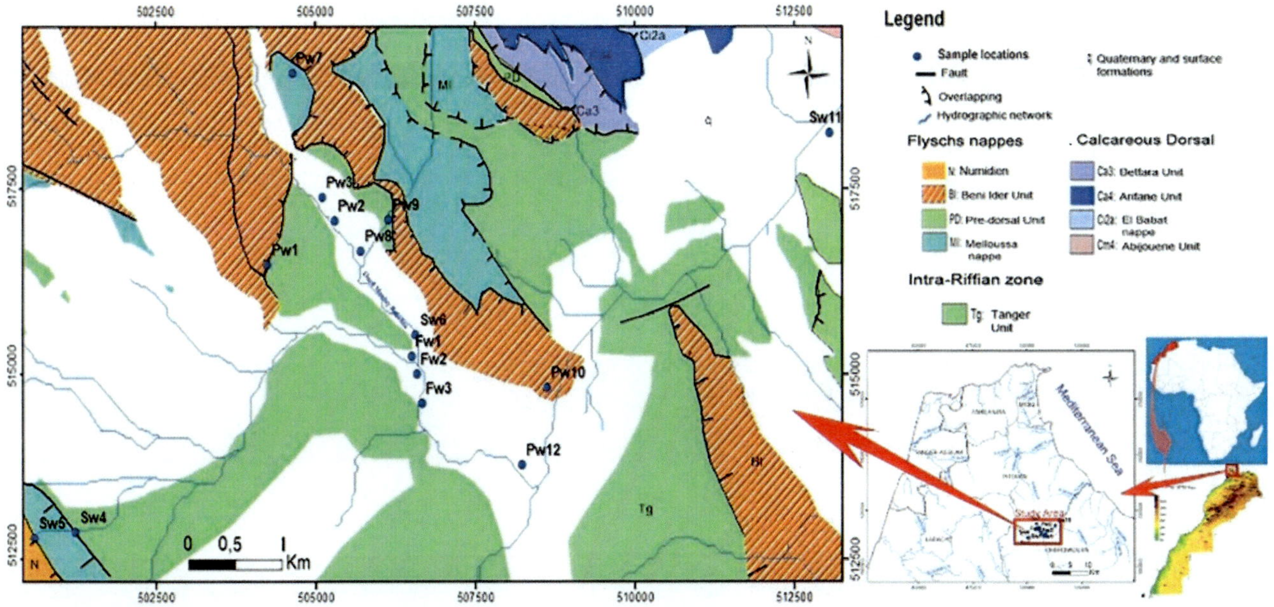

Fig. 1 Map of the study area showing the sample locations

(26.7%). The second class was consisted of good to fairly good water quality (73.3%), which can be used without any control for plant irrigation which was moderately salt tolerant (Batarseh et al., 2021). Bacteriological analyses (Table 1) reveal two groundwater classes: excellent water quality (73.3%) with coliform concentration inferior to 20 CFU/100 ml and good water quality (26.7%) with a concentration equal or superior to 20 CFU/100 ml (N.M: NORMES MAROCAINES, 1423).

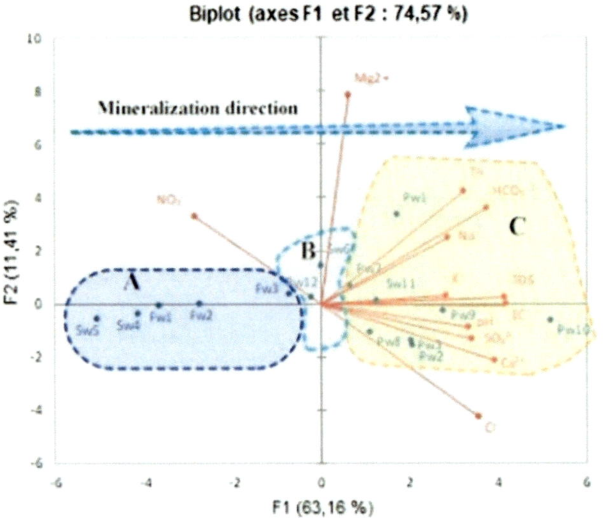

Fig. 2 Principal component analysis—conceptual structure map of the study area

4 Conclusion

The three geochemical facies characterize the aquifers in the study area. To satisfy the population needs in drinking and irrigation water uses of El Oued commune, chemical and bacteriological analyses evaluation were used. These methods reveal that the water was generally of excellent quality and exempt of Nitrate.

References

Akhtar, N., Ishak, MIS., Ahmad, MI., Umar, K., Yusuff, MS Md., Anees, MT., Qadir, A., & Almanasir, YK Ali. (2021). Modification of the water quality index (WQI) process for simple calculation using the multi-criteria decision-making (MCDM) method: A review. *Water, 13*(7), 905.

Allaoua, N., Hafid, H., & Chenchouni, H. (2024). An appraisal of physicochemical and bacteriological quality of groundwater resources in semiarid lands of Algeria. *Journal of Arid Land, 16*(2), 147–167.

Batarseh, M., Imreizeeq, E., Tilev, S., Al Alaween, M., Suleiman, W., Al Remeithi, AM., Al Tamimi, MK., & Al Alawneh, M. (2021). Assessment of groundwater quality for irrigation in the arid regions using irrigation water quality index (IWQI) and GIS-Zoning maps: Case study from Abu Dhabi Emirate, UAE. *Groundwater for Sustainable Development, 14*, 100611.

Blanco, E., & Donoso, G. (2021). *Global Water Security Issues (GWSI) 2020 Theme: The role of sound groundwater resources management and governance to achieve water security*. UNESCO.

Bouaroudj, S., Menad, A., Bounamous, A., et al. (2019). Assessment of water quality at the largest dam in Algeria (Beni Haroun Dam) and effects of irrigation on soil characteristics of agricultural lands. *Chemosphere, 219*, 76–88.

Cotruvo, J. A. (2017). 2017 WHO guidelines for drinking water quality: First addendum to the fourth edition. *Journal-American Water Works Association, 109*(7), 44–51.

Lamastra, L., Balderacchi, M., & Trevisan, M. (2016). Inclusion of emerging organic contaminants in groundwater monitoring plans. *MethodsX, 3*, 459–476.

N.M: NORMES MAROCAINES. (2002). *Bulletin officiel N° 5062 du 30 ramadan 1423 fixant les normes de potabilité à la consommation humaine*.

R.G.P.H. (2014). *Recensement Général de la Population et de l'Habitat Haut-commissariat au Plan du Maroc*.

Rodier, J., Legube, B., Merlet, N., Brunet, R., Mialocq, J. C., & Leroy, P. (2009). *L'analyse de l'eau*. (9e ed.). Paris. ISBN 978-2-10-054179-9.

Şener, Ş, Şener, E., & Davraz, A. (2017). Evaluation of water quality using water quality index (WQI) method and GIS in Aksu River (SW-Turkey). *Science of the Total Environment, 584*, 131–144.

Sunitha, V., & Reddy, B. M. (2022). Geochemical characterization, deciphering groundwater quality using pollution index of groundwater (PIG), water quality index (WQI) and geographical information system (GIS) in hard rock aquifer, South India. *Applied Water Science, 12*(3), 41.

Assessment of Groundwater Salinization Using Combined Hydrogeochemical Tools: The Case of Rhodope Aquifer (NE Greece)

Evangelos Tziritis, Ekaterini Sachsamanoglou and Maria Dolores Fidelibus

Abstract

The Rhodope coastal area (NE Greece) is an intensively cultivated agricultural region with profound signatures of water quality deterioration because of elevated salinity. Within the MEDSAL Project (www.medsal.eu) activities, a thorough study of hydrogeochemical characteristics was performed to assess the extent of groundwater salinization and identify its dynamics. A combined suite of hydrogeochemical tools was applied to the data of forty-six (46) groundwater samples collected in June 2020. Calculating salinization facies (SFs) allowed classifying samples according to their chloride content. The salinization phenomenon is evident since nearly 30% of groundwater samples result in brackish (B) or brackish saline (Bs) with $Cl- > 1000$ mg/L). Most saline samples (B and Bs) are distributed in two areas close to the coastline. Nevertheless, seawater intrusion is not the dominant salinization cause, as the effect from an additional diluted brine is profound. The fresh (F) and fresh-brackish (Fb) waters are located further inland than the B sample. However, they do not show any definite spatial pattern distribution. Considering the hydrogeological regime of the area, the F samples are likely to be significantly affected by the deeper (confined) aquifer layer of better qualitative conditions. Based on a series of binary plots of major elements versus Cl^-, we conclude that salinization may result from one single source or a combination of more sources from place to place. According to previous reports and surveys, additional cascading sources may include trapped saline lenses and irrigation water return.

E. Tziritis (✉) · E. Sachsamanoglou
ELGO-DIMITRA, Soil and Water Resources Institute (SWRI), 57400 Thessaloníki, Greece
e-mail: e.tziritis@swri.gr

M. Dolores Fidelibus
DICATECh Department, Politecnico di Bari, Via Orabona 4, 70125 Bari, Italy

Keywords

Groundwater salinization · Hydrogeochemistry · Salinization facies · Binary plots · Rhodope

1 Introduction

Understanding aquifer characteristics and connectivity is crucial for effective water management. Indeed, the investigation of interactions between shallow and deep aquifers provides insights into groundwater dynamics, vital for sustainable water use in coastal regions (Kallel et al., 2023). The Rhodope coastal area (RHO) is located in northeastern Greece, covering approximately 165 km^2 (Fig. 1). It mainly comprises agricultural land, which occupies more than 65%. The coastal area's geological formations include a Paleozoic metamorphic substrate (Rhodope massif) on the bottom of the sequence, composed of gneisses, amphibolites, marbles and migmatites. The substrate is overlaid by recent post-alpine formations of the Pliocene, Pleistocene and Quaternary ages. The more recent formations are of Holocene age and consist of non-consolidated deposits of alluvial origin characterized by high permeability. Two main aquifers are identified: a shallow semi-confined aquifer with an average thickness of 35 m and limited hydrogeological potential and an underlain thicker one (50–100 m), which is confined and hosts the regional groundwater reserves (Kallioras, 2008). According to Petalas and Lambrakis (2006), the two aquifers are hydraulically connected locally due to geomorphological and tectonic factors, thus, forming a unified aquifer system.

The irrational irrigation schemes applied in the last decades because of the intense agricultural activities led to a relevant groundwater drawdown, sometimes reaching up to 30 m or even more (Petalas & Lambrakis, 2006). As a result, saline water encroached on the coastal parts of

H. Chenchouni et al. (eds.), *Recent Advancements from Aquifers to Skies in Hydrogeology, Geoecology, and Atmospheric Sciences*, Advances in Science, Technology & Innovation, https://doi.org/10.1007/978-3-031-47079-0_15

Fig. 1 Spatial distribution of salinization facies with the dominant water types (hydrochemical facies)

the area. Groundwater salinization is justified by previous research (e.g. Kallioras, 2008; Petalas & Lambrakis, 2006) mainly from seawater intrusion owing to overexploitation of groundwater. However, the complex stratigraphy and further hydrogeochemical evidence denote that additional potential sources might exist whose overall intensity and impact are not yet fully understood.

2 Materials and Methods

A sampling campaign performed in June 2020 comprised 42 groundwater samples. In addition, 12 groundwater reference samples were further added based on the recorded data of the Greek National Monitoring Network; hence, 54 groundwater samples were considered in total. Twenty-nine parameters have been determined regarding the groundwater samples collected, including physicochemical parameters (measured in situ by portable probes), major/minor ions and trace elements. A flame photometer was used to analyze Na^+ and K^+, a spectrophotometer was used for NH_4^+, SO_4^{2}–, NO_3^- and B, and a flame atomic absorption spectrometer (AAS flame) was used for the analysis of Ca^{2+}, Mg^{2+}, Si and Zn. The titration method was used to analyze Cl– with neutralization of $AgNO_3$, and for the HCO_3^- with neutralization of HCl. The trace elements (Al, As, Ba, Cd, Cr, Cu, Fe, Li, Mn, Ni, Pb, Sb, Se, Sr, V) were analyzed with an AAS graphite furnace. Analytical accuracy was checked by calculating the Charge Balance Error (CBE) and was found $< 10\%$ for all samples. The groundwater samples were classified according to their Cl– content, based on a modification of the original hydrochemical facies classification performed by Stuyfzand (1989). Finally, the groundwater salinisation assessment has been performed by multiple binary plots of major and minor ions versus Chloride (e.g. Cl vs Ca, Na, Mg, K, SO_4, HCO_3, Li, Sr and B. Such molar ratios could provide useful information about the ongoing processes and potential salinization sources and are widely used in similar studies (e.g. de Montety et al., 2008).

3 Results and Discussion

The pH of the groundwater system is slightly alkaline with a medium (med) value of 7.5. The Specific Conductance (SpC) ranges from 612 to 19.900 μS/cm (med = 1.847 μS/cm), indicating the impact of groundwater salinization. Calcium (Ca^{2+}) is the most abundant cation in the groundwater water system, with a median (med) of 195.2 mg/L, followed by Na^+ (med = 124.2 mg/L), Mg^{2+} (med = 41.5 mg/L) and K^+ (med = 2.4 mg/L). The most anion is Cl^- (med = 458.3 mg/L), followed by HCO_3- (med = 227.6 mg/L), SO_4^{2-} (med = 49.4 mg/L) and NO_3- (med = 36.4 mg/L). Some noteworthy outliers concerning the measured trace elements in the groundwater samples were recorded for arsenic (68.8 μg/L) and boron (1657.7 μg/L). According to the modified salinization facies (SF) classification, 5 samples are classified as oligosaline (OH), 18 as fresh (F), 14 as fresh-brackish (Fb), 16 as brackish (B) and one as brackish saline (Bs). The spatial distribution of the SFs and the prevailing ions are displayed in Fig. 1. For reference, seven SW samples collected from the lagoons have been added and characterized as hypersaline (HS). The most brackish groundwater samples (Bs and B) are oriented across a linear orientation of the WNW-ESE direction, parallel to the coastline. The F and Fb are generally located further inland than the B samples; however, they are scattered without any profound spatial rationale in their distribution. Considering the hydrogeological regime of the area, the F samples are likely to be significantly affected by the deeper (confined) aquifer layer of better qualitative conditions. The OH groundwater samples are considered typical for the recharge waters.

The salinization phenomenon is evident in the groundwater resources of the coastal area, as nearly 30% of groundwater samples are characterized as brackish or brackish saline ($Cl^- > 1000$ mg/L). Based on a series of binary plots of major and minor elements versus Cl−, we may conclude that there is no single salinization source; however, seawater intrusion seems to be the dominant factor. According to previous reports and surveys, additional cascading sources may include trapped saline lenses and the impact or upconing of deep brines. Nevertheless, more sophisticated methods (e.g. isotopic analysis and modelling) are needed to verify the exact salinization sources and processes.

The spatial orientation of most saline samples (B and Bs) mainly occurs in two areas in SE and SW, close to the coastline, denoting that the higher salinity values are predominantly impacted by seawater intrusion. The most brackish groundwater samples (Bs and B) are oriented across a linear orientation of the WNW-ESE direction parallel to the coastline. The F and Fb are generally located further inland along the WNW-ESE direction than the B samples; however, they are scattered without any evident spatial pattern in their distribution. Considering the hydrogeological regime of the area, the F samples are likely to be significantly affected by the deeper (confined) aquifer layer of better qualitative conditions. The OH groundwater samples are considered typical for the recharge waters at W-NW of the RHO site. The salinity of B and Bs samples most probably originates mainly from seawater intrusion through two dominant axes: (a) in SE through the Ptelea and Elos lagoons and (b) in SW through Porto Lagos lagoon and Vistnida Lake. The temporal variation of some minor elements indicates that the western part is still affected by the lateral crossflows of the Kompsatos River. On the contrary, the eastern part is more impacted by salinization, with dominant seawater intrusion influences.

4 Conclusion

Various binary plots have been used to assess the salinization in Rhodope coastal aquifer. Over one salinization source and processes have been identified, which in most cases are overlapping. The dominant process is seawater intrusion (mainly near the coastline); however, the inland parts are mostly influenced by the upconing of a deeper diluted brine and the effect from the irrigation return flow, as suggested by the elevated nitrate values. The outcomes may be useful for mitigating the local scale's salinization phenomenon and planning managerial actions for sustainable groundwater resource management. However, the overall methodological framework may also be adopted in other Mediterranean regions subject to salinization and facilitate the identification of salinization sources and processes.

References

de Montety, V., Radakovitch, O., Vallet-Coulomb, C., et al. (2008). Origin of groundwater salinity and hydrogeochemical processes in a confined coastal aquifer: Case of the Rhône delta (Southern France). *Applied Geochemistry, 23*(8), 2337–2349. https://doi.org/10.1016/j.apgeochem.2008.03.011

Kallel, A., Barbieri, M., Rodrigo-Comino, J. (2023). *Selected Studies in Environmental Geosciences and Hydrogeosciences*. Springer, Cham. https://doi.org/10.1007/978-3-031-43803-5

Kallioras, A. (2008). *Groundwater resources management of aquifers subjected to seawater intrusion regime. The case of western coastal plain of the Prefecture of Rhodope* [PhD. Thesis]. Democritus University of Thrace (in Greek).

Petalas, C., & Lambrakis, N. (2006). Simulation of intense salinization phenomena in coastal aquifers—The case of the coastal aquifers of Thrace. *Journal of Hydrology, 324*, 51–64. https://doi.org/10.1016/j.jhydrol.2005.09.031

Stuyfzand, P. J. (1989). A new hydrochemical classification of water types, Regional Characterization of Water Quality. In *Proceedings of the Baltimore Symposium*. IAHS, Publ. no. 182.

Geostatistical Mapping of Piezometric Levels: Case Study of Chaouia Coastal Aquifer, Morocco

Saliha Najib, Ahmed Fadili, Khalid Mehdi, Joelle Riss and Najwa Hassou

Abstract

Located on the Moroccan Atlantic coast, coastal Chaouia aquifer, the region's subject of this study, constitutes the most extensive aquifer. It is composed mainly by a complex aquifer of geological formations from Ordovician-Cambrian, Cretaceous and Pliocene–Quaternary age, respectively. The geostatistical method by kriging, applied to piezometric data of 2009 and 2011, draws the evolution of water levels in these periods. The comparison of the obtained piezometric maps shows a seaward move of the potentiometric line 0 m, in 2011, which has been related to the importance of 2010 rainfall. The flow of water table is always oriented toward the sea with a convex trend of the piezometric curves at the vicinity of Oum Er-Rbia River, showing a permanent drainage phenomenon, proving that the piezometric level of the groundwater remains constantly above the level of the river water. Besides, piezometric chronicles recorded in the aquifer formations indicate that the piezometry of the water table is mainly related to possible changes in recharge rate under the rain.

Keywords

Geostatistical method · Piezometry · Aquifer · Chaouia coastal · Morocco

1 Introduction

Geostatistics, as a field of applied statistics, focuses on the analysis of spatially correlated data using the principles of regionalized variables theory (Hosseini et al., 2019). Geostatistics involves both analyzing and describing the spatial variability of a phenomenon, as well as making inferences and estimating unknown values of continuously distributed variables (Uyan & Cay, 2013).

The geostatistical kriging method is commonly applied in hydrogeology to estimate groundwater level (Ahmed & Marsily, 1987; Alley, 1993; Andres & Martin, 2005; Guekié Simo, 2012) and chemical contamination in groundwater research (Iskandar & Koike, 2010; Shamsudduha et al., 2009). The accomplishment of this method requires three steps: the first is variographic mapping, then model adjustment to each experimental variogram and finally the maps kriging.

The coastal aquifer is especially susceptible to degradation due to its proximity to seawater, in association with the intensive water demands that accompany high population densities in coastal areas. However, the distinctive feature of coastal aquifers is that freshwater within the aquifer is directly connected to marine water by an interface no thicker than a few meters.

The subject of this study is the coastal Chaouia, which serves as a significant hydrogeological unit in the Moroccan coastal Meseta, it is composed of a complex heterogeneous system comprising Plio-quaternary sandstone-limestone, Cretaceous marly limestone and altered shales from the Paleozoic (Bentayeb, 1972; Najib, 2014). The boundaries of the hydrogeological unit are located with the Oum Er-Rbia River, the Atlantic Ocean, the city of Casablanca and the

S. Najib (✉) · K. Mehdi
IRNE Team, MRI Laboratory, Polydisciplinary Faculty of Khouribga, Khouribga, Morocco
e-mail: sa.najib@usms.ma

S. Najib · A. Fadili · K. Mehdi
Marine Geosciences and Soil Sciences Laboratory, Faculty of Sciences, El Jadida, Morocco

A. Fadili
ERMAM Team, Department of Physics, Polydisciplinary Faculty of Ouarzazate, Ibnou Zohr University, Agadir, Morocco

J. Riss
I2M UMR 5295 Laboratory, ECE Department, Bordeaux University, Bordeaux, France

N. Hassou
VOMB Team, VMQB Laboratory Faculty of Sciences and Technologies, Mohammedia, Morocco

H. Chenchouni et al. (eds.), *Recent Advancements from Aquifers to Skies in Hydrogeology, Geoecology, and Atmospheric Sciences*, Advances in Science, Technology & Innovation, https://doi.org/10.1007/978-3-031-47079-0_16

Berrechid plain to the south-west, north-west, north-east and south-east respectively.

The only accessible water resource in the area under study is the unconfined groundwater system. Groundwater is the main resource for the population of the region. This resource has been marked by alternating periods of severe drought and a high demand for water, leading to a perturbation of the piezometric level due to the natural imbalance between marine and fresh groundwater (Najib et al., 2017). The purposes of this paper are (1) to establish the piezometric evolution of the unconfined aquifer; (2) to evaluate the influence of the rainfall on the groundwater level variations using geostatistical methods. For this reason, two measurement campaigns were used between the years April 2009 and June 2011.

2 Materials and Methods

Geostatistical modeling is performed in two main steps: variographic analysis and then estimation (Fig. 1). The variogram analysis is used to study the spatial structuring of the variable, the variogram map is constructed to detect possible directions of anisotropy, then the experimental variogram is calculated and finally a theoretical variogram model with widely known properties is chosen and fitted to the experimental variogram. The variogram charts are represented in polar coordinates in which each cell represents by a set of colors a variogram value for a direction and distance between points. The variogram map analyzes the spatial continuity of a variable in all directions in a 2D visualization space and indicates in particular possible directions of anisotropies.

The variogram is the most used function in geostatistics to describe the continuity of measured features (water level), it serves to quantify the correlation between two measurements of the piezometric level located one close to the other. The variogram distance quantifies the dissimilarity $\gamma(h)$ between data points divided by a specific variable or set of variables. Within a defined range of both distance and direction θ, If $Z(u_\alpha)$ and $Z(u_\alpha+h)$ represent pairs of samples, and $N(h)$ is the count of data pairs within this class, the experimental semi-variogram can be expressed as the average squared difference between the components of these data pairs (Goovaerts, 1999; Kigobe & Kizza, 2006), as shown in Eq. 1.

$$\gamma(h,\theta) = \frac{1}{2N(h,\theta)} \sum_{i=1}^{N(h,\theta)} [Z(u_\alpha) - Z(u_\alpha + h)]^2 \quad (1)$$

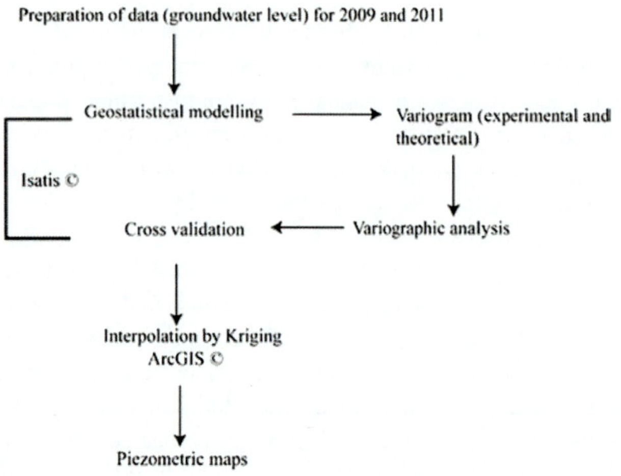

Fig. 1 Flowchart of the geostatistical method procedure

3 Results

3.1 Estimation of Water Level

The variographic map (Fig. 2) of water levels for May 2011 campaign indicates two directions of anisotropy i.e., N20° and N110° directions, which represent longitudinal and transverse flow respectively. The experimental water level variograms are calculated for these main anisotropy directions with an angular tolerance of 45°, in 15 time steps of 900 m with a distance tolerance equal to half the time step. The theoretical variogram model which is adjusted to these experimental variograms is non-stationary that corresponds to the spherical model expressed by Eqs. 2 and 3.

$$\gamma(h) = 500\left(1.5\frac{h}{11,000} + 0.5\left(\frac{h}{11,000}\right)^3\right) \text{ for direction of N20°} \quad (2)$$

$$\gamma(h) = 500\left(1.5\frac{h}{19,700} + 0.5\left(\frac{h}{19,700}\right)^3\right) \text{ for distance of N110°} \quad (3)$$

For this period of May 2011, the water level estimated by kriging shows a positive correlation with that measured in the field and a coefficient of determination R^2 equal to 0, 71 (Fig. 2). The H0 hypothesis of independence between measured and calculated piezometric levels is significantly rejected.

The variographic map (Fig. 3) of April 2009 campaign shows two directions of anisotropy i.e., N60° and N150°. The experimental variograms are calculated with an angular tolerance of 45°, in 10 steps of 1793 m with a tolerance on the distance equal to half the step. The theoretical variogram model that is adjusted to these experimental variograms is non-stationary and corresponds to the power model whose equations are expressed as follows (Eq. 4 and 5).

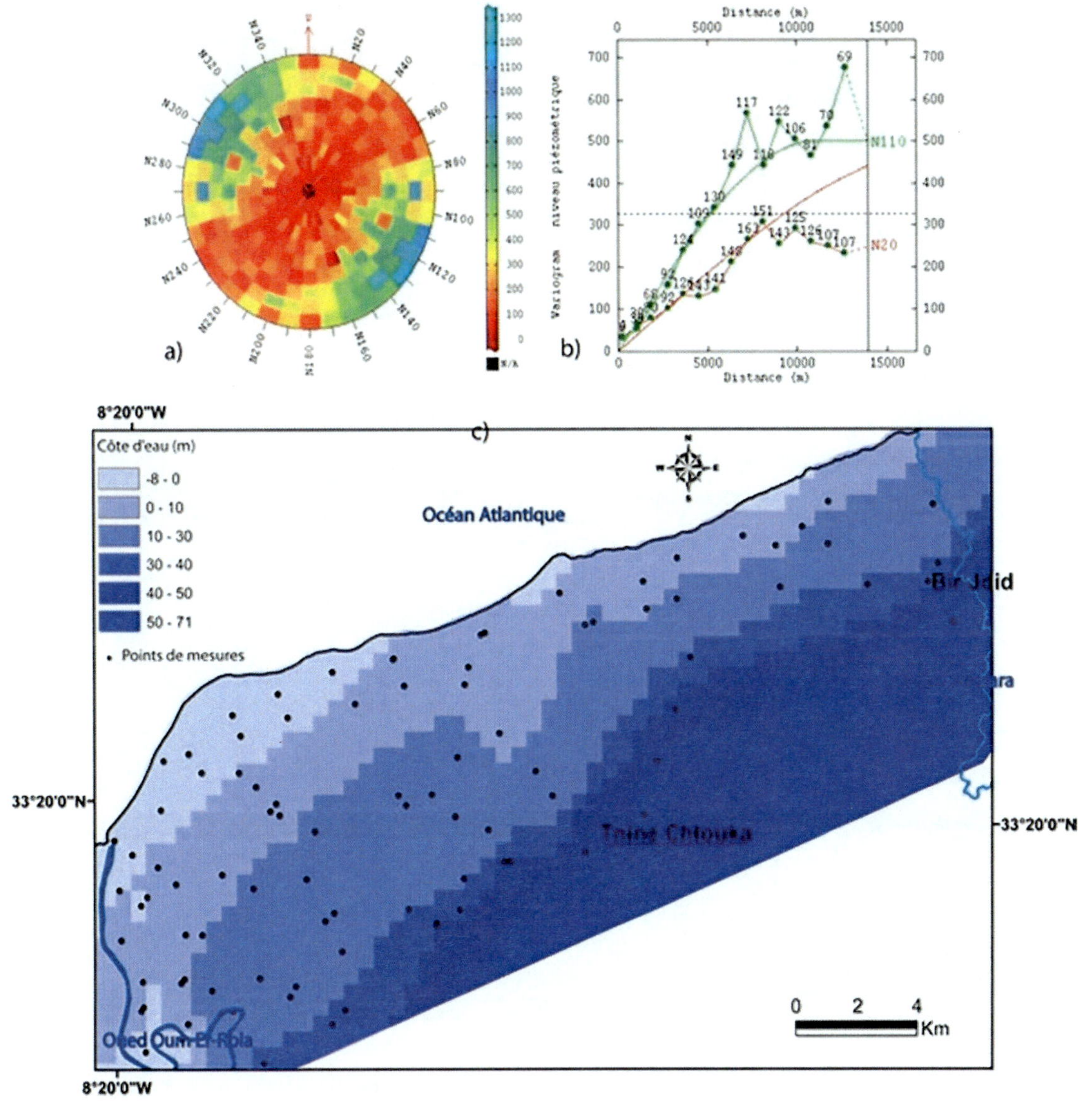

Fig. 2 Piezometric level maps established by geostatistical method on May 2011 **a** Variographic map, **b** experimental (dashed line) and adjusted theoretical (solid line) variogram of piezometric level in the main directions of anisotropy, **c** Kriging map

$$\gamma(h) = 3000\left(\frac{h}{8200}\right)^{1,8} \text{avec} 0 < \alpha < 2 \,\text{for direction of}\, \text{N}60° \tag{4}$$

$$\gamma(h) = 3000\left(\frac{h}{19,000}\right)^{1,8} \text{avec}\, 0 < \alpha < 2 \,\text{for distance of}\, \text{N}150° \tag{5}$$

The calculated level demonstrates a strong correlation with the measured level on the terrain, as indicated by a coefficient of determination R^2 of 0,95. The hypothesis H0 (independence between these two variables) is rejected in a significant way for the period of April 2009. A slight tendency toward an increase in the water table depth from downstream to upstream of the piezometric level is noted.

4 Discussion

The piezometric map of April 2009 shows a flow direction toward south-east except for the right bank of the Oum Er-Rbia River, indicating the existence of a communication between the groundwater and the river and manifested by the existence of curves spaced toward the wadi. The piezometric curves conserve their parallelism toward the Atlantic Ocean. The value and variations of the piezometric curves characterize the hydraulic gradient, as an important factor of the flow, it is about 5‰ in the north-east. On the other hand, in the downstream area (SW) between Tnine Chtouka and Azemmour (Fig. 2), the hydraulic gradient shows a value of 1‰ indicating a low permeability within the

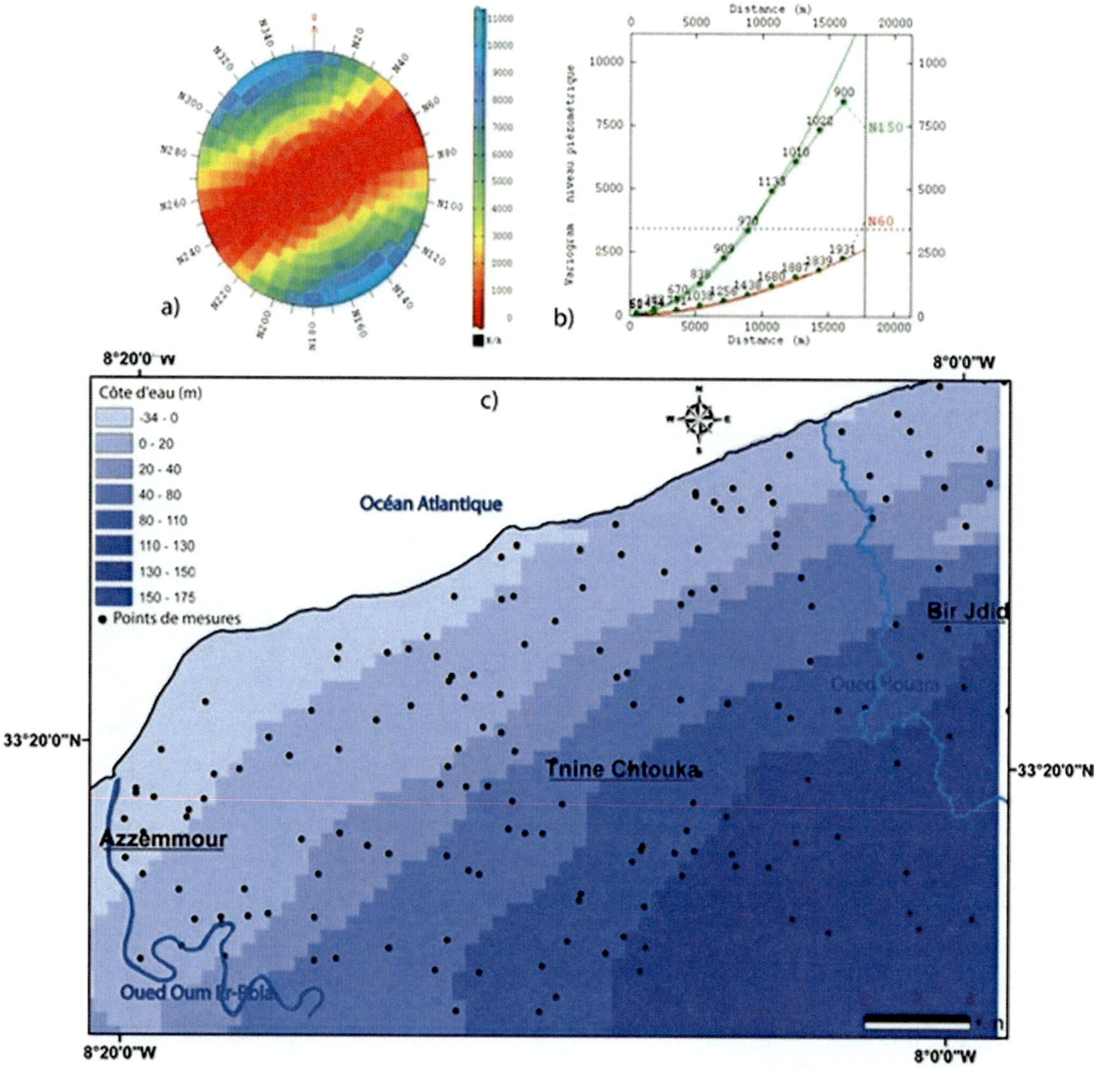

Fig. 3 Piezometric level maps established by geostatistical method on April 2009 **a** Variographic map, **b** experimental (dashed line) and adjusted theoretical (solid line) variogram of piezometric level in the main directions of anisotropy, **c** Kriging map

Cenomanian aquifer. The permeability of the aquifer terrains in the region is a determining factor in the groundwater flow variations.

The piezometric state, established for the period of May 2011, illustrates the same shape as the period of April 2009. The general flow direction is oriented toward the ocean, however, in the south-western part is directed toward the Oum Er-Rbia wadi (Najib et al., 2016). The obtained values of hydraulic gradient are not uniform. To the north-east the hydroisohypses are tight reflecting a high gradient of 3‰, and toward the SW they are spaced reflecting a low hydraulic gradient of 1‰. The comparison between piezometric maps from both periods shows an advance, during 2011, of the 0 m isopiez toward the ocean. This result is probably related to the important influence of rainfall recorded in 2010 (776 mm and 649 mm recorded respectively in the meteorological stations of Casablanca and El Jadida) on the water table recharge.

5 Conclusion

This study aimed to study the piezometric level and evolution of groundwater in the coastal Chaouia region geostatistics. During April 2009 period, the piezometry level is marked by an advance of the salt water toward land in the coastal sector between Azemmour and Tnine Chtouka. This situation is essentially due to the continuous solicitation of the aquifer and the irregularity of precipitations between 1997 and 2009 periods. The piezometric state of May 2011 underlines a regression of the zero line and an advance of the piezometric lines with respect to the sea. This finding is the consequence of the significant rainfall recorded in 2010. In the north-east zone, the water table is stable with piezometric lines parallel to the ocean and a flow direction towards the ocean.

References

Ahmed, S., & De Marsily, G. (1987). Comparaison of geostatistical methods for estimating transmissivity using data on transmissivity and specific capacity. *Water Resources Research, 23*(2), 1717–1737.

Alley, W. M. (1993). Geostatistical models. In *Regional ground-water quality* (pp. 87–108). Van Nostrand Reinhold.

Andres, A. S., & Martin, M. J. (2005). *Estimation of the water table for the Inland Bays watershed.* Delaware Geological Survey, Newark. No. 68.

Bentayeb, A. (1972). Etude hydrogéologique de la Chaouia Côtière avec essais de simulation mathématique en régime permanent, Maroc [doctoral dessertation].

Goovaerts, P. (1999). Geostatistics in soil science: State-of-the-art and perspectives. *Geoderma, 89*, 1–45. https://doi.org/10.1016/S0016-7061(98)00078-0

Guekié Simo, A. T. (2012). *Effets de la variation spatiale et temporelle des propriétés des terrains sur les défaillances des réseaux enterrés* (264 p) [Thèse de Doct.]. Univ. Bord.1, Bordeaux.

Hosseini, E., Gholami, R., & Hajivand, F. (2019). Geostatistical modelling and spatial distribution analysis porosity and permeability in the Shurijeh-B reservoir of Khangiran gas field in Iran. *Journal of Petroleum and Production Technology, 9*, 1051–1073.

Iskandar, I., & Koike, K. (2010). Distinguishing potential sources of arsenic released to groundwater around a fault zone containing a mine site. *Environment and Earth Science.* https://doi.org/10.1007/s12665-010-0727-8

Kigobe, M., & Kizza, M. (2006). Dealing with spatial variability under limited hydrogeological data. Case study: Hydrological parameter estimation in Mpigi-Wakiso. In *Proceedings from the international conference on advances in engineering and technology* (pp. 211–220). Elsevier.

Najib, S. (2014). Etude de l'évolution de la salinization de l'aquifère de la Chaouia côtière (Azemmour-Bir Jdid, Maroc): climatologie, hydrogéologie, hydrochimie et tomographie électrique [doctoral dessertation].

Najib, S., Fadili, A., Mehdi, K., Riss, J., Makan, A., & Guessis, H. (2016). Salinization process and coastal groundwater quality in Chaoui, Morocco. *Journal of Africain Earth Sciences, 115*, 17–31.

Najib, S., Fadili, A., Mehdi, K., Riss, J., & Makan, A. (2017). Contribution of hydrochemical and geoelectrical approaches to investigate salinization process and seawater intrusion in the coastal of Chaouia. *Journal of Contaminant Hydrology, 198*, 24–36.

Shamsudduha, M., Marzen, L. J., Uddin, A., Lee, M. K., & Saunders, J. A. (2009). Spatial relationship of groundwater arsenic distribution with regional topography and water-table fluctuations in the shallow aquifers in Bangladesh. *Environmental Geology, 57*, 1521–1535.

Uyan, M., & Cay, T. (2013). Spatial analyses of groundwater level differences using geostatistical modeling. *Environmental and Ecological Statistics, 20*, 633–646. https://doi.org/10.1007/s10651-013-0238-3

Assessing Extreme Monthly Runoff Over an Arid Basin Through Reanalysis Datasets

Pedro Rau, Fiorela Castillón, Kimberly Visitacion, Marcela Yeckle and Marco Cordova

Abstract

Water availability in arid basins is a serious concern and its quantification remains uncertain. Climate variability as the El Niño Southern Oscillation phenomenon (ENSO) can worsen this scenario, exacerbating flooding in some regions such as the Piura basin on the northern Peruvian Pacific coast. We analyzed four geospatial runoff datasets covering a 7622 km^2 basin area at ~50 km of resolution. They are two global datasets: Global Runoff (GRUN) and Linear Optimal Runoff Aggregate (LORA) based on both hydrological reanalysis; a Peruvian dataset named PISCO-Hym-GR2M based on satellite and in-situ conceptual water balance; and a semi empirical method called Rindex based only on in-situ conceptual water balance. Those monthly datasets were compared to historical observations over four selected ENSO events which accelerated and retarded the peak monthly runoff and reached up to 317 mm/month. GRUN, LORA, PISCO-Hym-GR2M and Rindex reached month predictions of 42%, 21%, 58% and 50%, respectively with an acceptable relative error below 50% during the December–May wet period. Using hydroclimatic match indices in a Taylor diagram, correlations coefficients over 0.75 were obtained for all products, the highest corresponding to the GRUN dataset. A reliable root mean square difference near 40 mm/month for GRUN and Rindex; and a reliable amplitude of the variations for PISCO-Hym-GR2M and Rindex around 80 mm/month were obtained. Most datasets underestimated runoff with shifts in peak monthly runoff timing related to spatial aggregation of input precipitation, being LORA the most accurate in matching peaks. Those products could help to study monthly runoff fluctuations, knowing their difficulty in prediction.

Keywords

Runoff · Extremes · GRUN · LORA · Peru

1 Introduction

The use of reanalysis products based on satellite, modeling and in-situ validation is gaining ground in hydrology because of its high spatiotemporal resolution. The information on these products offers an opportunity to explore their advantages and limitations over basins to understand streamflow variability better (Llauca et al., 2021). Arid basins are always misunderstood because of water balance complexities, mainly with climate forcing such as precipitation and evapotranspiration and at soil and subsoil reservoirs (Negm et al., 2020). Also, some regions present strong climate variability (e.g., El Niño phenomenon) where extreme runoff events can exacerbate basin water balance.

2 Methods

The four runoff datasets used and their sources are shown in Table 1. They were compared with monthly observed streamflow at Pte.Ñacara station in the Piura basin in Peru (see Fig. 1) through a Taylor diagram during four El Niño events: 1982–1983, 1991–1992, 1997–1998 and 1998–1999. Those events were the most destructive recorded El Niño events. Hydroclimatic characteristics of the basin are found in Hobeichi et al. (2019).

The relative error lets to compare the reliability of the products (X_p) with observed data (X_o) during El Niño events as Eq. 1. The prediction level (%) of each product was calculated taking the six rainiest months (Dec–May) with relative errors of <50% along the four El Niño events.

$$\left(X_p - X_o\right) * 100/X_o \tag{1}$$

P. Rau (✉) · F. Castillón · K. Visitacion · M. Yeckle · M. Cordova
Universidad de Ingeniería y Tecnología UTEC, CITA, Lima, Peru
e-mail: prau@utec.edu.pe

H. Chenchouni et al. (eds.), *Recent Advancements from Aquifers to Skies in Hydrogeology, Geoecology, and Atmospheric Sciences*,
Advances in Science, Technology & Innovation, https://doi.org/10.1007/978-3-031-47079-0_17

Table 1 Streamflow datasets and methods selected

Dataset	Resolution	Description	References
Global Runoff (GRUN)	0.5° × 0.5°	Reanalysis, runoff	Ghiggi et al. (2019)
Linear Optimal Runoff Aggregate (LORA)	0.5° × 0.5°	Reanalysis, runoff	Hobeichi et al. (2019)
PISCO-Hym-GR2M	Sub catchment	Satellite and In-situ, streamflow	Llauca et al. (2021)
Rindex method	Sub catchment	Conceptual, runoff	Rau et al. (2019)

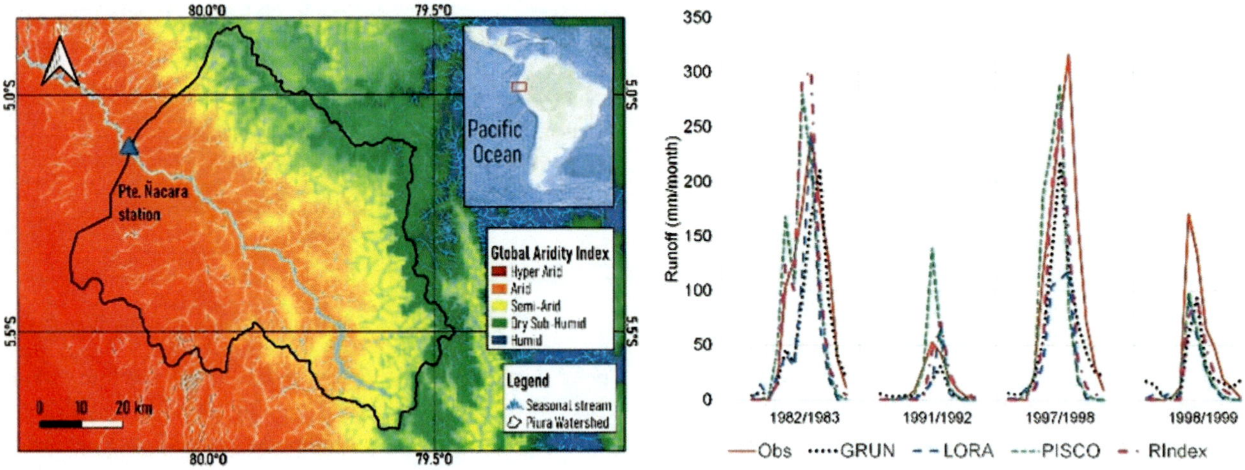

Fig. 1 Study area with aridity indexes from CGIAR (left). Runoff datasets comparison (right)

3 Results

3.1 Relative Error and Peaks

Table 2 shows the prediction level percentages of each dataset concerning the runoff observed at Pte.Ñacara station. The header of Table 2 shows the month with the highest streamflow value in each El Niño event (e.g., April during 1983); below them, each cell shows the relative error for the observed peak month and the lag in the maximum peak timing. For instance, GRUN reached a relative error of −65.2% in March 1992, and its maximum peak was observed 1 month after (+1) the one registered by the in-situ station.

Table 2 Relative error (%), prediction and month lag timing

Dataset	Prediction (%)	Peak 1983 (Apr)	Peak 1992 (Mar)	Peak 1998 (Apr)	Peak 1999 (Feb)
GRUN	42	−31.5 (+1)	−65.2 (+1)	−61.4 (−1)	−65.7 (+1)
LORA	25	−0.8 (0)	−67.4 (+1)	−62.6 (0)	−45.2 (0)
PISCO	58	−12.2 (−1)	164.6 (0)	−74.5 (−1)	−42.2 (0)
Rindex	50	23.4 (0)	−16.1 (+1)	−53.5 (−1)	−55.3 (+1)

Figure 1 shows the monthly runoff observations (Obs) and the differences between the peaks recorded by each product and the peaks of the in-situ station. The PISCO and Rindex products have the highest accuracy with predictions above 50%, while the LORA product is the lowest with a 25% success rate. Although LORA has the lowest percentage of prediction, it is the one that captures the month of the observed peak.

3.2 Datasets Comparison

Figure 2 shows a Taylor diagram comparing the proximity of the observed and predicted data from studied products. The LORA and GRUN datasets present the lowest centered root mean square error (RMS) values at 50 mm/month (green circles). On the other hand, the Rindex and PISCO

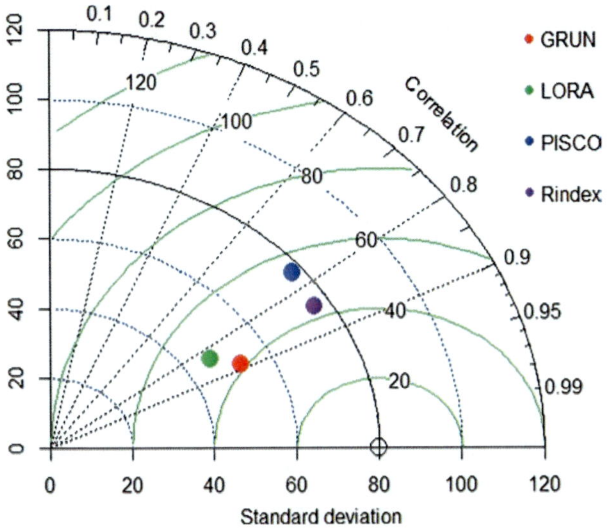

Fig. 2 Taylor diagram of predicted (colored circles) and observed runoff (white circle)

products give deviations very close to the observed values, compared to those of LORA and GRUN, which present dispersion values lower than those observed. All products show correlations >0.75.

4 Discussion

Piura basin presents limited spatialized data which makes hydrological studies at the basin scale oversimplified, especially in extreme episodes such as El Niño events. Despite the spatialized cover of reanalysis products, some limitations are commonly encountered over arid and high latitudinal regions according to their sources. For instance, GRUN usually sub-estimates precipitation in small catchments and is better suited to basins with low precipitation. Its performance against extreme events has been little studied (Ibarra et al., 2021). Our results are also concordant with sub-estimation for GRUN and LORA datasets. In contrast, PISCO and Rindex exhibited differences for events in 1983 and 1992, with a predominance of sub-estimation for 1998 and 1999.

Overestimation and sub-estimation could be managed in further studies. However, inaccuracies in predicting monthly timing peak runoff as a common issue in climatological studies could be challenging. The LORA dataset performed better than PISCO, GRUN and Rindex. These differences are mainly due to each product's timing of precipitation occurrences.

5 Conclusion

Reanalysis datasets showed an approach with a general underestimation to conceptual and observed-based methods in the Piura basin. The match over monthly timing peak runoff is an advantage. Even if hydrological reanalysis datasets could have limitations over arid lands and high latitudes, low latitudinal regions can benefit from its relative accuracy being an opportunity to explore them in regions with data scarcity.

Authors thank ProCiencia funding 084-2018 FONDECYT-BM-IADT-MU ePiura.

References

Ghiggi, G., Bourrel, L., Labat, D., Ruelland, D., Frappart, F., Lavado, W., Dewitte, B., & Felipe, O. (2019). GRUN: An observation-based global gridded runoff dataset from 1902 to 2014. *Earth System Science Data, 11*, 1655–1674.

Hobeichi, S., Abramowitz, G., Evans, J., & Beck, H. (2019). Linear Optimal Runoff Aggregate (LORA): A global gridded synthesis runoff product. *Hydrology and Earth System Sciences, 23*, 851–870.

Ibarra, D., David, C., & Tolentino, P. (2021). Evaluation and bias correction of an observation-based global runoff dataset using streamflow observations from small tropical catchments in the Philippines. *Hydrology and Earth System Sciences, 25*, 2805–2820.

Llauca, H., Lavado-Casimiro, W., Montesinos, C., Santini, W., & Rau, P. (2021). PISCO_HyM_GR2M: A model of monthly water balance in Peru (1981–2020). *Water, 13*(8), 1048.

Negm, A., Bouderbala, A., Chenchouni, H., & Barcelo, D. (2020). *Water Resources in Algeria-Part I: Assessment of Surface and Groundwater*. Springer, Cham.

Rau, P., Bourrel, L., Labat, D., Ruelland, D., Frappart, F., Lavado, W., Dewitte, B., Felipe, O. (2019). Assessing multi-decadal runoff (1970–2010) using regional hydrological modelling under data and water scarcity conditions in Peruvian Pacific catchments. *Hydrological Processes, 33*(1), 20–35.

Potential of Support Vector Machine Fed by ERA5 for Predicting Daily Discharge in the High Atlas of Morocco

Bouchra Bargam, Abdelghani Boudhar, Christophe Kinnard, Karima Nifa and Abdelghani Chehbouni

Abstract

In the high Atlas Mountains of Morocco, forecasting discharge is of crucial importance for water resource management but it remains a difficult task due to the scarcity of observations. This study presents meteorological re-analysis data as a promising alternative for feeding hydrological model where observation data are absent. Re-analysis data are used here as inputs to a 'Support Vector Regression' (SVR) algorithm to simulate the daily discharge over the Rheraya catchment in Tensift, Morocco. The SVR model is calibrated, and its accuracy is assessed using a time series from September 2003 to August 2016. The efficiency of daily predictions of streamflow is compared with the daily-observed discharge. SVR model fed by re-analysis data showed a good performance during the test period with an NSE and RMSE respectively equal to (0.88 and 0.89 m^3/s). This accuracy is mainly due to the high correlation between the discharge and the predictors. The SVR model fed by re-analysis data can be considered as a reliable tool to estimate daily discharge.

Keywords

ERA5 · Re-analysis data · Data scarcity · Daily discharge prediction · Artificial intelligence (AI) · SVM · SVR

B. Bargam (✉) · A. Boudhar · A. Chehbouni
Center for Remote Sensing Applications, Mohammed VI Polytechnic University, Salé, Morocco
e-mail: bouchra.bargam@um6p.ma

A. Boudhar · K. Nifa
Data4Earth Laboratory, Sultan Moulay Slimane University, Béni Mellal, Morocco

C. Kinnard
Département des Sciences de L'environnement, Centre de Recherche sur les Interactions Bassins Versants—Écosystèmes Aquatiques (RIVE), Université du Québec À Trois-Rivières, Trois-Rivières, QC, Canada

1 Introduction

In most arid and semi-arid Mediterranean regions, water resources are facing many challenges going from severe drought to the high demand for water (Giorgi & Lionello, 2008). An effective management and planning of water resources based on discharge forecasting is urgent. It is the case for the High Atlas Mountains in Tensift, Morocco, where water resources are under increasing pressure due to the reduction of precipitation and the expanding demand (Marchane et al., 2017; Tramblay et al., 2020). In the Tensift region, discharge is characterized by a high variability (Chaponnière et al., 2008, Zamrane et al., 2016), and the runoff generation mechanism is complex (Boudhar et al., 2009), which prevents to have accurate predictions. Several researchers have been tempted to predict discharge in this region, relying on conceptual or physically based modelling. In such conditions of data scarcity and the complexity of hydrological processes, applying hydrological models seems to be nonobvious. At the same time, Machine Learning (ML) techniques appear to be a promising and flexible tool for predicting discharge in Tensift region (Lange & Sippel, 2020). Throughout this study, we aim to evaluate a Machine Learning technique—Support Vector Regression (SVR)—in predicting daily discharge using (ERA5) re-analysis data, over the Rheraya sub-basin spanning the period September 2003 to August 2016. It is organized into two main sections: the first covers the SVM model as well as the data used. The second section presents the results and conclusions.

2 Materials and Methods

SVM Model

The Support Vector Machine (SVM) (Vapnik, 1995) is a supervised machine learning technique used in this study to address the regression problem of discharge simulation (SVR) but can also be used for classification problems

H. Chenchouni et al. (eds.), *Recent Advancements from Aquifers to Skies in Hydrogeology, Geoecology, and Atmospheric Sciences*, Advances in Science, Technology & Innovation, https://doi.org/10.1007/978-3-031-47079-0_18

(Support Vector Classification SVC) (Liu et al., 2021; Zhao et al., 2020). SVM works by mapping inputs into a high-dimensional space using a linear function and partitioning datasets by offset through an ideal hyperplane. Practically, SVM is implemented using a kernel function that can be polynomial, sigmoid, radial basis function (rbf).

Datasets

The dataset used in this study spanned the years 2003/2016 and it includes daily precipitation from ERA-5 re-analysis data that was averaged on Rheraya sub-basin, daily snow cover area (SCA) extracted from MODIS and averaged as well on the study area and daily observations of discharge obtained from the Tensift hydraulic water agency (ABHT). All variables are summarized in the Table 1.

Calibration

The SVR model was calibrated by partitioning the time series into two groups. From 2011 to 2016 is used to train the model, then the period 2003/2011 is used for testing. Hence, in the period of training which goes from 2011 to 2016, the SVR model is fed by both predictors and the corresponding target. This step consists of learning the different patterns between the target (the streamflow) and the predictors. Next, in the period of testing, the model is fed by unseen data. This means that SVR model will predict streamflow for the period 2003/2011 for which it has not seen the predictors before. Consequently, the accuracy of SVR model is assessed by comparing the predicted streamflow in the period of test with the observed one using two performance metrics NSE and RMSE.

Model Efficacy Indicators

For the accuracy of predictions, this study uses two performance indicators; the Nash Sutcliff efficiency (NSE, Eq. 1) and the Root Mean Squared Error (RMSE, Eq. 2)

$$\mathrm{NSE} = 1 - \frac{\sum_i (y_i - \hat{y}_i)^2}{\sum_i (y_i - \bar{y})^2} \tag{1}$$

$$\mathrm{RMSE} = \sqrt{\frac{\sum_{i=1}^{n} (y_i - \hat{y}i)^2}{n}} \tag{2}$$

Table 1 The used dataset

No	Variables	Index	Unit
1	Streamflow at day t	Qt	m^3/s
2	Precipitation	Pt	mm
5	Snow cover area	SCAt	%

y_i is the observed target, $\hat{y}_i$ is the predicted target, $\bar{y}$ is the mean observed target and n is the total count of observations.

3 Results

3.1 Input Selection

To select only the potential variables and avoid redundancy, the correlation between predictors and the discharge is analyzed throughout the cross-correlation (Fig. 1a) and the correlation matrix (Fig. 1b). Based on cross-correlation, these variables can be considered: P_t, P_{t-1}, SCA_t, SCA_{t-1}, SCA_{t-2}, Q_{t-1}, Q_{t-2}, Q_{t-3}. Then by analyzing the inter-correlation of these variables using the correlation matrix, only the following inputs are kept P_t, P_{t-1}, SCA_t and Q_{t-1}. In Rheraya sub-basin, the observed streamflow comes directly from rainfall and snowfall with a certain lag time besides the antecedent streamflow which explains the selected predictors.

3.2 Daily Discharge Simulation

It has to be mentioned that the summer period was deleted from the time series, to avoid dealing with the flood occurring from June to August. Figure 2 presents the scatter density of the predicted discharge against the observed one. SVR accuracy is expressed by NSE equal to 0.88 and RMSE equal to 0.89 m^3/s. SVR model showed a good enough simulation with a little underestimation especially for the peak flows. The SVR estimation of daily discharge is close to the observed one. As for the peak-flows, a slight difference between the prediction and the observation is noticed.

4 Discussion

SVR model showed a good performance during both training and test. This performance can be explained by the good correlation between discharge and the predictors especially the antecedent discharges. The hyper-parameters optimization is an important factor behind the accuracy of SVR model. A good enough optimization can help avoid the slight underestimation of the model. Besides, discharge measurements are subject to error which can affect the performance of SVR model. On the other hand, it was proved that precipitation from re-analysis data can be successfully used as inputs to predict daily discharge.

105
106
107

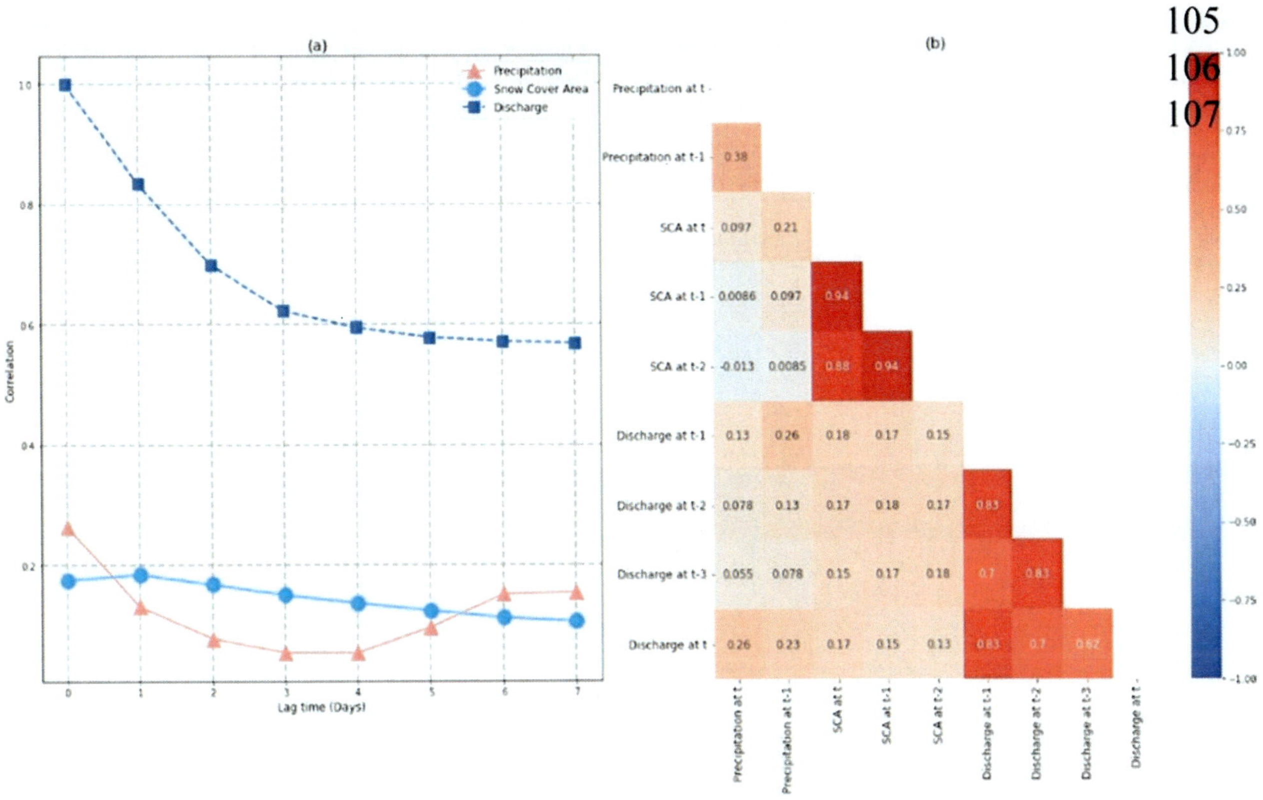

Fig. 1 **a** The correlation between the discharge and the predictors (precipitation, snow cover area and discharge) at a lag time of up to 7 days, **b** the correlation matrix for the selected predictors and the discharge

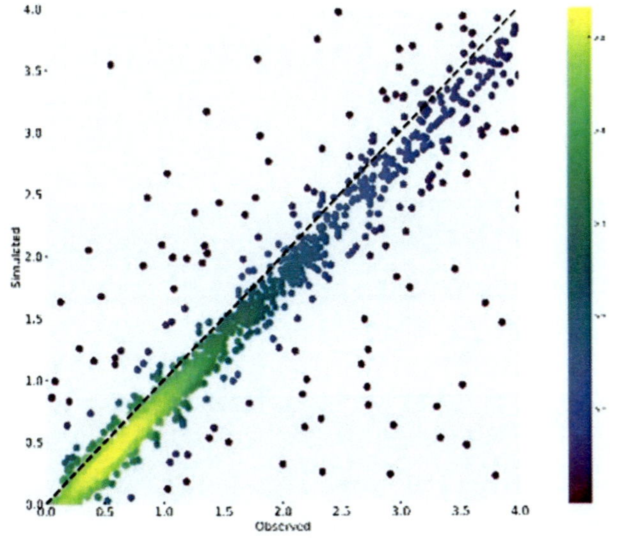

Fig. 2 The scatter density of the observed and simulated discharge (m^3/s) from 2003 to 2011 at Rheraya catchment using SVR model

5 Conclusion

In this study, the accuracy of SVR model fed by re-analysis data is evaluated for daily discharge simulation in Rheraya sub-basin. SVR model showed a good performance of prediction using P_t from ERA-5, SCA_t from MODIS and the observed discharge Q_t. It can be concluded that the SVR model can be successfully employed for daily estimation of discharge using re-analysis data.

References

Boudhar, A., Hanich, L., Boulet, G., Duchemin, B., Berjamy, B., & Chehbouni, A. (2009). Evaluation of the Snowmelt Runoff model in the Moroccan High Atlas Mountains using two snow-cover estimates. *Hydrological Sciences Journal, 54*(6), 1094–1113. https://doi.org/10.1623/hysj.54.6.1094

Chaponnière, A., Boulet, G., Chehbouni, A., & Aresmouk, M. (2008). Understanding hydrological processes with scarce data in a mountain environment. *Hydrological Processes, 22*(12), 1908–1921. https://doi.org/10.1002/hyp.6775

Giorgi, F., & Lionello, P. (2008). Climate change projections for the Mediterranean region. *Global and Planetary Change, 63*(2–3), 90–104. https://doi.org/10.1016/j.gloplacha.2007.09.005

Lange, H., & Sippel, S. (2020). Machine learning applications in hydrology (pp. 233–257). Springer. https://doi.org/10.1007/978-3-030-26086-6_10

Liu, D., Zhang, Y., Zhang, J., Xiong, L., Liu, P., Chen, H., & Yin, J. (2021). Rainfall estimation using measurement report data from time-division long term evolution networks. *Journal of Hydrology, 600*, 126530. https://doi.org/10.1016/j.jhydrol.2021.126530

Marchane, A., Tramblay, Y., Hanich, L., Ruelland, D., & Jarlan, L. (2017). Climate change impacts on surface water resources in the

Rheraya catchment (High Atlas, Morocco). *Hydrological Sciences Journal, 62*(6), 979–995. https://doi.org/10.1080/02626667.2017.1283042

Tramblay, Y., Koutroulis, A., Samaniego, L., Vicente-Serrano, S. M., Volaire, F., Boone, A., et al. (2020). Challenges for drought assessment in the Mediterranean region under future climate scenarios. *Earth-Science Reviews, 210*, 103348. https://doi.org/10.1016/j.earscirev.2020.103348

Vapnik, V. (1995). Support-vector networks. *IEEE Expert-Intelligent Systems and Their Applications, 7*(5), 63–72. https://doi.org/10.1109/64.163674

Zamrane, Z., Turki, I., Laignel, B., Mahé, G., & Laftouhi, N. E. (2016). Characterization of the interannual variability of precipitation and streamflow in tensift and ksob basins (Morocco) and links with the NAO. *Atmosphere, 7*(6), 84. https://doi.org/10.3390/atmos7060084

Zhao, G., Pang, B., Xu, Z., & Xu, L. (2020). A hybrid machine learning framework for real-time water level prediction in high sediment load reaches. *Journal of Hydrology, 581*, 124422. https://doi.org/10.1016/j.jhydrol.2019.124422

Toward a Better Consideration of Hydro-Meteorological Information for Flash-Flood Crisis Management Through Machine Learning Models

Salma Sadkou, Guillaume Artigue, Noémie Fréalle, Pierre-Alain Ayral, Séverin Pistre, Sophie Sauvagnargues and Anne Johannet

Abstract

Flash floods are among the deadliest natural hazards. With the increase of worldwide population, many more lives could be at risk. Hydrological modeling methods have significantly evolved in the last years. Also, new methods have emerged like machine learning. However, it is still challenging to anticipate and plan crisis management actions like evacuation. Crisis managers and hydrologists often work individually. Hydrologists will often forecast hydrological variables like discharge. Nevertheless, apart from big cities, crisis management plans are rarely based on this variable. The forecasts are, therefore, unlikely to be used regardless of their informational value. Thus, this study aims to identify relevant information for crisis management using different configurations of forecast bulletins. The study site is the flash-flood prone watershed of Gardon d'Anduze (545 km^2) in Anduze, Southern France. Floods can cause the discharge to increase from 100 l/s to 3000 m^3/s in about ten hours. Using artificial neural networks, water level and discharge were forecasted. From these forecasts, six different bulletins were proposed with variant types and levels of information like future rain, discharge at upstream watersheds, and threshold markers. To evaluate their relevancy, records of decisions, observation sheets, discussions, and surveys were used. The bulletins were then tested in a crisis management simulation setting with operational managers. The main findings are as follows: (1) the usefulness of reference to previous events depended on each crisis manager's personal experience, (2) discharge was difficult to interpret, (3) confidence in forecast bulletins was medium to high, (4) crisis managers were generally able to make decisions with satisfying degrees of anticipation regardless of the level of information, (5) they also preferred bulletins with the highest level of information due to an increase in visibility. This work showcased the importance of flash-flood forecast bulletins in crisis management and the reliability of neural networks forecasts.

Keywords

Flash flood · Forecasting · Neural networks · Crisis management · Gardon d'Anduze

1 Introduction

The Mediterranean area is particularly prone to flash floods. In France, these events have claimed the lives of 250 people between 1988 and 2015 and caused significant economic damages (Vinet et al. 2016). Thanks to an increase in computational capabilities, new methods have emerged in hydrological modeling like machine learning. Crisis management could benefit greatly from these advances. Nevertheless, this remains limited for two reasons. First, hydrologists and crisis managers tend to work separately. Second, hydrologists tend to forecast discharge, but this variable is rarely used in crisis management plans. Forecasts need to be adapted to the crisis management context for proper integration in this activity. The first step to achieve this objective is defining the managers' needs. For

S. Sadkou · G. Artigue · A. Johannet (✉)
HSM, Univ Montpellier, IMT Mines Ales, CNRS, IRD, Alès, France
e-mail: anne.johannet@mines-ales.fr

S. Sadkou · N. Fréalle · S. Sauvagnargues
LSR, IMT Mines Alès, Alès, France

P.-A. Ayral
UMR 7300 ESPACE, CNRS, Saint-Christol-Lez-Alès, France

S. Pistre
HSM, Univ Montpellier, CNRS, IRD, Montpellier, France

H. Chenchouni et al. (eds.), *Recent Advancements from Aquifers to Skies in Hydrogeology, Geoecology, and Atmospheric Sciences*, Advances in Science, Technology & Innovation, https://doi.org/10.1007/978-3-031-47079-0_19

this reason, crisis managers were invited to a workshop to evaluate flash-flood forecast bulletins for the Gardon of Anduze. The bulletins contained varying types of information and forecasts of water level and discharge made using artificial neural networks.

2 Materials and Methods

The scenario of the workshop is inspired by the flash-flood event of September 2002 in Anduze, France. The original timing was adapted. The workshop is organized in two sequences. The players were separated into two groups: the group of Anduze composed of the Mayor, the General Director of Services and two deputies and the other group composed of other professionals in Major Risk Prevention. The neural network model was designed by Bornancin-Plantier (2013) for water levels and following a method proposed by Kong-A-Siou et al. (2012) for discharges. The follow-up was carried using decision records, surveys, and observation sheets.

3 Results

3.1 Decision Records

The decisions varied sometimes significantly from a bulletin to another even within the same sequence. The types of decisions increased with the increase of data during sequence 1. The least amount of decision types corresponds to bulletin 4 (Table 1).

Table 1 Added data and types of decisions managers made by bulletin during each sequence, each new bulletin reuses information of the previous one

Sequence	Bulletin	Data	Decision types
Sequence 1: Beginning of water rise at 11:30 am (simulated time)	1	• Measured level • Median, maximum, and minimum forecasted level	Alert/inform the population Secure roads, building, etc. …
	2	• Threshold markers: submersion of submersible bridge, yellow, and orange vigilances (Vigicrues) • Average measured rain	Alert/inform the population Secure roads, building, etc. …
	3	• Future rain in Anduze • Measured level in two upstream watersheds	Alert/inform the population Secure roads, buildings, etc. … Shelter the population Evacuate
Sequence 2: 2:30 pm (simulated time) when the peak is almost reached	4	• Level is replaced by discharge variable	Secure roads, building etc. …
	5	• Level variable is reestablished • Threshold marker of red vigilance	Alert/inform the population Open emergency accommodation Shelter the population Secure roads, buildings etc. …
	6	• Potential flood zone map	Alert/inform the population Open emergency accommodation Shelter the population

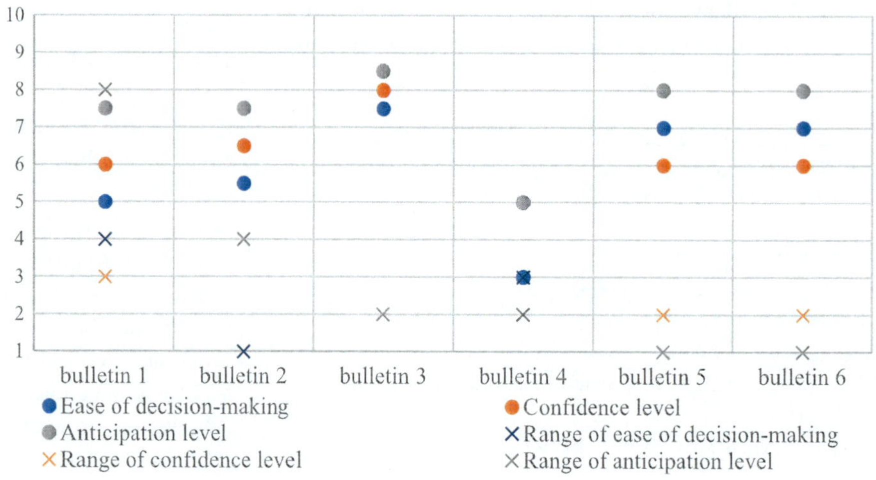

Fig. 1 Median and range of scores of three aspects in the surveys, on a scale of 1 to 10

3.2 Surveys and Rankings of Bulletins

At the end of each situation: bulletin-decision, managers completed a questionnaire. The evaluation scores of the situations are presented in Fig. 1. The ease of decision-making generally improved throughout the workshop. The confidence level was slightly above medium, while the anticipation level remained quite high. All of the scores dropped for bulletin 4. The ranges of the scores were also investigated. They are highest at the beginning as opposed to the end of the workshop (Fig. 1). As for the rankings, bulletins 5 and 6 were considered most appropriate.

3.3 Observations

During the workshop, many observations were made. For instance, the managers were interested in threshold markers not represented in the bulletin. In addition, if referencing past flash-flood events, recent ones should be privileged. Also, although the rain representation was difficult to understand initially, the managers became accustomed to it by the end of the workshop. Moreover, they expressed the need for some sort of "empirical validation" (e.g., reading a limnimetric scale). Finally, they wanted to be provided by as much meteo-hydrological data on upstream watersheds as possible.

4 Discussion

The workshop format is appropriate because it allows focus solely on the bulletins and adjustment to their format and content. Initially, the interpretations and reactions of the managers varied significantly probably due to their varying experience. They became more homogenous by the end of the workshop. The major event of September 2002 was not recognized. This demonstrates the importance of markers and additional information provided in the bulletins. Also, in addition to not being in the crisis management plans, discharge is not directly observable. Lastly, bulletins with the highest level of information were preferred and maybe due to an increased visibility.

5 Conclusion

This work showcased the importance of flash-flood forecast bulletins in crisis management and the reliability of neural networks' forecasts. Moreover, it was an opportunity to develop a methodology to evaluate forecast bulletins. These bulletins should be enriched with ample and relevant information identified to aid in decision-making. In the future, to further enhance the bulletins, artificial neural networks could be used to predict variables specific to crisis management. The returns from managers could be more quantitative to limit interpretations, possible in the present format of the answers.

References

Bornancin-Plantier, A. (2013). *Conception de modèles de prévision des crues éclair par apprentissage artificiel* (PhD, Paris 6). http://www.theses.fr/2013PA066015

Kong-A-Siou, L., Johannet, A., Valérie, B. E., & Pistre, S. (2012). Optimization of the generalization capability for rainfall–runoff modeling by neural networks: The case of the Lez aquifer (southern France). *Environmental Earth Sciences, 65*(8), 2365–2375. https://doi.org/10.1007/s12665-011-1450-9

Vinet, F., Boissier, L., & Saint-Martin, C. (2016). Flashflood-related mortality in southern France: First results from a new database. *E3S Web of Conferences*, *7*, 06001. https://doi.org/10.1051/e3sconf/20160706001

Analysis of the Hydrological Regime of the Congo River in Brazzaville and Its Impact on Navigation

Edouard P. S. Konzi, Janny M. Ciabembi and Levy S. Ayissou

Abstract

Our study consists of analyzing the hydrological behavior of the Congo River to detect the anomalies observed which have negatively impacted navigation for several years. The databases of the Research Institute for Development (RID) and the Excel files of the Joint Waterway Maintenance Service (JWMS) were used, namely (IL Consulting Engineers in Study for the dredging of Brazzaville-Hydraulic study report. FED, 2006) for different parameters: (i) daily water heights from 1972 to 2006; (ii) monthly discharges from 1902 to 2005; (iii) water levels from 1953 to 1995; (iv) gauging results from 1982 to 1994. These results reflect the degradation of the hydrological behavior of this arm of the Congolese river on the right bank, with negative consequences on navigation. The negative consequences of this phenomenon require preventive dredging. In addition, studies should be carried out to help promote a significant inflow of water in the main channel which runs alongside the port of Brazzaville during low flow.

Keywords

Gauging · Hydracces · Rating curve · Raising · Low flow

E. P. S. Konzi (✉)
Department of Physics and Applied Sciences, National Pedagogical University, BP 8815, Kinshasa, Democratic Republic of Congo
e-mail: edouard.konzi@upn.ac.cd

J. M. Ciabembi
Mechanical Section, Higher Institute of Applied Techniques, BP 6593, Kinshasa, Democratic Republic of Congo

L. S. Ayissou
Joint Waterways Maintenance Service, BP 179, Brazzaville, Republic of Congo

1 Introduction

The studies by Laraque et al. (2001), Laraque and Georges (2005) and Laraque and Adèle (2007) on the hydroclimatic evolution of the two major tributaries of the right bank of the Congo River (Oubangui, Sangha) note that the hydrological regime of these two tributaries remains deficient, despite a slight recovery observed. In addition, Bouka and Mpounza (2009), who map the annual average isohyets of the Republic of the Congo for the periods (1951–1970, 1971–1980, 1971–2000), observe that less and less rainfall in the past. These authors also show that to this climate change must be added the degradation of the soils which impacts not only on the flows, but also on the siltation of the diversions of the rivers of the right bank as well as of the Congo River in Brazzaville low tide. In this paper, using the database of the Excel file of the Joint Waterways Maintenance Service (IWMS) of Brazzaville and that of the Research Institute for Development (RID), we have analyzed the distribution flows that pass through the right bank with regard to the flows of the left bank. We also studied the evolution of water levels in Brazzaville as well as the period of risk to navigation. We have broadened our conclusion by proposing the resumption of the measurement of flows in different channels of the navigation route, by studying the dynamics of sandbanks and the impact on the environment.

2 Materials and Methods

The staff gauge installed at the Beach of Brazzaville by the (JWMS) was used to read the water level, every day in the morning. The reconstruction of missing limnimetric data from 1997 by Laraque and Adèle (2007) provided us with a homogeneous hydrological chronicle from 1902 to 2005. Correlation with surrounding hydrometric stations by multiple regression (Lallemant, 2018) made it possible to reconstruct missing data from 1997. The analysis and

H. Chenchouni et al. (eds.), *Recent Advancements from Aquifers to Skies in Hydrogeology, Geoecology, and Atmospheric Sciences*, Advances in Science, Technology & Innovation, https://doi.org/10.1007/978-3-031-47079-0_20

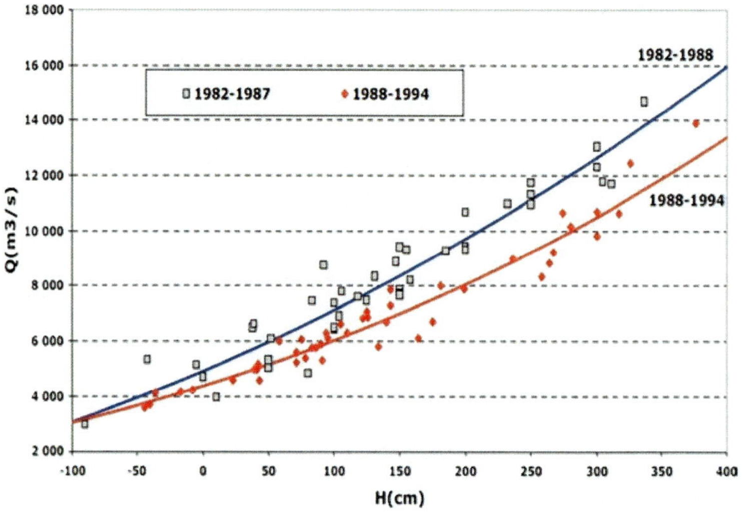

Fig. 1 Evolution of the height–discharge relationship at the height of Brazzaville, gray points (height–discharge, 1982–1988) in blue (rating curve 1982–1988). Red dots (stage-discharge, 1988–1994), rating curve (1988–1994) in red

interpretation of the results and the various graphs have been made possible by relying on the calculations carried out using the basic management and operating software of Hydracces (2004) hydrological data of the Research Institute for Development (RID). All these features allowed us to analyze and interpret the different graphs found in this study.

3 Results

Given that the chronicle of flow measurements stopped in 1994, it can be estimated that during periods of low water, through the Congolese arm on the right bank, it passes between 12 and 15% of the total flow of the river (this depends the severity of the low water level) and this proportion of the total flow which can rise to more than 22–23% during floods (IL Consulting Engineers, 2006; Konzi, 2014). The distribution of the gaugings in two periods made it possible to establish two rating curves (IL Consulting Engineers, 2006),

- From 1982 to 1987, the calibration curve equation is as follows:

$$Q_2 = 2.08 \times (H_m + 8.9)^{3.01} \tag{1}$$

- From 1988 to 1994, the rating curve is given by the following relationship:

$$Q_2 = 304.3 \times (H_m + 8.9)^{3.01} \tag{2}$$

In the light of Fig. 1 below, the evolution of the rating curve of the arm of the river on the right bank during the two periods shows the raising of its average bed.

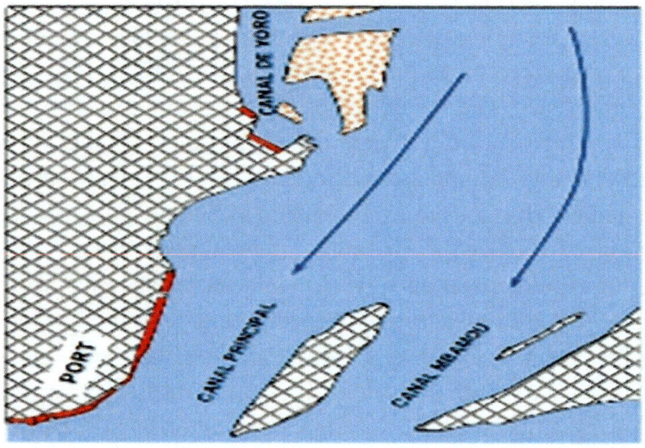

Fig. 2 Blue arrow on the left indicates the main channel and the Mbamou channel (IL Consulting Engineers, 2006; Konzi, 2014)

In addition, the analysis of flow measurements taken between 1982 and 1986 suggests a trend toward an increase in the hydraulicity of the arm of the right bank at low water in 1985, with regard to the series of gaugings carried out in 1982, 1983, and 1984 (Sogreah and Soreni, 1982, 1983). It should be noted that the evolution of the observed runoff does not represent a trend over a longer period, with regard to the analysis of the gaugings from 1982 to 1994. At the port of Brazzaville, the flow of the Congolese arm on the right bank is shared between the Mbamou canal and the main channel passing through the port (Fig. 2).

It will be noted that before 1988, the flows in the main channel were always higher than the flows in the Mbamou channel (Fig. 3). This trend in flows was reversed from 1988, when the flows of the Mbamou canal became higher than those passing through the main channel until 1994. This is due to a gradual raising of the passes on the right bank of the arm of the Congo River by sandbanks.

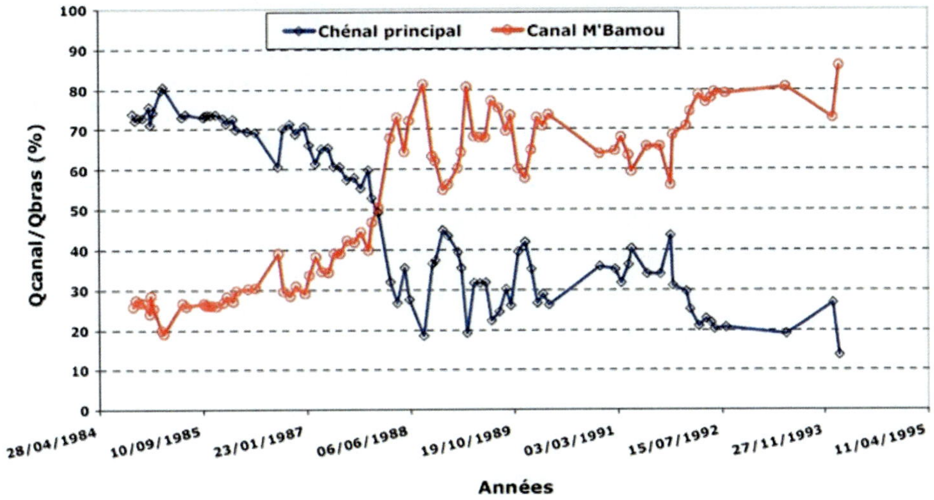

Fig. 3 Evolution of the flow distribution of the right bank arm of the Congo River in the two main channels. The blue curve represents the flow in the main channel. The red curve, the flow in the Mbamou channel

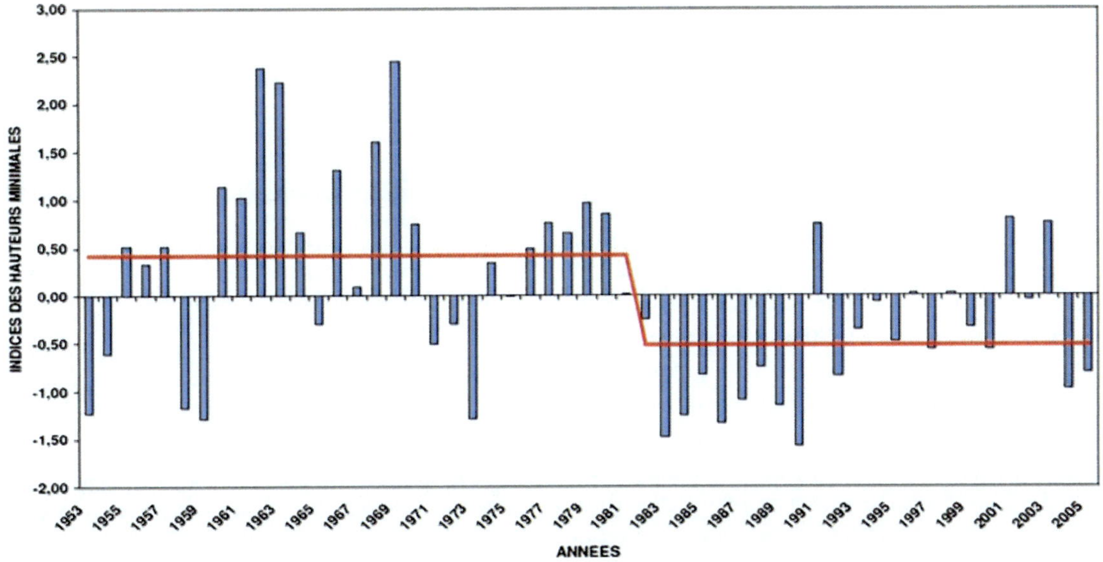

Fig. 4 Evolution of minimum annual heights at Beach. The blue bars, their minimum heights. The red curve represents their average

4 Discussion

Since the minimum annual heights are the most important parameter for navigation and this, during low water, two periods are observed (Fig. 4): (i) a period of less severe low water levels with an average minimum height of 0.40, from 1953 to 1981; (ii) a period of much more severe low water levels from 1983 to 2003 with an average minimum height of −0.25

In general, navigation is compromised during low water because of the severity of the low water compared to the draft of vessels in operation. For the JWMS, <100 cm away, access to the port of Brazzaville is handicapped by sandbanks. According to the series of water levels from 1953 to 2005, there are on average 85 days, during which the water level is <100 cm (level above which navigation is done in complete safety). This also depends on the state of silting up of the passes. Based on Fig. 5 below, we see that since 1980, the number of days of risk to navigation has exceeded the average for the series, reaching 110 days on average per year.

It should be noted that for navigation, in addition to the height of water, this also depends on the importance of deposits likely to significantly limit the draft available to boats.

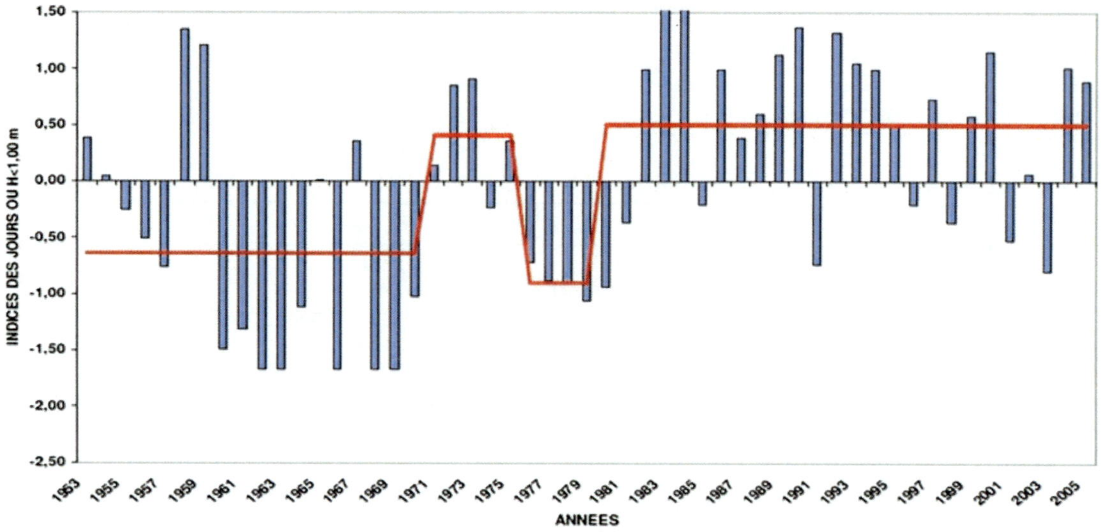

Fig. 5 Evolution of the number of days when the water level is less than 100 cm. The blue bars indicate the indices for the number of days. The red curve represents their average

5 Conclusion

In this opus which discussed the hydrological regime of the Congo River and its impact on navigation, we will retain that: (i) low water levels have become more severe over the past 25 years given that the average values of minimum water heights have dropped by 157% with a negative impact on navigation (Data from the JWMS Excel file); (ii) The number of days in the year when the Beach scale reading shows <100 cm increased by 90% between the period 1953–1980 and that going from 1981 to 2005. In addition, the proportion of flows from the right bank of the river arm, which represents 21% of the total flow of the Congo River during the period 1982 to 1987, fell by 18% on average during the period 1988–1994. The cessation of gauging from 1994 did not confirm this trend. The future of navigation in the main channel is likely to deteriorate further if corrective measures are not considered to pass a significant flow through the main channel, where ships transit. We also propose that an environmental study be conducted to assess its impact on this ecosystem of about 30 km.

References

Bouka, C. B., & Mpounza, M. (2009). Impacts of current and expected climate change in the Republic of Congo, SIFEE Colloquium, Cc and Evaluation and Env, Niamey.

Hydracces. (2004). Hydrological database management and exploitation software, in 4 languages (English, Spanish, French, Portuguese). www.upl.ird.fr/hybam/outils/hydraccess/en.htm

IL Consulting Engineers. (2006). Study for the dredging of Brazzaville-Hydraulic study report. FED.

Konzi, P. S. E. (2014). Estimation of solid transport in rivers: test case at Pool-Malebo on the Congo River (right bank). Doctoral Thesis, National Pedagogical University, Kinshasa.

Lallemant, D. (2018). The risks and resilience of cities: How disasters can catalyse transformative change. In: Vasbinder, J.W. (Ed) *Disrupted Balance: Society at Risk*. World Scientific Publishing, pp. 65–71. https://doi.org/10.1142/9789813239227_0008

Laraque, A., & Georges, A. (2005). Congo Basin (Congo Stations at Beach-Brazzaville, Sangha at Ouesso, Oubangui at Bangui).

Laraque, A., & Adèle, A. (2007) Critical inventory and homogenization of Congo Basin hydrological data. IRD154 LMTG/U 019, OBHI, Fort de France

Laraque, A., Mahé, G., Orange, D., & Marieu, B. (2001). Spatiotemporal variations in hydrological regimes within Central Africa during the XXth century. *Journal of Hydrology, 245*, 104–117.

Sogreah & Soerni. (1982). Study of the improvement of access to the port of Brazzaville: Report of the second mission (Agence Transcongolaise des Communications/Direction des Voies Navigables).

Sogreah & Soerni. (1983). Study of the improvement of access to the port of Brazzaville: Report of the third mission (Agence Transcongolaise des Communications/Direction des Voies Navigables).

Improving Agro-ecosystems Resilience in Water Limited Areas Through Smart Irrigation Management: Economic Analysis at the Farm Level

Hacib El Amami, Ali Chebil and Insaf Mekki

Abstract

The expansion of intensive irrigation in the central part of Tunisia has introduced the challenge of groundwater sustainability as water table is lowering at the rate of 1–2 m per year. More productive and resilient agriculture requires changes in natural resources management, particularly water and energy. Smart water management for precision irrigation in agriculture is essential for enhancing crop yield and efficient use of water, improving irrigation scheduling, decreasing farming costs while sustaining the environment. It also contributes to make agriculture more resilient to future water-related shocks and global economic changes. Within Sustainable Production in Water Limited Areas of Mediterranean Agro-Systems (SUPROMED) project, we have selected two groups of farms situated in the same areas with similar environmental conditions. In the first sample, the irrigation of selected crops (wheat, oat, olive, pistachios and almond trees) was scheduled by MOPECO model according to soil characteristics, crop types, weather conditions and irrigation network characteristics. For the second, the farmers were left to perform the prevailing water irrigation management. The main objective of this study was to assess the MOPECO irrigation scheduling at the farm level. Agronomic yields, water productivity, farmer's income and gross margin resilience towards market prices variability were used as indicators to compare such management to the prevailing one. Results showed that smart irrigation management allowed better yields and water use productivity, comparative to conventional management. The farmer's incomes were also higher. Results also showed that under smart management the gross margin of selected crops become more resilient towards market prices variability leading thus to better economic viability of farming systems in the water limited environments.

Keywords

Smart management · Groundwater · Productivity · Resilience · Irrigation

1 Introduction

Smart-irrigation management is regarded as a valuable approach to increase crops yields and decrease costs while contributing to saving water and improving energy use by scheduling the amount of irrigation according to soil characteristics, crop types and weather conditions (Fellah et al., 2018; Kamienski et al., 2021). Water is saved by scheduling the amount of irrigation according to local data on soil characteristics, crop types, and weather conditions, thus improving the crops yields while decreasing the cost of irrigation and energy use (Levidow, 2014; Vories et al., 2021).

The aim of this work was to analyse how smart irrigation management can help agriculture become more resilient to future water-related shocks and economic global change (increasing inputs costs and/or falling market prices). The focus has been on the potential impact resulting in utilizing smart irrigation management on agronomic yield, farmers' income, water and energy productivity as well as the resilience of crops' gross margin towards market prices variability.

H. El Amami (✉) · A. Chebil · I. Mekki
Water and Forests, National Research Institute for Rural Engineering, Carthage University, Tunis, Tunisia
e-mail: hacib.amami@gmail.com

H. Chenchouni et al. (eds.), *Recent Advancements from Aquifers to Skies in Hydrogeology, Geoecology, and Atmospheric Sciences*, Advances in Science, Technology & Innovation, https://doi.org/10.1007/978-3-031-47079-0_21

2 Methodology

The study area (Sidi-Bouzid) is located in the central parts of Tunisia where surface water is scarce and has random occurrences. Agriculture is primarily based on intensive irrigation using groundwater. In this area, groundwater level is falling quickly at a rate of 1–2 m annually (Ben Guessim, 2015; Hemdi, 2022). In order to assess the impact of smart irrigation management, two groups of farms, having similar characteristics, were selected. At the level of the first group, called "Treated group", farmers receive advices as the volume of water and the time of irrigation is scheduled by MOPECO model based on local data (soil characteristics, types of crops, phonological stages, and local weather conditions). At the level of the second group, referred to as the "control group", farmers do not receive advices about water irrigation management. They were left to perform the prevailing water management practice in the area. Seven crops were monitored during two years (2020/2021 and 2021/2022). They include wheat, onion, oat, olive trees, almond, and pistachios.

Detailed information relative to all process of production was collected, and several indicators were used to assess the impact of innovative water irrigation.

3 Results

3.1 Yields Increase and Water Productivity

Results showed significant differences between indicators obtained at the level of treated group and control group in terms of yields, water use, and water productivity. As an example, the irrigation management, based on MOPECO model, saved water, for onion crop, by 20% and gave 65% more yield as compared to that of control farmers group. For almond trees, the water saved is 32%, and the yield has increased by 49%. These results reveal that innovative irrigation management based on local data combined with best agricultural practices could increase water productivity and improve grain yields.

3.2 Energy Use

Specific energy shows the amount of energy spent to produce one unit of marketable product (Kwh/kg). As shown in Table 1, the specific energy was lower for crops managed according to MOPECO irrigation scheduling. That means less energy was used on treated plots to produce the same quantity of wheat (1 kg), comparative to "control" plots.

Table 1 also showed that the percentage of energy cost in the total output value is lower, compared to conventional management. For onion crop conducted under DSS smart irrigation, around of 5% of total output value is spent on energy input; that is 0.05 TND for each Dinar of product sold. This value is 8.9% for the one conducted under conventional management.

3.3 Increase of Farmer's Income

The impact of smart water management on farmer's incomes is analysed by comparing participants and non-participants, using the difference in differences (DID) method. Table 2 shows that the double difference, which is the difference between the two mean farm income differences (7172–2100 TND), is then 5072 TND; indicating thus an increase in participant farmer's income.

This indicates that active participation in the SUPROMED project and the adoption of innovative water management enables the participants to generate more income by optimizing water and energy use.

3.4 Resilience Towards Market Prices Fluctuations

Break-even price estimates the point at which a farm's gross income is equal to its total variable costs. Results in Table 3 revealed that the sensitivity of gross margin to price fluctuations is the highest at the level of control farms plots.

The gross margin of crops conducted under innovative irrigation management is much less sensitive.

Table 1 Specific energy and its share in the total product value

	Onion		Wheat		Oat		Pistachios	
	TF	CF	TF	CF	TF	CF	TF	CF
Specific energy (Kwh/kg)	0.024	0.048	0.39	0.173	0.093	0.145	1.42	2.6
Energy cost /Total production value (TND*) (%)	5	8.9	10	20	8	11.6	4.8	6.25

TF treated farms, *CF* control farms
* 1 TND ≈ 0.33 €

Table 2 Impact of smart water irrigation on farmer's incomes (TND)

Groups	Change in farmer's income		
	Before (2019)	After (2022)	Difference
Participants	32,600	39,772	7172
Non-participants	26,850	28,950	2100
Difference between groups	5750	10,822	5072

Table 3 Resilience towards prices fluctuations: Break-even analysis

Crops	Average market price (TND/kg)	Break-even price	
		Supromed plots	Control plots
Wheat	1.2	0.430	0.596
Oat	0.9	0.570	0.720
Onion	0.3	0.113	0.152
Olive	1.785	0.597	0.710
Pistachios	25	4.460	4.562
Almond	5.5	5.077	3.720

4 Discussion

Empirical results suggested that smart irrigation management enhances yields, water productivity and increased the obtained gross margin of each monitored crop. They also promote resilience of farming systems against market variability prices. From this point of view, the new irrigation management can be considered as promising method to achieve sustainable water use in drylands. Many economic analyses revealed that advanced irrigation scheduling system could decrease energy and water usage, while improving profitability (Lecina, 2015; Fellah et al., 2018; Vatta et al., 2018; Bazaluk et al., 2022, Liu & Smith, 2020; Liu & Liakos, 2020).

However, it is important to recognize the existence of some barriers to the successful adoption of DSS smart irrigation as such system is very knowledge intensive.

To overcome these barriers and facilitate the more effective implementation of smart irrigation management in Tunisia, farmers need to be better supported through the extension advice and the multiplication of training sessions in order to improve their knowledge and building capacities in this field.

5 Conclusion

This study aimed to identify the potential socio-economic impact resulting in utilizing innovative irrigation management at the farm level. Results showed that innovative management allows higher yield, lower production cost as well as irrigation cost, comparative to conventional management. Results also showed that under smart management farmers' incomes become more resilient towards market prices variability leading thus to better economic viability of the farms.

Acknowledgements The research was funded by the SUPROMED Project, PRIMA, Program H2020.

References

Bazaluk, O., Havrysh, V., Nitsenko, V., Mazur, Y., & Lavrenko, S. (2022). Low-cost smart farm irrigation systems in kherson province: Feasibility study. *Agronomy, 12*, 1013. https://doi.org/10.3390/agronomy12051013

Ben Guessim, A. (2015). For a new vision of the underground water management. In *Workshop on: Valorization of Scientific Research Achievements in Water Sector*, March 17th-18th, Gammart, Tunisia.

Fellah, S., Khiari, A., Kribaa, M., Arar, A., & Chenchouni, H. (2018). Effect of water regime on growth performance of Durum Wheat (*Triticum durum Desf.*) during different vegetative phases. *Irrigation and Drainage, 67*(5), 762–778.https://doi.org/10.1002/ird.2289.

Hemdi, M. (2022). Situation des ressources en eau dans le Gouvernorat de Sidi Bouzid. Regional seminar on: The adoption of smart irrigation technologies in scheduling irrigation in water limited areas, June, 6–8 Gafsa, Tunisia.

Kamienski, C., Soininen, J. P., Taumberger, M., Dantas, R., Toscano, A., Cinotti, T. S., Maia, R. F., & Torre Neto, A. T. (2021). Smart water management platform: IoT-based precision irrigation for agriculture. *Sensors, 2019*(19), 276. https://doi.org/10.3390/s19020276

Lecina, S. (2015). Farmerless profit-oriented irrigation scheduling strategy for solid sets. II: Assessment. *Journal of Irrigation, and Drainage Engineering, 142*(4), 04015068. https://doi.org/10.1061/(ASCE)IR.1943-4774.0000988

Levidow, L., Zaccaria, D., Maia, R., Vivas, E., Todorovic, M., & Scardigno, A. (2014). Improving water-efficient irrigation: Prospects and difficulties of innovative practices. *Agricultural Water Management, 146*, 84–94. https://doi.org/10.1016/j.agwat.2014.07.012

Liu, Y., & Liakos, V. (2020). Economic risk analysis of modern irrigation scheduling method on cotton production in Georgia. In *Selected Paper prepared for presentation at the Southern Agricultural Economics Association (SAEA) Annual Meeting*, Louisville, Kentucky, February 1–4, 2020.

Liu, Y., & Smith, A. R. (2020). Economic risk analysiz of modern irrigation scheduling method on cotton production in Georgia. In *Selected Paper prepared for presentation at the Southern Agricultural Economics Association (SAEA) Annual Meeting*, Louisville, Kentucky, February 1–4, 2020.

Vatta, K., Sidhu, R. S., Lall, U., Birthal, P. S., Taneja, G., Kaur, B., Devineni, N., & Alister, C. M. (2018). Assessing the economic impact of a low-cost water-saving irrigation technology in Indian Punjab: The tensiometer. *Water International, 43*(2), 305–321. https://doi.org/10.1080/02508060.2017.1416443

Vories, E., O'Shaughnessy, S., Sudduth, K., Evett, S., Andrade, M., & Drummond, S. (2021). Comparison of precision and conventional irrigation management of cotton and impact of soil texture. *Precision Agriculture, 2021*(22), 414–431. https://doi.org/10.1007/s11119-020-09741-3

Hydrologic Response to Land Use Change in the Kifisos Experimental Sub-Basin, Athens, Greece

Evgenia Koltsida, Nikos Mamassis and Andreas Kallioras

Abstract

This study investigated the impact of land use change on mean annual discharge in an experimental sub-basin in the metropolitan area of Athens, Greece, using the Soil and Water Assessment Tool (SWAT) model. The study site, the NW sub-basin of the Kifisos river basin, experienced an increase in artificial surfaces from 42.1 km^2 in 1990 to 71.6 km^2 in 2018. Therefore, the selected study site is an excellent case study to quantify the impact of urbanization on water balance components. The simulated discharge was compared to four land use scenarios (i.e. land use maps: 1990, 2000, 2006, and 2012) under the same climate conditions. Results showed that from 1990 to 2018: (i) the expansion of urban areas led to an increase of discharge by 8.7%, as water yield increased and evapotranspiration decreased, (ii) under the same climate conditions, runoff changes were observed during the late dry season, and (iii) the impact of land use change was more noticeable in the suburban sub-basins with significant urbanization. Overall, the outcomes of this study provide essential knowledge of the hydrological processes of an urban/peri-urban system which will be useful for planning future development strategies in the region.

Keywords

SWAT · Land use change · Hydrological response · Urbanization

E. Koltsida (✉) · A. Kallioras
School of Mining and Metallurgical Engineering, Laboratory of Engineering Geology and Hydrogeology, National Technical University of Athens, Athens, Greece
e-mail: ekoltsida@metal.ntua.gr

N. Mamassis
School of Civil Engineering, Laboratory of Hydrology and Water Resources Development, National Technical University of Athens, Athens, Greece

1 Introduction

Land use changes (physically or artificially) significantly impact the physical characteristics of a catchment and cause local decreases or increases in infiltration, canopy interception, evapotranspiration, and surface runoff. These impacts provide valuable knowledge for land use management and urban planning. The SWAT model (Arnold et al., 2012) has been successfully applied to estimate the impacts of urbanization (Zhou et al., 2013) and to investigate the effects of land use changes on runoff (Lin et al., 2015) and sediment yield (Mendonça dos Santos et al., 2020). However, the applications for estimating the impact of land use changes on daily discharge are limited. The main objective of this study is to investigate the effect of human-caused land use change on discharge at a daily scale in an experimental sub-basin. The model (i.e. baseline scenario: land use map of 2018) was calibrated for 2018 and validated for 2019 using the SUFI-2 algorithm in the SWAT-CUP program (Abbaspour et al., 2007). The land use map of 2018 was used for the construction of the baseline scenario, and the simulated discharge was compared to four different land use scenarios with decreased urbanization (i.e. land use maps: 1990, 2000, 2006, and 2012) under the same climate conditions to estimate the changes in the hydrological components caused by each scenario. This study will provide a detailed assessment of the impact of land use change on the hydrological processes at a critical daily scale and establish a calibrated tool for future modelling applications and urban development strategies.

2 Materials and Methods

The input data include a digital elevation model, a soil map, a land use map, and hydrological and meteorological data. The meteorological data were obtained from the National Observatory of Athens (NOA) (Lagouvardos

H. Chenchouni et al. (eds.), *Recent Advancements from Aquifers to Skies in Hydrogeology, Geoecology, and Atmospheric Sciences*, Advances in Science, Technology & Innovation, https://doi.org/10.1007/978-3-031-47079-0_22

et al., 2017), and the discharge data were provided by the Open Hydrosystem Information Network (Mamassis et al., 2021). Land use maps (1990–2018) were obtained from the European project Corine Land Cover (2020). The Corine classes were modified according to SWAT land use categories in the model's database. The three main land use categories are artificial surfaces (i.e. urban fabric, industrial, commercial, and transport units), agricultural areas (i.e. arable land, pastures, and heterogeneous agricultural areas), and forests and semi-natural areas (i.e. forests, scrub, and herbaceous vegetation associations). In the SWAT model, Penman–Monteith method was used to estimate potential evapotranspiration, and the curve number (CN) method was selected to calculate surface runoff.

3 Results and Discussion

The hydrological response to land use changes was estimated by importing the land use maps from 1990 to 2018 (Fig. 1) into the SWAT model under fixed climate conditions. In order to assess the impact of land use changes, the same set of parameters of the original calibration was used. The discharge outputs of each scenario were compared to the outputs of the baseline scenario to estimate the discharge change rate. Figure 2 displays the spatial distribution of surface runoff variation over the period 1990–2018. In total, the artificial surfaces increased from 30.1% in 1990 to 51.2% in 2018, which led to a discharge increase of 8.79% for the whole catchment. The sub-basins 7, 20, and 13, in the suburban areas with the highest urbanization rate, showed the most significant change in surface runoff. Figure 3 presents average monthly discharge change rate. The discharge change rate was relatively low from 1990 to 2000 and higher from 2006 to 2012 due to the rapid increase in the urban areas in the study area. In all scenarios, an increase in discharge was observed during the wet months and a decrease during the dry months.

Furthermore, in order to estimate the spatial impact of land use changes on the hydrological components, a correlation analysis was performed in 25 sub-basins over the period 1990–2018. High correlation was observed between the surface runoff change and the increase of the artificial surfaces. Lower correlations were observed between surface runoff change and agricultural areas, forests, and semi-natural areas. In particular, the increase of the artificial areas had a high correlation with surface runoff ($R=0.81$), water yield ($R=0.60$), and a lower correlation with actual evapotranspiration ($R=-0.38$), lateral flow ($R=-0.20$), soil water ($R=-0.50$), percolation ($R=-0.55$), and groundwater ($R=-0.55$). Overall, the results indicated that the increase of artificial surfaces (i.e. urbanization) was the major cause of land use change affecting the hydrological components.

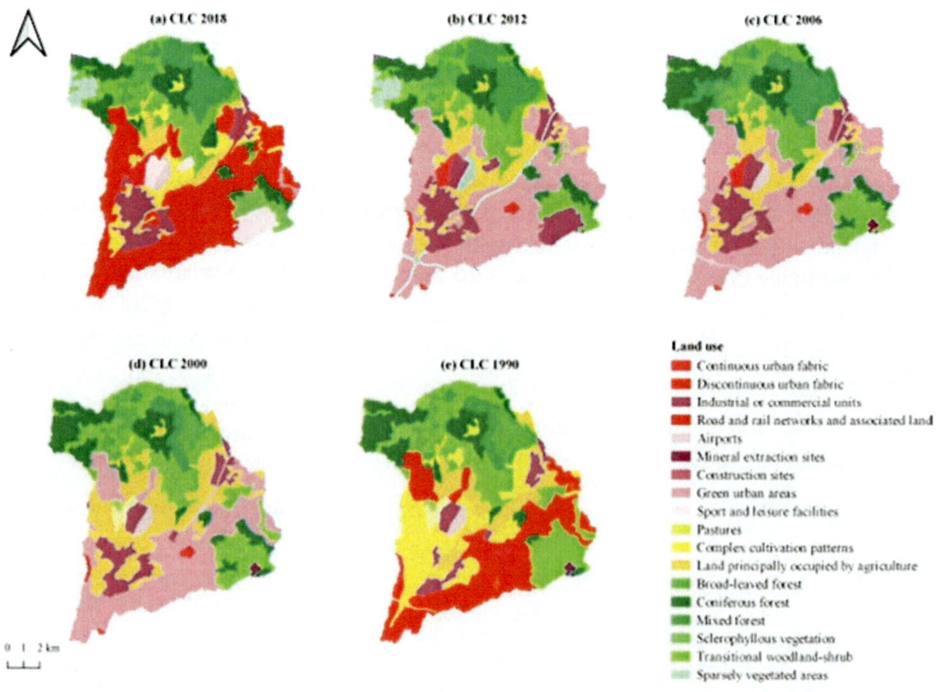

Fig. 1 Spatial distribution of land use over the years **a** 2018, **b** 2012, **c** 2006, **d** 2000, and **e** 1990

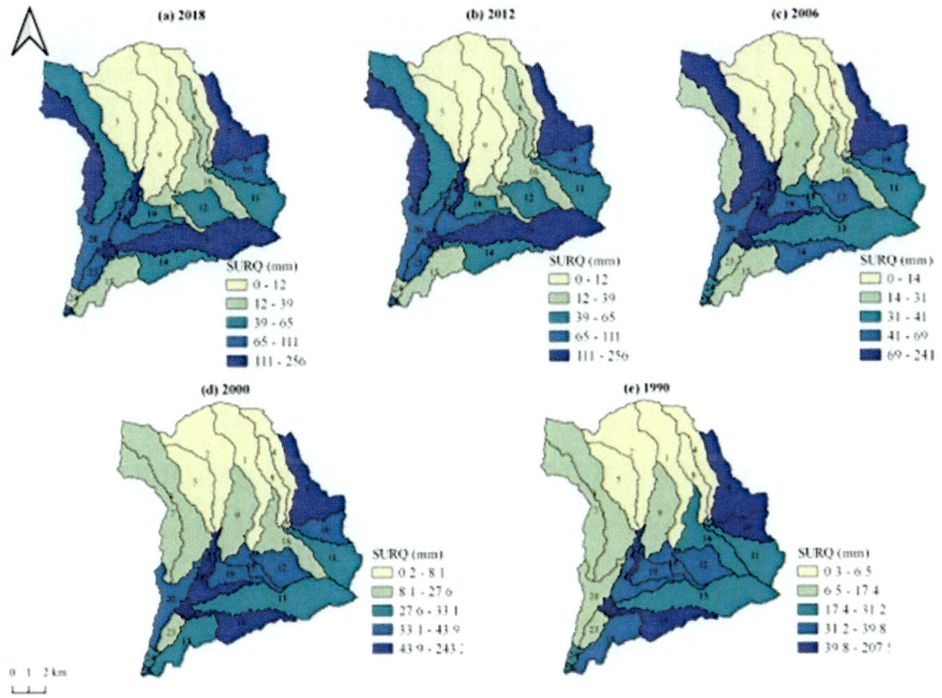

Fig. 2 Spatial distribution of surface runoff (SURQ) over the years **a** 2018, **b** 2012, **c** 2006, **d** 2000, and **e** 1990

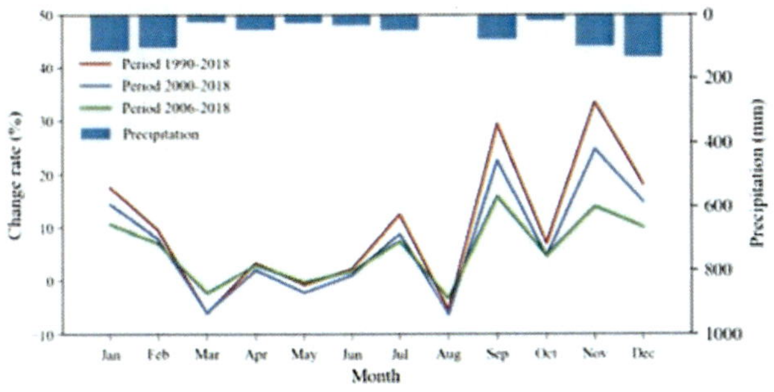

Fig. 3 Mean monthly discharge change rate (%) from 1990 to 2018

4 Conclusion

This study investigated the impact of land use change on mean annual discharge from 1990 to 2018. The SWAT model was used to calculate discharge using the land use maps of 1990, 2000, 2006, 2012, and 2018. The simulated discharge of each year was compared to the simulated discharge of the baseline scenario of 2018.

Results indicated that the artificial areas increased from 30.1% in 1990 to 51.2% in 2018 in the whole study area. The land use changes were observed in the lower part of the study area close to the metropolitan area. A correlation analysis was used to identify the major cause of land use change affecting discharge for each sub-basin. The results showed that surface runoff was mainly correlated to changes in artificial surfaces ($R > 0.8$). Due to the rapid increase in artificial areas from 1990 to 2018, the mean annual discharge increased by 8.79%. This study provides essential knowledge for the local planner to use in urban planning and future development.

Acknowledgements The research work was supported by the Hellenic Foundation for Research and Innovation (HFRI) under the HFRI PhD Fellowship grant (Fellowship Number: 1586).

References

Abbaspour, K. C., Yang, J., Maximov, I., Siber, R., Bogner, K., Mieleitner, J., Zobrist, J., & Srinivasan, R. (2007). Modelling hydrology and water quality in the pre-alpine/alpine Thur watershed using SWAT. *Journal of Hydrology, 333*, 413–430. https://doi.org/10.1016/j.jhydrol.2006.09.014

Arnold, J. G., Moriasi, D. N., Gassman, P. W., Abbaspour, K. C., White, M. J., Srinivasan, R., Santhi, C., Harmel, R. D., Van Griensven, A., Van Liew, M. W., et al. (2012). SWAT: model use, calibration, and validation. *Transactions of the ASABE, 55*, 1491–1508. https://doi.org/10.13031/2013.42256

Corine Land Cover (CLC). (2020). Land use data. https://land.copernicus.eu/. Accessed on 15 Dec 2020.

Lagouvardos, K., Kotroni, V., Bezes, A., Koletsis, I., Kopania, T., Lykoudis, S., Mazarakis, N., Papagiannaki, K., & Vougioukas, S. (2017). The automatic weather stations NOANN network of the National Observatory of Athens: Operation and database. *Geoscience Data Journal, 4*, 4–16. https://doi.org/10.1002/gdj3.44

Lin, B., Chen, X., Yao, H., Chen, Y., Liu, M., Gao, L., & James, A. (2015). Analyses of landuse change impacts on catchment runoff using different time indicators based on SWAT model. *Ecological Indicators, 58*, 55–63. https://doi.org/10.1016/j.ecolind.2015.05.031

Mamassis, N., Mazi, K., Dimitriou, E., Kalogeras, D., Malamos, N., Lykoudis, S., Koukouvinos, A., Tsirogiannis, I., Papageorgaki, I.,

Papadopoulos, A., et al. (2021). OpenHinet: a synergistically built, national-scale infrastructure for monitoring the surface waters of Greece. *Water, 13*, 2779. https://doi.org/10.3390/w13192779

Mendonça dos Santos, F., Proença de Oliveira, R., & Augusto Di Lollo, J. (2020). Effects of land use changes on streamflow and sediment yield in Atibaia River Basin—SP, Brazil. *Water, 12*, 1711. https://doi.org/10.3390/w12061711

Zhou, F., Xu, Y., Chen, Y., Xu, C.-Y., Gao, Y., & Du, J. (2013). Hydrological response to urbanization at different spatio-temporal scales simulated by coupling of CLUE-S and the SWAT model in the Yangtze River Delta region. *Journal of Hydrology, 485*, 113–125. https://doi.org/10.1016/j.jhydrol.2012.12.040

Application of a Decision Support Tool to Assist Water Governance Within a Water-Stressed Area: Case of the Souss Basin, Morocco

Oumaima Attar, Youssef Brouziyne, Lhoussaine Bouchaou, Yassine Ait Brahim and Abdelghani Chehbouni

Abstract

The Souss basin in Morocco is distinguished by an uneven surface water supply and steady overexploitation of groundwater resources. In recent years, the basin experienced constant disrupted water balance, due mostly to year-to-year variable weather conditions and rising water demand. This situation is projected to affect the region's economic development and threaten the prospects of many economic sectors. Therefore, this study was conducted to analyze the water balance of the Souss basin as well as to develop a decision support tool (DSS) to assist water sector managers in sustainable water governance. MODSIM 8.1, a generic river basin management decision support system, for water allocation modeling designed as a computer-aided tool, was selected for that study. Once calibrated and validated over the period from 1990 to 2019 using recorded data about physical processes and hydraulic infrastructures, the model is used to simulate water supply and demand in the Souss basin. The findings indicate that the basin's water resources are vulnerable, and dams' inputs in the basin have undergone significant fluctuation under both changing climate and increasing water demand in the growing area. In addition, there is a significant amount of unmet demand across all demand sites. As result, the average total unmet demand for surface water from reservoirs in irrigated areas reached 201 mm^3 between 1990 and 2019, and the monthly average demand increased by 55% in the dry season, compared to the demands in the rest of the year. The significant amount of unmet demand across all demand sites suggests that demands are satisfied by the withdrawal of water from groundwater resources. Using MODSIM, the adopted simulation approach has shown to be a useful decision support tool to understand water resources planning challenges. Water managers require reliable tools to represent the basin's various trade-offs with regards to water resources. Thus, an investigation improving the representation of groundwater/surface water interaction in operations modeling is required to enhance the evaluation of the consequences of different uses at the basin level.

Keywords

Souss basin · DSS · MODSIM · Water allocation modeling · Climate change

1 Introduction

In recent years, researchers and policymakers have increasingly highlighted the importance of water security (Chenchouni et al., 2022; Grey et al., 2013). The Souss basin, which is located in central-western Morocco, has a significant amount of economic potential that is based on its natural resources, agricultural, fishing, and tourism industries. The agricultural industry, which has experienced remarkable growth in recent decades, is a significant source of socio-economic support and integration for the local community in the area. The management of complex basin

O. Attar (✉) · L. Bouchaou · Y. Ait Brahim · A. Chehbouni
International Water Research Institute, Mohammed VI Polytechnic University (UM6P), 43150 Benguerir, Morocco
e-mail: oumaima.attar@um6p.ma

Y. Brouziyne
International Water Management Institute (IWMI), MENA Office, Giza 12661, Egypt

L. Bouchaou
Laboratory of Applied Geology and Geo-Environment, Faculty of Sciences, Ibn Zohr University, Cité Dakhla, BP 8106, 8000 Agadir, Morocco

A. Chehbouni
Institut de Recherche Pour Le Développement (IRD), Unité Mixte de Recherche (UMR), Centre d'études Spatiales de La Biosphère (Cesbio), 31400 Toulouse, France

H. Chenchouni et al. (eds.), *Recent Advancements from Aquifers to Skies in Hydrogeology, Geoecology, and Atmospheric Sciences*, Advances in Science, Technology & Innovation, https://doi.org/10.1007/978-3-031-47079-0_23

systems, such as the Souss basin in Morocco, necessitates the development of effective decision support technologies that are capable of assessing system components in a completely integrated manner. The research measures the existing water demands and evaluates the influence that recent developments in water resources management have had on the water balance of four agricultural perimeters located within the Souss basin. The simulation was constructed by adapting and modifying the MODSIM generalized river basin network flow model.

2 Study Area

The Souss basin is located in the central-western part of Morocco (Fig. 1). The High Atlas Mountains are in the north of the basin, the Anti-Atlas Mountains are in the south, the Siroua Massif in the east, and the Atlantic Ocean are in the west of the basin. There is a large temporal and geographical range of fluctuation in the amount of precipitation that falls within the basin. The average annual precipitation ranges from 600 mm in the north, on the summits of the High Atlas, to 150 mm in the south, in the eastern part of the Anti-Atlas. Temperatures range from 14 °C in the region of the High Atlas in the north to 20 °C in the region of the Anti-Atlas in the south, with the highest daily temperature reaching 49 °C in the summer and the lowest temperature in the winter falling to −3 °C (Abahous et al., 2017; Ait Brahim et al., 2017; Seif-Ennasr et al., 2017). Water for the basin's dams comes from the streamflow created by winter and spring rains and melting in the spring. Under climate change, the most pessimistic simulations expect a reduction in the dam storage (Seif-Ennasr et al., 2016).

3 Modeling Approach

In this study, MODSIM (Labadie, 2010) is used to develop the Souss basin system. The basin was represented with nodes and links. The nodes represent the reservoirs (dams), demands, and confluences of rivers. The links represent the sections of the rivers or channels linking the dams with the consumption demands. The three reservoir nodes represent the three dams, namely Aoulouz Dam, Imi-lkhengu Dam, and Abdelmoumen Dam. The G1 perimeter, as well as the modern and traditional Issen perimeters, the Elguerdane perimeter, and the P12-P13 perimeter are the agricultural perimeters that are represented by the irrigation demand nodes.

The model was calibrated using the basin physical data and hydrologic processes. Manual evaluation of the data over time was used to evaluate the calibration and validation. Trial-and-error was used to adjust the settings and compare observed and simulated outputs. The method was to estimate network parameters and data and then verify that expected flows fit observed data trends and general tendency. The approach included creating the best estimate of network characteristics and then ensuring that simulated matched observed trends.

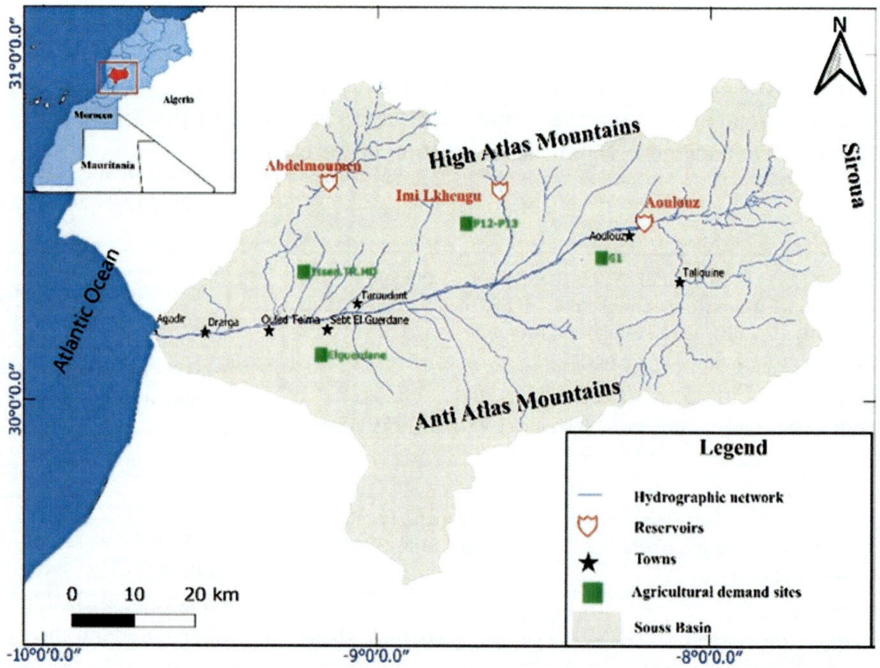

Fig. 1 Souss basin localization including gauging stations, topography, and streamflow network

4 Results and Discussion

4.1 Streamflow Simulation

Figure 2 shows one of the results of calibration and validation results. The figure compares the result of simulated and observed discharge hydrographs for the calibration (1990–2009) and validation (2010–2019) period at the inputs of Abdelmoumen dams. According to the linear regression results, the model appears to be effective in terms of monthly inflows to the dams. The MODSIM model's efficiency in terms of R^2 at monthly inflows to the dams for the calibration and validation period is shown in the figure for Abdelmoumen dam.

The findings reveal that the basin's water supplies are vulnerable and that water allocations in the basin have fluctuated significantly as a result of both climate change and increased water demand in the rapidly expanding region of the basin.

The calibration and testing of the model's monthly simulation have revealed that the model is capable of properly simulating the measured flows in the system. The development of well-calibrated models was a significant accomplishment of this research.

4.2 Water Demand

Results of agricultural demands for water are shown in Fig. 3. During the wet seasons, the average monthly demand falls to 300 mm^3, whereas during the hot season it rises to 550 mm^3, representing a 55% increase in comparison with the demands seen during the rest of the year. The number presents a comparison of the water demand and the shortages that are recorded each month by all of the perimeters. The figure presents an illustration of the demand for water that cannot be met. The higher levels of demand throughout the summer months have led to an increase in the level of scarcity that existed at this time. The

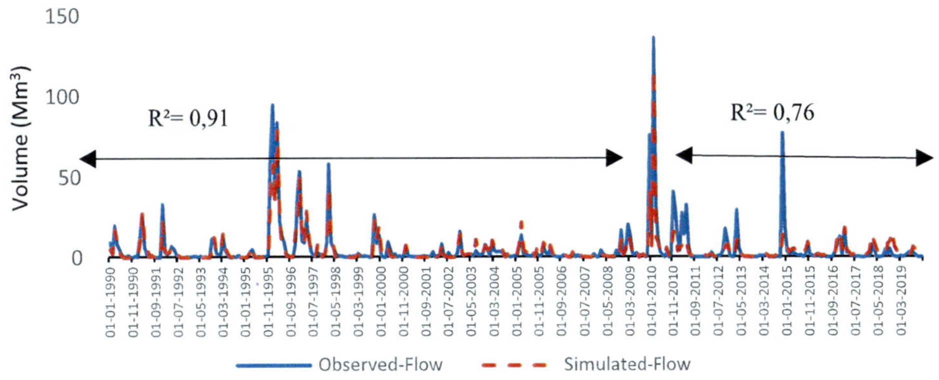

Fig. 2 Monthly simulated and observed dams' inputs for calibration and validation periods at Abdelmoumen dam

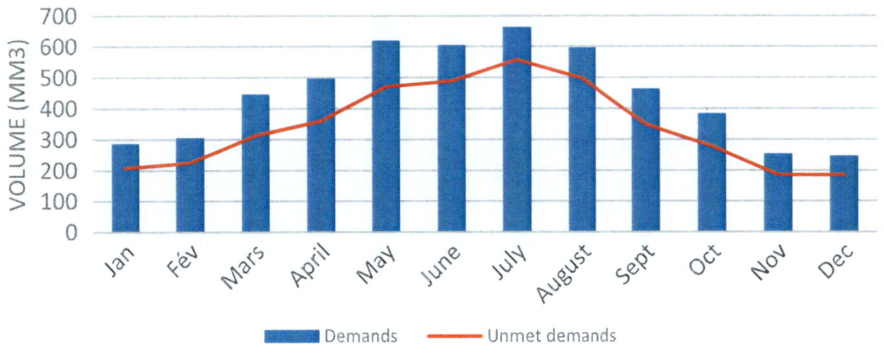

Fig. 3 Monthly distribution of demand and unsatisfied demand during the simulation period (1990–2019)

agricultural sector in the area is the one that uses the most water, which is why the region has the worst water shortage issues during the warm months (May–Sep). The fact that these months fall during the dry seasons provides an explanation for this phenomenon. In the same way, the amount of groundwater that is pumped out for the purpose of irrigation will continually rise during this time in response to the rising demand.

5 Conclusions

The analysis of inflows was performed with the model at the dams' inputs and the water demands of four parameters in the basin. The inflows to the Abdelmoumen reservoir were successfully modeled. Using linear regression, the analysis shows good results for the inflows recorded at the dams. The model simulated flows were well matched to the observed hydrograph volumes for the gauging station installed at Oued Souss. The results indicate that the monthly average demand increases by 55% in the hot season, compared to the demands of the rest of the year. The unmet water demands will be probably satisfied by the uncontrolled withdrawal of water from groundwater resources. Therefore, we suggest that further research needs to be conducted to evaluate the various water management strategies, with the goal of mitigating the negative effects of both water withdrawal and climate change and providing useful information for integrated water resource management among all relevant stakeholders.

Acknowledgements This work is carried out in part with the Project CHARISMA with assistance of Hassan II Academy of Sciences and Techniques and the Project PRIMA-AGREEmed financed by MESRSI (Morocco).

References

Abahous, H., Siffeddine, A., Bouchaou, L., Ronchail, J., ElMorjani, Z. E. A., Ait Brahim, Y., & Kenny, L. (2017). Inter-annual variability of precipitation in the Souss Massa region and linkage of the North Atlantic Oscillation. *Journal of Materials and Environmental Science, 9*(7), 622–627.

Ait Brahim, Y., Saidi, M. E. M., Kouraiss, K., Sifeddine, A., & Bouchaou, L. (2017). Analysis of observed climate trends and high resolution scenarios for the 21st century in Morocco. *Journal of Materials and Environmental Science, 8*(4), 1375–1384.

Chenchouni, H., Chaminé, H. I., Khan, M. F. (Eds.) (2022). *New Prospects in Environmental Geosciences and Hydrogeosciences*. Cham, Springer.

Grey, D., Garrick, D., Blackmore, D., Kelman, J., Muller, M., & Sadoff, C. (2013). Water security in one blue planet: twenty-first-century policy challenges for science. *Philosophical Transactions of the Royal Society A: Mathematical, Physical and Engineering Sciences, 371*(2002), 20120406. https://doi.org/10.1098/rsta.2012.0406.

Labadie, J. W. (2010). MODSIM: Decision support system for river basin (p. 6). Department of Civil and Environmental Engineering at CSU

Seif-Ennasr, M., Malki, M., Bouchaou, L., Beraaouz, E. H., Hirich, A., Zaaboul, R., ElMorjani, Z. E. A., Choukr-Allah, R., & Nrhira, A. (2017) Assessment of global change impacts on groundwater resources in Souss-Massa basin. In O. Abdalla, A. Kacimov, M. Chen, A. Al-Maktoumi, T. Al-Hosni, & I. Clark (Eds.), *Water resources in arid areas: The way forward* (pp. 295–318). Springer. https://doi.org/10.1007/978-3-319-51856-5_8

Seif-Ennasr, M., Zaaboul, R., Hirich, A., Caroletti, G. N., Bouchaou, L., El Morjani, Z. E. A., Beraaouz, E. H., McDonnell, R. A., & Choukr-Allah, R. (2016). Climate change and adaptive Water management measures in Chtouka Aït Baha region (Morocco). *Science of the Total Environment, 573*, 862–875. https://doi.org/10.1016/j.scitotenv.2016.08.170

Biogeochemistry, Geobiology and Geoecology

Oceanographic Investigation for Sustainable Development of Algaculture with an Ecosystem Approach in Sidi Rahal Marine Coast (Morocco)

Abdelmaoula El Meraoui, Chaymae Najimi, Kaoutar Jamous, Nouh Lahmam, Miriam Wahbi, Hassan Nhhala, Omar El Kharki and Mustapha Maatouk

Abstract

The "lower Chaouia" or coastal Chaouia region, which is the most agricultural part of the coastal Chaouia, is a sub-Atlantic plain that has developed along the Atlantic Ocean. To ensure the integration of sustainable aquaculture in this region, a study was conducted to identify the specific richness of the environment. It consisted of the mapping of physicochemical parameters and a sedimentary study for the exploration of potential algaculture sites. To conduct this study, a database was built from a combination of satellite data. Besides, field measurements were also carried out during the oceanographic investigation missions of the study area to understand the temporal variability of the various parameters with the help of geographic information system (GIS) tools. The treatment and analysis of all the data collected revealed important macroalgal biodiversity in the Sidi Rahal Chatai area. Indeed, in this relatively rich environment, where the maritime zone (located between the coastline and the rocky strip offshore) which remains covered with seawater, with a depth of water greater than 30 cm, is home to more than 22 species of marine macroalgae. The particle size study of the sediments shows that the bottom consists of the rocky sand type which promotes the fixation of algae. The actual variability of some environmental parameters confirmed the suitability of this area for algae cultivation. For this reason, algaculture is an alternative solution for preserving algae. It is an environmentally friendly activity that can even limit the effects of chemical and organic pollution at the local level and mitigate the effects of climate change a socio-economic activity for the development of the region.

Keywords

Algaculture · Sedimentary study · Sustainable aquaculture · Satellite data · GIS

1 Introduction

The use of macroalgae is very important in several fields: human food, breeding, agriculture, cosmetics, and pharmaceutical on a world scale (Leandro et al., 2019). Among these countries, Morocco is considered the third largest producer of agar–agar, extracted mainly from *Gelidium sesquipedale* (Givernaud et al., 2005). This natural resource shows signs of overexploitation (ElMtili et al., 2013). In addition, the growing demand for raw materials in the gelling industry in Morocco has led to a significant interest in recent years in other *agarophytes* such as *gracilaria* to compensate the overexploitation and insufficient natural stocks.

Morocco has implemented a specific strategy to meet this demand. This ambition cannot be achieved, particularly for marine aquaculture, without a good choice of appropriate sites.

In order to accomplish that, the selection of potential areas is an important aspect to ensure before the establishment of any aquaculture exploitation. The environmental, physical, and socio-economical aspects of the sites must be

A. El Meraoui (✉) · K. Jamous · M. Wahbi · H. Nhhala · O. El Kharki · M. Maatouk
Applied Geosciences Research and Development Laboratory, FSTT, Abdelmalek Essaadi University, Tetouan, Morocco
e-mail: abdelmaoula.elmeraoui@etu.uae.ac.ma

C. Najimi
Laboratory LGCE, Superior School of Technology, Mohammed V University, Salé, Morocco

N. Lahmam
Laboratory EMRN, FSTT, Abdelmalek Essaadi University, Tetouan, Morocco

H. Nhhala
Expert and Consultant in Technical Sciences and Aquaculture Engineering, Tangier, Morocco

H. Chenchouni et al. (eds.), *Recent Advancements from Aquifers to Skies in Hydrogeology, Geoecology, and Atmospheric Sciences*, Advances in Science, Technology & Innovation, https://doi.org/10.1007/978-3-031-47079-0_24

carefully evaluated before proceeding with the establishment of an algal farm. The present study aims to to select favorable areas for the cultivation of mainly *gracilaria* and other commercially valuable algal species in Sidi Rahal Chattai.

2 Materials and Methods

Water samples were collected from six stations of Sidi Rahal Chatai which is a coastal town in the Casablanca-Settat region of Morocco. The physicochemical hydrology and biology were assessed in this study. A portable multiparameter was used on-site to measure temperature, turbidity, salinity, conductivity, and dissolved oxygen from six stations S1, S2, S3, S4, S5, and S6. Besides, nitrates, orthophosphates, and chlorophylls from two stations (S1 and S4) were analyzed by spectrophotometry at the laboratory of the Higher School of Technology of Sale Morocco according to The protocol published by the water analysis book Control and interpretation (Rodier & Legube, 2016).

On the other hand, bathymetry, sediment sampling for particle size analysis and fund type identification was also conducted at each of the six sites on the low tide of white-water to understand the background nature of the whole site.

Samples of different types of algae were collected in the study area and rinsed with seawater to remove debris and small crustaceans. Food and Agriculture Organization of the United Nations guide to the identification of species to fish (Fischer et al., 1987), the algaebase site (http://www.algaebase.com), the National Natural Heritage Inventory (INPN) site (https://inpn.mnhn.fr/) and the Word Register of Marine Spices WoRMS site (http://www.marinespecies.org/) was used to identify different species their taxonomies. The samples were observed in the hours following the harvest, as some elements useful for recognition degrade quickly. Identification is based primarily on the morphology of these species, specifically on the color of the species, the type of Dallas and its ramification, and the absence or presence of the stem and its size. Small algae were placed in flat basins filled with seawater. For some species, microscopic examination is often indispensable for the recognition and consists of mounting a piece of the thallus between slide and slide in a drop of seawater. Often, a transverse or longitudinal section (made with a razor blade) was necessary for the identification of certain species. The magnification used in our case is 40–400 times.

3 Results

3.1 Bottom Sediment Survey and Bathymetry of Sidi Rahal Chatai

Figure 2 shows the portions of this marine area that remain covered with seawater, with a depth of water of more than 30 cm, located between the shoreline and the offshore rocky strip. The areas found at low tide are rocky outcrops or rocky platforms with altimetry varying in elevation from sea level at low tide. This site has generally irregular terrain, characterized by crevices of different shapes, sizes and depths, wide cracks, and more or longer grooves in the form of labyrinths. These discovered parts present a large floral diversity (algae, phanerogams, etc.) which local women collect.

3.2 Marine Macroalgal Biodiversity of Sidi Rahal Chatai

The area of Sidi Rahal Chatai reveals very important macroalgal biodiversity. In this relatively rich environment, 22 species of marine macroalgae were found. The latter consists essentially of three taxonomic groups: Rhodophyceae, Pheopheae, and Chlorophyceae (Table 1). Rhodophyceae or red algae are most diverse in the study area with 11 species divided into five families (Table 1), including *Gelidium sesquipedale* and *Chondracanthus acicularis*, which belong to the families of Gelidiaceae and Gigartinaceae, respectively. These two species are among the most abundant macroalgae species harvested by women from a cooperative in the area (Fig. 1).

The Phaeophyceae or the brown algae are present with nine species divided into five families with varied morphologies (Table 1), among the latter the Sargassaceae are the most abundant covering a very large area of the zone forming a Sargasso alga (Fig. 1). The macroalgal specific richness in Chlorophyceae or green algae is estimated at two species: *Codium tomentosum* and *Ulva lactuca* belonging to the families of Codiaceae and *Ulvaceae,* respectively (Table 1).

3.3 Water Features from Sidi Rahal Chatai

The results of the various parameters analyzed for the six stations studied are summarized in Table 2, of which we note that there is no significant variation in values except

Table 1 Species of algae present in the area of Sidi Rahal

Species of algae	
Codium tomentosum	*Gracilaria gracilis*
Ulva lactuca	*Gracilaria multipartita*
Sargassum muticum	*Gracilaria verrocusa*
Sargassum vulgare	*Mastocarpus stellatus*
Bifurcaria bifurcata	*Chondrus crispus*
Cystoseira caespitosa	*Chondracanthus acicularis*
Cystoseira amentacea	*Gelidium corneum*
Saccorhiza polyschides	*Gelidium sesquipedale*
Padina boryana	*Ceramium virgatum*
Cutleria multifida	*Chondria scintillans*
Fucus vesiculosus	*Porphyra umbilicalis*

of different parameters except for dissolved oxygen, which varies between 6.18 mg of O_2/L as the minimum value recorded in station S1 and 15.14 mg of O_2/L as the maximum value recorded in station S4. This variation at this low tide can be explained by the stagnation of water.

3.4 Water Features from Sidi Rahal Chatai

The granulometric analysis makes it possible to distinguish the granulometric classes of the sand obtained. Indeed, the different classes of arenites (sands and rocks) are separated according to the classification scale adopted by Cailleux (1952).

The facies obtained is also of the rocky sand type (Fig. 2).

4 Discussion

The study of the large-scale area through field missions was very beneficial to improve the results, and field missions were necessary for the acquisition of in-depth data. The use of satellite data is exclusively for the interpretation and validation of remote sensing information; these results only help us eliminate areas unsuitable for algaculture but will not allow us to reliably select areas suitable for the culture of the selected species.

The investigation of area has been investigated mainly because of its hydrodynamic capacity, which perfectly meets the requirements of the cultivation technique adopted, due to the presence of an offshore rocky strip that makes the site well protected from tides and strong winds coming from the sea and that also prevents the waves from moving toward the coast and will therefore ensure the stability of the facility as well as ease of monitoring and the harvesting of algae. From an economic point of view, this band has a significant effect on the costs of the charges required for the structure of the crop.

The study of the real variability of certain environmental parameters analyzed in situ confirmed the suitability of this area for the cultivation of *gracilaria* algae. Thus, the existing macroalgal biodiversity in the area of which the selected species was one of the factors confirming its potential for algaculture since the availability of local stocks of species

Fig. 1 Geographical location and sampling points in the study area

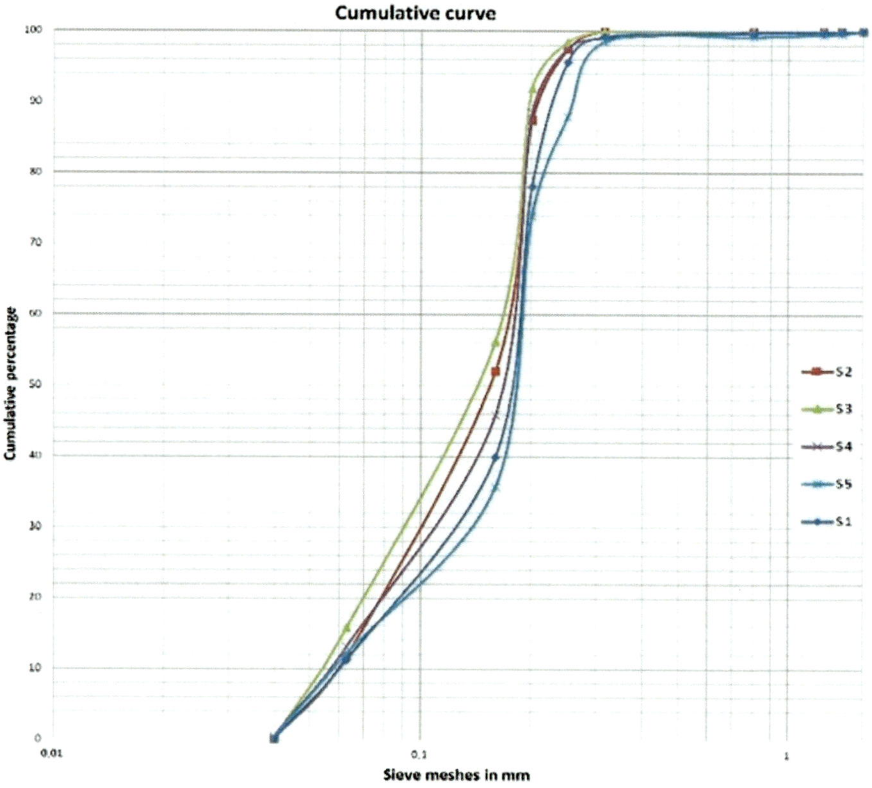

Fig. 2 Grain size analysis cumulative percentage curve

Table 2 Water quality in the study area

Parameters	Water sampling stations						
	S1	S2	S3	S4	S5	S6	
Chlorophyll (mg/L)	0.031			0.042			Analyzed in laboratory
Nitrate (mg/L)	1.7472			1.1849			
Orthophosphate (mg/L)	0.4370			0.4185			
Temperature (°C)	22.5	21.3	24.2	23.9	22.9	23.7	Measure with multiparameter sondes in situ
Dissolved oxygen (mg/L)	6.18	6.15	6.60	12.44	11.40	12.34	
SPC (g/L)	55.5	55.4	55.7	49.64	55.8	56.8	
Conductivity (mS/cm)	51,882	51,836	52,463	48,674	54,472	55,266	
Salinity (ppt)	36.84	36.81	36.99	32.49	37.02	37.78	
pH	8.67	8.63	8.60	8.99	8.94	8.9	

to be cultivated on the site is a good indicator that the conditions The site's ecological features are conducive to the growth and development of the species, which approves the results of the study of the effects of physicochemical parameters and nutrients in the same area (Mouttaqui, et al., 2022).

5 Conclusions

To improve the results, we recommend incorporating additional criteria to further refine the areas obtained, such as the assessment of economic costs through the calculation of distances to potential markets and a roadmap to ensure that the selected site is also accessible by land.

In conclusion, the great interest and economic value of the *Gracillaires*, which are considered sources of the most valuable *phycocolloids* of the world market, must be well protected and preserved. This is done through the sustainable use of its resources. For this, algaculture is a solution for the preservation of algae, it is an activity respectful of the environment, and it is likely to adapt even to mitigate the effects of chemical and organic pollution at the local level and to mitigate the effects of climate change.

The monitoring of the parameters in a regulatory way is necessary in order to guarantee a better production of this algae and all the changes that may appear in the environments.

References

Algaebase. http://www.algaebase.com

Cailleux, A. (1952). Morphoskopische Analyse der Geschiebe und Sandkörner und ihre Bedeutung für die Paläoklimatologie. *Geologische Rundschau, 40*, 11–19.

El Mouttaqui, S., Bekkali, M., Ouazzani, K. C., Baali, A., Zidane, H., Bourhim, N., & Rezzoum, N. E. (2022). The effect of physico-chemical parameters and nutrients on the bioecology of *Chondracanthus acicularis* (Roth) Fredericq at three sites in the Moroccan Atlantic Ocean. *Egyptian Journal of Aquatic Biology Fisheries., 26*(4), 273–287.

ElMtili, N., Fakihi Kachkach, F. Z., & El Harchi, M. (2013). Les algues marines: Nouvelle potentialité économique pour le Maroc. Quelle stratégie biotechnologique? *Cahiers UAE, 8–9*, 1–7.

Fischer, W., Bauchot, M. L., & Schneider, M. (eds.) (1987) Fiches FAO d'identification pour les besoins de la pêche. (rev. 1). Méditerranée et mer Noire. Zone de pêche 37. Vol. II. Commission des Communautés Européennes and FAO, Rome

Givernaud, T., et al. (2005). Mapping and biomass estimation for a harvested population of *Gelidium sesquipedale* (Rhodophyta, Gelidiales) along the Atlantic coast of Morocco. *Phycologia, 44*(1), 66–71.

INPN (Inventaire National du Patrimoine Naturel). https://inpn.mnhn.fr/

Leandro, A., Pereira, L., & Gonçalves, A. M. M. (2019). Diverse applications of marine macroalgae. *Marine Drugs, 18*(1), 17.

Rodier, J., & Legube, B. (2016). L'analyse de l'eau. Contrôle et interprétation, avec la collaboration de MERLET Nicole.

WoRMSWord Register of marine Spices. http://www.marinespecies.org/

Assessment of Soil Factors Influencing Productivity of Fish Ponds Under Two Contrast Agro-ecological Regions

Asim Kumar Bhowmick, Gunindra Nath Chattopadhyay, Kapil Deo Sah and Dipak Sarkar

Abstract

The bottom soils of fish ponds not only influence the availability of different nutrient elements to benefit the primary food organisms but also control many important bio-geo-chemical reactions in the pond environment exerting significant influence on the pond health and, in turn, the fish growth. However, little attention is generally paid to various productivity attributing soil properties while undertaking any aquaculture operation in various soil zones. We assessed the relative effects of different soil factors on productivity of the fish ponds situated under two contrast agro-ecological regions of West Bengal, India, viz. hot-dry sub-humid (HDSH) region represented by red and lateritic soils and hot-moist sub-humid (HMSH) region exemplified by alluvial soils with the primary objectives of identifying the productivity limiting soil properties of the fish ponds of these two regions and understanding their impacts on productivity. For this, the magnitude of primary fish food production in 85 and 110 no of ponds in HMSH and HDSH regions, under culture of composite fish species, were studied in the form of net primary productivity (NPP) and correlated with relevant properties of those pond soils. NPP values were considerably lower in the fish ponds under HDSH region as compared to those under HMSH region. Statistical correlation and multiple regression analyses showed pH and available N and P status to be the key soil factors to influence the NPP of pond water in case of hot-moist region. For the hot-dry region, pH, organic C, available N and P, and clay content emerged as the major factors influencing the productivity. Among all these properties, available P was observed to be the most important soil factor influencing the net primary productivity of pond water in both the agro-ecological regions and was followed by soil pH and available N. Together, these three soil properties could explain 83 and 72% of variability in primary productivity of the fish ponds under hot-moist and hot-dry regions, respectively. The results suggest that the aquaculture practices in these two agro-ecological regions need specific management of these three soil properties for attaining better fish production.

Keywords

Fish pond productivity · Contrast agro-ecological regions · Pond soils · Productivity influencing soil factors · Key soil properties

A. K. Bhowmick
Bathanberia Srinibas H.S. School, Kolaghat, West Bengal 721134, India

G. N. Chattopadhyay (✉)
Seacom Skills University, Kendradangal, West Bengal 731236, India
e-mail: gunin.c2010@gmail.com

K. D. Sah · D. Sarkar
Ex-ICAR-National Bureau of Soil Survey and Land Use Planning, Kolkata 700091, India

1 Introduction

The productivity of any natural fish culture system depends largely on the nature of its bottom soil which has been classically designated as the "Chemical laboratory of a fish pond" by Hickling (1971). The role of bottom soils in influencing availability of different nutrient elements to primary producers and, thereby, the fishes through various food chains has been discussed (Boyd, 1995; Ouali et al., 2018). In addition, the importance of the soil phase with regard to various other physical, chemical, and biological attributes of aquatic productivity is also well documented (Delince, 1992). Such significance of the pond soils in controlling different factors of pond productivity results in considerable variations in the yield levels of the fish ponds

H. Chenchouni et al. (eds.), *Recent Advancements from Aquifers to Skies in Hydrogeology, Geoecology, and Atmospheric Sciences*, Advances in Science, Technology & Innovation, https://doi.org/10.1007/978-3-031-47079-0_25

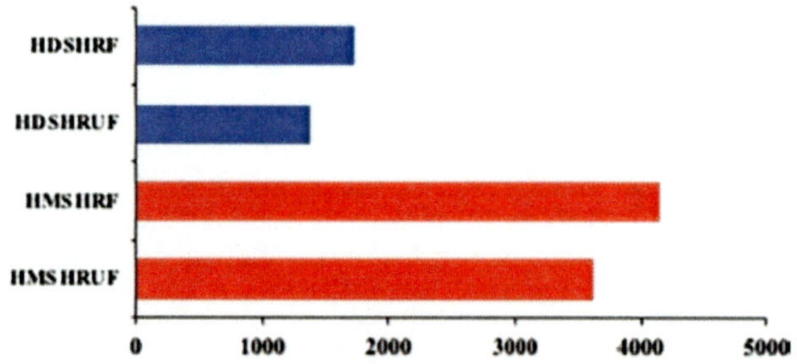

Fig. 1 Average fish yield (kg ha$-^1$) in ponds under hot-moist and hot-dry sub-humid regions treated with and without fertilizers (Bhowmick, 2008)

situated under different agro-ecological regions (AER) having varying soil properties. As discussed by Boyd et al. (2002); Chattopadhyay (2018) and others, the bottom soils of fish ponds are generally developed on the sub-surface soils through gradual transportation and sedimentation of the surface soils along with varying levels of organic matter accumulated largely from un-utilized food organisms, manures, and supplementary feeds. Nature and properties of the bottom soils in the fish ponds under different AERs thus closely reflect those of their upland soils. However, limited information is so far available on this aspect. While carrying out an initial study on productivity of fish ponds located under hot-dry sub-humid (HDSH) region represented by red and lateritic soils in comparison with those under a contrast hot-moist sub-humid (HMSH) region dominated by alluvial soils, the first author observed the yield rates in the fish ponds under HDSH region to be significantly lower in comparison with those in HMSH region (Fig. 1).

t_{tab} was 2.145 at 14 d.f., at 0.005 level of significance while t_{cal} values for unfertilized and fertilized ponds were 56.57 and 63.90 respectively indicating the differences in fish yield levels to be significantly different for the two AERs. To explain this behavior, we undertook a detailed work program with the primary objective of identifying the key soil factors responsible for variations in pond productivity under these two contrast AERs. It has been expected that the results of the study will help in developing some efficient soil specific pond productivity management options for these AERs.

2 Materials and Methods

For this study, soil and water samples were collected from 85 and 110 no of fish ponds of HMSH and HDSH regions of West Bengal, India, which were under culture of composite fish species with varying food habits leading to larger utilization of wide spectrum of food organisms in the pond environment. The HMSH region was largely represented by alluvial soils, while red and lateritic soils were the predominant soil group for the HDSH region. Variations in soil properties in these two AERs emerged from differences in soil forming processes, climatic factors, and presence of parent materials. At least 5 no. of submerged surface soil samples were collected from each pond with the help of a soil sampler and then mixed thoroughly as a composite sample for every pond. These soil samples were processed and analyzed for different properties, viz. soil texture, particle and bulk density, water holding capacity, pore space, pH, electrical conductivity, cation exchange capacity, exchangeable Ca^{2+}, Mg^{2+}, Na^{+}, and K^{+}, easily oxidizable organic C, easily mineralizable N, available P, K, Fe, Cu, Mn, and Zn by following standard methods, as compiled by Chattopadhyay (2022). From each pond under study, surface water samples were collected and analyzed for net primary productivity (NPP) by estimating the dissolved oxygen (DO) contents of pond water at spot using a specific time interval during day time (Chattopadhyay, 2022). The increment in DO values due to photosynthesis was then converted in terms of photosynthetic C by following Odum (1973). The variations in the soil and water properties of the fish ponds under the two agro-ecological regions were assessed statistically by using the test for equality of two population means (Sahu, 2007). In this study, the tabulated T value (the value of $T_{\alpha/2}$) was calculated to be 1.96 at $\alpha = 0.05$. In this test, if the calculated T value is greater than the tabulated T value, we can reject the null hypothesis and accept the alternative hypothesis, and the values differ significantly. In addition, the pond soil properties were correlated with the dependent NPP of pond water samples. The identified parameters were further subjected to multiple regression analysis to find the relative contributions of different soil properties to NPP of pond water and, thereby, the fish pond productivity of the two AERs under study.

3 Results

General physical properties of the pond soils from the two AERs have been presented in Table 1. The HMSH region showed significantly higher silt and clay contents in the pond soils over those in HDSH AER. Textural grouping indicated the pond soils of HMSH region to be largely under clayey category and those under HDSH region exhibited mostly loamy texture, dominated by sandy clay loam group. Bulk density, particle density, and water holding capacity values were also significantly higher in the pond soils under HMSH region.

The electro-chemical properties of the pond soils (Table 2) indicated the bottom soils under HMSH AER to show

Table 1 Some physical properties of the pond soils under two agro-ecological regions

Properties	Hot-moist sub-humid region		Hot-dry sub-humid region		
	Range	Average	Range	Average	*T* value
Sand (%)	10.2–58.9	32.3	48.3–90.4	63.8	– 17.40
Silt (%)	8.4–53.8	23.0	2.7–22.8	15.5	5.37
Clay (%)	12.3–54.2	45.7	2.5–23.4	20.7	19.00
Bulk density (g cc^{-1})	0.87–1.34	1.2	0.88–1.10	1.0	13.70
Particle density (g cc–^{1})	2.31–2.72	2.60	2.10–2.32	2.22	29.03
Water holding capacity (%)	49.0–77.1	62.0	40.2–47.6	45.7	19.94
Pore space (%)	21.0–35.7	29.7	38.7–44.9	41.6	– 26.51

T_{tab} value for the study was 1.96. *T* values above the tab value indicate significant difference

higher pH and electrical conductivity (EC) values over those in HDSH region (Table 2). Cation exchange capacity (CEC) and concentrations of exchangeable Ca^{++} and Mg^{++} were also more in such soils. However, amount of exchangeable K^{+} and Na^{+} values, which were observed to be comparatively lesser in occurrence, were statistically at par in the pond soils under both of the regions.

The availability of the major nutrients in the pond soils, e.g., organic C, N, P, and K were in higher order in the pond soils of HMSH region (Table 3). Among the four micronutrients, availability of Cu and Zn were found to be higher in HMSH region. However, Fe and Mn, which showed larger occurrence under both the AERs, did not show significant variation between the two regions.

Analyses of the water samples revealed the net primary productivity (NPP) values of the fish ponds situated in the HMSH region to be significantly higher than those under HDSH region indicating the natural productivity of the fish ponds to be much higher under HMSH region (Table 4).

Table 2 Electro-chemical properties of the pond soils under two agro-ecological regions

Pond soil properties	Hot-moist sub-humid region		Hot-dry sub-humid region		
	Range	Average	Range	Average	*T* value
pH	5.9–8.0	7.1	5.2–7.4	6.2	10.52
Electrical conductivity (dSm^{-1})	0.7–5.0	2.0	0.2–2.4	1.2	5.89
Cation exchange capacity (Cmol (P^{+}) kg^{-1})	6.1–20.3	16.8	4.0–18.6	12.8	6.98
Exchangeable Ca^{++} (Cmol (P^{+}) kg–^{1})	2.1–8.5	6.6	1.8–8.6	4.1	9.51
Exchangeable Mg^{++} (Cmol (P^{+}) kg^{-1})	1.9–5.6	4.3	1.0–6.6	3.2	5.82
Exchangeable K^{+} (Cmol (P^{+}) kg–^{1})	0.5–2.5	0.8	0.1–1.8	0.8	0.00
Exchangeable Na^{+} (Cmol (P^{+}) kg^{-1})	0.6–3.6	1.2	0.4–3.1	1.0	1.76

T_{tab} value for the study was 1.96. *T* values above the tab value indicate significant difference

Table 3 Availability of plant nutrients in pond soils under two agro-ecological regions

Properties	Hot-moist sub-humid region		Hot-dry sub-humid region		
	Range	Average	Range	Average	*T* value
Easily oxidizable organic C (%)	0.2–2.2	1.1	0.1–2.0	0.8	3.86
Easily mineralizable N (mg kg–^{1})	84.0–205.0	162.3	61.0–165.0	141.0	4.74
Available P (mg kg^{-1})	7.8–22.7	15.3	2.5–12.9	9.1	12.13
Available K (mg kg–^{1})	42.0–134.0	109.0	37.0–119.0	84.0	07.21
DTPA extractable Cu (mg kg^{-1})	1.3–6.7	3.9	0.9–2.9	1.9	12.34
DTPA extractable Fe (mg kg–^{1})	11.4–81.4	29.5	15.5–90.6	34.7	– 01.80
DTPA extractable Mn (mg kg^{-1})	10.4–87.6	27.6	9.8–95.8	47.6	– 06.15
DTPA extractable Zn (mg kg–^{1})	0.9–10.7	6.8	0.8–8.5	4.9	05.42

T_{tab} value for the study was 1.96. *T* values above the tab value indicate significant difference

Table 4 Net primary productivity values of pond water under two agro-ecological regions

Net primary productivity (mg C m^{-3} h$-^{1}$)	Hot-moist sub-humid	Hot-dry sub-humid
Range	352.1–1024.8	163.0–507.1
Average	704.1	342.0
T value	14	

T_{tab} value for the study was 1.96. *T* value above the tab value indicates significant difference

4 Discussion

The pond soils under HMSH AER appeared to be heavier than those under HDSH region owing to dominance of finer textured particles in the alluvial soils of HMSH region. This property, in conjunction with higher bulk density values, resulted in increased water holding capacity of these soils helping them to retain larger volume of water in the fish ponds of HMSH region. This behavior also tended to restrict leaching of plant nutrients from such pond environment, as compared to the ponds under HDSH region.

General electro-chemical properties of the pond soils have been presented in Table 2. Importance of neutral to slightly alkaline pH levels of water and soil in maintaining the pond productivity has long been emphasized due to better availability of plant nutrients and also occurrence of congenial conditions for the physiological activities of the living components in the ponds under such pH range (Boyd, 1995, Chattopadhyay, 2018). In this study, pH values of the pond soils under HMSH region remained mostly under neutral to alkaline range with a few samples showing acidic pH values. On the other hand, the pond soils under HDSH region were largely under acidic range which is not likely to support good aquaculture. Occurrence of acidic pH in pond soils of red and lateritic soils of HDSH region has been discussed by Chattopadhyay and Banerjee (2005) and has been attributed as a major productivity limiting constraint for such ponds. EC, which represents the amount of soluble salts in the system, showed comparatively higher values in the ponds of HMSH region. However, excepting very marginal cases, EC values of pond soils under both the AERs did not appear to be too high or low to influence the pond productivity. CEC values of the pond soils also occurred in comparatively higher ranges in the pond soils of HMSH region. The relevance of CEC for several other soil properties like nutrient retention, lime requirement, use efficiency of added fertilizers, etc., are well known (Mandal & Chattopadhyay, 1990). Such variation in CEC values in the pond soils under the two contrast AERs may, therefore, be considered to be an important indicator of their productivity potentials. Higher concentration of exchangeable bases was observed in the pond soils of HMSH region with Ca and Mg being the significantly dominant bases. Such increased base saturation in the HMSH pond soils was reflected in their neutral to alkaline pH values.

Conversion of a substantial amount of un-utilized fish food organisms to soil organic matter has been discussed by Delince (1992). In this study also, larger production of fish food organisms in the pond system under HMSH region (Table 4) resulted in increased organic C content in the pond soils (Table 3). On the other hand, occurrence of both primary fish food organisms and soil organic C were of lower order in the ponds under HDSH region.

Importance of different plant nutrients in supporting the primary productivity and, thereby, the growth of fishes has been discussed by Boyd (2018). In this study, availability of three major nutrients, viz. N, P, and K, maintained higher status of these nutrients under HMSH region, as compared to those under HDSH region (Table 4). Several factors like heavy texture, high CEC, congenial pH range, and higher amount of organic matter contributed to better availability of these nutrients in these pond soils and to induce higher productivity in the fish ponds under HMSH region over those under HDSH region. On the other hand, availability of P was substantially low in the pond soils under HDSH region. This is likely to be due to greater fixation of P with Fe and Al under the prevailing acidic soil conditions of such ponds (Chattopadhyay & Banerjee, 2005). High concentration of soluble Fe and predominance of acidity in the bottom soils under HDSH region have been observed in this study also. Among the micronutrients studied, availability of Cu and Zn were higher in the pond soils of HMSH region while Fe and Mn occurred in high concentrations in both the AERs. Inherent properties of red and lateritic soils under HDSH region in maintaining higher concentrations of Fe and Mn are well known.

NPP denotes the net amount of organic matter available to the fishes through photosynthesis after meeting the demand of C for respiration of the food organisms (Delince, 1992). Occurrence of more congenial environment in the pond soils of HMSH region resulted in significantly higher NPP values in such pond water over those in HDSH AER. Since NPP generally exhibits a close relationship with fish growth (Boyd, 1984), the higher NPP values in the pond water of HMSH region indicated better productivity potential of these ponds over those under HDSH region.

Regression analysis of the results showed available soil P to be the most important soil factor determining the NPP values in both the AERs and was followed by soil pH (Table 5). Available N, which was in low order in the pond soils under both the agro-ecological regions, emerged as another important contributor to primary productivity of these fish ponds. In addition, clay content and organic C status, which were low in the pond soils under HDSH region, also appeared to influence the productivity of the

Table 5 Regression equations for different soil properties with relation to NPP values

Equation No.	Hot-moist sub-humid region	R^2
1	NPP = 293.27 + 2.46(Av.P)	0.42
2	NPP = 174.21 + 1.98(Av.P) + 13.91(pH)	0.72
3	NPP = 193.08 + 2.52(Av.P) + 14.08(pH) + 0.31(Av.N)	0.83
	Hot-dry sub-humid region	R^2
1	NPP = 322.91 + 3.80(Av.P)	0.37
2	NPP = 286.59 + 2.57(Av.P) + 1.73(pH) + 0.31	0.60
3	NPP = 205.42 + 12.60(Av.P) + 2.53(pH) + 26.81(Av.N)	0.72
4	NPP = 282.62 + 2.38(Av.P) + 1.45(pH) + 23.28(Av.N) + 16.83(clay)	0.76
5	NPP = 175.71 + 2.43(Av.P) + 1.31(pH) + 10.98(Av.N) + 17.99(clay) + 16.51(O.C.)	0.81

NPP—Net primary productivity of water (mg C m^{-3} $h-^{1}$). Av. N, P, and K indicate available status of respective nutrient (mg kg^{-1}). O.C. and clay contents are in percent

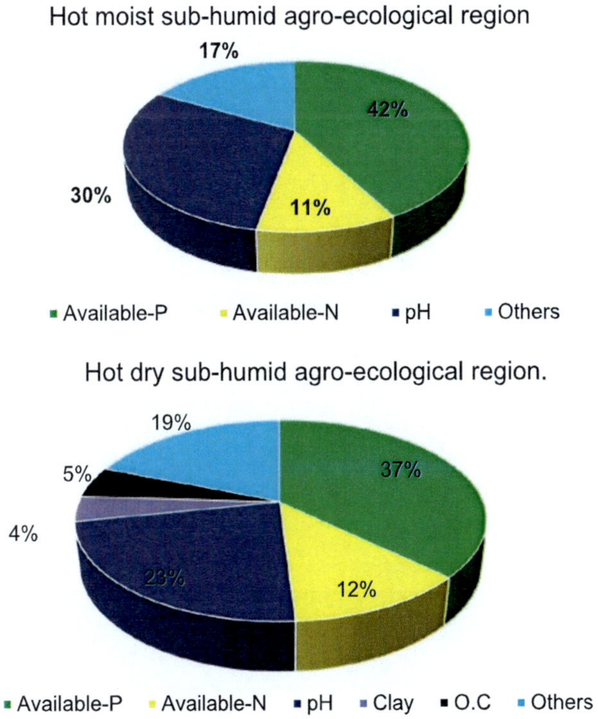

Fig. 2 Percent contributions of different soil properties on NPP of pond water under two AERs

fish ponds of HDSH AER significantly. Together, availability of P, pH, and available N of the pond soils contributed to 83 and 72% variation in pond productivity under HMSH and HDSH regions, respectively (Fig. 2). Further inclusion of clay and O.C. values with these properties helped to increase the predictability in variation of NPP of the fish ponds of HDSH region to 81%.

5 Conclusion

This study on identification of soil factors influencing the productivity of the fishponds under two contrast agro-ecological regions showed available P, pH, and available N status to be the key drivers for the NPP of pond water under both the hot-moist and hot-dry sub-humid regions. For the hot-dry region, in addition to these three soil properties, organic C and clay content emerged as two other soil factors which showed significant influence on pond productivity. Among all these properties, available P was observed to be the most important soil factor for both the agro-ecological regions and was followed by soil pH and available N, respectively. Together, these three soil properties could explain 83 and 72% of variability in net primary productivity of the fish ponds under hot-moist and hot-dry region, respectively (Fig. 2). The results suggest that for better productivity management for the fish ponds under these two agro-ecological regions, specific management of these key soil properties needs to be given greater attention.

Acknowledgements The authors are thankful to Indian Council of Agricultural Research for supporting the study. They also acknowledge the helps from Dr. Ramyajit Mondal and Ms. Moupiya Roy during the preparation and revision of the manuscript.

References

Banerjee, A., Chattopadhyay, G. N., & Boyd, C. E. (2010). Soil system-based approach: A tool for fish pond fertilization. *Better Crops Plant Food, 94*(1), 22–24.

Bhowmick, A. K. (2008). *Utilization of soil resources for assessment of pond productivity under two contrasting agro-ecological regions of West Bengal* (Ph.D. thesis). Visva-Bharati University.

Boyd, C. E. (1984). *Water quality in warm water fish ponds* (p. 369). Auburn University.

Boyd, C. E. (1995). *Bottom soils, sediment and pond aquaculture* (pp. 366). Chapman and Hall.

Boyd, C. E. (2018). Aquaculture pond fertilization. *CAB Reviews, 13*, 002. https://doi.org/10.1079/pavsnnr201813002

Boyd, C. E., Wood, C. W., & Thujai, T. (2002). *Aquaculture pond bottom soil management*. https://www.researchgate.net/publication/242592208

Chattopadhyay, G. N. (2018). Soil and water quality management for sustainable aquaculture. In S. D. Tripathi, W. S. Lakra, & N. S. Chadha (Eds.), *Aquaculture in India* (pp. 355–389). Narendra Publishing House.

Chattopadhyay, G. N. (2022). *Chemical analysis of fish pond soil and water* (p. 132). Daya Publishing House.

Chattopadhyay, G. N., & Banerjee, A. (2005). Soil system-based productivity management of fish ponds in red and lateritic soil zones. *World Aquaculture, 36*(2), 57–61.

Delince, G. (1992). *The ecology of the fish pond ecosystem* (p. 1230). Kluwer Academic Publishers.

Hickling, C. F. (1971). *Fish culture* (p. 225). Faber and Faber.

Mandal, L. N., & Chattopadhyay, G. N. (1990). Chemical and electrochemical environment in fish ponds and its impact on aquaculture. In G. K. Manna & B. B. Jana (Eds.), *Impact of environment on animals and aquaculture* (pp. 79–84).

Odum, E. P. (1973). *Fundamentals of ecology*. W.B. Saunders Company.

Ouali, N., Belabed, B. E., & Chenchouni, H. (2018). Modelling environment contamination with heavy metals in flathead grey mullet *Mugil cephalus* and upper sediments from north African coasts of the Mediterranean Sea. *Science of the Total Environment, 639*, 156–174. https://doi.org/10.1016/j.scitotenv.2018.04.377

Sahu, P. K. (2007). *Agriculture and applied statistics-1* (pp. 185–270). Kalyani Publishers.

Vass, K. K., Chattopadhyay, G. N., & Jana, B. B. (2017). Soil, fishery and aquaculture. In H. Pathak, S. K. Sanyal, & P. N. Takkar (Eds.), *State of Indian agriculture—Soil* (pp. 165–182). National Academy of Agricultural Sciences.

Compliance with WHO and FAO Standards for Treated Water from the Nouaceur and Mediouna's Wastewater Treatment Plants for Reuse in Watering and Irrigation

Meryem Zarri, Samah Ait benichou, Abdelilah Fahde, Fouad Amraoui and Mohamed Tahiri

Abstract

For a decade, the preservation of water resources has become a major environmental concern in Morocco, through the National Sanitation Program which aims to collect and treat wastewater efficiently by using valuable wastewater treatment plants (WWTP) for the reuse of cleaned wastewater in agricultural irrigation, urban watering, and recovery of rivers. The acute drought that has hit Morocco and the Mediterranean basin in the last five decades has pushed the Moroccan government to issue a new law on water saving and the reuse of purified water: law 10–95. To be reused, the treated water must comply with the standards imposed by the law 10–95 and the WHO. The Casablanca region with 4 million inhabitants evacuates more than 350,000 m^3 of wastewater into the sanitary network. Only 100,000 pass through wastewater treatment plants installed in the peri-urban area. The cities of Casablanca and Mohammedia use an outlet that discharges 250,000 m^3 into the Atlantic Ocean without extensive treatment. The cities of Mediouna and the airport area have wastewater treatment plants based on membrane reactor technologies, and activated sludge, respectively, generate purified water with different purification yields. In this context, we studied the conformity of cleaned water in wastewater treatment plants in the peri-urban environment of Casablanca using the WHO and FAO standards for irrigation waters by analyzing different physicochemical and bacteriological parameters. Our partnership with 'Lyonnaise des Eaux of Casablanca' allowed us to take regular samples from two main WWTPs, surrounding Casablanca City (Mediouna and Nouaceur airport zone) and make a quality monitoring and comparative study of treatment degree of the two WWTPs. The follow-up interests: Temperature, pH, electrical conductivity (EC), chemical oxygen demand (COD), biochemical oxygen demand (BOD5), NH_4^+, NO_3^-, NKT, PO_4^{3-}, and fecal coliforms. The physicochemical and bacteriological analyses of samples taken during the second half of the year 2021 showed that at the level of the station of Mediouna the registered abatement rate reached 90.9% for COD, 99.25% for BOD5, 99.1% for total suspended solids (TSS), 95.59% for Total Kjeldahl Nitrogen, and 80.20% for orthophosphate. For the airport zone, the registered abatement rate reached 90.2% for COD, 98.41% for BOD5, 99.02% for TSS, 80.07% for orthophosphate, and 83% for Total Kjeldahl Nitrogen. At the same time, the level of physicochemical quality (pH and EC) and hygienic quality for the two WWTPs are found to comply with WHO and the FAO intended for watering and irrigation.

Keywords

Treated wastewater · Physicochemical parameters · Bacteriological parameters · Reuse of cleaned wastewater · Agriculture irrigation

M. Zarri (✉) · M. Tahiri
Laboratory of Organic Synthesis, Extraction and Valorization, Ain Chock's Sciences Faculty, Hassan II University of Casablanca, Casablanca, Morocco
e-mail: zarrimeryem30@gmail.com

F. Amraoui
Laboratory of Geosciences Applied to Development Engineering, Ain Chock's Sciences Faculty, Hassan II University of Casablanca, Casablanca, Morocco

S. A. benichou · A. Fahde
Laboratoire de Biologie et d'écologie Animales, Ain Chock's Sciences Faculty, Hassan II University of Casablanca, Casablanca, Morocco

H. Chenchouni et al. (eds.), *Recent Advancements from Aquifers to Skies in Hydrogeology, Geoecology, and Atmospheric Sciences*, Advances in Science, Technology & Innovation, https://doi.org/10.1007/978-3-031-47079-0_26

1 Introduction

Agriculture, as the largest water consumer globally, confronts varied challenges in water quality and availability (Shortle et al., 2021). Research efforts are focused on addressing these issues, including the potential use of treated wastewater for irrigation. The wastewater treatment plants (WWTP) in urban and peri-urban areas provide opportunities for the valorization of treated water resources (Chenchouni et al., 2023; Moussaoui et al., 2024). In Morocco, agriculture is the largest consumer of water resources. These resources, depending on the regions from which they come, and their possible contact with pollution sources, have very diverse characteristics. Moreover, given the decrease in water supply observed over the last decades, farmers, especially in continental regions, are interested in using wastewater. Thus, water quality standards for irrigation have been established to protect them (Jin & Young, 2001; Belghyti et al., 2009). Many research efforts are directed towards addressing these problems. The wastewater treatment plants in the peri-urban area of Casablanca treat a very large volume of wastewater from the city of Mediouna and the Nouaceur airport area. These stations produce treated water (Rassam et al., 2012; Arisily & Hajji, 2020). The present work is part of the valorization of treated water from wastewater treatment plants in the peri-urban area of Casablanca.

2 Materials and Methods

2.1 Description of the Study Site

Mediouna located in the northwestern part of the region and east of the province of Casablanca. It currently has a population of about 171,822 inhabitants, with 30.2% of the population living in rural areas. The area of the province is 234 km^2, divided into 2 urban and 3 rural communes. For wastewater treatment plant of Mediouna has been programmed as part of the Lydec liquid sanitation scheme in Casablanca (Lyonnaise des Eaux de Casablanca (LYDEC), 2010). Launched in April 2013, the WWTP has a capacity of 40,000 equivalent inhabitants, with a future extension, by 2030, of 80,000 equivalent inhabitants. The daily treatment capacity of the plant is 2800 m^3 which can reach, in the future around 2017, 3800 m^3 with a peak flow of 300 m^3/h (Lyonnaise des Eaux de Casablanca (LYDEC), 2013). This WWTP (Lyonnaise des Eaux de Casablanca (LYDEC), 2011) adopts a compact system of intensive treatment of RBM type which allows a drastic elimination of suspended solids and a disinfection/hygienization of the treated water which can be reused in agricultural irrigation. The treatment process implemented starts with an automatic 6 mm screen followed by a desanding/degreasing or preliminary clarification (PC) with oil and grease removal. Nouaceur is a commune in the Casablanca-Settat region of Morocco. Its population was 12,696 inhabitants. The Mohammed-V airport of Casablanca is located there, and it is part of the province of Nouaceur, which bears the same name. Since 2010, ONDA has launched the construction of its new WWTP to meet the wastewater treatment needs of the entire area of the airport Mohammed V/Nouaceur until 2020. This activated sludge plant, commissioned in August 2013, has two treatment streams (Moutaib et al., 2014, 2005). One for domestic wastewater with the possibility of reusing the treated water and the other for industrial wastewater with a prospect of reusing the treated water. At the same time, the ONDA has started to encourage and oblige the industrialists of the aeropole to treat their wastewater before discharging it into the sewerage system (ONDA, 2007). Both cities have a population growth rate of about 10% per year; one of the highest rates in the country (general national census of 2014). They concentrate a large number of economic housing projects, as well as tourism projects.

2.2 Materials and Methods

All analyzes were carried out according to the standardized methods (Rodier et al., 2016, and Afnor) at the Laboratory 'Organic Synthesis, Extraction and Valorization'. During our study cycle, in-situ measurements and samples were taken monthly during the period from December 2021 to August 2022. The in-situ measurements of temperature, pH, and electrical conductivity were performed using a portable multi-parameter (Consort, type C 535). For the physicochemical analyses, the water samples were taken in the field in polyethylene bottles and transported cold (4 °C) in a cooler to the laboratory. The physicochemical analyses of the water were carried out on the parameters listed in the table below (Table 1) according to the standards recommended by AFNOR or those approved by Rodier et al. (2016).

The present study examined treated water collected from Mediouna (M) and Nouaceur (N) stations located in the Casablanca-Settat region of Morocco. The samples taken in December 2021 until August 2022 were stored in the laboratory in a cool, dry, and dark place.

3 Results

3.1 Physicochemical Characterization of Treated Water

Table 2 presents the results of some parameters of the complete characterization of the treated water issue from Mediouna and Nouaceur wastewater treatment plants.

Table 1 Measuring equipment and analytical methods used

Parameters	Methods	Unit	Reference
total suspended solids (TSS)	Filtration method on Whatman GFC filter (0.45 μm)	mg/L	Reference 1825-047
Chemical oxygen demand (COD)	Oxidation method with potassium dichromate on LCK kits and reading with DR 5000 spectrophotometer	mg/L	NM 03.7.054/2012
Biochemical oxygen demand (BOD5)	Manometric method with Oxitop VELP Scientifica	mg/L	NM ISO 5815-1/2012
Orthophosphates	Ammonium molybdate method	mg/L	NM ISO 10304-2/2009
Fecal coliforms	Filtration on membrane and culture on agar medium	(1/1000 mL)	NF EN ISO 9308-1

Table 2 Results of physicochemical and bacteriological analyses of treated water

WWTPs	Parameter	Treated water				
		Minimum	Average	Maximum	WHO	FAO
Mediouna	T (°C)	19.8	23 ± 1	27.0	<35	0–30
	pH	6.5	7.0 ± 0.1	8.3	6.5–8.4	6.5–8.4
	EC (μS/cm)	2670	3300	4538	<7680	>3000
	COD (mg/L)	22.6	42.0 ± 0.1	59	250	37.83–61.33
	BOD5 (mg/L)	0.6	3.0 ± 0.1	6.7	120	6–13.8
	TSS (mg/L)	1.7	3.0 ± 0.2	5.9	No standard	2.45–5.72
	Orthophosphates (mg/L)	7.8	9.0 ± 0.1	18.4	6.4–30.64	2.59–13.83
	Fecal coliforms (1/100mL)	433	600	841	1000	No standard
Nouaceur	T (°C)	20.8	21 ± 1	27.4	30	0–30
	pH	7.3	7.0 ± 0.1	8.4	6.5–8.4	6.5–8.4
	EC (μS/cm)	2918	2900	3322	7680	>3000
	COD (mg/L)	16	29.0 ± 0.1	41	250	37.83–61.33
	BOD5 (mg/L)	0.2	2.0 ± 0.1	5.1	120	6–13.8
	TSS (mg/L)	2.1	3.0 ± 0.2	4.4	No standard	2.45–5.72
	Orthophosphates (mg/L)	6.25	10.0 ± 0.1	13.8	6.4–30.64	2.59–13.83
	Fecal coliforms (1/100mL)	200	500	700	1000	No standard

WHO World Health Organization; *FAO* Food and Agriculture Organization

4 Discussion

For both WWTPs, the temperature, pH, EC, and TSS in the treated water are well below the limit recommended by WHO and FAO. The low TSS load of the WWTPs is due to the treatment systems adopted by the WWTPs which eliminate almost all the TSS. COD is an indication of the rate of elimination of oxidizable matter in the treated waters. They are of good quality and largely in conformity with WHO and FAO and can even be reused in agricultural irrigation. For BOD5 the great performance in terms of organic matter removal of our plant is due to the good performance of the bacterial biomass and the growth of certain bacteria specialized in the assimilation and mineralization of the specific organic matter in aerobic conditions (Amori, 2022). Even if the load of orthophosphates in treated water seems low compared to the needs of market gardening, for example, the regular supply of these elements by treated water could accelerate the growth of crops and thus ensure good agricultural yields, which makes the reuse of water in irrigation very promising (Boudiabi et al., 2023). The bacteriological pollution of treated water is evaluated through the research and the enumeration of fecal coliforms whose presence in high number indicates a contamination of fecal origin and probably a presence of pathogenic microorganisms (Guemmaz et al., 2020). The preliminary results of the physicochemical characterization show that the majority of these parameters respect WHO and FAO standards for treated water, which confirms that we can reuse treated water in watering and irrigation. At the end of this evaluation of the degree of physicochemical pollution, we conclude that all the parameters studied, in particular the parameters of organic pollution (BOD5, COD, and TSS) place the waters of the cities of Mediouna, Nouaceur, in the range of concentration less important to low according to Metcalf & Eddy et al. (2014). This is related to the dilution of organic matter (soluble or suspended) due to the high-water consumption per capita.

5 Conclusion

Most of the values of the physicochemical analyses of the treated water are perfectly in line with the Moroccan quality grid for water intended for irrigation in terms of pH, temperature, electrical conductivity, TSS, etc. This good physicochemical and microbial quality of the treated wastewater is a testimony to the efficiency of the systems adopted by the plants, which protects the receiving environment and allows the treated water to be reused for irrigation of the region's agricultural land. This reuse may contribute to reduce, in the future, the pressure of pumping water from the local groundwater.

References

Amori, P. N. (2022). Reclamation and agricultural reuse of mixed industrial and domestic wastewater treated by aerobic biological process and ozonation. Doctoral thesis, University of Surrey. https://doi.org/10.15126/thesis.900338

Arisily, T., & Hajji, A. (2020). Technologie de séchage solaire des boues des stations d'épuration des eaux usées et son impact sur la gestion des boues au Maroc. *Revue Marocaine des Sciences Agronomiques et Vétérinaires*, *8*(1), 57–66.

Belghyti, D., El Guamri, Y., Ztit, G., et al. (2009). Caractérisation physico-chimique des eaux usées d'abattoir en vue de la mise en œuvre d'un traitement adéquat: cas de Kénitra au Maroc. *Afrique Science*, *5*(2), 61730. https://doi.org/10.4314/afsci.v5i2.61730

Boudiabi, S., Ababsa, N., & Chenchouni, H. (2023). Sewage and sewage treatment. In R. Brinkmann (Ed.), The Palgrave handbook ok global sustainability (pp. 719–745). Palgrave Macmillan, Cham.

Chenchouni, H., Chaminé, H.I., Zhang, Z., et al. (2023). *Recent Research on Hydrogeology, Geoecology and Atmospheric Sciences*. Springer, Cham. https://doi.org/10.1007/978-3-031-43169-2

Guemmaz, F., Neffar, S., & Chenchouni, H. (2020). Physicochemical and bacteriological quality of surface water re-sources receiving common wastewater effluents in drylands of Algeria. In: Negm, A., et al. (Eds). *Water Resources in Algeria-Part II: Water Quality, Treatment, Protection and Development* (pp 117–148). Springer, Cham. https://doi.org/10.1007/698_2019_400

Jin, L., & Young, W. (2001). Water use in agriculture in China: importance, challenges, and implications for policy. *Water Policy*, *3*(3), 215–228. https://doi.org/10.1016/S1366-7017(01)00015-0

Lyonnaise des Eaux de Casablanca (LYDEC). (2010). *EIE-STEP Mediouna* 128.

Lyonnaise des Eaux de Casablanca (LYDEC). (2011). *Document YT - Suez Environnement – Dégréement* 18.

Lyonnaise des Eaux de Casablanca (LYDEC). (2013). *Rapport interne* 86.

Metcalf & Eddy, Tchobanoglous, G., Stensel, H.D., Tsuchihashi, R., Burton, F., Abu-Orf, M., Bowden, G., & Pfrang, W. (2014). *Wastewater Engineering: Treatment and Resource Recovery*. McGraw Hill Education: New York, NY, USA.

Moutaib, Z., Hebabaze, S., Brand, C., & Chlaida, M. (2014). Wastewater of the airport Mohammed V platform (Casablanca, Morocco): physico-chemical characterization and treatment with activated sludge. *International Journal of Innovation and Applied Studies*, *9*(2), 725-733.

Moussaoui, M., Ababsa, N., Bougoufa, H., et al. (2024). Climatic drivers of wastewater treatment efficiency of wastewater treatment facilities. In: Çiner, A., Barbieri, M., Khan, M.F., et al. (Eds). *Recent Research on Environmental Earth Sciences, Geomorphology, Soil Science and Paleoenvironments*. Springer, Cham. https://doi.org/10.1007/978-3-031-48754-5_28

ONDA. (2007). Réalisation d'une nouvelle station de traitement des eaux usées de la zone aéroportuaire de Nouasser/Aéroport MOHAMMED V, mémoire de dimensionement (45 p).

Rassam, A., Chaouch, A., Bourkhiss, B., & Bourkhiss, M. B. (2012). Performances de la dégradation de la matière organique par lagunage aéré dans la station d'épuration des eaux usées de la ville d'Oujda (Maroc oriental). *Bulletin de la Société Royale des Sciences de Liège*, *81*, 121–125. https://popups.uliege.be/0037-9565/index.php?id=3892

Rodier, J., Legube, B., & Merlet, N. (2016). *L'analyse de l'eau* (10e ed). Dunod, Paris.

Shortle, J., Ollikainen, M., & Iho, A. (2021). *Water Quality and Agriculture: Economics and Policy for Nonpoint Source Water Pollution*. Palgrave Macmillan, Cham. https://doi.org/10.1007/978-3-030-47087-6

Analysis of Perchlorate in Aquatic Food and Evaluation of Human Exposure

Mohamed Mahmud El-Mounja, Antonietta Rizzo, Chiara Telloli, Elena Marrocchino and Carmela Vaccaro

Abstract

Chemical contaminants have become a concern in terms of food security, as they immediately affect the quality of food (food safety) and can affect the availability and nutritional access to that resource (food security) over a long period. For this reason, the objective of this research work is to analyze perchlorate in aquatic food, an anthropogenic contaminant could affect human health. Samples were analyzed using micro-Raman spectroscopy. Analyzes of mussel samples collected in the Goro lagoon show the presence of perchlorate, but fortunately in very short concentration.

Keywords

Perchlorate · Aquatic food · Raman spectroscopy

1 Introduction

To ensure food security for a specific population, it is crucial to protect the environment and the natural habits of animals and plants that contribute to the food chain (Meemken & Qaim, 2018; Ouali et al., 2018). Within this dynamic system, the water quality is important to monitor and secure the water quality (Bhagwat, 2019; Belhouchet et al., 2024). Given the widespread occurrence of chemical contaminants in foodstuffs and their serious health risks (Almoselhy, 2021), the prevention of food contamination is a public health priority. In recent years a subset of harmful organic chemicals, mostly of anthropogenic origin, for example, perchlorate, has been of concern (Weber et al., 2019). Perchlorate (ClO_4-) is an environmental pollutant (Cao et al., 2019) that is growing due to its occurrence in water and its health effects to thyroid uptake of iodine and production of hormones (Lisco et al., 2020). There are geochemical conditions that make high solubility of ClO_4^- in the aquatic environment (Pleus & Corey, 2018). Perchloric acid and most perchlorate salts readily dissolve in water, generating the perchlorate anion (ClO_4-). Although a strong oxidizing agent, the perchlorate anion is persistent in the environment, due to the high activation energy associated with its (abiotic) reduction to chlorate (ClO_3^-). In addition, due to its relatively low charge density, it does not form complexes with metals like other anions. For this reason, its ionic state does not readily sorb to environmental media. This combination of solubility, stability, and mobility create the potential for both localized and area-wide impacts of toxicological interest (Urbansky, 2002). The focus of this work is to analyze different kinds of mussels and shell samples collected in the Adriatic Sea (northeast of Italy) to identify the presence of perchlorate using Raman spectroscopy (Kuruvinashetti et al., 2020; Luo et al., 2023).

2 Materials and Methods

Different kinds of mussels and shell samples were collected in the Goro lagoon (Adriatic Sea, northeast of Italy), and after having been cleaned, they have been observed with micro-Raman spectroscopy, as a rapid screening tool for qualitative analysis.

Raman spectrum of both mussel and shell samples was obtained by BWTEK i-Raman plus spectrometer equipped with a 785 nm laser at the Traceability Laboratory of ENEA

M. M. El-Mounja
ARSN—Mauritanian National Authority of Radiation Protection, Nuclear Security, and Safety, Nouakchott, Mauritania

A. Rizzo · C. Telloli
ENEA, FSN-SICNUC-TNMT, Bologna, Italy

E. Marrocchino (✉) · C. Vaccaro
Department of Environmental and Prevention Sciences, University of Ferrara, Ferrara, Italy
e-mail: mrrlne@unife.it

H. Chenchouni et al. (eds.), *Recent Advancements from Aquifers to Skies in Hydrogeology, Geoecology, and Atmospheric Sciences*, Advances in Science, Technology & Innovation, https://doi.org/10.1007/978-3-031-47079-0_27

in Bologna (Emilia-Romagna Region, Italy). The measurement parameters, acquisition time, the number of repetitions, and laser energy, have been selected for each sample to maximize the signal-to-noise ratio. For each spectrum, a reference acquisition with the same parameters was previously carried out to subtract the instrumental background.

3 Results

First of all, with micro-Raman spectroscopy, standard perchlorate is observed to better know the position of the peak of perchlorate in the spectrum, which is at 928 cm$-^1$, using a standard reference material certified.

Then, knowing the position of the perchlorate peak, mussel and shell samples were observed. Figure 1a shows the spectrum of the mussel in which no presence of perchlorate was observed, or if present it has a very low concentration and is below the detection limit of the instrument. On the opposite in the shell samples, a perchlorate peak is present. Both sides of the shell were analyzed: the inner (Fig. 2b) and outer shell (Fig. 2c). The difference between the two sides is very low, but in both, perchlorate peak was visible: around 3% for the inner shell and around 6.25% for the outer shell.

4 Discussions

Comparing the recent shell of the mussels collected for this research work and the shell lived and collected in the past period, it is clear the changing of the perchlorate.

Figure 2 shows two different kinds of samples analyzed at micro-Raman spectroscopy: Fig. 2a shell samples coming from Terra del Fuego in Argentina (Pacific Ocean) and dated by the carbon 14 method to 6500 BC (Iron age); Fig. 2b shell samples from Tunisia (Mediterranean area) and dated by the carbon 14 method to 5000 BC. Both figures show lower values of perchlorate present in shells, with respect to the recent shell, as shown in Fig. 1. The peaks are very low, but there are around 1.5% both for mussels form Terra del Fuego and from Tunisia. The decreasing levels

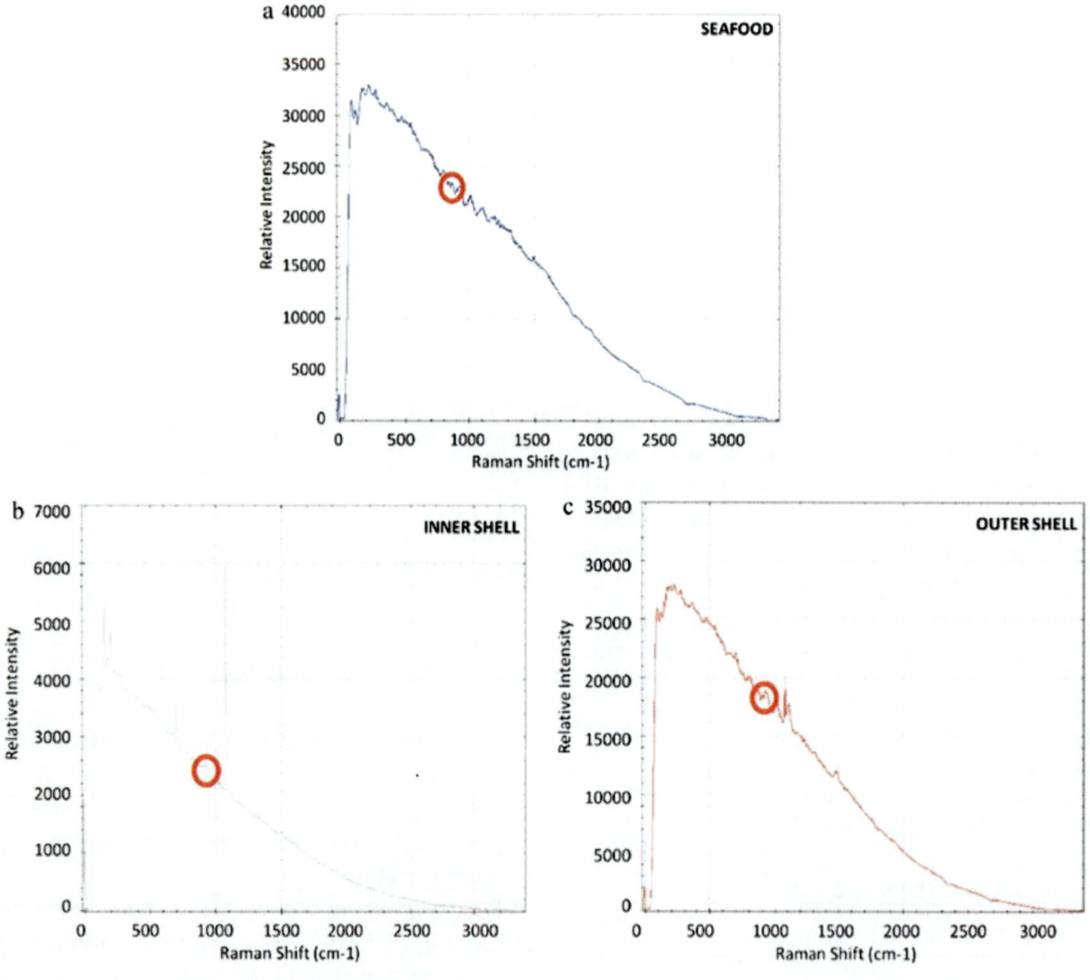

Fig. 1 Raman spectra of **a** mussel; **b** inner shell; **c** outer shell

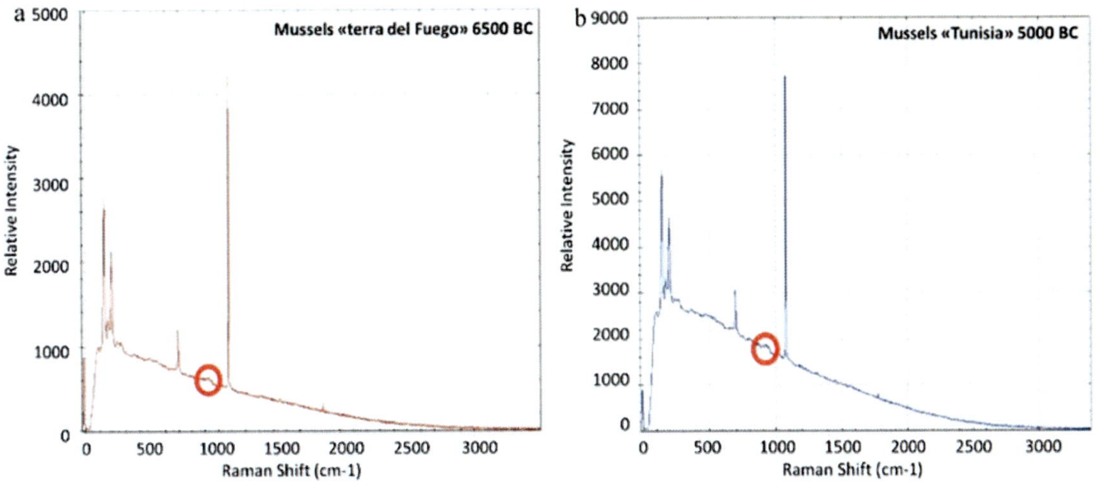

Fig. 2 Raman spectra of **a** mussels sampled in the Terra del Fuego (Argentina, 6500 BC); **b** mussels sampled in Tunisia (5000 BC)

of perchlorate observed in ancient shell samples compared to recent ones suggest a notable change in environmental contamination over time. This highlights the importance of ongoing efforts to monitor and mitigate chemical pollutants in aquatic ecosystems, emphasizing the need for proactive measures to safeguard water quality and public health (Pleus & Corey, 2018; Zieritz et al., 2022).

5 Conclusion

Micro-Raman spectroscopy was used a rapid screening tool for qualitative analysis of perchlorate. Analyzes of mussel samples collected in the Goro lagoon do not show a high presence of perchlorate, but comparing this kind of samples with samples collected and lived in the past period (6500 and 5000 BC), it is clear that the perchlorate has increased in recent years. This indicates that anthropogenic watering also affects the food security of aquatic species.

References

Almoselhy, R. I. M. (2021). A review on health risks from processing contaminants in edible oils and foodstuffs. *International Journal of Advanced Technology and Science Research.*

Belhouchet, N., Inal, A., Nait-Mohand, H. (2024). Assessment of water-sediment-biota pollution in coastal ecosystems of Algeria. *Regional Studies in Marine Science, 70*, 103355.

Bhagwat, V. R. (2019). Safety of water used in food production. In *Food safety and human health* (Chap. 9, pp. 219–247).

Cao, F., Jaunat, J., Sturchio, N., Cancès, B., Morvan, X., Devos, A., Barbin, V., & Ollivier, P. (2019). Worldwide occurrence and origin of perchlorate ion in waters: A review. *Science of the Total Environment, 661*, 737–749.

Kuruvinashetti, K., Zhang, Y., Li, J., & Kornienko, N. (2020). Shell isolated nanoparticle enhanced Raman spectroscopy for renewable energy electrocatalysis. *New Journal of Chemistry, 44*, 19953.

Lisco, G., De Tullio, A., Giagulli, V. A., De Pergola, G., & Triggiani, V. (2020). Interference on iodine uptake and human thyroid function by perchlorate-contaminated water and food. *Nutrients, 12*, 1669.

Luo, Y., Naidu, R., & Fang, C. (2023). Raman imaging towards in-situ visualisation of perchlorate adsorption. *Water Research, 229*, 119510.

Meemken, E. M., & Qaim, M. (2018). Organic agriculture, food security, and the environment. *Annual Review of Resource Economics, 10*, 39–63.

Ouali, N., Belabed, B. E., & Chenchouni, H. (2018). Modelling environment contamination with heavy metals in flathead grey mullet *Mugil cephalus* and upper sediments from north African coasts of the Mediterranean Sea. *Science of The Total Environment, 639*, 156–174.

Pleus, R. C., & Corey, L. M. (2018). Environmental exposure to perchlorate: A review of toxicology and human health. *Toxicology and Applied Pharmacology, 358*, 102–109.

Urbansky, E. T. (2002). Perchlorate as an environmental contaminant. *Environmental Science and Pollution Research, 9*, 187–192.

Weber, R., Bell, L., Watson, A., Petrlik, J., Paun, M. C., & Vijgen, J. (2019). Assessment of pops contaminated sites and the need for stringent soil standards for food safety for the protection of human health. *Environmental Pollution, 249*, 703–715.

Zieritz, A., Sousa, R., Aldridge, D. C. (2022). A global synthesis of ecosystem services provided and disrupted by freshwater bivalve molluscs. *Biological Reviews, 97*(5), 1967–1998.

Degraded Arid Soil Reclamation for Cotton Cultivation Using Organic Waste Amendments

Reginald Kogbara, Helmi Hamdi, Ali Al-Sharshani, Osman Abdalla, Udeogu Onwusogh and Sabah Solim

Abstract

Qatar is one of the most fertilizer-dependent countries due to challenging soil and climatic conditions. The country strives toward self-sufficiency in agricultural production in alignment with the Qatar National Vision 2030. Hence, this work investigates the potential of utilizing nutrient-rich resources that are currently wasted for the reclamation of degraded arid soils to support the cultivation of industrial crops such as cotton (*Gossypium* spp.). Two abundant organic wastes, industrial biosludge and cow dung compost, were employed as soil amendments at a 3% application rate on a silty loam soil with relatively high salinity (electrical conductivity = 5.60 dS/m) and compared with conventional chemical fertilization. Cotton (May 344 variety) was then grown on the biowaste-amended soils in lysimeters for ten months (March through January) spanning through the hot season in Qatar, with the average temperature ranging from 19 to 37 °C. Soil properties and plant growth characteristics, including soil metal concentrations, days to germination and flowering, plant height, and cotton yield, were determined at set periods. The results indicated that different from the chemical fertilizer treatment, the organic amendments led to a significant release of potassium eight months after planting, roughly twice the concentration available at the initial sampling period. In all treatments, soil magnesium and iron concentrations generally increased, while phosphorus and zinc decreased over time. There was generally no significant difference in the concentrations of metals analyzed such as chromium, copper, nickel, and zinc between soils amended with the organic wastes and chemical fertilizer. The concentrations of metals were below the regulatory limits for sewage sludge applied to soils. The days to germination were 2, 9, and 11, while the days to flowering were 61, 92, and 77 for the cow dung compost, biosludge, and fertilizer treatments, respectively. The average cumulative plant heights were 74, 65, and 63 cm, while the average cumulative cotton boll yield was 7.3, 5.4, and 2.6 tons/ha, respectively, in the cow dung compost, biosludge, and fertilizer treatments. The results demonstrate that the organic amendments, especially cow dung compost, can help reclamation of degraded/saline arid soils under the described pedo-climatic conditions.

Keywords

Cotton · Cow dung compost · Industrial biosludge · Organic waste amendments · Plant growth parameters

1 Introduction

Qatar is one of the most fertilizer-dependent countries (World Bank, 2016) due to challenging soil and climatic conditions. Hence, as the country strives toward self-sufficiency in agricultural production in alignment with the Qatar National Vision 2030, using biosludge and other organic wastes as soil enhancers presents a unique opportunity to utilize a

R. Kogbara (✉)
Texas A&M University at Qatar, Education City, P.O. Box 23874, Doha, Qatar
e-mail: regkogbara@cantab.net

Izmir Institute of Technology, Gulbahce, Urla, 35430 Izmir, Turkey

H. Hamdi
Center for Sustainable Development, Qatar University, Doha, Qatar

A. Al-Sharshani · U. Onwusogh · S. Solim
Qatar Shell Research & Technology Center, QSTP LLC, Doha, Qatar

O. Abdalla
Department of Agricultural Research, Ministry of Municipality & Environment, Doha, Qatar

H. Chenchouni et al. (eds.), *Recent Advancements from Aquifers to Skies in Hydrogeology, Geoecology, and Atmospheric Sciences*, Advances in Science, Technology & Innovation, https://doi.org/10.1007/978-3-031-47079-0_28

nutrient-rich wasted resource. Thus, this work investigates the potential of utilizing two abundant organic wastes rich in nutrients, industrial biosludge and cow dung compost, for the reclamation of degraded arid soils to support the cultivation of industrial crops such as cotton (*Gossypium* spp.). Cotton was chosen among other industrial crops as it thrives in warm and humid climates so could survive in Qatar's climate.

Organic wastes benefit crops by efficiently using organic nitrogen and phosphorus, which are released slowly throughout the growing season, allowing for better nutrient absorption (Boudjabi & Chenchouni, 2021; Boudjabi et al., 2019; Kogbara et al., 2020a, 2020b). However, using these biowastes may accumulate toxic metals and other contaminants in soils depending on their origins (Kogbara et al., 2020b). There is a paucity of literature on similar studies as cotton is not usually grown in Qatar, which has a hyper-arid desert climate. Several studies have shown that the ideal temperature for cotton growth and development ranges from 20 to 32 °C (Reddy et al., 1992). Hence, this study aimed to investigate the use of organic waste amendments to reclaim degraded arid soil for cotton production under Qatar's prolonged hot climatic conditions and to evaluate changes in soil properties and the effect on cotton performance.

2 Materials and Methods

2.1 Lysimeters and Soil

The experiments were conducted in 9 field lysimeters (3 replicates per treatment), each with 9 m^2 surface area (3 m by 3 m) and a depth of 1 m. The lysimeters were filled with soil of silty loam texture (9% gravel, 27% sand, 54% silt, and 10% clay). Table 1 shows key properties of the degraded saline soil. The soil had an in situ electrical conductivity of 5.60 mS/cm determined using an EC 1210 soil EC meter (*Nieuwkoop BV*, Netherlands), and 0.68% NaCl content.

2.2 Soil Amendments

Three nutrient sources were compared for fertilization of the soil, namely industrial biosludge, cow dung compost and an inorganic fertilizer. The properties of the biosludge and cow dung compost used can be found elsewhere (Khater, 2015; Kogbara et al., 2020a). The biosludge and cow dung compost was applied at 3% dosage of the soil mass in a given lysimeter (i.e., 115 ton/ha). The inorganic fertilizer was applied using chemical fertilizers to supply the required levels of N, P, and K for cotton. Specifically, urea and K_2SO_4 were applied at 200 kg/ha divided into three equal doses to supply 92 and 86 kg/ha of N and K, respectively, while 150 kg/ha of triple super phosphate supplied 30 kg/ha P.

2.3 Sowing, Irrigation, and Analytical Methods

About 45 cotton seeds (May 344 variety) were sown in the nutrient-amended soils. These were grown for 10 months (March through January) spanning through the long hot and dry season in Qatar. The lysimeters were irrigated at set periods, on average 9 days per month for about 6 min with a bubbler flow rate of 4.75 L/min. Soil properties and plant growth characteristics were determined using methods documented in previous related publications (Epa, 2018; Goodman & Dale Monks, 2003; Rayment & Lyons, 2011).

2.4 Data Analysis

This involved simple statistics such as mean and standard deviation. Analysis of variance (ANOVA) was used to test significant differences at the 5% level.

3 Results and Discussion

The soil and plant performance in the three treatments are shown in Table 2 and Fig. 1. Table 2 compares the elemental concentration of the soils at the initial stage (before seed sowing) and at the final growth stage (8 months after planting). Figure 1 shows plant yield and height over the 10-month period, and days to germination and flowering.

The results show that different from the inorganic fertilizer treatment, the organic amendments led to a significant release of K eight months after planting, roughly twice the concentration available at the initial stage, and K is critical

Table 1 Key physicochemical properties of the soil used for cotton cultivation

C/N ratio	pH	NaCl (%)	Element content (mg/kg)							
			Cl^-	Al	Cr	Cu	Fe	Mn	Ni	K
113:1	7.56	0.68	4100	69,643	51.6	12.6	32,225	298	61.9	8666
			Ca	Mg	Na	Zn	Cd	S	B	Pb
			212,083	566.8	6735	40.0	<0.5	1598	58.5	2.8

Table 2 Average key soil elemental concentrations over an 8-month period

Treatment	Elemental concentration[a] (mg/kg)									
	Initial	Final	Initial	Final	Initial	Final	Initial	Final	Initial	Final
	Total nitrogen		Total phosphorus		Potassium		Magnesium		Iron	
BS	300	500	2793	1164	4292	11,355	39,807	96,371	35,419	75,179
CDC	335	503	830	681	3578	7427	34,147	58,886	24,081	111,314
Fertilizer	310	155	739	708	6643	7121	47,028	59,761	32,052	76,937
	Calcium ($\times 10^3$)		Chromium		Copper		Nickel		Zinc	
BS	231	368	48	121	61	0.2	77	81	224	102
CDC	183	232	93	137	43	34	100	78	136	62
Fertilizer	191	212	117	137	34	0.2	79	78	111	57

BS: biosludge, CDC: cow dung compost, fertilizer: inorganic fertilizer
[a]Concentrations were determined for 0–30 cm soil depth

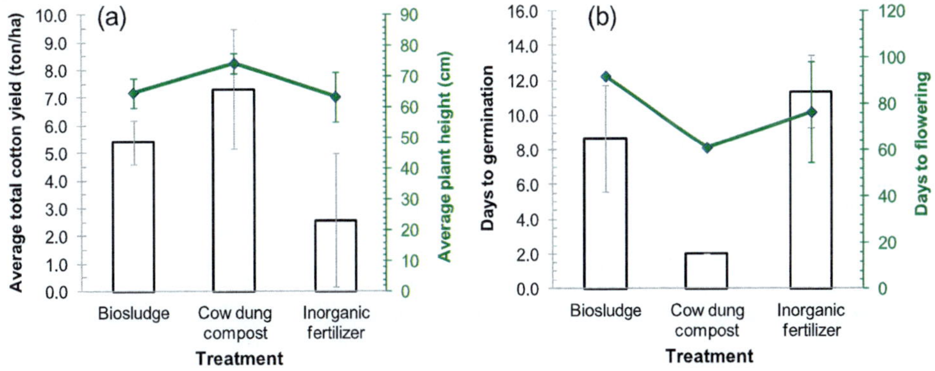

Fig. 1 Plant performance in the treatments in terms of **a** average cotton yield and plant height, and **b** number of days to germination and flowering, over the 10-month growth period. *Note* The columns represent the average total cotton yield on the primary *y*-axis, the dots represent the average plant height on the secondary *y*-axis, while the error bars represent the standard deviation of three replicates

for cotton growth. This corroborates the findings of a previous study in which the application of organic manure significantly increased K release in the soil in contrast to inorganic fertilizer. This was attributed to the dissolution of some K-containing soil minerals through organic acids resulting from decomposition of the organic amendments (Bader et al., 2021). The total nitrogen concentration in the soil increased between the initial and final stages for the treatments with organic amendments but decreased for the inorganic fertilizer treatment. This confirms observations made in previous studies on the quick loss of nitrogen from inorganic fertilizers (possibly through leaching and denitrification) from the soil compared to its slow release by organic amendments such as biosludge and compost (Ayotamuno et al., 2009; Kogbara et al., 2020a, 2020b).

In all treatments, soil Ca, Mg, and Fe concentrations generally increased, while P and Zn decreased over time. Similar increases in Ca, Mg, and Fe concentrations have been recorded in soils amended with organic and inorganic fertilizers (Amjadian et al., 2021). The decreases in P and Zn, however, differ from the work of Amjadian et al. (Amjadian et al., 2021) as different types of fertilizers and soils were considered in both studies. There was generally no significant difference in the concentrations of metals analyzed such as Cr, Cu, Ni, and Zn between soils amended with the organic wastes and inorganic fertilizer at the 5% probability level. This differs from observations in a previous study in which farmyard manure showed significantly higher concentrations of Cr, Cu, and Zn than NPK fertilizer (Singh et al., 2010). The concentrations of the metals were below the regulatory limits (< 300 for Cr, Ni and Zn, and < 100 mg/kg for Cu) for sewage sludge applied to soils (Boudjabi & Chenchouni, 2023; GSO, 1997).

The average cumulative cotton yield was 7.3, 5.4, and 2.6 tons/ha, while the average cumulative plant heights were 74, 65, and 63 cm, respectively, in the cow dung compost, biosludge, and fertilizer treatments. The days to germination were 2, 9, and 11, while the days to flowering were 61, 92, and 77 for the cow dung compost, biosludge, and fertilizer treatments, respectively. These results differ from a recent study in which mineral fertilizer gave superior cotton growth and yield compared to cow-dung-based compost on loamy sand soils (Tovihoudji, et al., 2022). Thus, the observations here show that the organic amendments thrive better for the specific (saline silt loam) soil type considered.

4 Conclusion

The performance of the treatments in reclaiming the degraded soil for cotton production in Qatar's hot climate was in the order, cow dung compost > biosludge > inorganic fertilizer. The results show that the organic amendments, especially cow dung compost, can better supply needed nutrients than inorganic fertilizer and help reclamation of degraded/saline arid soils under the described pedo-climatic conditions.

Acknowledgements Funding provided through the NPRP project (NPRP12S-0205-190045: Industrial crop production in Qatar with biosolids and treated sewage effluent and industrial water) from the Qatar National Research Fund (QNRF—a member of The Qatar Foundation) is gratefully acknowledged. The statements made herein are solely the responsibility of the authors.

References

Amjadian, E., Zeinodini, A., & Doğan, H. (2021). Effect of fertilizer management systems on growth and balance of nutrients in wheat cultivation. *Central Asian Journal of Plant Science Innovation, 1*(2), 56–69. https://doi.org/10.22034/CAJPSI.2021.02.01

Ayotamuno, J. M., Kogbara, R. B., & Agoro, O. S. (2009). Biostimulation supplemented with phytoremediation in the reclamation of a petroleum contaminated soil. *World Journal of Microbiology & Biotechnology, 25*(9), 1567–1572. https://doi.org/10.1007/s11274-009-0045-z

Bader, B. R., et al. (2021). Potassium availability in soil amended with organic matter and phosphorous fertiliser under water stress during maize (*Zea mays* L.) growth. *Journal of the Saudi Society of Agricultural Sciences, 20*(6), 390–394. https://doi.org/10.1016/j.jssas.2021.04.006

Boudjabi, S., & Chenchouni, H. (2021). On the sustainability of land applications of sewage sludge: How to apply the sewage biosolid in order to improve soil fertility and increase crop yield? *Chemosphere, 282*, 131122. https://doi.org/10.1016/j.chemosphere.2021.131122

Boudjabi, S., & Chenchouni, H. (2023). Comparative effectiveness of exogenous organic amendments on soil fertility, growth, photosynthesis and heavy metal accumulation in cereal crops. *Heliyon, 9*(4), e14615. https://doi.org/10.1016/j.heliyon.2023.e14615

Boudjabi, S., Kribaa, M., & Chenchouni, H. (2019). Sewage sludge fertilization alleviates drought stress and improves physiological adaptation and yield performances in Durum Wheat (*Triticum durum*): A double-edged sword. *Journal of King Saud University—Science, 31*, 336–344. https://doi.org/10.1016/j.jksus.2017.12.012

Goodman, B., & Dale Monks, C. (2003). Farm demonstrations method for estimating cotton yield in the field for use by extension agents and specialists. *Journal of Extension, 41*(6). https://archives.joe.org/joe/2003december/iw3.php

GSO. (1997). GSO 901 Standard: Municipal solid waste compost. GCC Standardization Organization (GSO), Riyadh, Saudi Arabia.

Khater, E.-S.G. (2015). Some physical and chemical properties of compost. *International Journal of Waste Resources, 5*(1), 72–79.

Kogbara, R. B., et al. (2020a). Effect of gas-to-liquid biosludge on soil properties and alfalfa yields in an arid soil. *Journal of Cleaner Production, 250*, 119524. https://doi.org/10.1016/j.jclepro.2019.119524

Kogbara, R. B., et al. (2020b). Recycling industrial biosludge for buffel grass production in Qatar: Impact on soil, leachate and plant characteristics. *Chemosphere, 247*, 125886. https://doi.org/10.1016/j.chemosphere.2020.125886

Rayment, G. E., & Lyons, D. J. (2011). *Soil chemical methods—Australasia*. CSIRO Publishing.

Reddy, K. R., Reddy, V. R., & Hodges, H. F. (1992). Temperature effects on early season cotton growth and development. *Agronomy Journal, 84*(2), 229–237. https://doi.org/10.2134/agronj1992.00021962008400020021x

Singh, A., Agrawal, M., & Marshall, F. M. (2010). The role of organic vs. inorganic fertilizers in reducing phytoavailability of heavy metals in a wastewater-irrigated area. *Ecological Engineering, 36*(12), 1733–1740. https://doi.org/10.1016/j.ecoleng.2010.07.021

Tovihoudji, G. P., et al. (2022). Profitability and agronomic potential of cotton (*Gossypium hirsutum* L.) under biochar-compost-based amendments in three agroecological zones of northern Benin. *Frontiers in Sustainable Food Systems*, 492, https://doi.org/10.3389/fsufs.2022.1036133

US EPA. (2018). Test method 6010D (SW-846): Inductively coupled plasma-optical emission spectrometry (ICP-OES), Revision 5. United States Environmental Protection Agency (US EPA), Washington.

World Bank. (2016). *World development indicators 2016—3.2: Agricultural inputs*. World Bank. https://doi.org/10.1596/978-1-4648-0683-4

Effects of Date Palm Residues Derived Biochar on GHG Emissions and NO_3-N Leaching in Urea-Fertilized Desert Soil

Khaled Alotaibi, Saud Aloud, Hattan Alharbi and Abdullah Al-Modaihsh

Abstract

Biochar (BC) addition to soil has received a growing interest worldwide due to its numerous agricultural and environmental benefits, including nutrient retention in soil and reduction of greenhouse gas emissions. The aim of this study was to evaluate the impact of date palm residue (DPR) and its BC on greenhouse gases (CO_2 and N_2O) emissions and NO_3-N leaching following the addition of urea fertilizer to a desert soil. The BC was produced at four temperatures: 300, 400, 500 and 600 °C (BC300, BC400, BC500 and BC600) and mixed with the soil, followed by urea addition and incubated for 40 days. The CO_2 and N_2O were measured periodically on day 1, 3, 6, 10, 15, 20, 25, 30, 35 and 40, whereas NO_3-N in leachate was measured weekly up to five weeks. Cumulative CO_2 emissions were the greatest in DPR followed by BC300 treatment, whereas the BC600 treatment exhibited the lowest CO_2 emissions. The urea alone treatment showed the highest N_2O emission, but this was significantly reduced with any other treatment, with the BC300 treatment providing the most pronounced effect on N_2O reduction. The DPR followed by BC300 and BC400 remained the most effective treatments in reducing the amount of leached NO_3-N during the whole period of incubation.

Keywords

Biochar · Carbon dioxide · Nitrous oxide · Nitrate leaching · Desert soil

1 Introduction

Date palm trees generate huge amount of residues, with its majority being disposed of by incineration, and this can have adverse impacts on the environment. However, a better benefit of date palm residues (DPR) may be achieved by converting these materials to biochar (BC), having important characteristics that can increase nutrient retention abilities and therefore improve nutrient use efficiency. Biochar has shown enormous agricultural and environmental benefits (Cayuela et al., 2014; Gul et al., 2015; Spokas et al., 2012; Xie et al., 2015; Boudjabi et al., 2023). However, the magnitude effects of BC on agricultural and environmental systems largely depend on many factors including biochar- and soil-related properties (Ababsa et al., 2023; Tag et al., 2016; Tomczyk et al., 2020). Addition of BC with fertilizers may stimulate GHGs emissions or increase nitrate levels in water derived from soil leachates, leading to environmental degradation. This study aimed at evaluating the impact of DPR and their biochar on greenhouse gases (CO_2 and N_2O) emissions and NO_3-N leaching following the addition of urea fertilizer to a desert soil.

2 Materials and Methods

Two experiments were designed to meet the objectives of the study. The first experiment was set up to study the impact of BC addition on CO_2 and N_2O emissions. In this experiment, 50 g air-dried soil was weighed into a 100-mL plastic vial and mixed with one rate (50 t ha$-^1$) of each BC (BC300, BC400, BC500 and BC600) in addition to feedstock (FS). Urea fertilizer was applied at 200 kg N ha^{-1}.

K. Alotaibi (✉) · S. Aloud · A. Al-Modaihsh
Department of Soil Science, College of Food and Agriculture Sciences, King Saud University, P.O. Box 2460, Riyadh 11451, Saudi Arabia
e-mail: khalotaibi@ksu.edu.sa

H. Alharbi
Department of Plant Protection, College of Food and Agriculture Sciences, King Saud University, P.O. Box 2460, Riyadh 11451, Saudi Arabia

H. Chenchouni et al. (eds.), *Recent Advancements from Aquifers to Skies in Hydrogeology, Geoecology, and Atmospheric Sciences*, Advances in Science, Technology & Innovation, https://doi.org/10.1007/978-3-031-47079-0_29

A treatment without BC (only urea) was included as a control (Cont). Each treatment was replicated 4 times. The vial containing the soil was placed into a 5-L flask equipped with rubber septa to allow for gas measuring using a Gasmet FTIR gas analyzer. The evolved CO_2 and N_2O were measured at intervals: 1, 3, 6, 10, 15, 20, 25, 30, 35 and 40 days. The second experiment was designed to study the impact of the same treatments on NO_3- leaching. In this experiment, 500 g air-dried soil was mixed with the BC treatments and packed into a PVC column (20 cm height) and leached once a week for a period of 5 weeks. The NO_3^- was measured in the leachate. The PROC MIXED procedure of SAS was used to test the effects of biochar treatments, incubation time and their interaction between CO_2 and N_2O emissions and nitrate content in the leachate. Treatments means were separated using the Student–Newman–Keuls test. The treatment effect was declared significant at $P<0.05$.

3 Results and Discussion

Treatment, time and their interaction had a significant effect on carbon dioxide (CO_2) emissions (Fig. 1). Cumulative CO_2 emissions were the greatest in FS followed by BC300 treatment, whereas the BC600 treatment exhibited the lowest CO_2 emissions (Fig. 1). Similarly, Figueiredo et al. (2019) found that sewage sludge biochar produced at 300 and 400 °C increased CO_2 emissions, whereas biochar produced at 500 °C reduced the emissions. Zimmerman et al. (2011) also found that the CO_2 evolution increased with decreasing pyrolysis temperature. The urea alone treatment showed the highest N_2O emission, but this was significantly reduced with any other treatment, with the BC300 treatment providing the most pronounced effect on N_2O reduction (Fig. 2). This is consistent with a previous study which found reduced N_2O emissions after co-application of fertilizer and biochar (Grutzmacher et al., 2018). The impact of treatments on NO_3-N content in leachate was significant at any week, but the impact was more evident after the first two weeks (Fig. 3). The FS followed by BC300 and BC400 remained the most effective treatments in reducing the amount of leached NO_3-N during the whole period of incubation. This is in line with a recent study where biochar addition to sandy soil decreased NO_3 leaching (Purkaystha et al., 2022). Overall, biochar was able to reduce N_2O emissions and NO_3-N content in leachate compared to the control, but it was not as effective as DPR. This can be related to the higher soluble C in DPR compared to the biochar, especially those produced at higher temperatures (> 500 °C) which is assumed to be more recalcitrant, as shown by its limited effects on CO_2 emissions rates. Long-term studies are required to verify the results of the current study and to further understand the mechanisms by which this specific BC can affect N dynamics under fertilized desert soil conditions.

4 Conclusion

The effects of date palm residues biochar on CO_2 and N_2O emissions and NO_3-N leaching varied according to the production temperature. Lower temperature BCs increased CO_2 emission and reduced NO_3-N leaching, whereas high temperature BC did not reduce NO_3-N content in the leachate and did not stimulate CO_2 emission. BCs produced at lower temperature are expected to have soluble C for microbial utilization and thereby improving microbial activity.

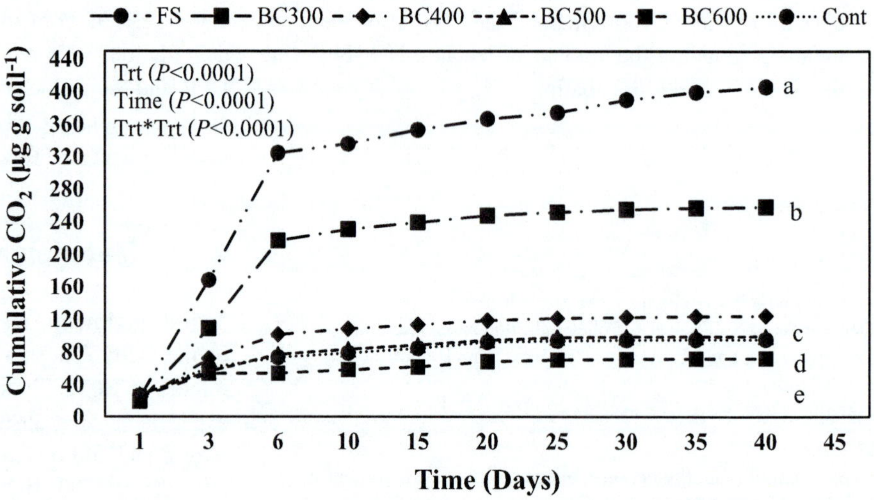

Fig. 1 Effects of BC addition on carbon dioxide (CO_2) emission from urea-fertilized desert soil. Treatments sharing the same letter are not significantly different at $P<0.05$

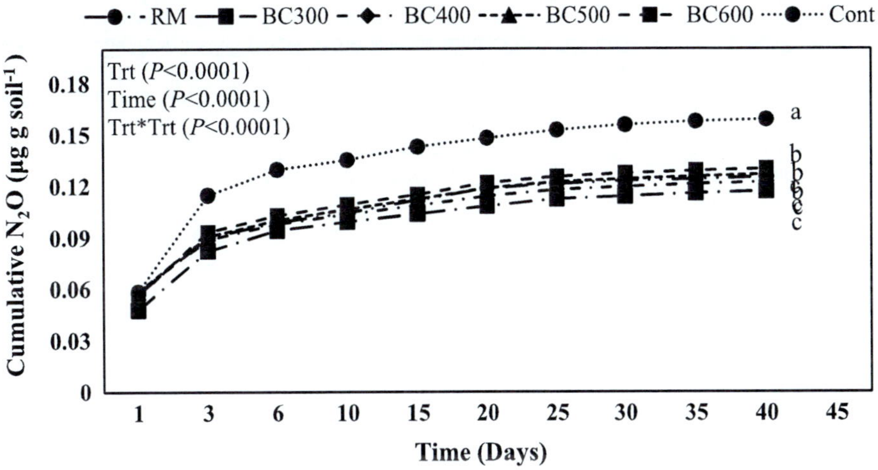

Fig. 2 Effects of BC addition on nitrous oxide (N_2O) emission from urea-fertilized desert soil. Treatments sharing the same letter are not significantly different at $P<0.05$

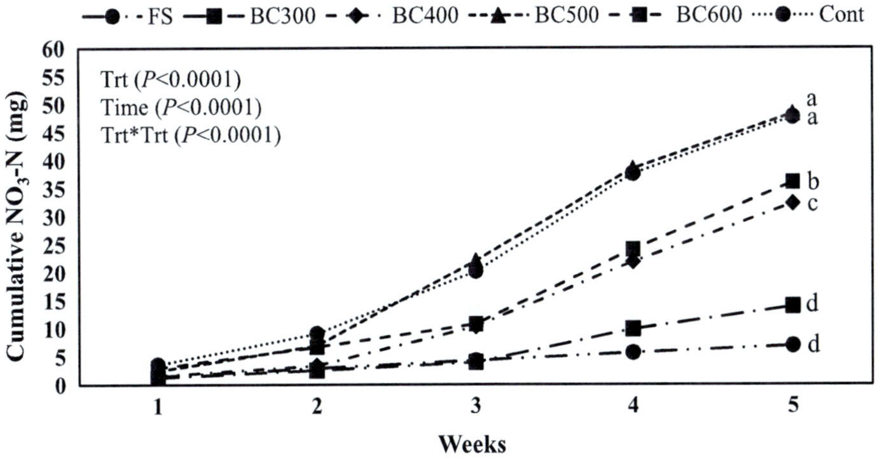

Fig. 3 Effects of BC addition on cumulative NO3-N content in leachate from urea-fertilized desert soil over five weeks of incubation. Treatments sharing the same letter are not significantly different at $P<0.05$

Acknowledgements This project was funded by the National Plan for Science, Technology and Innovation (MAARIFAH), King Abdulaziz City for Science and Technology, Kingdom of Saudi Arabia, Award Number (13-ENV1102-02).

References

Ababsa, N., Boudjabi, S., & Chenchouni, H. (2023). Biochar amendments changed soil properties and improved cereal crop growth under salt stress. *Journal of Soil Science and Plant Nutrition, 23*(4), 4912–4925. https://doi.org/10.1007/s42729-023-01453-7

Boudjabi, S., Ababsa, N., & Chenchouni, H. (2023). Enhancing soil resilience and crop physiology with biochar application for mitigating drought stress in durum wheat (*Triticum durum*). *Heliyon, 9*(12), e22909. https://doi.org/10.1016/j.heliyon.2023.e22909

Cayuela, M. L., Van Zwieten, L., Singh, B. P., Jeffery, S., Roig, A., & Sánchez-Monedero, M. A. (2014). Biochar's role in mitigating soil nitrous oxide emissions: A review and meta-analysis. *Agriculture, Ecosystems & Environment, 191*, 5–16. https://doi.org/10.1016/j.agee.2013.10.009

Figueiredo, C. C. D., Coser, T. R., Moreira, T. N., Leão, T. P., Vale, A. T. D., & Paz-Ferreiro, J. (2019). Carbon mineralization in a soil amended with sewage sludge-derived biochar. *Applied Sciences, 9*(21), 4481. https://doi.org/10.3390/app9214481

Grutzmacher, P., Puga, A. P., Bibar, M. P. S., Coscione, A. R., Packer, A. P., & de Andrade, C. A. (2018). Carbon stability and mitigation of fertilizer induced N_2O emissions in soil amended with biochar. *Science of the Total Environment, 625*, 1459–1466. https://doi.org/10.1016/j.scitotenv.2017.12.196

Gul, S., Whalen, J. K., Thomas, B. W., Sachdeva, V., & Deng, H. (2015). Physico-chemical properties and microbial responses in biochar-amended soils: Mechanisms and future directions. *Agriculture, Ecosystems & Environment, 206*, 46–59. https://doi.org/10.1016/j.agee.2015.03.015

Purkaystha, J., Prasher, S., Afzal, M. T., Nzediegwu, C., & Dhiman, J. (2022). Wheat straw biochar amendment significantly reduces nutrient leaching and increases green pepper yield in a less fertile soil. *Environmental Technology & Innovation, 28*, 102655.

Spokas, K. A., Cantrell, K. B., Novak, J. M., et al. (2012). Biochar: A synthesis of its agronomic impact beyond carbon sequestration. *Journal of Environmental Quality, 41*, 973–989. https://doi.org/10.2134/jeq2011.0069

Tag, A. T., Duman, G., Ucar, S., & Yanik, J. (2016). Effects of feedstock type and pyrolysis temperature on potential applications of biochar. *Journal of Analytical and Applied Pyrolysis, 120*, 200–206.

Tomczyk, A., Sokołowska, Z., & Boguta, P. (2020). Biochar physicochemical properties: Pyrolysis temperature and feedstock kind effects. *Reviews in Environmental Science and Bio/technology, 19*, 191–215.

Xie, T., Sadasivam, B. Y., Reddy, K. R., Wang, C., & Spokas, K. (2015). Review of the effects of biochar amendment on soil properties and carbon sequestration. *Journal of Hazardous, Toxic, and Radioactive Waste, 20*, 1–14.

Zimmerman, A. R., Gao, B., Ahn, M. Y. (2011). Positive and negative carbon mineralization priming effects among a variety of biochar-amended soils. *Soil Biology and Biochemistry, 43*(6):1169–1179. https://doi.org/10.1016/j.soilbio.2011.02.005

The Spatial Structure of Vegetation Cover of Abrau Peninsula (Northwestern Caucasus)

Maxim Bocharnikov, Galina Ogureeva and Elena Suslova

Abstract

The uniqueness of the Abrau peninsula is determined by a high level of floristic and phytocoenotic diversity and the relict nature of the vegetation cover of the shiblyak—broad-leaved type of altitudinal zonality. For this mountain territory, a complex spatial organization is reviled. In this research, a cartographic method for the evaluation of the vegetation cover of a mountain territory has been approved by identifying its spatial structure. The complex analysis of field data (geobotanical descriptions) and remote sensing data (multispectral satellite images, digital elevation model) made it possible to obtain a digital cartographic model of vegetation for the key area using maximum likelihood classification with etalons and discriminant analysis. The compiled large-scale (s. 1:50,000) inventory map of the vegetation reflects the actual vegetation cover, represented by a diversity of hemixerophytic sub-Mediterranean and mesophytic nemoral communities. The change in the basic communities in the altitudinal spectrum determines the identification of two belts. The lower seaside belt of hemixerophytic forests, sparse forests, and shrub communities is expressed at heights of 0–150 (200) m a.s.l. It is characterized by the development of pistachio-juniper (*Juniperus excelsa*, *Pistacia mutica*) forests and sparse forests, oak (*Quercus pubescens*) forests, as well as serial vegetation with shrub and herb-dwarf shrub petrophyte-herb communities. The belt of mesophytic broad-leaved forests is located at heights of 150 (200)–450 m a.s.l. Pine-oak (*Quercus petraea*, *Pinus kochiana*) and broad-leaved (*Carpinus betulus*, *Tilia begoniifolia*, *Acer campestre*) forests dominate in it. The important patterns in the spatial structure of the belt are expressed within the altitudinal belts. The structure of vegetation communities is formed under the conditions of several groups of relief forms: watershed surfaces, slopes, seismic forms, erosion forms, and marine forms. The most valuable results of the investigation are connected with reviling the gradients of phytocoenotic diversity. Along with altitudinal gradient, the position on different levels of catena has significant role in its determination. The identified spatial patterns are important for understanding the formation of Northwestern Caucasus ecosystems as a part of Mediterranean region.

Keywords

Phytocoenotic diversity · Ecosystem · Remote sensing · Mapping · Mountain territories

1 Introduction

The vegetation cover of Abrau peninsula is reflected on a number of small-scale maps. It makes it possible to identify basic vegetation formations. There is also experience in large-scale mapping of vegetation in some territories of the Utrish reserve in the central part of Abrau peninsula: the northern macroslope of the Navagir range (Ogureeva & Bykova, 2013), Lobanova and Topolnaya gorges (Suslova & Rets, 2013). Necessity of understanding of spatial structure of vegetation cover requires expanding the research area and mapping according to the principles of ecological–phytocoenotic classification of vegetation.

The complexity of mountain vegetation as an object of mapping is determined by the diversity of conditions (the influence of bioclimatic altitudinal gradient, dynamic substrate). The spatial structure of the vegetation cover of the mountain area is determined by the diversity of rocks, the influence of the sea, insolation, and the redistribution of

Elena Suslova—Deceased.

M. Bocharnikov (✉) · G. Ogureeva · E. Suslova
Lomonosov Moscow State University, Leninskie Gory, 1, 119991 Moscow, Russia
e-mail: maxim-msu-bg@mail.ru

H. Chenchouni et al. (eds.), *Recent Advancements from Aquifers to Skies in Hydrogeology, Geoecology, and Atmospheric Sciences*, Advances in Science, Technology & Innovation, https://doi.org/10.1007/978-3-031-47079-0_30

temperature and moisture supply at each altitudinal level. Additional differentiation in the vegetation cover is introduced by different disturbances, mainly cutting and fires. It is reflected in the development of short- and long-term-derived vegetation communities. The complexity of the organization of the vegetation cover determines to distinguish heterogeneous structures (mainly of geomorphogenic genesis): phytocatenas, exposure combinations of phytocatenas, slope combinations, petrophytic series.

2 Materials and Methods

The vegetation mapping is based on the integration of field materials and Earth remote sensing data. At the first stage, more than 300 geobotanical descriptions were carried out along a network of ecological and topological profiles on different elements of mountain territories from the watershed of the Navagir range to the Black Sea coast (from 0 to 450 m a.s.l.).

The second stage of the research is related to the classification of vegetation communities according to the identification of the ecological specificity of the identified syntaxonomic units. An analysis of the typological diversity of communities was carried out on the basis of ecological–coenotic and geographic–genetic classifications of vegetation (Neshataev, 1987; Sochava, 1980). Communities were classified according to the predominant bioecogroups.

The third stage of the research is related to selection and analysis of satellite images. The principal analysis scheme is based on relationships between the reflectance values of the described vegetation communities in different spectral bands and vegetation indices (Landsat-8), as well as the values obtained from the digital elevation model (ASTER GDEM), and the position of the communities in the hierarchical structure of typological diversity (Popov, 2018). The spectral information of vegetation communities directly contains data about their composition and structure.

At the fourth stage of the research, the evaluation of the spatial patterns of vegetation diversity was carried out by Shannon and Simpson indices. The last stage represents final inventory vegetation map. The compiled map is the basis for the analysis of vegetation structure of Abrau peninsula.

3 Results

Typological diversity of vegetation of Abrau peninsula contains hemixerophytic sub-Mediterranean formations and mesophytic broadleaf forests. The first include oak (*Quercus pubescens*) forests, pistachio-juniper (*Juniperus excelsa*, *Pistacia mutica*) forests and sparse forests, petrophytic shrub and herb communities. The second contain oak (*Quercus petraea*), linden (*Tilia begoniifolia*), hornbeam (*Carpinus betulus*), maple (*Acer campestre*), ash (*Fraxinus excelsior*), and pine (*Pinus kochiana*) forests (Fig. 1).

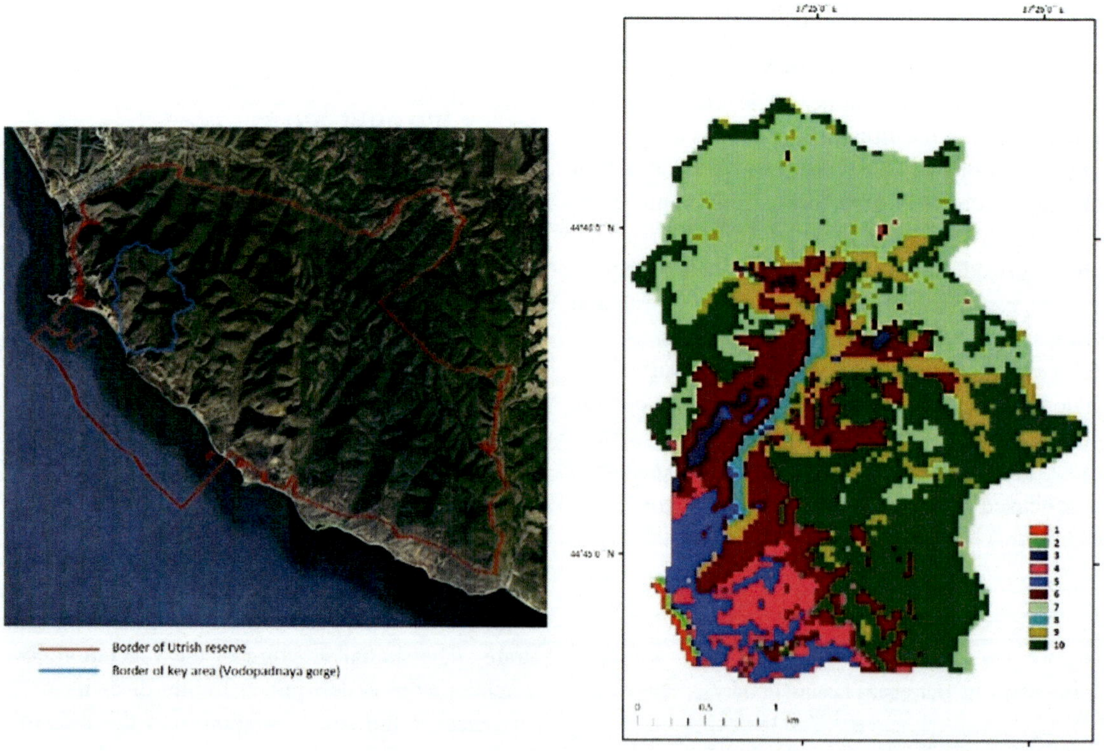

Fig. 1 Cartographic model of vegetation of the key area (Vodopadnaya gorge)

Vegetation subdivisions.

1. Halophytic series of communities.
2. Grass–petrophytic series of communities.
3. Shrub series of communities.
4. Oak (*Q. pubescens*)—pistachio (*P. mutica*)—juniper (*J. excelsa*) sparse forests.
5. Pistachio (*P. mutica*)—juniper (*J. excelsa*) forests.
6. Oak (*Q. pubescens*) hornbeam (*Carpinus orientalis*) forests.
7. Oak (*Q. petraea*) and pine (*P. kochiana*)—oak forests.
8. Hornbeam (*C. betulus*) forests.
9. Hornbeam (*C. betulus*)—maple (*A. campestre*)—linden (*Tilia begoniiflia*) forests.
10. Ash (*F. excelsior*) forests.

The spatial structure of the phytocoenotic diversity of the key area, identified by Shannon and Simpson indices, reveals significant relationships with orographic conditions for absolute altitude and slope exposure. There is a trend for an increase in the diversity of communities with a decreasing in altitude and with the change of slopes from shadow to light exposures. According to the altitudinal gradient, the lower belt has a higher level of diversity than the upper one. The increasing of diversity is expressed from eluvial to accumulative positions of slopes (Table 1).

The surfaces of watersheds are characterized by the lowest level of phytocoenotic diversity with a minimum species richness. The communities in the lower parts of the slopes and bottoms of the valleys are more diverse. Conditions of increased moisture contribute to the development of polydominant broad-leaved forests. They are formed in the bottom of the valleys (the lower altitudinal belt) and in transaccumulative positions (the upper belt). A regular increase in the diversity of communities is expressed from the upper to the lower positions of the catenas.

4 Discussion

The botanical and geographical specificity of Abrau peninsula is determined by the lower belt. It is formed by heterogeneous structures with the participation of communities of hemixerophytic sub-Mediterranean vegetation formations on a relatively small area. It is developed at altitudes of 0–150 (200) m. Along with a high level of biodiversity (Litvinskaya, 2004), it is characterized by a complex spatial structure (Ogureeva et al., 2020).

The belt of mesophytic broad-leaved forests is developed at altitudes of 150 (200)–450 m. It has a lower level of species richness and the complexity of the spatial structure of the vegetation cover. The vegetation is based on homogeneous communities of broad-leaved forests that occupy all forms of relief and catena levels.

Spatial patterns in the distribution of phytocoenotic diversity are associated with a set of conditions, the gradients of which are expressed between different landforms and due to changes in morphometric characteristics. The dynamic status of communities is also important. The closest relationships are found with absolute altitude, slope exposure, as well as in connection with the catenary structure of the territory. It is characterized by the complex spatiotemporal structure of mountain ecosystems (Morley et al., 2019; Qiu et al., 2012).

The major diversity is formed in the lower belt with a heterogeneous typological composition and structure on coastal insolated slopes and seismic relief forms. The change in temperature along the altitudinal gradient is associated with key patterns that determine the differentiation of vegetation between altitudinal belts. Within altitudinal belts, the spatial pattern of diversity is dominated by gradients of two key complex factors. The first factor is also related to the temperature indicators, which cause an increase in the

Table 1 General scheme of phytocoenotic diversity in the structure of mountain territory of Abrau peninsula (a case study in the key area, Vodopadnaya gorge)

		Levels of phytocatena		
		Shadow exposure slopes	Light exposure slopes	
Altitudinal belts	Belt of mesophytic broadleaf forests	Eluvial Transit Accumulative		↓
	Belt of hemixerophytic sub-Mediterranean formations			
		→		Phytocoenotic diversity gradients

diversity on the slopes of light exposures. The development of sparse forest, shrub and herb communities is related to them. The second factor is determined by an increase in moisture in the accumulative positions of catenas. The lower parts of the slopes have a higher level of diversity along the altitudinal spectrum.

5 Conclusion

As a result of the vegetation map compiling, important regularities in the spatial organization of the vegetation cover of Abrau peninsula were determined. The lower belt is expressed at altitudes of 0–150 (200) m. Hemixerophytic sub-Mediterranean forests, sparse forests, and shrub communities form this belt. It is characterized by an increased level of floristic and phytocoenotic diversity, a complex spatial structure associated with the contrast of ecotopes in the conditions of development of different morphostructures. The upper belt is located at altitudes of 150 (200)–450 m. Mesophytic broad-leaved forests form this belt. The most important patterns in its spatial structure are associated with the development of communities at different levels of the catena and at slopes of different exposure and steepness.

References

Litvinskaya, S. A. (2004). *Vegetation of the Black Sea coast of Russia (Mediterranean enclave)*. Krasnodar.

Morley, P. J., Donoghue, D. N. M., Chen, J., & Jump, A. S. (2019). Quantifying structural diversity to better estimate change at mountain forest margins. *Remote Sensing of Environment, 223*, 291–306.

Neshataev, Yu. N. (1987). *Methods of analysis of geobotanical materials*. Nauka.

Ogureeva, G. N., Bocharnikov, M. V., & Suslova, E. G. (2020). Structure of the botanical diversity of the Utrish-Tuapse variant of the Crimean-Novorossiysk Orobiom. *Arid Ecosystems, 10*(4), 261–268.

Ogureeva, G. N., & Bykova, P. A. (2013). Cenotic diversity of vegetation cover of the northern macroscline of the Navagir ridge. *Biodiversity of the Utrish State Nature Reserve, 1*, 77–90.

Popov, S. Yu. (2018). Geobotanical map of the Pinega State Nature Reserve. *Vegetation Mapping*, 3–18.

Qiu, B., Zhong, M., Zeng, C., Tang, Z., & Chen, C. (2012). Effect of topography and accessibility on vegetation dynamic pattern in Mountain-hill Region. *Journal of Mountain Science, 9*, 879–890.

Sochava, V. B. (1980). *Geographical aspects of the Siberian taiga*. Nauka.

Suslova, E. G., & Rets, E. P. (2013). Vegetation of the Abrau Peninsula and its display on a large-scale geobotanical map. *Biodiversity of the Utrish State Nature Reserve, 1*, 67–76.

Influence of Long-Term Mineral Fertilization on Soil Microbiota, Organic Matter Content and CO_2 Emissions

Gergana Kuncheva, Galin Ginchev and Iliana Ivanova

Abstract

The objective of this study is to analyse the effect of different long-term fertilization treatments on soil microbiological and chemical properties and CO_2 emissions on cropped soils. The fertilization treatments were N, P, K, NP, NK, PK, NPK, 0, which have been applied constantly for 64 years on four agricultural crops (wheat, barley, corn and beans) in rotation, in two replicates. The study was conducted on Haplic Chernozem soil type. The highest organic carbon content was observed in the treatments NPK (1.54%) and NP (1.53%), and the lowest in the control (0.97%). On average, of all measurements, CO_2 emissions were largest in NP (96.28 kg ha$-^{1}$ day^{-1}), NPK (93.13 kg ha$-^{1}$ day^{-1}) and PK (79.40 kg ha$-^{1}$ day^{-1}) treatments. Strong positive correlation of mineral nitrogen with electrical conductivity, strong negative correlation with pH and moderate positive with soil organic carbon content were found. The amount of available potassium had the strongest effect on the count of actinomycetes and microbial biomass carbon. There was a positive correlation of CO_2 emissions with soil organic carbon stocks and mineral nitrogen content. Long-term fertilization with NPK and NP led to enhanced accumulation of organic matter and higher available nutritious, in comparison to unfertilized soil.

Keywords

Long-term fertilization experiment · Soil microbiota · Organic matter content · CO_2 emissions

G. Kuncheva (✉)
ISSAPP "Nikola Poushkarov", Bansko Shousse Street 7, 1331 Sofia, Bulgaria
e-mail: g.kuncheva@issapp-pushkarov.org

G. Ginchev · I. Ivanova
Institute of Agriculture and Seed Science "Obraztsov Chiflik", Ruse, Bulgaria

1 Introduction

Many long-term fertilizer experiments were established to provide information on the amounts and forms of fertilizers to use to increase crop yields, of plant nutrition and soil fertility (Johnston & Poulton, 2018). In the current climate change context, another key goal of agriculture is optimizing fertilization management in terms of greenhouse gas emission reduction and improvement of soil carbon sequestration (Boudjabi & Chenchouni, 2022). To achieve climate change mitigation and to ensure food security, it is essential to understand how fertilizing strategies impact the dynamics of soil organic carbon (SOC) (Chenchouni & Neffar, 2022; Xiang et al., 2022). The aim of the present study is to establish the influence of long-term mineral fertilization on the soil chemical and microbiological properties, the SOC storage and the soil CO_2 emissions.

2 Materials and Methods

A long-term fertilization experiment was carried out in the experimental field of IASS Obraztsov Chiflik, region Ruse, Danube plain on highly leached Haplic Chernozem. The experiment was established 64 years ago. The experiment is positioned on area, of 1 ha. The size of the trial plots is 100 m^2, and the harvest plots—60 m^2. The applied methods of fertilization are N, P, K, NP, NK, PK, NPK, 0, and they had been brought to the same places for 64 years on four agricultural crops (wheat, barley, maize and beans), in rotation, in two replicates. Nitrogen rates for wheat and corn were N—150 kg ha$-^{1}$; for beans N—50 kg ha^{-1} and for barley N—100 kg ha$-^{1}$. Phosphorus—120 kg ha^{-1} and potassium—70 kg ha$-^{1}$ were applied for all crops. For comparison, the unfertilized control N0P0K0 is maintained. Phosphorus and potassium fertilizers are applied once, before the main tillage, and nitrogen fertilizers—once after the emergence of the crops, as an early spring feeding.

H. Chenchouni et al. (eds.), *Recent Advancements from Aquifers to Skies in Hydrogeology, Geoecology, and Atmospheric Sciences*, Advances in Science, Technology & Innovation, https://doi.org/10.1007/978-3-031-47079-0_31

The soil type, highly leached Haplic Chernozem, on which the experiment was carried out, is characterized by low organic carbon content—1.04%, mineral N—21.5–23.6 mg kg^{-1}, available forms of P_2O_5—3.6–4.9 mg 100 g$-^{1}$ and available K_2O—24–26 mg 100 g^{-1} in the 0–40 cm layer. The soil texture is Silty Clay Loam (SiCL) with 10% sand (2–0.05 mm), 60% silt (0.05–0.002 mm) and 30% clay (< 0.002 mm), and the soil pH 5.5–6.0. During the spring of 2021, samples were taken for chemical and microbiological analyses from all treatments and two replicates of the experiment, for all crops. CO_2 emissions were measured monthly during the growing season of four crops (6 times). Opaque, plastic chambers, attached to an infrared gas analyser, were used to record soil emissions. Ammonium and nitrate nitrogen were determined according to Kjeldal, and acetate–lactate method for the determination of available phosphorus and potassium forms was used. Microbial biomass carbon (Cmic) measurements was determined by the substrate-induced respiration method (SIR). The number of main groups of soil microorganisms (10^6 CFU g^{-1}) were determined on agar plates: on MPA for saprophytic bacteria, on starch ammonium agar for actinomycetes and on Chapek medium for microscopic fungi. Total soil organic carbon content (SOC, %) was determined by the Tjurin's method by wet digestion.

3 Results

A two-way analysis of variance was conducted, and the results showed that different long-term fertilization treatments had a significant effect on pH, available forms of potassium, soil organic carbon (SOC), microbial biomass carbon. For the rest of the agrochemical and microbiological indicators, the influence of crop is significant. The differences between agrochemical and microbiological parameters between crops were more pronounced than between treatments, because of the different periods of fertilizers application and crops development (Table 1).

The highest electrical conductivity, mineral nitrogen and the lowest pH were reported for the treatments N, NP, NK, NPK (Table 2; Fig. 1). The highest organic carbon content was observed in the treatments NPK (1.54%) and NP (1.53%), and the lowest in the control (0.97%). The long-term fertilization had a significant effect on SOC, as well as on Cmic (Table 1). Cmic is the highest in NK, K, PK treatments.

The highest number of saprophytic bacteria was reported for the treatments N, P, NK, NPK. In the unfertilized control about 6 times lower number of saprophytic bacteria was recorded, compared to these combinations of fertilization. The unfertilized soil had the lowest numbers of all groups of microorganisms, as well as low amounts of available nutritional elements, EC and microbial biomass carbon. CO_2 emissions were increased in NP (96.28 kg ha^{-1} day$-^{1}$), NPK (93.13 kg ha^{-1} day$-^{1}$) and PK (79.40 kg ha^{-1} day$-^{1}$) treatments.

4 Discussion

Compared to natural ecosystems, SOC stocks in agrarian ecosystems are more complex and dynamic, due to often intensive and repeated disturbances associated with management practices such as fertilizer applications (Gai et al., 2018). The impact of N fertilization on organic matter stocks is controversial. Synthetic N fertilizer may reduce SOC stocks because it enhances organic carbon mineralization (Mulvaney et al., 2009; Russell et al., 2009), but some reports indicate that N fertilization increases SOC stocks (Mahal et al., 2019). Higher C storage was observed in soils fertilized with NPK, N, P and K, compared with unfertilized soil (Zhao et al., 2017). In our study long-term NP, NPK and P treatments enhanced SOC content and it was lowest in the N treatment and unfertilized soil. It was estimated a positive moderate correlation between SOC stocks and mineral nitrogen content ($r=0.54$), EC ($r=0.64$), and negative with pH ($r=-0.56$).

Table 1 Two-way analysis of variance of measured parameters

Parameter	Factor	F-ratio	P-value	Parameter	Factor	F-ratio	P-value	Parameter	Factor	F-ratio	P-value
pH	t	6.97	0.0002	Avail. P	t	1.19	0.35	Sapr. bacteria	t	0.99	0.4674
	c	37.53	0.0000		c	5.88	0.0045		c	2.48	0.0888
EC	t	1.14	0.3759	Avail. K	t	3.77	0.0085	Actinomycetes	t	1.26	0.3143
	c	5.01	0.0089		c	0.7	0.5604		c	6.24	0.0034
SOC	t	6.85	0.0003	Cmic	t	11.67	0.00000	Fungi	t	0.95	0.493
	c	0.33	0.8069		c	1.65	0.2076		c	10.75	0.0002
Min. *N*	t	1.25	0.3223	CO_2	t	1.74	0.15				
	c	4.33	0.0159		c	5.4	0.0065				

Factors—t-treatment; c-crop. Parameters—EC (μS cm$-^{1}$), pH, SOC (%), mineral N (mg kg^{-1}), available forms of P_2O_5, K_2O (mg 100 g$-^{1}$), CO_2 emissions (kg ha day^{-1}), Cmic (mg kg$-^{1}$), main groups soil microorganisms—CFU (colony forming units) × 10^6 g$-^{1}$ abs. dry soil

Table 2 Average by measurements EC (μS cm^{-1}), pH, SOC (%), mineral N (mg kg^{-1}), available forms of P_2O_5, K_2O (mg 100 g^{-1}), CO_2 emissions (kg ha day$-^1$), Cmic (mg kg^{-1}), main groups soil microorganisms—CFU (colony forming units) × 10^6 g^{-1} abs. dry soil

Treatment	N	P	K	NP	NK	PK	NPK	O
EC	43.23ab	24.70ab	25.70ab	44.55ab	40.70ab	29.50ab	51.82b	16.40a
pH	5.42ab	5.65cd	5.70cd	5.28ab	5.25a	5.50bc	5.30ab	5.74d
SOC	1.13ab	1.40cd	1.22bc	1.53d	1.25bc	1.18abc	1.54d	0.97a
Min. N	79.07ab	26.02ab	26.01ab	78.18ab	84.93ab	37.54ab	105.45b	20.79a
Avail. P	36.16b	37.91b	27.98ab	31.32ab	21.85a	33.55ab	36.30b	32.85ab
Avail. K	56.81a	70.45ab	84.93b	52.20a	85.50b	87.77b	76.24b	71.14ab
CO_2 emissions	60.74ab	62.69abc	53.56a	96.29ab	63.70ab	79.41abc	93.13c	74.00abc
Saprophytic bact.	13.07a	13.30a	9.49a	7.45a	11.91a	7.22a	12.21a	2.75a
Actinomycetes	0.72ab	0.58ab	0.79ab	0.40a	0.74ab	0.93b	0.51ab	0.36a
Microsc. fungi	0.09a	0.08a	0.08a	0.12a	0.11a	0.08a	0.09a	0.07a
Cmic	253.66cd	210.46bc	281.84de	186.01ab	315.66e	293.11de	169.10ab	152.19a

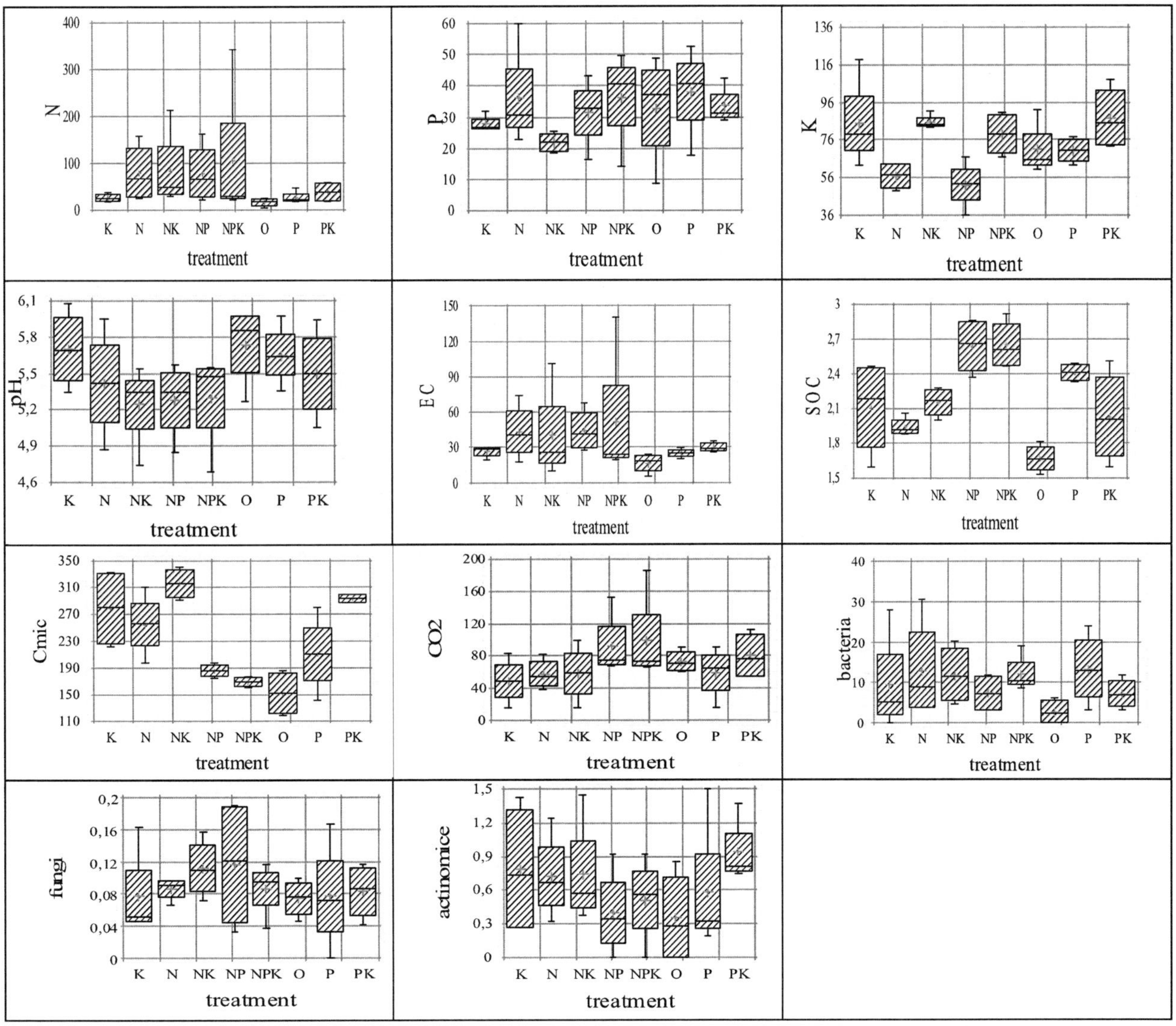

Fig. 1 Microbiological and chemical parameters, EC (μS cm$-^1$), pH, SOC (%), mineral N (mg kg^{-1}), available forms of P_2O_5, K_2O (mg 100 g$-^1$), CO_2 emissions (kg ha day^{-1}), Cmic (mg kg$-^1$), main groups soil microorganisms—CFU (colony forming units) × 10^6 g$-^1$ abs. dry soil

In our study, Cmic had a very high correlation ($r = 0.91$) with the amount of actinomycetes, and available potassium content ($r = 0.49$). Multiple regression analysis shows a relationship between Cmic and number of actinomycetes and fungi ($R^2 = 90.9$). The amount of fungi had moderate correlation with mineral nitrogen content ($r = 0.64$), and the number of actinomycetes had positive moderate correlation ($r = 0.56$) with available potassium. In research of Luo et al. (2015), long-term N fertilization declined Cmic and dehydrogenase activity, while combined NP, NPK fertilization increased these parameters.

Emissions of greenhouse gasses increased with higher N input (Saeed et al., 2022) and the cumulative CO_2 fluxes under long-term NP and NPK treatments were enhanced in comparison with zero and NK fertilization (Zhai et al., 2011). In our experiment, the CO_2 emissions were increased in the NP, NPK treatments compared to K, N, P fertilization treatments and the control.

5 Conclusion

Long-term fertilization with NPK and NP on Haplic Chernozem led to higher accumulation of organic matter and available nutritional elements, while unfertilized control had decreased amounts of most groups of microorganisms, available forms of nitrogen, phosphorus and potassium and less accumulation of soil organic carbon. CO_2 fluxes in NP and NPK treatments were highest and had positive correlation with SOC and mineral nitrogen content.

Acknowledgements The authors acknowledge funding of research activities received from the National Science Fund under grant agreement КП-06 Н 46/1.

References

Boudjabi, S., & Chenchouni, H. (2022). Soil fertility indicators and soil stoichiometry in semi-arid steppe rangelands. *CATENA, 210*(3), 105910. https://doi.org/10.1016/j.catena.2021.105910

Chenchouni, H., & Neffar, S. (2022). Soil organic carbon stock in arid and semi-arid steppe rangelands of North Africa. *CATENA, 212*(1), 106004. https://doi.org/10.1016/j.catena.2021.106004

Gai, X., Liu, H., Liu, J., Zhai, L., Yang, B., Wu, S., Ren, T., Lei, Q., & Wang, H. (2018). Long-term benefits of combining chemical fertilizer and manure applications on crop yields and soil carbon and nitrogen stocks in North China Plain. *Agricultural Water Management, 208*, 384–392. https://doi.org/10.1016/j.agwat.2018.07.002

Johnston, A. E., & Poulton, P. R. (2018). The importance of long-term experiments in agriculture: Their management to ensure continued crop production and soil fertility; the Rothamsted experience. *European Journal of Soil Science, 69*(1), 113–125. https://doi.org/10.1111/ejss.12521.PMID:29527119;PMCID:PMC5832307

Luo, P., Han, X., Wang, Y., Han, M., Shi, H., Liu, N., & Bai, H. (2015). Influence of long-term fertilization on soil microbial biomass, dehydrogenase activity, and bacterial and fungal community structure in a brown soil of northeast China. *Annales de Microbiologie, 65*, 533–542. https://doi.org/10.1007/s13213-014-0889-9

Mahal, N. K., Osterholz, W. R., Miguez, F. E., Poffenbarger, H. J., Sawyer, J. E., Olk, D. C., Archontoulis, S. V., & Castellano, M. J. (2019). Nitrogen fertilizer suppresses mineralization of soil organic matter in maize agroecosystems. *Frontiers in Ecology and Evolution, 7*, 59. https://doi.org/10.3389/fevo.2019.00059

Mulvaney, R. L., Khan, S. A., & Ellsworth, T. R. (2009). Synthetic nitrogen fertilizers deplete soil nitrogen: A global dilemma for sustainable cereal production. *Journal of Environmental Quality, 38*, 2295–2314. https://doi.org/10.2134/jeq2008.0527

Russell, A. E., Cambardella, C. A., Laird, D. A., Jaynes, D. B., & Meek, D. W. (2009). Nitrogen fertilizer effects on soil carbon balances in Midwestern US agricultural systems. *Ecological Applications, 19*, 1102–1113. https://doi.org/10.1890/07-1919

Saeed, Q., Zhang, A., Mustafa, A, Sun, B., Zhang, S., & Yang, X. (2022). Effect of long-term fertilization on greenhouse gas emissions and carbon footprints in northwest China: A field scale investigation using wheat-maize-fallow rotation cycles. *Journal of Cleaner Production, 332*, 130075, https://doi.org/10.1016/j.jclepro.2021.130075. ISSN 0959-6526

Xiang, Y., Cheng, M., Wen, Y., & Darboux, F. (2022). Soil organic carbon sequestration under long-term chemical and manure fertilization in a cinnamon soil, Northern China. *Sustainability, 14*, 5109. https://doi.org/10.3390/su14095109

Zhai, L., Liu, H., Zhang, J., Huang, J., & Wang, B. (2011). Long-term application of organic manure and mineral fertilizer on N_2O and CO_2 emissions in a red soil from cultivated maize-wheat rotation in China. *Agricultural Sciences in China, 10*(11), 1748–1757. https://doi.org/10.1016/S1671-2927(11)60174-0

Zhao, H., Sun, B., Lu, F., et al. (2017). Roles of nitrogen, phosphorus, and potassium fertilizers in carbon sequestration in a Chinese agricultural ecosystem. *Climatic Change, 142*, 587–596. https://doi.org/10.1007/s10584-017-1976-2

Evapotranspiration and Maize Productivity Characteristics Under Soil Erosion Control Technologies on Sloped Terrains

Gergana Kuncheva, Iliana Ivanova, Milena Kercheva, Viktor Kolchakov and Evgeni Enchev

Abstract

The aim of the study was to assess the influence of water erosion and the applied erosion control technologies on components of soil water balance and crop productivity indicators during a three-year field experiment on Epicalcic Chernozem. Four treatments were tested: conventional technology, applied along the slope (control); conventional technology applied across the slope; surface mulching; and minimum tillage with vertical mulching with manure. The volume of surface water runoff was measured by a stationary method. The actual evapotranspiration was calculated according to the FAO 56 procedure. The water runoff was reduced in the treatments grown minimum tillage systems. On average for three years, the runoff was reduced by 435.8 m^3 ha^{-1} for conventional technology applied across to the slope, by 524.6 m^3 ha^{-1} for the treatment with surface mulching and by 848.2 m^3 ha^{-1} for the minimum tillage with vertical mulching, compared to the control treatment, grown along the slope. The average grain yields for the three years range from 6026.3 kg $ha-^1$ in the control treatment to 7018.3 kg ha^{-1} in the treatment with application of minimum tillage and vertical mulching. The increase in evapotranspiration in the erosion control treatments was reduced by the lower evaporation due to the plant residues on the soil surface in minimum tillage and surface mulching in the initial crop stage, despite the increased transpiration because of the lower runoffs and higher soil moisture content. This reflected in a better plant development and higher yields in maize production under these technologies.

Keywords

Actual crop evapotranspiration · Water erosion · Erosion control technologies · Surface water runoff

1 Introduction

Water erosion is a soil degradation process with significant negative impact on natural and agricultural ecosystems. It leads to the formation of surface water runoff, significantly reduces infiltration and soil water storage on sloping terrains (Arar & Chenchouni, 2014; Darouich et al., 2022). The erosion control technologies for crop production have the potential to reduce the negative consequences of these processes (Bagagiolo et al., 2018). The long use of Epicalcic Chernozems (Kastanozems, WRB) lead to the deterioration of the structure of these types of soils, observed in arable land due to changes in their mechanical, microaggregate and structural composition and soft, soil-forming rock, susceptible to degradation and erosion (Teoharov et al., 2015). The aim of the study was to assess the influence of soil erosion and the applied erosion control technologies on components of soil water balance and crop productivity indicators during a three-year field experiment on Epicalcic Chernozem.

2 Materials and Methods

A three-year (2017–2019) study was conducted, on the territory of the experimental field near the village of Tarstenik, Ruse region, North Bulgaria, Danube plain (25.9005 °E 43.6783 °N, 114 m a.s.l.), under rainfed conditions, on

G. Kuncheva (✉) · M. Kercheva · V. Kolchakov
ISSAPP "Nikola Poushkarov", Bansko Shousse Street 7, 1331 Sofia, Bulgaria
e-mail: G.Kuncheva@Issapp-Pushkarov.Org

I. Ivanova
Institute of Agriculture and Seed Science "Obraztsov Chiflik", Ruse, Bulgaria

E. Enchev
University of Ruse "Angel Kanchev", Studentska Street 8, 7017 Ruse, Bulgaria

H. Chenchouni et al. (eds.), *Recent Advancements from Aquifers to Skies in Hydrogeology, Geoecology, and Atmospheric Sciences*, Advances in Science, Technology & Innovation, https://doi.org/10.1007/978-3-031-47079-0_32

moderately eroded Epicalcic Chernozem, on slope with inclination of 5° (8.7%). The treatments of the experiments were as follows: CT—conventional technology, applied along the slope—control; CTT—conventional technology, applied across the slope; SM—surface mulching with manure (4500–5000 kg ha$-^1$), applied across the slope; MTVM—technology for minimum tillage, including vertical mulching with manure and main tillage without reversing the layer—loosening, instead of ploughing. In the MTVM treatment, pre-sowing was performed across the slope and vertical (internal soil) mulching with manure was carried out according to a band scheme (distance between slots 1.4 m and interval between the strips in the field 3 m) at a depth of 0.40 m.

The surface water runoff and eroded soil were measured by a stationary method with runoff sites, with area of 75 m^2 and runoff collecting chambers. The leaf area indices (LAI) in 5th, 9th leaf and tasselling stages and yields were measured in each treatment by leaf parameters method.

The FAO56 (Allen et al., 1998) procedure had been used to calculate the reference evapotranspiration (Et_o), crop evapotranspiration (Et_c) and actual evapotranspiration (Et_a). The crop evapotranspiration, Et_c, was calculated by multiplying the reference crop evapotranspiration, Et_o, by a crop coefficient, K_c. Actual evapotranspiration (Et_a) was calculated using a general soil water balance equation and by using a water stress coefficient K_s (water stress coefficients) by adjusting K_c. For different growth stages, calculation of K_c was performed. The crop coefficients in the initial phase—$K_{c_{ini}}$ for SM and MTVM treatment, were reduced for every 10% plant residues or mulch material with 5% (Allen et al., 1998). The application of surface mulching in treatment SM was reflected to coefficients in initial stage, which was reduced by 15%. The application of direct sowing and the presence of plant residues in MTVM treatment were also reflected in a 15% reduction in K_c for initial growth stage.

3 Results

Precipitation has an uneven distribution in the studied region. During the growing season of maize, most of the rainfalls occur in April, May and June, while after mid-July there is a drought, which often extends to September. This also determines the reduction of Et_a in the late season stage of the crop development (Table 1; Fig. 1).

The daily values of reference (Et_o), crop (Et_c) and actual (Et_a) evapotranspiration are shown in Fig. 1. Six erosive rainfalls were registered during 2017 and 2018, and three in 2019. The application of the SM and MTVM technologies led to a reduction of the surface water runoff on average for three years by 21.1 mm and 45.4 mm, respectively, for the growing season. This was due to changed direction of the tillage operations, minimum tillage instead of ploughing, formation of slits, plant residues and surface mulching (Table 2).

A higher evapotranspiration was estimated for the treatments CTT, SM and MTVM, compared to the control treatment, due to reduced surface water runoff. The treatments, grown conventionally across the slope and with surface mulching, differ in reduced runoff and soil surface coverage, which reduced evaporation in the initial phase of the cultivated crop. Hereof evapotranspiration was slightly lower in SM than in CTT treatment. In the MTVM treatment, due to the minimum tillage, plant residues remained on the soil surface, evaporation was reduced in the initial phase. After this stage, the reduced surface water runoff leads to a higher Et_a.

The yields were the lowest in the treatment grown conventionally, along the slope. Better plant development and yields was observed in the erosion control treatments SM and MTVM (Table 2; Fig. 2).

Table 1 Average for three years crop coefficients (K_c), reference evapotranspiration (Et_o, mm), crop evapotranspiration (Et_c, mm), actual evapotranspiration (Et_a, mm) for maize growing stages

Maize growing stage	K_c		Et_o, mm	Et_c, mm	Et_a, mm			
	CT, CTT	SM, MTVM			CT	CTT	SM	MTVM
Initial stage	0.530	0.402	117.5	60.1	60.1	60.1	51.1	51.1
Development stage	0.530–1.224	0.402–1.224	194.5	172.6	166.7	167.7	162.0	164.3
Mid-season stage	1.224	1.224	286.3	350.4	108.8	124.3	134.2	162.9
Late season stage	0.451	0.451	132.4	176.8	0.5	0.7	0.8	1.3
	$K_{c_{ini}} = 0.33$–0.63; $K_{c_{end}} = 0.33$–0.54 St. dev $K_{c_{ini}} = 0.024$; $K_{c_{mid}} = 0$; $K_{c_{end}} = 0.023$		ANOVA: Eta ini, p=0.0012 Eta dev. NS; Eta mid NS; Eta late NS					

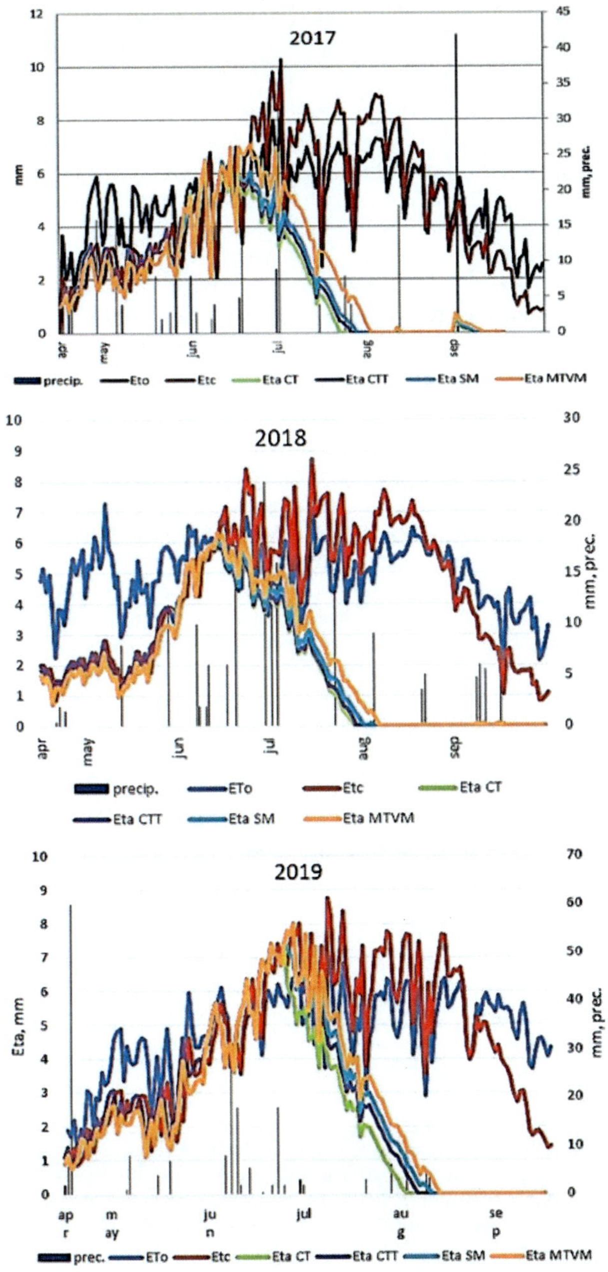

Fig. 1 Reference evapotranspiration Et_0 (mm), crop evapotranspiration, Et_c (mm), actual evapotranspiration, Et_a (mm) for treatments CT, CTT, SM, MTVM

4 Discussion

The mechanical manipulation of the soil affects soil properties, evapotranspiration and crop production (Busari et al., 2015; Kuncheva et al., 2022). A number of studies establish that the conventional tillage leads to an increased soil erosion and loss of soil organic carbon, aggregate stability and soil water content (Tadesse et al., 2021). The conservation tillage practices have the potential to reduce the surface water runoff and improve soil water infiltration processes (Feyereisen et al., 2008). Some of these practices may result in reduction of non-beneficial loss of water (evaporation), which does not necessarily translate into yield benefit (Fellah et al., 2018; Irmak et al., 2019). Many studies have shown a higher soil water storage under the conservation tillage compared to the conventional tillage systems (Tadesse et al., 2021), but there are also studies (Kováč et al., 2005) that show the opposite results—a higher soil moisture content under the conventional tillage than under conservation tillage.

In this experiment, the runoff volume was reduced in the treatments with surface mulching SM and minimum tillage and vertical mulching MTVM, on average 1.4 and 3.4 times. Plant residues on the soil surface and mulching lead to a reduction of soil evaporation, despite the increased transpiration, due to the better crop development and the increased soil water reserve. Overall, this reflected in maize development and grain yields (for SM increased with an average of 10% and for MTVM with 15%).

5 Conclusion

As a result of the research, the following conclusions can be made: (1) the applied soil protection technologies (SM and MTVM) lead to a reduction of surface water runoff and reduced evaporation in the initial crop stage; (2) The application of these technologies leads to the improvement of water reserves in the soil, which was reflected in a better plant development and higher yields in maize production under these technologies on sloped terrains.

Table 2 Seasonal volume of the surface water runoff, mm and evapotranspiration (mm), maize yields (kg ha$-^{1}$)

Treatments	Surface water runoff, mm				Actual evapotranspiration, mm				Yields, kg ha^{-1}			
Year	CT	CTT	SM	MTVM	CT	CTT	SM	MTVM	CT	CTT	SM	MTVM
2017	93.3	70.4	62.5	23.2	320.5	335.5	330.9	378.9	5723	6041*	6326*	6677*
2018	57.4	49.7	41.0	23.2	318.1	322.5	318.4	342.3	6105	6426***	6720***	7122***
2019	41.7	31.4	25.6	9.9	368.5	399.4	396.8	416.2	6925	7256***	7561***	7969***
	ANOVA: HSD[0.05] = 24.12 HSD[0.01] = 29.7, $p < 0.001$				ANOVA: NS				2017 HSD[0.05] = 38.3, HSD[0.01] = 45.0; 2018 HSD[0.05] = 66.9 HSD[0.01] = 101.4; 2019 HSD[0.05] = 72,0 HSD[0.01] = 109.8			

$* p \leq 0.05$; $** p \leq 0.01$; $*** p \leq 0.001$

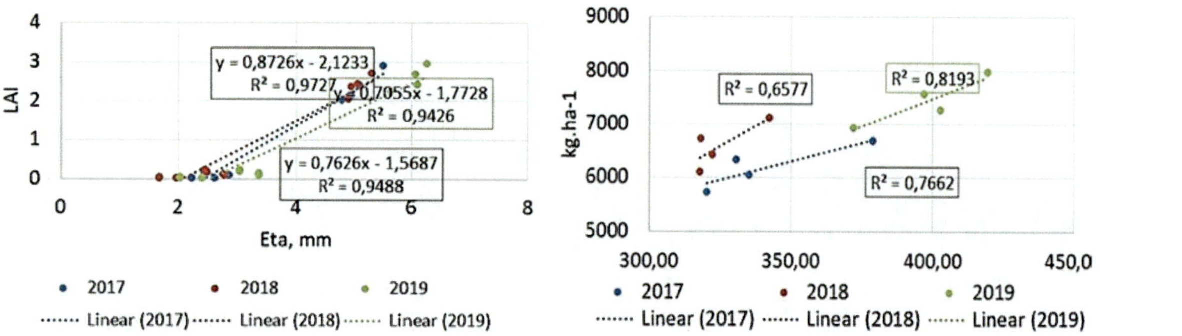

Fig. 2 Dependence of LAI on daily Et_a (mm) in stages for 5th, 9th leaf and tasselling (**a**), and of yields (kg ha$-^1$) on seasonal Et_a (mm) for treatments CT, CTT, SM, MTVM (**b**)

Acknowledgements The authors acknowledge funding of research activities received from the National Science Fund under grant agreement КП-06 Н 46/1 2020 (project "Efficiency of erosion control agrotechnologies for improvement of soil quality and water regime and mitigation of greenhouse gas emissions").

References

Allen, R. G., Pereira, L. S., Raes, D., & Smith, M. (1998). Crop evapotranspiration—Guidelines for computing crop water requirements. FAO Irrigation and drainage paper 56. FAO—Food and Agriculture Organization of the United Nations, Rome. https://www.fao.org/3/X0490E/x0490e00.htm#ContentsAuthor

Arar, A., & Chenchouni, H. (2014). A "simple" geomatics-based approach for assessing water erosion hazard at montane areas. *Arabian Journal of Geosciences, 7*(1), 1–12. https://doi.org/10.1007/s12517-012-0782-4

Bagagiolo, G., Biddoccu, M., Rabino, D., & Cavallo, E. (2018). Effects of rows arrangement, soil management, and rainfall characteristics on water and soil losses in Italian sloping vineyards. *Environmental Research, 166*, 690–704. https://doi.org/10.1016/j.envres.2018.06.048

Busari, M. A., Kukal, S. S., Kaur A., Bhatt, R., & Dulazi, A. A. (2015). Conservation tillage impacts on soil, crop and the environment. *International Soil and Water Conservation Research, 3*, 119–129. https://doi.org/10.1016/j.iswcr.2015.05.002

Darouich, H., Ramos, T. B., Pereira, L. S., et al. (2022). Water use and soil water balance of Mediterranean vineyards under rainfed and drip irrigation management. Evapotranspiration partition and soil management modelling for resource conservation. *Water, 14*, 554. https://doi.org/10.3390/w14040554

Fellah, S., Khiari, A., Kribaa, M., et al. (2018). Effect of water regime on growth performance of Durum Wheat (*Triticum durum* Desf.) during different vegetative phases. *Irrigation and Drainage, 67*(5), 762–778. https://doi.org/10.1002/ird.2289

Feyereisen, G., Strickland, T., Bosch, D. D., Truman, C., Sheridan, J., & Potter, T. (2008). Curve number estimates for conventional and conservation tillage in the Southwest coastal plain. *Journal of Soil and Water Conservation, 63*(3), 120–128. https://doi.org/10.2489/jswc.63.3.120

Irmak, S., Kukal, M. S., Mohammed, A.T., & Djaman, K. (2019). Disk-till vs. no-till maize evapotranspiration, microclimate, grain yield, production functions and water productivity. *Agricultural Water Management, 216*, 177–195. https://doi.org/10.1016/j.agwat.2019.02.006

Kováč, K., Macák, M., & Švančárková, M. (2005). The effect of soil conservation tillage on soil moisture dynamics under single cropping and crop rotation. *Plant, Soil and Environment, 51*, 124–130. https://doi.org/10.17221/3564-PSE

Kuncheva, G. S., Ivanova, I. I., Dimitrov, P. D., & Enchev, E. T. (2022). Evapotranspiration, development and productivity of wheat under different tillage systems on sloping terrains. In: 8th International conference on energy efficiency and agricultural engineering (EE&AE), pp 1–4. https://doi.org/10.1109/EEAE53789.2022.9831273

Tadesse, H. K., Moriasi, D. N., Gowda, P. H., et al. (2021). Comparison of evapotranspiration and biomass simulation in winter wheat under conventional and conservation tillage systems using APEX model. *Ecohydrology & Hydrobiology, 21*(1), 55–66. https://doi.org/10.1016/j.ecohyd.2020.08.003

Teoharov, M., Shishkov, T., Hristov, B., et al. (2015). Chernozems in Bulgaria—Systematic, specific features and problems. *Soil Science Agrochemistry and Ecology, 48*(3–4), 3–9. https://soilscience-bg.org/page/download.php?articleID=50

Biodiversity and Conservation of Sub-Mediterranean Landscapes of the North-Western Caucasus

Marina Petrushina, Maxim Bocharnikov and Alexandra Andreeva

Abstract

Sub-Mediterranean landscapes with high biodiversity, rare and endemic species of flora and fauna are unique for Russia, being formed on the Black Sea coast with intensive and prolonged anthropogenic effect, in the most northeastern area of Mediterranean ecosystems of Europe. The main object of this study was the Utrish Nature Reserve, which was organized in 2010 to protect these landscapes in the North-Western Caucasus. It became a model area for large-scale complex study and monitoring of sub-Mediterranean forest landscapes under current climatic changes, including their restoration after previous anthropogenic effect. Large-scale landscape and botanical maps, compiled on the field complex descriptions on sample sites, repeated observations on the model plots with special attention to vegetation, and the analysis of remote sensing data were used to identify factors of formation, structure and dynamics of vegetation and landscapes. The relationship of vegetation to abiotic conditions was determined by statistical methods. The main factors and basic regularities of the spatial organization of vegetation and landscapes, and phytocoenotic diversity were revealed on the basis of landscape and botanical maps (1:25,000 and 1:50,000) and transects. The dominance of slope geosystems and their maximum diversity, altitudinal zonality with formation of sub-Mediterranean landscapes at 0–250 m a.s.l., the asymmetry of geosystems are typical for the landscape structure, which determines the heterogeneity of vegetation. Coastal landscapes are distinguished by the more complex spatial structure and high biodiversity with rare and relict species, including a high participation of Mediterranean species. The sea–land contact zone with the predominance of abrasion types of shores is characterized by dynamism, which leads to the destruction of part of the coastal landscapes. Vegetation restoration after recreational use and fires, including a major fire in 2020, was determined based on the study of monitoring plots. Rapid formation of herb and shrub layers began after the fire, the structure of which had both similar features in different communities and differed in its specifics. The results of the research formed the basis for expanding the boundaries of the reserve in 2021 at the expense of coastal geosystems that were not previously included and the organization of monitoring.

Keywords

Sub-mediterranean landscapes · Biodiversity · Structure · Dynamics · Utrish nature reserve

1 Introduction

The sub-Mediterranean landscapes of the North-Western Caucasus stretch along the Russian Black Sea coast between the cities of Anapa and Tuapse within the low mountain ridges and are characterized by high biodiversity (more than 800 vascular plants), rich in relict species of flora and fauna. They are vulnerable to the intensification of modern anthropogenic impact and climate change with growing of average annual temperatures and the number of extremely hot days (Bogdanovich et al., 2021), which increase the risk of fires, as it was in 2020. These landscapes were previously studied mainly while compiling medium- and small-scale maps for the entire Caucasus or its regions (Beroutchashvili, 1995). Large-scale landscape and botanical researches were carried out for some area, which results became a basis for justification the creation of the Utrish State Nature Reserve in 2010, the only one

M. Petrushina (✉) · M. Bocharnikov · A. Andreeva
Lomonosov Moscow State University, Leninskie Gory, 1, 119991 Moscow, Russia
e-mail: mnpetrushina@mail.ru

H. Chenchouni et al. (eds.), *Recent Advancements from Aquifers to Skies in Hydrogeology, Geoecology, and Atmospheric Sciences*, Advances in Science, Technology & Innovation, https://doi.org/10.1007/978-3-031-47079-0_33

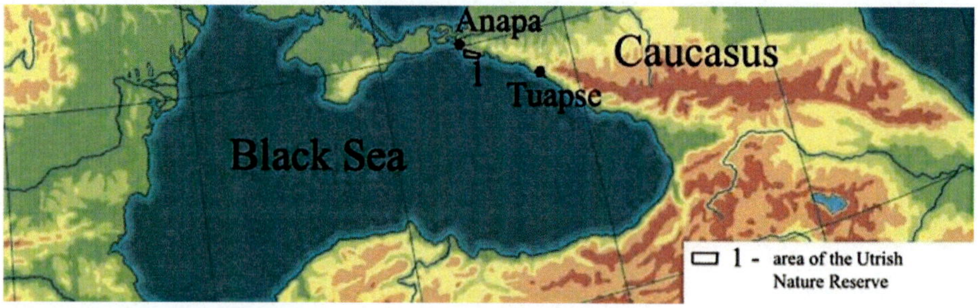

Fig. 1 Location of the study area

of this status to protect the sub-Mediterranean landscapes in the study region (Leontyeva et al., 2009) (see Fig. 1). The main goal of our study was to reveal the factors and basic patterns of vegetation and geosystem diversity of the Reserve as a key area for monitoring and conservation of unique landscapes for Russia.

2 Materials and Methods

Field measurements on the transects from the watershed to the local basis of erosion across slopes of different exposures and distance from the sea and valley bottoms, as well as digital elevation models and remote sensing data of different years (Landsat-8), were used to compile large-scale landscape and botanical maps (1:25,000) as the basis to identify factors and patterns of biodiversity formation. About 400 descriptions on the sample plots within transects included the characteristics of vegetation—species composition of communities and their morphometric indicators (height, crown closeness, projective cover, etc.) and ecotopic conditions—relief forms (type, altitude, steepness and exposure), rocks (lithology, mechanical composition and surface stony) and soils (type, thickness and features of horizons). Special attention was paid to the fixation of rare, protected and requiring protection plant species.

The spatial distribution of species diversity was determined based on multivariate regression analysis of field data and multispectral satellite images of Landsat-8. This method is widely used in research at the present time (Cavender-Bares et al., 2020).

Repeated observations at selected monitoring sites since 2013 were carried out to study vegetation dynamics, including regenerative successions after previous anthropogenic impact (logging, recreation) and fires.

Statistical analyses (correlation analysis by nonparametric Spearman rank correlation coefficient and analysis of variance (ANOVA)) were used to determine the closeness of the relationship between phytocoenotic characteristics of vegetation and ecotopic conditions (Puzachenko, 2004).

3 Results

Based on the compiled maps, it was revealed that the spatial structure of landscapes and vegetation is heterogeneous and varies within altitudinal zones. Sub-Mediterranean landscapes are formed at the altitudes below 200–250 m a.s.l. on the northern slopes of the main low ridges with dominance of hemixerophilic oak-hornbeam (*Quercus pubescens* Willd., *Carpinus orientalis* Mill.) forests and at altitudes 250–300 m on the southern exposed slopes with pistachio-juniper (*Pistacia mutica* Fisch. & S.A.Mey, *Juniperus excelsa* M.Bieb., *J. foetidissima* Willd., and *J. oxycedrus* L.) and juniper sparse forests in combination with shrub thickets and petrophytic herb communities at altitudes 0–150 (200) m. At higher altitudes, they are replaced by the landscapes of mesophilic forests. The most complex structure of landscapes with high floristic and phytocoenotic diversity is typical for the southern seaside slopes, especially for the coastal zone, where the number of Mediterranean plant species reaches 50% in pistachio-juniper forests and 20% of the total flora. The age of individual junipers is more than 350 years. The unique pine forests (*Pinus brutia* Ten.) form a mosaic of small plots closely to the sea. The deterioration of vegetation was noted as a result of a fire in 2020. The sea–land zone is also distinguished by prevailing abrasive shores, the dynamics of which contributes to the reduction of the area of coastal sub-Mediterranean landscapes.

Observations at monitoring sites showed the restoration of vegetation, mainly herbage and shrubs (*Jasminum fruticans* L., *Ruscus aculeatus* L., *Rhus coriaria* L.), with woody growth of *Quercus pubescens, Pistacia mutica, Fraxinus oxycarpa* Willd.

Statistical analysis revealed a strong positive correlation between the plant diversity and distance from the sea and altitude. It also increases from eluvial to accumulative positions in phytocatena.

4 Discussion

Landscape and vegetation diversity of the study area is defined mainly by southern sea position, highly dissected relief, complicated by different seismic gravitational forms, heterogeneity of rocks and anthropogenic effect. In conditions of low precipitation during the growing season, the role of moisture in formation and functioning of vegetation increases, which is associated with the exposure and shape of the slopes, the position in the catena, the altitude and distance from the sea.

After the creation of the Reserve, as well as the inclusion of part of the coastal landscapes in its borders in 2021, the restoration and improvement of vegetation, including Mediterranean species, began. Plant successions with the rapid appearance of herbage and shrubs with separate woody growth were also noted after the fire in 2020. It destroyed pine and juniper forests of the Reserve and neighboring territories on an area of about 120 hectares, including individual juniper specimens in other plant communities with better survival of pistachio, oak and hornbeam. Restoration of vegetation had common features in different communities and differed in specifics. The predominance of *Rhus coriaria* was typical to the burning sites in the pine forest with increasing of *Ruscus aculeatus L.* and *Jasminum fruticans* in the pistachio-juniper and pine with juniper sparse forests. The appearance of individual pine sprouts and oak growth renewal was also noted. The floristic diversity on the second year after the fire was up to 26 species (on the area of 100 m^2), including some Mediterranean species. Pistachio undergrowth was found in another pistachio-juniper forest that burned down in 2009, unlike juniper, which indicates a longer recovery time.

5 Conclusion

In spite of rather small area (near 9910 ha of land), the significance of the Utrish Reserve in the conservation of sub-Mediterranean landscapes and their complex studies is great. It is characterized by high biodiversity due to the complexity of landscape structure, especially of the southern exposed slopes and the coastal zone. The Reserve is also distinguished by significant areas of geosystems with a recovering structure after recent recreation impact and repeated effects of fires that increased its role as a background for monitoring neighboring territories with intensive and varied uses. The results of the research formed the basis for expanding the boundaries of the Reserve in 2021 by some coastal geosystems and the organization of monitoring. Further studies of the diversity and dynamics of vegetation and landscapes, including at monitoring sites, will allow us to determine trends in their changes under modern climatic fluctuations and to organize protection more correctly inside and around the Reserve.

References

Beroutchashvili, N. L. (1995). *The Caucasus: Landscapes, models, experiments*. Tbilisi state University Press.

Bogdanovich, A. Yu., Lipka, O. N., Krylenko, M. V., Andreeva, A. P., & Dobrolyubova, K. O. (2021). Climate threats in the North-West Caucasus Black Sea coast: modern trends. *Fundamental and Applied Climatology, 7*(4), 44–70.

Cavender-Bares, J., Gamon, J. A., & Townsend, P. A. (2020). *Remote sensing of plant biodiversity*. Springer Nature.

Leontyeva, O. A., Suslova, E. G., Pereshkolnik, S. L., & Petrushina, M. N. (2009). To create Utrish reserve at the Abrau Peninsula. In Ozhan, E. (Ed.). MEDCOAST 09 (Vol. 2, pp. 249–260) (2009).

Puzachenko, Yu. G. (2004). *Mathematical methods in ecological and geographical research*. Academy.

Mercury Concentrations in Coastal Heath Forests of Peninsular Malaysia

Francis Q. Brearley and Jamilah Mohd Salim

Abstract

Mercury (Hg) is a globally distributed bioaccumulative neurodegenerative pollutant. The main way Hg enters the ecosystem is via stomatal uptake of gaseous Hg and subsequent sequestration into the soil via leaf litterfall. Thus, determining litterfall Hg concentrations will help our understanding of a key flux in Hg biogeochemistry. Concentrations of Hg in terrestrial ecosystem compartments are not well known in South-east Asia, so to rectify this deficit we determined Hg concentrations in litterfall and soil in two contrasting forest types in Terengganu, Peninsular Malaysia. We collected litterfall and soils (bulk: < 2 mm and fine: < 63 µm) from forest and heath vegetation types over coarse sandy soils and assessed Hg by an acid digestion (HCl and HNO_3) and hydride generation ICP–OES technique. Litterfall Hg was 42.6 ± s.e. 2.17 ng g–^{1}, and bulk soil Hg was 10.1 ± 2.30 ng g^{-1}, whereas the fine soil fraction was 173 ± 20.9 ng g–^{1}. There were correlations between Hg and macronutrient elements (C, N, P, K, Ca, Mg and S) in soil but not leaves. Total Hg return to the soil via litterfall was estimated as 16.8 µg Hg m^{-2} yr–^{1} which compares well with other global studies. Overall, our research implies that this 'unpolluted' location in Malaysia has Hg input comparable to other global studies, but we note that other forests in this region with greater biomass may return more Hg to the soil.

Keywords

Litterfall · Mercury · Pollution · Soil fractionation · Tropical rain forest

F. Q. Brearley (✉)
Manchester Metropolitan University, Manchester M1 5GD, UK
e-mail: f.q.brearley@mmu.ac.uk

J. Mohd Salim
Universiti Malaysia Terengganu, 21030 Kuala Nerus, Terengganu, Malaysia

1 Introduction

Mercury (Hg) is a bioaccumulative and neurodegenerative pollutant distributed around the world. The main input (<90%) of Hg to terrestrial ecosystems is through vegetation uptake of gaseous Hg via stomata and subsequent sequestration into the soil via leaf litterfall (Zhou et al., 2021). Thus, determining leaf and litterfall Hg concentrations will help our understanding of this key flux in Hg biogeochemistry. Zhou et al. (2021) and Wang et al. (2016) compiled a global data set on leaf and litterfall Hg concentrations, respectively, and found that only about 2% of all measurements were from South-east Asia marking a major knowledge gap which is important to fill given the large population in this region. Soil is also a key stock for terrestrial Hg (Bishop et al., 2020); for many metals, as the soil particles decrease in size through a sand–silt–clay gradient, metal concentrations increase but this has been less studied for Hg than other metals (Fernández-Martínez et al., 2014; Gómez-Armesto et al., 2018). Whilst studies on soil Hg concentrations have been conducted at sites contaminated by gold mining, particularly across Indonesia (Barkdull et al., 2019; Limbong et al., 2005; Tomiyasu et al., 2020), we have little data on Hg concentrations from 'unpolluted' South-east Asian locations (Brearley et al., 2023). Therefore, in this work, we provide details on the Hg concentrations in litterfall and soil from an 'undisturbed' site in Peninsular Malaysia to help fill this gap in our knowledge of Hg cycling in this region.

2 Materials and Methods

The study location was the Jambu Bongkok Forest Reserve (4°54′N, 103°22′E), Terengganu, Malaysia, which has coastal forest and heath vegetation on sandy former dune soils. The soils on the dune ridges are Spodosols with a high proportion of coarse sand and acidic (also known as BRIS soils)

H. Chenchouni et al. (eds.), *Recent Advancements from Aquifers to Skies in Hydrogeology, Geoecology, and Atmospheric Sciences*, Advances in Science, Technology & Innovation, https://doi.org/10.1007/978-3-031-47079-0_34

and considered low in nutrients (Roslan et al., 2010); they are generally confined to the east coast of Peninsular Malaysia. Two locations in two contrasting sites were sampled: a forest dominated by *Shorea materialis* (Dipterocarpaceae), and heath vegetation which is a more stunted vegetation type on the former dune ridges (Jamilah et al., 2104).

Samples were collected in January 2015 (wetter season). The soil was dug from up to 15 cm depth; at each site, three replicates were taken with replicated collections separated by at least three metres. Litterfall was collected in traps left in place for two weeks. Bulk soil pH was measured in a 1:2.5 soil:deionised water slurry after 1-h equilibration with a Sartorius PB-11 pH meter. Soils were fractionated by passing through a 63-μm sieve to obtain the fine soil fraction. Total C and N were measured using a Leco TruSpec (bulk soil<2 mm and leaf litter) or a Vario ELcube (fine soil<63 μm). The samples were homogenized and digested in 7.5 ml HNO_3 plus 2.5 ml HCl in a CEM Mars Xpress microwave in sealed PTFE tubes. For quality control, strawberry leaves (LGC7162) and light sandy soil (Metranal-32) certified reference materials were digested alongside the samples with recoveries of 96% and 106%, respectively. Digests were analysed on a Thermo Scientific iCAP6300 ICP–OES using a hydride generation technique (Brearley et al., 2023; Lofthouse, (no date)). All other elements were determined in 'normal' running mode.

3 Results

The two vegetation types did not differ greatly in their litterfall or soil nutrient concentrations, but the forest soil was slightly more acidic by 0.2 pH units and had less potassium (Table 1). Litterfall Hg was 42.6 ± s.e. 2.17 ng g^{-1}, and bulk soil Hg was 10.1 ± 2.30 ng g$-^{1}$, whereas the fine soil fraction was an order of magnitude greater at 173 ± 20.9 ng g^{-1} (Fig. 1). If we consider an estimated leaf litterfall of 6 Mg ha yr$-^{1}$ for a typical heath forest (Dent et al., 2006; Miyamoto et al., 2007; Proctor et al., 1983), then this equals Hg deposition of 42 g Hg ha^{-1} yr$-^{1}$ or 16.8 μg Hg m^{-2} yr$-^{1}$. There were strong correlations between bulk and fine soil Hg and all macronutrient concentrations (mean $\rho = 0.78$), but these were not observed for leaves.

4 Discussion

The litterfall Hg concentrations reported here are very similar to the global median of 46 ng g^{-1} (Wang et al., 2016) which is roughly twice that of live leaves (Zhou et al., 2021). However, it was lower than the mean for tropical broadleaved trees of 56 ng g$-^{1}$ (Wang et al., 2016). It is not clear whether this higher value for tropical leaves is

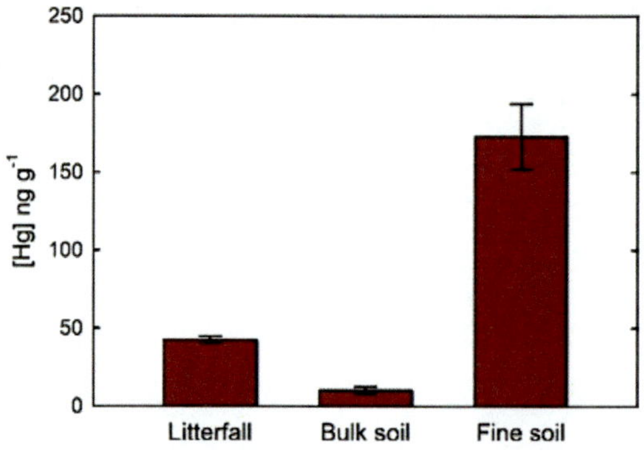

Fig. 1 Litterfall and bulk (<2 mm) and fine (<63 μm) soil mercury concentrations at the Jambu Bongkok Forest Reserve in Peninsular Malaysia. Values are mean ± standard error

consistent across all tropical biomes as the data set is dominated by data from South America where concentrations were particularly high. More locally, litterfall Hg concentrations compared well with those from Setiu wetlands on the Terengganu coast (east Peninsular Malaysia) with a range between about 25 and 45 ng g^{-1}; concentrations did not differ between seasons, but they were approximately doubled in old/decaying leaves compared to live actively photosynthetic leaves (Siau et al., 2021). Similarly, on the west coast of Peninsular Malaysia, mangrove senescing leaf Hg concentrations were similar at 33.9 ng g$-^{1}$, despite the area being more influenced by human activities (Wolswijka et al., 2020). As the soil at our study site is dominated by coarse sand, this is one reason for a clear increase in the fine (<63 μm) soil fraction that had greater carbon concentrations and hence more negative charges for binding with Hg (Huang et al., 2020). The strong correlations between Hg concentrations and carbon (and other macronutrients) in soils confirm the interaction between soil carbon and Hg. There are little data with which to compare our bulk soil data with from terrestrial environments in the region, but soil from Kinabalu Park in East Malaysia had a greater concentration with a mean of 277 ng g^{-1} across four soil depths (plus the litter layer) (Brearley et al., 2023).

Mercury flux via litterfall was estimated to be 16.8 μg m$-^{2}$ yr^{-1} which was also similar to the global median of Wang et al. (2016) of 15.3 μg m$-^{2}$ yr^{-1}, but the correct estimation of this value is dependent upon (i) accurate measurement of annual litterfall mass and (ii) assumption of no seasonal changes in litterfall Hg concentrations. We note that the annual litterfall mass in our studied heath forest will be lower than larger-statured lowland evergreen rain forests that may be around 10 Mg$-^{1}$ ha^{-1} yr$-^{1}$ (Niiyama et al., 1992), so assuming comparable leaf and litterfall Hg concentrations, Hg fluxes will be greater in more typical lowland evergreen rain forests of Indo-Malaya. Additional studies should examine other potential Hg fluxes such as throughfall, and root and wood decomposition that will improve our understanding of Hg biogeochemical cycling in the region with implications for human health.

5 Conclusion

We show there that Hg concentrations in litterfall and soil in this 'unpolluted' Malaysian site are fairly close to the global mean, but note that biomass and litterfall were lower than many other larger-statured forests in the region. There are little other data from similar South-east Asian sites, and, thus, we provide some of the first measurements of aspects of the Hg cycle from this region helping in our understanding of Hg biogeochemistry.

Table 1

Litterfall and soil (bulk < 2 mm and fine < 63 μm) pH and macronutrient concentrations at the Jambu Bongkok Forest Reserve in Peninsular Malaysia. Values are mean ± standard error.

	Soil (<2 mm)		Soil (<63 μm)		Litterfall	
	Forest	Heath	Forest	Heath	Forest	Heath
pH	3.88 ± 0.02	4.09 ± 0.07	–	–	–	–
C (%)	7.71 ± 2.36	7.79 ± 0.99	27.4 ± 3.15	32.0 ± 0.59	47.7 ± 0.61	48.4 ± 0.62
N (%)	0.36 ± 0.11	0.37 ± 0.05	1.23 ± 0.15	1.50 ± 0.04	1.02 ± 0.04	0.99 ± 0.04
P (mg kg^{-1})	35.8 ± 11.9	34.8 ± 1.3	523 ± 117	677 ± 50	419 ± 46.7	424 ± 46.8
K (mg kg–^{1})	25.8 ± 8.67	33.5 ± 1.84	504 ± 68.2	828 ± 76.2	1180 ± 81.2	1160 ± 152
Ca (mg kg^{-1})	104 ± 45.2	112 ± 14.3	1630 ± 387	2140 ± 271	5760 ± 464	5540 ± 427
Mg (mg kg–^{1})	93.2 ± 32.4	52.0 ± 3.96	1340 ± 289	1060 ± 123	2220 ± 98.8	2610 ± 137
S (mg kg^{-1})	98.7 ± 33.4	96.4 ± 6.12	1630 ± 293	2160 ± 197	1270 ± 45	1170 ± 84.5

References

Barkdull, N. M., Carling, G. T., Rey, K., & Yudiantoro, D. F. (2019). Comparison of mercury contamination in four Indonesian watersheds affected by artisanal and small-scale gold mining of varying scale. *Water, Air and Soil Pollution, 230*(9), 214.

Bishop, K., Shanley, J. B., Riscassi, A., et al. (2020). Recent advances in understanding and measurement of mercury in the environment: terrestrial Hg cycling. *Science of the Total Environment, 721*, 137647. https://doi.org/10.1016/j.scitotenv.2020.137647

Brearley, F. Q., Sellan, G., McKendry, D., Sumail, S., & van der Ent A. (2023). Soil mercury along an elevation gradient in northern Borneo. In A. Kallel, M. Barbieri, J. Rodrigo-Comino, H., et al. (Eds.), *Selected studies in environmental geosciences and hydrogeosciences* (pp. 99–102). Springer, Cham. https://doi.org/10.1007/978-3-031-43803-5_22

Dent, D., Bagchi, R., Robinson, D., Majalap-Lee, N., & Burslem, D. F. R. P. (2006). Nutrient fluxes via litterfall and leaf litter decomposition vary across a gradient of soil nutrient supply in a lowland tropical rain forest. *Plant and Soil, 288*(1–2), 197–215.

Fernández-Martínez, R., Loredo, J., Ordóñez, A., & Rucandio, I. (2014). Mercury availability by operationally defined fractionation in granulometric distributions of soils and mine wastes from an abandoned cinnabar mine. *Environmental Science: Processes & Impacts, 16*(5), 1069–1075.

Gómez-Armesto, A., Bibián-Núñez, L., Campillo-Cora, C., Pontevedra-Pombal, X., Arias-Estévez, M., & Nóvoa-Muñoz, J. C. (2018). Total mercury distribution among soil aggregate size fractions in a temperate forest podzol. *Spanish Journal of Soil Science, 8*(1), 57–73.

Huang, B., Yuan, Z., Li, D., Zheng, M., Nie, X., & Liao, Y. (2020). Effects of soil particle size on the adsorption, distribution, and migration behaviors of heavy metal(loid)s in soil: a review. *Environmental Science: Processes & Impacts, 22*(8), 1596–1615.

Jamilah, M. S., Nur-Faiezah, A. G., Siti Kehirah, A., Siti Mariam, M. N., Razali, M. S. (2014). Woody plants on dune landscape of Terengganu, Peninsular Malaysia. *Journal of Tropical Forest Science, 26*(2), 267–274.

Limbong, D., Kumampung, J., Ayhuan, D., Arai, T., & Miyazaki, N. (2005). Mercury pollution related to artisanal gold mining in north Sulawesi Island, Indonesia. *Bulletin of Environmental Contamination and Toxicology, 75*(5), 989–996.

Lofthouse, S. (no date). Determination of mercury using a basic vapour system and the iCAP 6000 Series ICP. Thermo Scientific Sales Team Note.

Miyamoto, K., Rahajoe, J. S., Kohyama, K., & Mirmanto, E. (2007). Forest structure and primary productivity in a Bornean heath forest. *Biotropica, 39*(1), 35–42.

Niiyama, K., Ripin, A., Yasuda, M., Sato, T., Zamah Shari, N. H. (2019) Data paper: Long-term litter production in a lowland dipterocarp forest, Peninsular Malaysia from 1992 to 2017. *Ecological Research* 34(1), 30.

Proctor, J., Anderson J M, Fogden S C L, Vallack H W. (1983). Ecological studies in four contrasting lowland rain forests in Gunung Mulu National Park, Sarawak. II. Litterfall, litter-standing crop and preliminary observations on herbivory. *Journal of Ecology, 71*(1), 261–283.

Roslan, I., Shamshuddin, J., Fauziah, C. I., & Anuar, A. R. (2010). Occurrence and properties of soils on sandy beach ridges in the Kelantan-Terengganu plains, Peninsular Malaysia. *Catena, 83*(1), 55–63.

Siau, Y. F., Le, D. Q., Suratman, S., Jaamana, S., Tanaka, K., & Kotaro, S. (2021). Seasonal variation of total mercury transfer through a tropical mangrove food web, Setiu Wetlands. *Marine Pollution Bulletin, 162*, 111878.

Tomiyasu, T., Hamada, Y. K., Bransono, C., Kodamatani, H., Matsuyama, A., Imura, R., Hidayati, N., & Rahajoe, J. S. (2020). Mercury concentrations in paddy field soil and freshwater snails around a small-scale gold mining area, West Java, Indonesia. *Toxicology and Environmental Health Sciences, 12*(1), 23–29.

Wang, X., Bao, Z., Lin, C.-J., Yuan, W., & Feng, X. (2016). Assessment of global mercury deposition through litterfall. *Environmental Science and Technology, 50*(16), 8548–8557.

Wolswijka, G., Satyanarayana, B., Le, D. Q., Siau, Y. F., Ali, A. N. B., Saliu, I. S., Fisol, M. A. B., Gonnelli, C., Dahdouh-Guebas, F. (2020). Distribution of mercury in sediments, plant and animal tissues in Matang Mangrove Forest Reserve, Malaysia. *Journal of Hazardous Materials,* 387, 121665.

Zhou, J., Obrist, D., Dastoor, A., Jiskra, M., & Ryjkov, A. (2021). Vegetation uptake of mercury and impacts on global cycling. *Nature Reviews: Earth and Environment, 2*(4), 269–284.

Atmospheric and Oceanographic Sciences

Improved Navigation Based on Received Signal Quality Monitoring (RSQM)

Madad Ali Shah, Arif Hussain, Syed Hadi Hussain Shah, Izhar Hussain and Hina Magsi

Abstract

The global navigation satellite system (GNSS) provides an easy, efficient, and cost-effective way to determine position, time, and direction anywhere around the globe. However, standard GNSS can provide users with highly accurate positioning information in clear open sky environments only and often lacks consistency in maintaining required navigation performance thresholds in the urbanized environments. In such operating environments, the quality of GNSS signals is significantly degraded due to multipath (MP) and non-line-of-sight (NLOS) reception which can ultimately result in inaccurate estimation of position and navigation parameters. The MP/NLOS reception remains a potential obstacle to the practical realization of GNSS in highly urbanized environments, and the severity of the effect varies greatly with the type of environment. Thus, in order to meet stringent requirements defined for practical applications of GNSS, the characterization of signal disruption which could significantly affect a GNSS performance is highly important. In this paper, an adaptive signal quality monitoring (RSQM) method is proposed which can monitor the quality of signal and anticipate the possible degradation by investigating the temporal variations in GNSS features. The proposed RSQM method performs rigorous quality tests on GNSS fundamental features, i.e., carrier-to-noise ratio (CNR) and range residuals, and then ranks the quality of signal into three categories, i.e., best, mediocre, and worse. Leveraging the redundancy of range measurements, the RSQM selects navigation signals having the best quality and discards or de-weights the poor quality signals. To validate the performance of the proposed RSQM algorithm, a detailed study on the performance of a multi-GNSS receiver in the quad-constellation mode has been carried out on a pre-surveyed track. The experimental results show that RSQM method can accurately detect and categorize the signal quality which can help in improving the navigation accuracy by suppressing the MP/NLOS effects.

Keywords

GNSS · Multipath · Accuracy · CNR

1 Introduction

Accurate, seamless, and reliable navigation has become an integral part of nearly every intelligent transportation system (ITS). The significance of accurate and precise navigation information is obvious for certain applications such as transportation, precision agriculture, and location-based services, but certain safety–critical applications, i.e., autonomous driving architectures, intelligent transportation systems, and geo-fencing majorly, rely on accurate positioning and precise navigation to carry out essential operations (European, 2020; Michael Dooley et al., 2015). The global navigation satellite system (GNSS) is the long-established satellite-based navigation system providing positioning, navigation, and timing (PNT) services across the globe. In recent times, the GNSS technology has advanced at a breakneck pace, and the system has grown from the GPS, only system, to four independent global constellations including the Russian GLONASS, European Galileo, and

M. A. Shah · H. Magsi
Department of Electrical Engineering, Sukkur IBA University, Sukkur, Pakistan
e-mail: madad@iba-suk.edu.pk

A. Hussain (✉) · I. Hussain
The Benazir Bhutto Shaheed University of Technology and Skill Development Khairpur Mirs, Khairpur, Pakistan
e-mail: arif@bbsutsd.edu.pk

S. H. H. Shah
Muhammad Ali Jinnah University, Karachi, Pakistan

H. Chenchouni et al. (eds.), *Recent Advancements from Aquifers to Skies in Hydrogeology, Geoecology, and Atmospheric Sciences*, Advances in Science, Technology & Innovation, https://doi.org/10.1007/978-3-031-47079-0_35

Chinese BeiDou, leading to a remarkable boost in satellite availability. Furthermore, GNSS modernization has also enabled signal transmission at multiple frequencies, increasing capability in terms of signal acquisition, atmospheric correction, multipath detection, resilience to interference, and precise positioning (Hussain et al., 2020; Joubert et al., 2020). Despite remarkable advancements and boost in satellite availability, a standalone GNSS often lacks consistency in maintaining required navigation performance thresholds in urban environments due to the serious impact of signal blockage, multipath, and NLOS reception. In such environmental contexts, the navigation signals are impaired, faded, and sometimes completely blocked which can ultimately result in inaccurate estimation of positioning solutions and degraded navigation performance (Ioannides et al., 2016; Zidan et al., 2020), although there has been considerable research into improving the navigation performance in urban environments by detecting and mitigating the NLOS/multipath effects at various stages (Gao & Groves, 2018; Marais et al., 2020; Pirbhulal et al., 2019). However, these mitigation models work fairly well in some environments but are of limited effectiveness in others. Keeping this in view, this paper presents an adaptive signal quality monitoring (RSQM) method which can monitor the quality of signal and anticipate the possible degradation by investigating the temporal variations in GNSS features. The proposed RSQM method performs rigorous quality tests on GNSS fundamental features, i.e., carrier-to-noise ratio (CNR) and range residuals, and then ranks the quality of signal into three categories, i.e., best, mediocre, and worse. Finally, leveraging the redundancy of range measurements, the RSQM selects navigation signals having the best quality and discards the poor quality signals.

2 Methodology and Field Experimentation

The field experimentation has been carried out in the dynamic mode to acquire the navigation signals and detect inconsistencies in the required navigation performance by driving an instrumented vehicle inside the urbanized city Sukkur, Pakistan. The entire track is 18 km long and covers all types of working environments, i.e., dense urban, urban, and suburb as shown in Fig. 1a. The route is further categorized into three different tracks, i.e., T1, T2, and T3, based on the signal reception conditions. Figure 1b is a 5-km long track, T1, and contains clear environment with minimal degradations. The second track, T2, is 7 km long having highly urbanized environment, which contains the high walls buildings, trees, and traffic disturbances as shown in Fig. 1c. The last track has the similar characteristics as of T1.

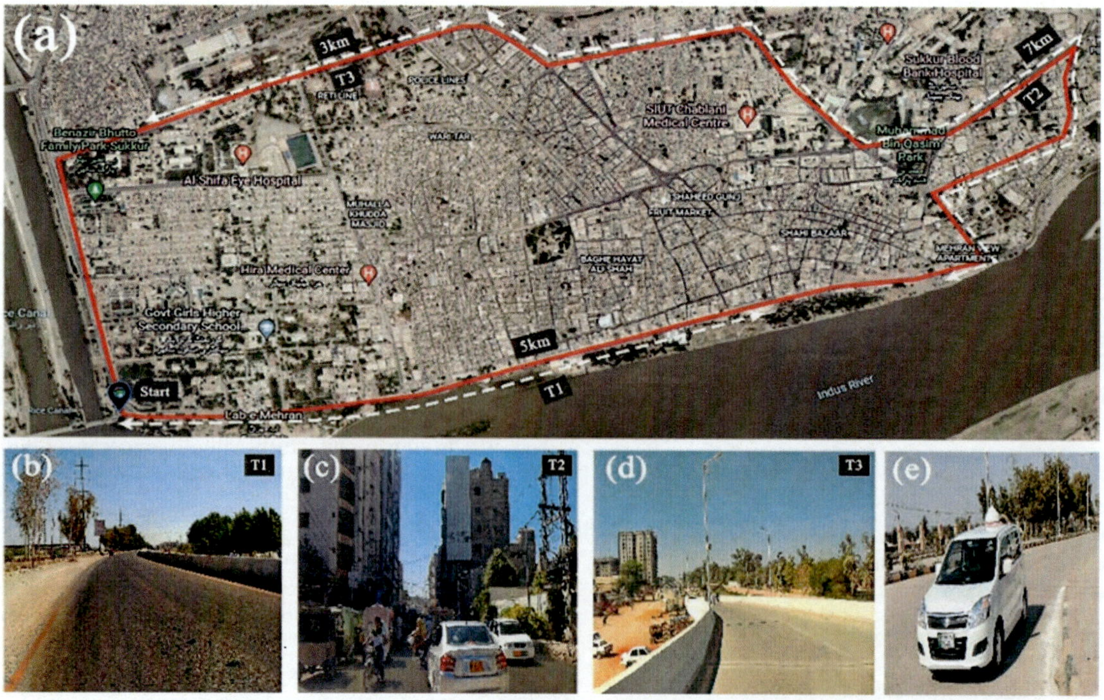

Fig. 1 The entire route containing three distinct tracks T1, T2, and T3 for the field experimentation and signal acquisition using instrumented test vehicle

3 Results

The performance assessment results of GNSS receiver including satellite availability (SA) that indicates the total number of satellites tracked by GNSS receiver at particular time and root mean square (RMS) error which is the estimated position error are shown in Fig. 2. The significant variations can be observed in the SA and RMS when moving from one track to another. On the first track T1, averagely more than 30 satellites were acquired and tracked which dropped to 23 on the track T2, due to changing environmental contexts. Similar characteristics in the satellite availability were observed at the T3 having open sky views. However, the notable deviations in the RMS errors were observed on T1 and T2 tracks, i.e., the average and maximum RMS errors increased to 6.7 m and 27.8 m on T2 track which were 2.3 m and 6.7 m, respectively, on T1 track. This level on inaccuracy may be problematic for many safety–critical applications. Although adequate satellite availability was observed in all the three tracks, RMS error is still problematic, and this may be due to the poor quality of the navigation signals contaminated by multipath/NLOS reception.

In order to further investigate the signal quality and anomalies that occurred due to MP/NLOS, we have carefully selected two satellites in different environments (open sky and degraded environment) and compared the carrier-to-noise ratio (CNR), elevation angle, and range residuals (RR) as shown in Fig. 3. Figure 3a shows the CNR, (b) is the elevation angle, and (c) represents the RR of G32 and

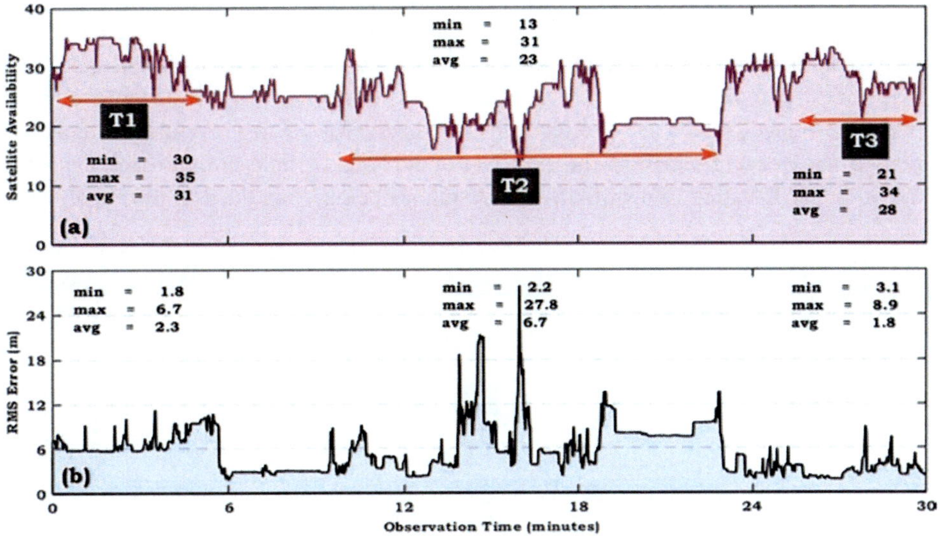

Fig. 2 Performance analysis of multiconstellation GNSS in moving mode. **a** Satellite availability, **b** RMS error

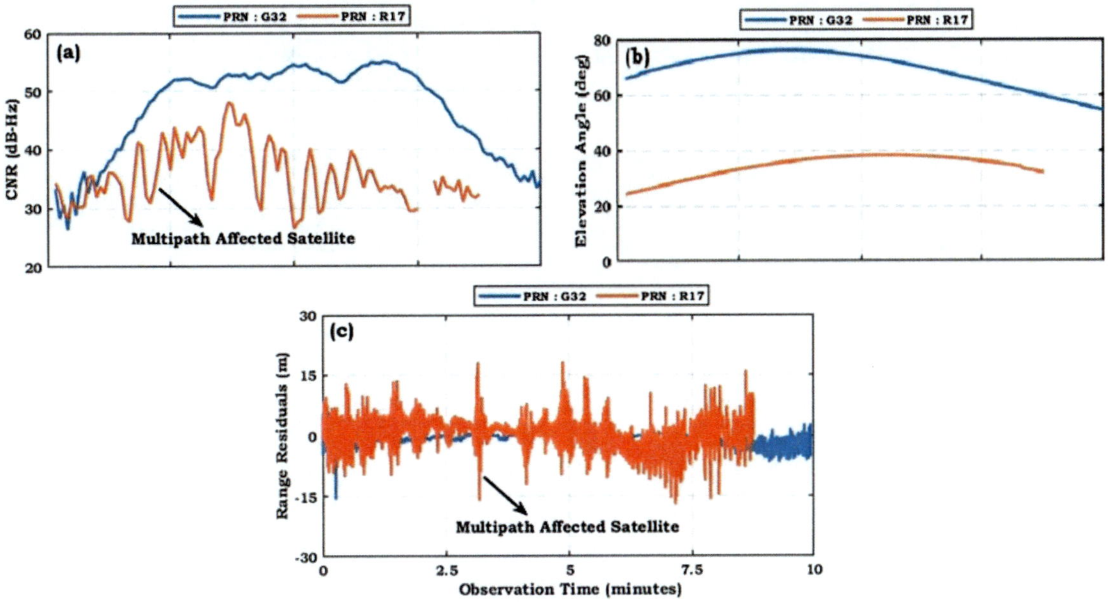

Fig. 3 GNSS satellite characteristics in open sky and degraded environment for G32 and R17. **a** Carrier-to-noise ratio (CNR), **b** elevation angle, and **c** range residuals

Table 1 The signal quality characterization model of RSQM

S. No.	CNR (dB)	Range residual (m)	Signal quality
01	CNR ≥ 40	RR ≤ | 1 |	Best
02	25 ≤ CNR < 40	| 1 | < RR ≤ | 5 |	Mediocre
03	CNR < 25	RR > | 5 |	Worst

R17 satellites for an observation period of 10 min. It can be observed from Fig. 2a and b that the CNR of the PRN G32 satellite varied above 35 (dB-Hz) during the entire observation time and at higher elevation angles. G32 being the best quality signal satellite and contributes to lower values of range residuals deviating between −1 m and + 1 m. It can be noted that all three signal quality parameters have a negligible effect of MP/NLOS anomalies on G32 due to no obstructions and complicated surrounding buildings. In the case of T2 track (degraded), number of satellites are greatly disturbed due to complex building structures, traffic, and trees. One of those satellite named as R17 is shown in Fig. 3 as orange colored and is affected by multipath. We observed the frequent variations in the CNR of R17, and most of the CNR varied below 40 dB-Hz. Despite of being higher elevation angles, range residuals of R17 increased to 15 m and deviated between −20 m and 15 m. Hence, it can be said that MP/NLOS significantly disturbs the signal quality in urban environment.

Based on the detailed analysis of GNSS performance and signal quality, it is established that quality of signal is greatly affected by MP/NLOS in highly urbanized environment. Keeping this in view, we have proposed an algorithm received signal quality monitoring (RSQM) that performs rigorous quality tests on GNSS fundamental features, i.e., CNR and range residuals. The RSQM identifies best quality signals and discards/masks the poor quality signals that can affect the position error using following model (Table 1).

Once the signal quality is detected and identified, then RSQM method adaptively updates the satellite masking level to discard or utilize the signal for navigation solution based on detected quality. After implementing the RSQM algorithm, the better performance is observed in the urbanized environment specifically on T2 as shown in Fig. 4. The average RMS error is reduced to 4.1 m which was around 7 m previously without using RSQM. Hence, RSQM can help in improving the navigation performance by suppressing the degradation effects.

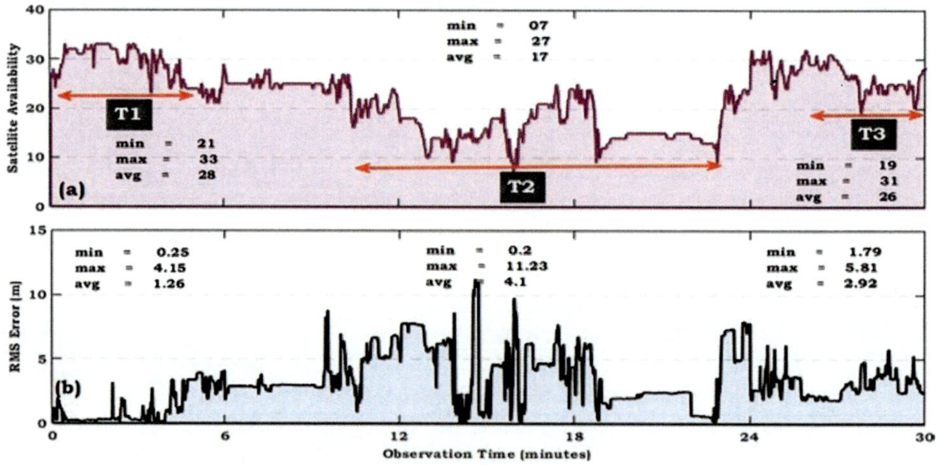

Fig. 4 Performance analysis of multi-constellation GNSS in moving mode after RSQM. **a** Satellite availability, **b** RMS error

4 Conclusion

This paper has investigated the navigation signal quality by performing rigorous quality tests on the GNSS signals acquired and tracked on various tracks. Then, an adaptive RSQM method is proposed to improve the navigation performance by suppressing the degradation effects.

References

European, G. N. S. S. (2020). GNSS market report. https://doi.org/10.2878/565013

Gao, H., & Groves, P. D. (2018). Environmental context detection for adaptive navigation using GNSS measurements from a smartphone. *Navigation: Journal of the Institute of Navigation, 65*(1), 99–116. https://doi.org/10.1002/navi.221

Hussain, A., Ahmed, A., Magsi, H., & Tiwari, R. (2020). Adaptive GNSS receiver design for highly dynamic multipath environments. *IEEE Access, 8*, 172481–172497. https://doi.org/10.1109/ACCESS.2020.3024890

Ioannides, R. T., Pany, T., & Gibbons, G. (2016). Known vulnerabilities of global navigation satellite systems, status, and potential mitigation techniques. *Proceedings of the IEEE, 104*(6), 1174–1194. https://doi.org/10.1109/JPROC.2016.2535898

Joubert, N., Reid, T. G. R., & Noble, F. (2020). Developments in modern GNSS and its impact on autonomous vehicle architectures. *IEEE Intelligent Vehicles Symposium proceeding*, 2029–2036. https://doi.org/10.1109/IV47402.2020.9304840

Marais, J., Kazim, S. A., Cocheril, Y., & Meurie, C. (2020). Multipath and NLOS detection based on the combination of CN0 values and a fish-eye camera. In *2020 European Navigation Conference ENC 2020*, November 2020. https://doi.org/10.23919/ENC48637.2020.9317408

Michael Dooley, G. B., Pirjanian, P., Romanov, N., Chiu, L., Di Bernardo, E., & Stout, M. (2015). Application of localization, positioning & navigation systems for robotic enabled mobile products.

Pirbhulal, S., Pombo, N., Felizardo, V., Garcia, N., Sodhro, A. H., & Mukhopadhyay, S. C. (2019). Towards machine learning enabled security framework for iot-based healthcare. In *Proceedings of the international conference on sensing technology*, ICST, Dec 2019, (vol. 2019-Decem). https://doi.org/10.1109/ICST46873.2019.9047745

Zidan, J., Adegoke, E. I., Kampert, E., Birrell, S. A., Ford, C. R., & Higgins, M. D. (2020). GNSS vulnerabilities and existing solutions: A review of the literature. *IEEE Access, 4*, 1–1. https://doi.org/10.1109/access.2020.2973759

Space Analog's Searching to Improve Deterministic Forecasting Using Analog Ensemble Method Over Morocco

Badreddine Alaoui, Driss Bari and Yamna Ghabbar

Abstract

The forecasting of surface weather parameters is a key stone to every weather and climate application, and it is mainly conducted by deterministic approaches, where several uncertainty sources are faced. These limitations enhanced the reflection toward using ensemble forecasting methods, which provides helpful tools for decision-making. In this work, we aim to develop a low computational cost ensemble forecasting approach based on the analog ensemble method (AnEn), to predict six surface parameters (T2m, WS10m, WD10m, RH2m, SURFP, and MSLP) at 10 airports of Morocco for 24 forecast hours. For this goal, we use hourly observations and forecasts from the operational mesoscale numerical model AROME covering the 4-year period (2016–2019). The latter was split into training (2016–2018) and testing (2019) periods. In AnEn, the selection of analogs commonly considers only historical data for each grid point in the study domain closest to the observation site. Herein, we propose two novelties: firstly, a new weighting strategy for predictors where we use three machine learning algorithms (linear regression, XGBoost, and random forest) to assign predictors' weights. Secondly, since AnEn requires larger training dataset to enhance chances to find best analogs, we extend the search space by integrating neighboring grid points. Thus, the analog detection is based here on 16 nearest grid points. As a result of the combination of these two techniques, it is found that the machine learning weighting strategies proved an improvement of performances in bias and root mean square error for different lead times and locations. Since the new space neighboring strategy maximizes the chances to find the best analogs, clear improvements were perceived for most airports. However, performances remain geographically dependent. In some airports, where topography is heterogeneous, applying this new analog searching strategy might lead to some worsening since weather conditions can vary at a hectometric scale.

Keywords

Analog ensemble · Machine learning · Space neighborhood · AROME · Space analog search

1 Introduction

Weather surface parameters play a key role in all weather and climate applications. For most world weather centers, predicting those parameters is based on determinism which raises several limitations and uncertainty sources (Eckel & Mass, 2005; Orrell et al., 2001; Paimazumder & Mölders, 2009). Ensemble forecasting provides a reliable alternative since it covers all the potential prediction scenarios (Leith, 1974), but compared to determinism, the latter requires high computational capacities. Analog ensemble method is one of the most effective and low-cost ensemble methods. This method has shown high success in several applications (Alessandrini et al., 2018; Davò et al., 2016; Delle Monache et al. 2013). The fundamental idea behind is to look for all the similar weather situations to the current one in a set of past forecasts, analogs, provided by the same numerical weather prediction (NWP) model. For a given location, whenever the most analogous past situations are defined, their associated observations construct our

B. Alaoui (✉) · D. Bari
Direction Générale de la Météorologie, CNRM, Casablanca, Maroc
e-mail: alaoui.badreddine.abe@gmail.com

B. Alaoui · Y. Ghabbar
Hassania School of Public Works, LaGeS/MoNum, Casablanca, Maroc

H. Chenchouni et al. (eds.), *Recent Advancements from Aquifers to Skies in Hydrogeology, Geoecology, and Atmospheric Sciences*, Advances in Science, Technology & Innovation, https://doi.org/10.1007/978-3-031-47079-0_36

ensemble. For all the applications above, the search space of the analogs is only restricted to the past forecasts' series for single grid points (Clemente-Harding et al., 2018). Sperati et al. (2017), suggested a new approach where they extend AnEn application to a 2D grid instead of specific locations. The suggested method shows high profits regarding gridded probabilistic products and also outperforms the European Center of Medium Range Weather Forecasting (ECMWF) ensemble prediction system (EPS) up to the third day of forecast. On the other hand, Sha et al. (2022) proposed a new hybrid approach combining AnEn and convolutional neural networks to post-process spatial precipitation forecasts. The latter study underlined the skillfulness of this hybrid approach in producing more realistic spatial precipitation patterns and precipitation intensity.

In this work, we suggest two novelties: conversely to the traditional analog ensemble method where we search for analogs independently in a unique location which is the closest grid point to the studied station. We will extend the search space of analogs to cover the 16 closest grid points to the studied location. Once the best analogs are defined, their associated observation from the studied location constructs our ensemble. On the other hand, for all the previous applications of the AnEn method, predictors were assigned constant weights which is physically inconvenient. Newly, in this work, for each predictand in the similarity metric, predictors will be assigned weights depending on their importance coefficient related to the predictand. The weighting process is achieved by three machine learning algorithms: linear regression, XGBoost, and random forest.

2 Data and Methods

For this study, hourly observations and forecasts' datasets from 2016 to 2019 have been used. Forecasts are available at the nearest 16 grid points of the 10 studied stations. These forecasts come from the mesoscale operational model AROME (Seity et al., 2011) with a resolution of 2.5 km and 90 vertical levels with the first one starting at about 5 m. The studied locations are well distributed and represent different geographic and climatic zones all across Morocco. The basic idea behind the analog ensemble method is to find all the similar weather situations to the current forecast in the past forecasts' archives. This process is based on a set of predictors and a similarity metric. The similarity measure used by authors in Delle Monache et al. (2013) is defined as follows:

$$\|F_t, A_{t'}\| = \sum_{i=1}^{N_v} \frac{w_i}{\sigma_i} \sqrt{\sum_{j=-\tilde{t}}^{\tilde{t}} (F_{i,t+j} - A_{i,t'+j})^2}, \tag{1}$$

where F_t is the current NWP deterministic forecast valid at the future time t at a station location; $A_{t'}$ is an analog (past forecast) at the same location and with the same forecast lead time but valid at a past time t'; N_v and w_i are the number of physical variables used in the analogs' search and their weights, respectively; σ_i is the standard deviation of the time series of past forecasts of a given variable i at the same location and forecast lead time; $\tilde{t}$ is equal to half the number of additional times over which the metric is computed; and $F_{i,t+j}$ and $A_{i,t'+j}$ are the values of the forecast and the analog in a time window with j from $-\tilde{t}$ to $+\tilde{t}$ for a given variable i.

In this work, our goal is to predict the six fundamental weather surface parameters: temperature (T2m) and relative humidity (RH2m) at 2 m, wind speed and direction (WS10m, WD10m) at 10 m, and surface and mean sea level pressures (SURFP and MSLP). For each of the studied variables, the predictors' set includes past forecasts of the target variable in addition to the 5 remaining parameters. However, instead of looking at the best analogs only at the studied station nearest grid point (AnEn_1GP), we extend the space of search, to include the 16 nearest points to the target location (AnEn_16GP (see Eq. 2). The best analogs verify the following condition:

$$\min_{gp=1,16} F_t^{gp}, A_{t'}^{gp}, \tag{2}$$

By this way, we enlarge our training dataset and multiply its size by 16. For the same training period (2016–2018), we search for the most analogous weather conditions at each of the 16 NWP grid points past forecasts' series. Once the most analogous situations are defined, their verifying observations at the studied location (i.e., center of the grid or the closest NWP grid point to the station) construct the ensemble forecasts. Moreover, we will use a slightly changed version of the analog's similarity measure in Delle Monache et al. (2013), where predictors' weights are equal to their normalized importance coefficients relative to the predictand (Alaoui et al., 2022). These weights are defined by three machine learning algorithms: linear regression (LR), random forest (RF), and XGBoost (XGB) (see Fig. 1). The constant equal weights basic (DM13) version is used also as a benchmark.

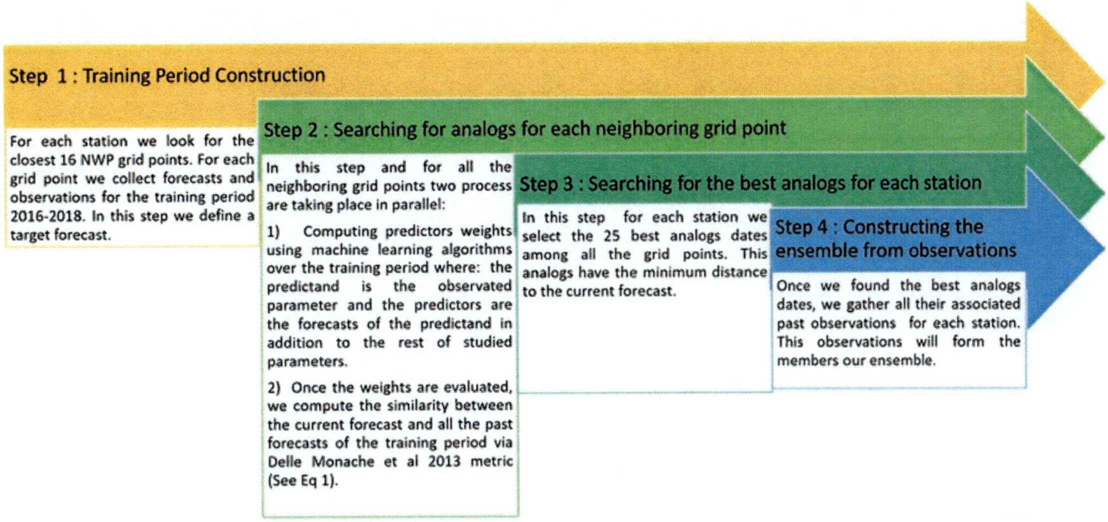

Fig. 1 Analog ensemble spatial extension general flow chart

To evaluate the performance of each configuration setup, the mean systematic error (bias) and root mean square error (RMSE) are used. In fact, the forecast error, between the output f_i and the observation o_i (the truth), can be expressed as $e_i = f_i - o_i$ where bias $= \overline{e_i}$ and RMSE $= \sqrt{\overline{e_i^2}}$.

3 Results

In this section, we will introduce the main findings of this work and we will evaluate the performances of both AnEn configurations with single and 16 neighboring grid points. Through this verification process, we will use two main metrics: the bias and RMSE (see section below).

In Fig. 2, we draw the AnEn single point (AnEn_1GP) and 16 neighboring grid points (AnEn_16GP) versions' bias and RMSE for T2m and RH2m. The bias expresses the mean systematic error between the ensemble mean and the observation. This score is averaged over all the studied stations and is plotted then for each forecast lead time.

For RH2m, both versions of AnEn generally outperform AROME during night and morning lead times, showing a bias that goes from 0 to −2%. Opposed to the AnEn_1GP version, with the spatial extension of analogs' search space, AnEn_16GP improves further bias particularly for early morning and day lead times. Besides, AnEn generally shows the best RMSE than AROME during the night. A slight improvement associated with AnEn_1GP is noticed during day lead times compared to AnEn_16GP.

For T2m, AnEn_16GP draws the same bias magnitude as AROME for morning and night lead times. However, both versions are outperformed by AROME during day lead times for both bias and RMSE. Findings in Fig. 2 show high sensitivity of AnEn performances to lead times. AnEn_16GP is performing well only during night and morning lead times; however, in day lead times, the latter is outperformed by AROME. Machine learning weighting strategies assessed slight improvement of RMSE for most studied parameters when compared to the basic DM13 constant weights.

From Fig. 3, we notice that AnEn displays high values of bias gains for WS10m (more than 100%). For RMSE, gains exceed 10% for most locations. Those gains are found mainly in coastal stations. However, AnEn draws few losses for some interior and south locations. In these locations, NWP performs well and draws good bias and RMSE which explains the deterioration carried by AnEn. Indeed, AnEn performances are highly sensitive to geography. For instance, the latter performs well, in locations surrounded by dense observation networks and where topography is quite homogeneous (coastal regions in northwest of Morocco as example). However, for mountainous and interior regions, performances are getting worse.

4 Discussion and Conclusion

In this work, we tested two novelties: firstly, we assigned new weights to each predictor depending on its importance coefficients to the predictand for three AI algorithms. Secondly, we extended the search space of analogs to 16 neighboring NWP grid points to the studied location instead of the nearest grid point.

Results show high sensitivity of AnEn performances to time and space. Indeed, AnEn displays best performances only for morning and night lead times. In day lead times,

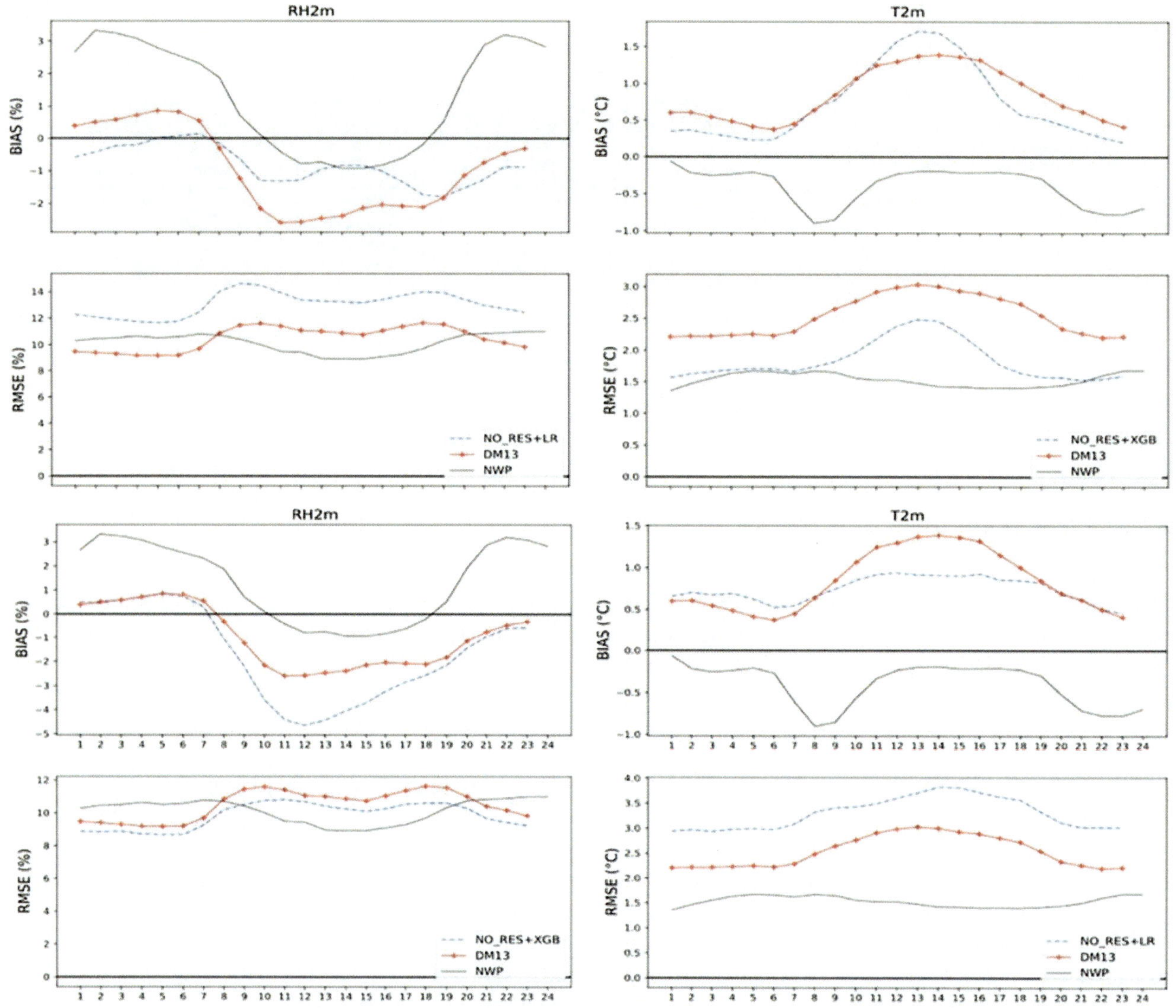

Fig. 2 T2m, RH2m, bias, and RMSE (averaged over all locations) as function of lead time for AnEn_16GP (up) and AnEn_1GP (down), where AnEn best weighting algorithm in dashed blue line, AROME in black solid line, and DM13 in x red

performances are similar or worse than that of the operational NWP AROME model.

However, using new weighting strategies helps to improve AnEn RMSE. From a geographic standpoint, AnEn with 16 nearest grid points (AnEn_16GP) displays high gains in terms of bias and RMSE for most parameters. However, for few locations, some worsening is noticed. Finding analogs of good quality depends directly on the length of the training datasets (Horton et al., 2012). Thus, extending analogs' search space to spatial neighbors enlarges the training dataset; hence, it improves AnEn performances. Furthermore, including calibration methods to correct bias in AnEn outputs is highly recommended (Alessandrini et al., 2019).

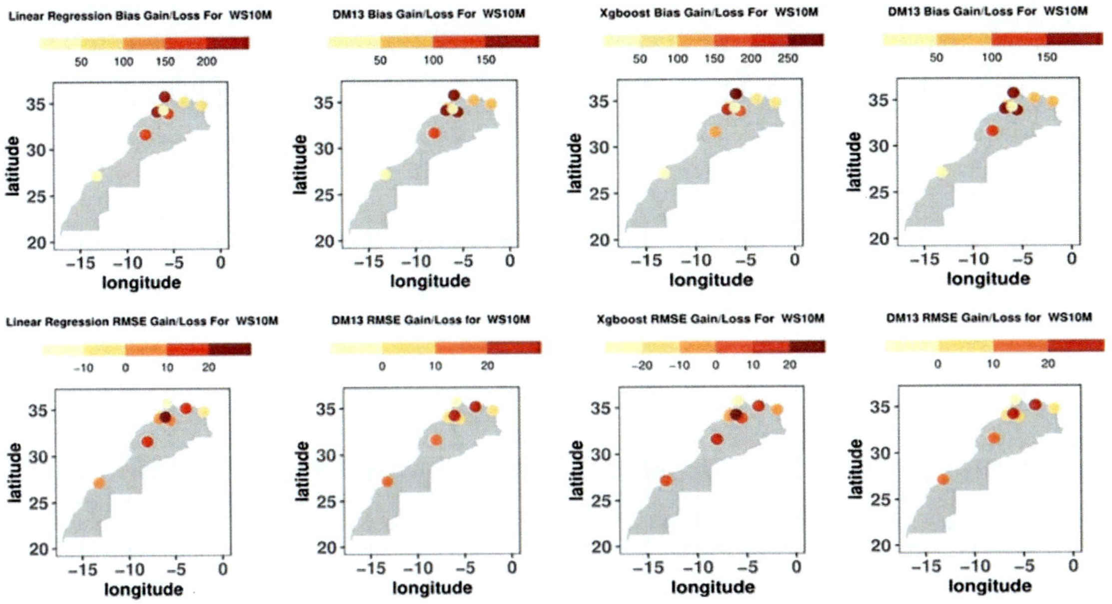

Fig. 3 WS10m, bias (up), and RMSE (down) gains/losses in percent for all the studied stations and the best weighting configurations (linear regression for single point and XGBoost for 16 neighboring points). This is shown for both experiments: AnEn_1GP (left) and AnEn_16GP (right)

References

Alaoui, B., Bari, D., & Ghabbar, Y. (2022). Surface weather parameters forecasting using analog ensemble method over the main airports of Morocco. *Journal of Meteorological Research, 36*(6), 866–881.

Alessandrini, S., Delle Monache, L., Rozoff, C. M., & Lewis, W. E. (2018). Probabilistic prediction of tropical cyclone intensity with an analog ensemble. *Monthly Weather Review, 146*

Alessandrini, S., Sperati, S., & Delle Monache, L. (2019). Improving the analog ensemble wind speed forecasts for rare events. *Monthly Weather Review, 147*(7), 2677–2692.

Clemente-Harding, L., Young, G. S., Cervone, G., Delle Monache, L., & Haupt, S. E. (2018). Toward a search space extension for the analog ensemble technique to improve probabilistic predictions, vol. 2018.

Davò, F., Alessandrini, S., Sperati, S., Delle Monache, L., Airoldi, D., & Vespucci, M. T. (2016). Post-processing techniques and principal component analysis for regional wind power and solar irradiance forecasting. *Solar Energy, 134*

Delle Monache, L., Eckel, F., Rife, D., Nagarajan, B., & Searight, K. (2013). Probabilistic weather prediction with an analog ensemble. *Monthly Weather Review, 141*, 3498–3516

Eckel, F. A., & Mass, C. (2005). Aspects of effective mesoscale, short-range ensemble forecasting. *Weather and Forecasting, 20*(3), 328–350.

Horton, P., Jaboyedoff, M., Metzger, R., Obled, C., & Marty, R. (2012). Spatial relationship between the atmospheric circulation and the precipitation measured in the western Swiss Alps by means of the analogue method. *Natural Hazards and Earth System Sciences, 12*(3), 777–784.

Leith, C. E. (1974). Theoretical skill of monte carlo forecasts. *Monthly Weather Review, 102*(6), 409–418.

Orrell, D., Smith, L., Barkmeijer, J., & Palmer, T. N. (2001). Model error in weather forecasting. *Nonlinear Processes in Geophysics, 8*(6), 357–371.

Paimazumder, D., & Mölders, N. (2009). Theoretical assessment of uncertainty in regional averages due to network density and design. *Journal of Applied Meteorology and Climatology, 48*, 1643–1666

Seity, Y., Brousseau, P., Malardel, S., Hello, G., Bénard, P., Bouttier, F., Lac, C., & Masson, V. (2011). The AROME-France convective-scale operational model. *Monthly Weather Review*, 139.

Sha, Y., Gagne, D. J., II., West, G., & Stull, R. (2022). A hybrid analog-ensemble–convolutional-neural-network method for post-processing precipitation forecasts. *Monthly Weather Review, 150*(6), 1495–1515.

Sperati, S., Alessandrini, S., & Delle Monache, L.: Gridded probabilistic weather forecasts with an analog ensemble. *Quarterly Journal of the Royal Meteorological Society, 143*(708), 2874–2885.

Global TEC Response to the November 2021 Geomagnetic Storm

Abdollah Masoud Darya, Muhammad Shaikh, Ilias Fernini and Hamid AlNaimiy

Abstract

This work reports the global ionospheric total electron content (TEC) distribution using global ionosphere maps (GIM) and Swarm satellite observations one day before and one day after the November 2021 geomagnetic storm. The November 2021 storm was the first intense storm of solar cycle 25, resulting from a cannibalistic coronal mass ejection, which is a combination of two coronal mass ejections. The main features observed were the enhancement of the equatorial ionization anomaly for up to 8 h post-storm sudden commencement (SSC), and the depletion of TEC along the equator compared to the post-storm day. Additionally, a hemispherical asymmetry was observed 8 h post-SSC in the form of an enhancement of TEC along the high-latitude northern hemisphere and depletion along the high-latitude southern hemisphere compared to the pre-storm day. This asymmetry lasted for more than 8 h. The maximum upper ionosphere (above the F2 layer peak) TEC observed by Swarm satellites started decreasing 6 h post-SSC and reached levels lower than that of pre-storm day. Additionally, the mean TEC of the upper ionosphere observed by Swarm satellites was lower for the entire post-storm day than the pre-storm day. This feature contradicts what was observed from GIM, where an enhancement of TEC was observed post-storm, particularly along low latitudes, and northern hemisphere high latitudes. This decrease of TEC in the upper ionosphere can be attributed to the lowering of the post-storm ionospheric F-region height on the dayside of the globe, as reported in previous work and confirmed empirically in this work.

Keywords

Ionosphere · Total electron content · GIM · Swarm

1 Introduction

The period of time when a long-lasting and sufficiently intense interplanetary electric field leads to an intensified ring current is defined as a geomagnetic storm (Gonzalez et al., 1994). Studying the ionospheric disturbances induced by geomagnetic storms is a topic of interest due to the ionosphere's vital role in satellite navigation and radio communication (Mansilla & Zossi, 2022). Furthermore, different geomagnetic storms have varying effects on the ionosphere, and as such, more studies are required to improve our understanding of the Sun–Earth interaction (Blagoveshchensky et al., 2018). One way to study the solar-ionospheric interaction resulting from space weather events is to analyze the distribution and variation of the total electron content (TEC) of the ionosphere (Davies & Rüster, 1976).

Recent studies on the TEC response to geomagnetic storms have targeted the August 2018 geomagnetic storm (Akala, 2021; Imtiaz et al., 2021; Mansilla & Zossi, 2020, 2022; Shah et al., 2021). The enhancement of the equatorial ionization anomaly (EIA) after the geomagnetic storm was reported by Shah et al. (2021); Akala (2021), Imtiaz et al., (2021), as a result of the prompt penetration electric field during the main phase of the storm (Kumar & Kumar, 2020; Picanço, et al., 2021). Additionally, an enhancement of TEC in equatorial and low latitude post-storm was reported in Akala (2021), Imtiaz et al., (2021) and was attributed to the strong enhancement of thermospheric O/N_2. They also reported the presence of a hemispherical asymmetry, where the northern hemisphere was observed with higher TEC during the recovery phase of the storm. This asymmetry was reported previously in Astafyeva (2020) and was attributed to the seasonal asymmetry in the neutral mass density

A. M. Darya (✉) · M. Shaikh · I. Fernini · H. AlNaimiy
SAASST, University of Sharjah, Sharjah, UAE
e-mail: abdollah.masoud@ieee.org

H. Chenchouni et al. (eds.), *Recent Advancements from Aquifers to Skies in Hydrogeology, Geoecology, and Atmospheric Sciences*, Advances in Science, Technology & Innovation, https://doi.org/10.1007/978-3-031-47079-0_37

and high-latitude plasma distributions. The asymmetries in the geomagnetic field were also a contributing factor. The magnetic field strength difference between the northern and southern polar caps at ionospheric heights resulted in larger ion drift velocities in the northern hemisphere (Astafyeva, 2020). It is also worth noting that the opposite was observed during the June 2015 storm, where an increase in TEC was observed in the southern hemisphere, and a decrease in the northern hemisphere (Mansilla, 2018; Şentürk, 2020). In this work, we study the global short-term variation of TEC one day before and one day after the November 2021 geomagnetic storm. First, we utilize the TEC observed by ground-based receivers and represented by the global ionosphere maps (GIM). We also include TEC observed by Swarm satellites to study the upper ionosphere, above the F2 peak height.

2 Data and Methodology

This work compares the global TEC distribution one day before and one day after the storm sudden commencement (SSC) time. From this point onwards, we refer to the pre-storm day: from 20 UTC, 2nd of November, to 19 UTC, 3rd of November. We also refer to the post-storm day: from 20 UTC, 3rd of November, to 19 UTC, 4th of November. The final GIM data for these two periods of time were provided by the Center for Orbit Determination in Europe (CODE),[1] by combining data from around 300 global navigation satellite systems (GNSS) stations. We also used TEC data for the same period generated by the polar-orbiting Swarm satellites Alpha, Bravo, and Charlie (A, B, and C) (Friis-Christensen et al., 2008). Both Swarm A and C were orbiting at a mean altitude of 440 km, while Swarm B was orbiting at 512 km during the study period. As per the recommendation of the European Space Agency (ESA),[2] we only considered Swarm vertical TEC observations with an elevation of more than 50°.

3 Results and Discussion

On the second of November 2021, an M1.7-class solar flare was released by sunspot AR2891. The coronal mass ejection (CME) released in this event merged with another slower CME released earlier and formed a cannibalistic CME (Gopalswamy, 2001). The SSC time was at around 20:20 UTC, as represented by the red line in Fig. 1.

The onset of the storm occurred with the almost instantaneous increase of plasma flow speed from 460 to 700 km/s, as well as the interplanetary magnetic field (IMF) magnitude average exceeding 10 nT, and 20 nT soon after. Half an hour later, at around 20:50 UTC, the interplanetary electric field (E_y) increased to over 10 mV/m, and the symmetric disturbance field in H (SYM/H) magnetic field peaked at over 50 nT. The SYM/H then started decreasing until it reached a minimum of -117 nT around 16:20 h post-SSC. Since the storm had a peak Dst index of less than -100 nT, it can be considered as an intense storm (Gonzalez et al., 1994). Storms of this magnitude are classified by the National Oceanic and Atmospheric Administration (NOAA) as G3 category storms (Hanslmeier, 2002). This was the first G3 category storm in solar cycle 25.

In Fig. 2, we present the bi-hourly GIM in terms of the vertical TEC represented by TEC units (TECU). These TEC values represent the difference found by deducting the pre-storm TEC from the post-storm TEC.

Figure 2a–d shows the formation of two lobes separated at the equator as a result of the enhancement of the equatorial ionization anomaly (EIA), which is a feature that was reported during the August 2018 geomagnetic storm (Akala, 2021; Imtiaz et al., 2021; Shah et al., 2021). This feature is illustrated by the clear decrease of TEC along the equator, and the increase seen at low latitudes (±20°).

In Fig. 2e–h, the intensity of the two lobes resulting from the EIA is seen to decline, compared to Fig. 2a–d. However, we start seeing the presence of a hemispherical asymmetry in terms of TEC, where in the southern hemisphere, especially under −50° latitude, we observe a decrease of TEC post-storm ranging from 0 to −20 TECU, compared to pre-storm time. On the other hand, the northern hemisphere, particularly above +50° latitude, witnesses an increase in TEC post-storm ranging from 0 to +10 TECU, compared to pre-storm time. This asymmetry was reported previously during the August 2018 geomagnetic storm (Akala, 2021; Imtiaz et al., 2021) and was attributed to the seasonal asymmetry in the neutral mass density and high-latitude plasma distributions (Astafyeva, 2020). The opposite was observed during the June 2015 storm, where an increase in TEC was observed in the southern hemisphere, and a decrease in the northern hemisphere (Mansilla, 2018; Şentürk, 2020). It is important to note that the hemispherical asymmetry observed in Fig. 2e–h occurred during the main phase of the storm, since the SYM/H is yet to reach its minimum around 10 h later, whereas in Astafyeva (2020), the asymmetry was observed during the recovery phase of the storm.

In Fig. 2i–l, the gradual recovery of the TEC values to their pre-storm values can be seen, except for a small region at low latitudes in the northern hemisphere that shows a difference of +30 TECU compared to pre-storm time. Furthermore, the hemispherical asymmetry of TEC is also

[1] These data are publicly available at: {http://cddis.nasa.gov/archive/gnss/products/ionex}.

[2] https://earth.esa.int/eogateway/missions/swarm/product-data-handbook.

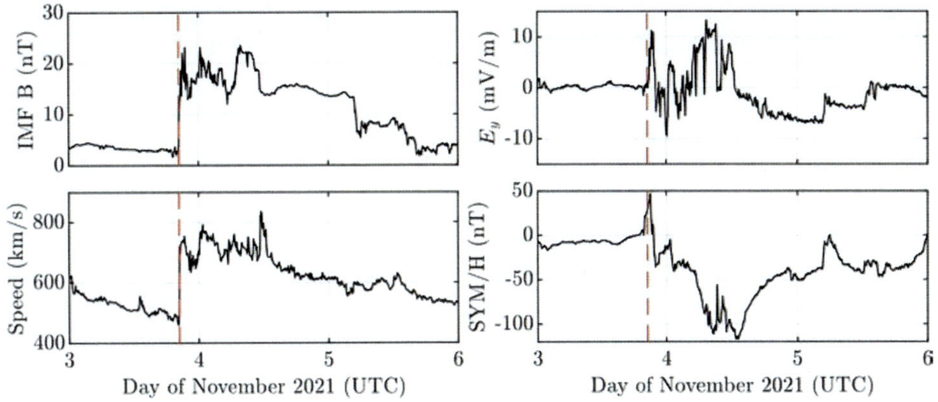

Fig. 1 The 5-min averaged magnetic field and solar wind parameters from November 3, 2021 to November 6, 2021. These parameters are freely available at: omniweb.gsfc.nasa.gov/form/omni_min.html. The dotted red line corresponds to the SSC time

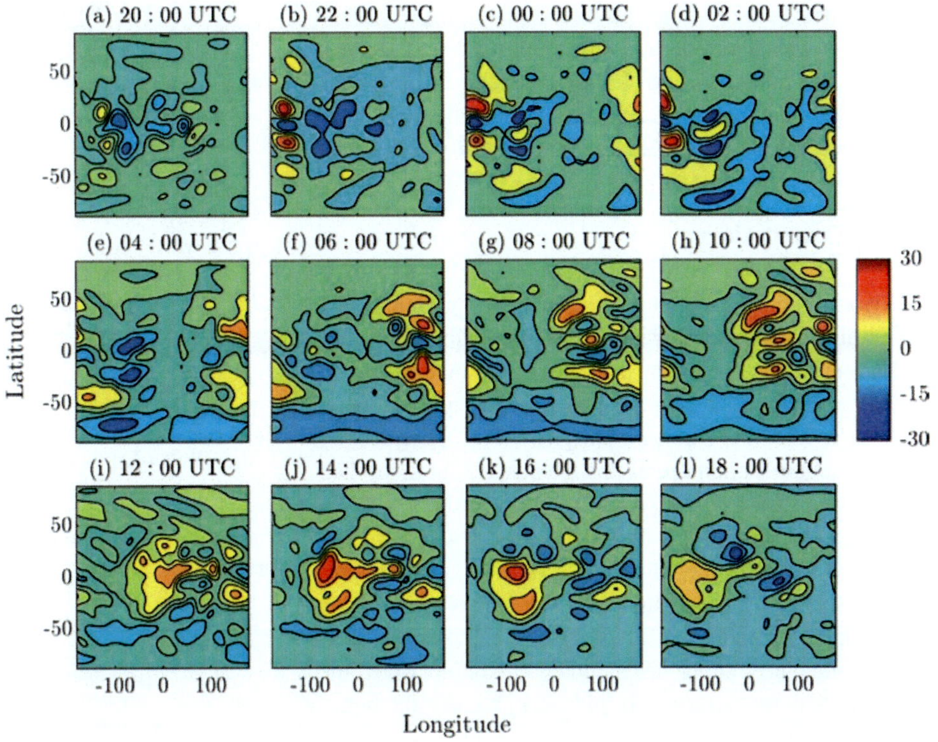

Fig. 2 The bi-hourly TEC difference, i.e., post-storm minus pre-storm TEC. For example, **a** shows the difference between the TEC for 11/03/2021 20:00 UTC (post-storm) and 11/02/2021 20:00 UTC (pre-storm). The color bars represent TECU

subsiding, with the TEC difference between the north and south hemispheres ranging from + 5 to −5 TECU.

In Fig. 3, we present the maximum and mean pre-storm and post-storm vertical TEC observed by Swarm satellites. These observations have been included in this work to study the upper ionosphere TEC (above the F2 peak) and compare it to the TEC observations made by ground-based GNSS receivers and presented by GIM.

In Fig. 3, we notice that the maximum TEC is higher than that of the pre-storm day 6 h post-SSC. However, from 12 to 18 h post-SSC, the maximum TEC observed by all Swarm satellites indicates a decrease in TEC post-storm, compared to pre-storm. The maximum TEC returned to its pre-storm levels 24 h post-SSC.

What is interesting about the mean TEC trends in Fig. 3 is that they were higher for the entire pre-storm day than the post-storm day. Additionally, they can be seen to decline from −24 h pre-SSC, reaching a minimum at + 18 h post-SSC. It is important to note that this feature was not seen in GIM observations (Fig. 2), where a clear enhancement of TEC was observed post-storm, especially at low latitudes. This reduction of plasma density was modeled previously in Fuller-Rowell et al. (2002) and was attributed to the lowering of the ionospheric F-region peak height on the dayside of the globe after a geomagnetic storm. Our work empirically confirms the existence of this reduction. Additionally, a hemispherical asymmetry was observed 8 h post-SSC in the form of an enhancement of TEC along the high-latitude

Fig. 3 The maximum and mean TEC observed by Swarm A, B, and C ± 24 h from SSC

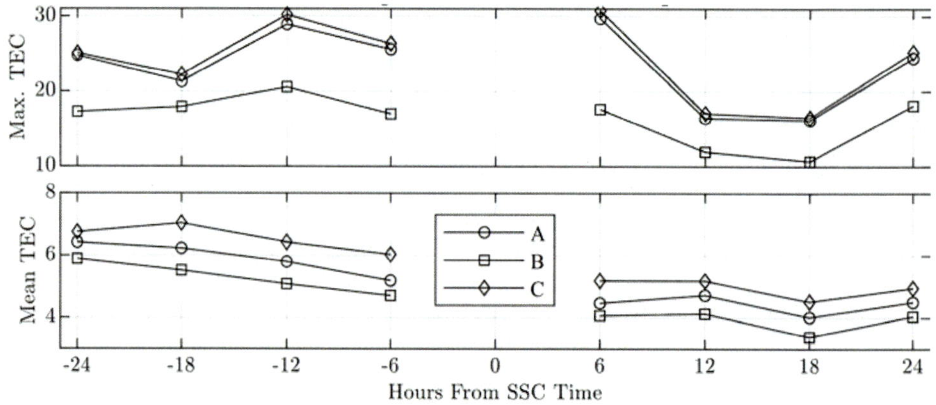

northern hemisphere and depletion along the high-latitude southern hemisphere compared to the pre-storm day. This asymmetry was also reported during the August 2018 geomagnetic storm (Akala, 2021; Imtiaz et al., 2021) and is opposite to what was observed during the June 2015 geomagnetic storm (Mansilla, 2018; Şentürk, 2020).

4 Conclusion

This work studied the global TEC distribution using GIM and Swarm satellite observations one day before and after the November 2021 geomagnetic storm. The main observed features were the enhancement of the EIA for up to 8 h post-SSC. The maximum upper ionosphere TEC observed by Swarm satellites started decreasing 6 h post-SSC and reached levels lower than that of pre-storm day. Furthermore, the mean TEC of the upper ionosphere observed by Swarm satellites was lower for the entire post-storm day than the pre-storm day. This feature contradicts what was observed from GIM, where an enhancement of TEC was observed post-storm, particularly along low latitudes. In future work, the two main features observed in this work, namely the hemispherical asymmetry, and the decrease of TEC post-storm in the upper ionosphere above the F2 peak will be studied in terms of plasma electrodynamics.

References

Akala, A. O., Oyedokun, O. J., Amaechi, P. O., et al. (2021). Solar origins of August 26, 2018 geomagnetic storm: Responses of the interplanetary medium and equatorial/low-latitude ionosphere to the storm. *Space Weather, 19*(10).

Astafyeva, E., Bagiya, M. S., Förster, M., & Nishitani, N. (2020). Unprecedented hemispheric asymmetries during a surprise ionospheric storm: A game of drivers. *Journal of Geophysical Research: Space Physics, 125*(3).

Blagoveshchensky, V., Maltseva, O., & Sergeeva, M. (2018). Impact of magnetic storms on the global TEC distribution. *Annales Geophysicae, 36*(4), 1057–1071.

Davies, K., & Rüster, R. (1976). Theoretical studies of storm effects in ionospheric total electron content. *Planetary and Space Science, 24*(9), 867–872.

Friis-Christensen, E., Lühr, H., Knudsen, D., & Haagmans, R. (2008). Swarm–an Earth observation mission investigating geospace. *Advances in Space Research, 41*(1), 210–216.

Fuller-Rowell, T. J., Millward, G. H., Richmond, A. D., & Codrescu, M. V. (2002). Storm-time changes in the upper atmosphere at low latitudes. *Journal of Atmospheric and Solar-Terrestrial Physics, 64*(12–14), 1383–1391.

Gonzalez, W. D., Joselyn, J. A., Kamide, Y., et al. (1994). What is a geomagnetic storm? *Journal of Geophysical Research: Space Physics, 99*(A4), 5771–5792.

Gopalswamy, N., Yashiro, S., Kaiser, M. L., Howard, R. A., & Bougeret, J. L. (2001). Radio signatures of coronal mass ejection interaction: Coronal mass ejection cannibalism?. *The Astrophysical Journal, 548*(1), L91.

Hanslmeier, A. (2002). The NOAA space weather scales. *The Sun and Space Weather*, 193–200.

Imtiaz, N., Hammou Ali, O., & Rizvi, H. (2021). Impact of the intense geomagnetic storm of August 2018 on the equatorial and low latitude ionosphere. *Astrophysics and Space Science, 366*(11), 1–14.

Kumar, S., & Kumar, S. (2020). Equatorial ionospheric TEC and scintillations under the space weather events of 4–9 September 2017: M-class solar flares and a G4 geomagnetic storm. *Journal of Atmospheric and Solar-Terrestrial Physics, 209*.

Mansilla, G. (2018). Ionospheric response to the magnetic storm of 22 June 2015. *Pure and Applied Geophysics, 175*(3), 1139–1153.

Mansilla, G., & Zossi, M. (2020). Longitudinal variation of the ionospheric response to the 26 August 2018 geomagnetic storm at equatorial/low latitudes. *Pure and Applied Geophysics, 177*(12), 5833–5844.

Mansilla, G., & Zossi, M. (2022). Ionospheric response to the 26 August 2018 geomagnetic storm along 280° E and 316° E in the South American sector. *Advances in Space Research, 69*(1), 48–58.

Picanço, G. A. D. S., Denardini, C. M., Nogueira, P. A. B., et al. (2021). Equatorial ionospheric response to storm-time electric fields during two intense geomagnetic storms over the Brazilian region using a Disturbance Ionosphere indeX. *Journal of Atmospheric and Solar-Terrestrial Physics, 223*, 105734.

Şentürk, E. (2020). Investigation of global ionospheric response of the severe geomagnetic storm on June 22–23, 2015 by GNSS-based TEC observations. *Astrophysics and Space Science, 365*(7), 1–15.

Shah, M., Abbas, A., Ehsan, M., et al. (2021). Ionospheric-thermospheric responses in south America to the August 2018 geomagnetic storm based on multiple observations. *IEEE Journal of Selected Topics in Applied Earth Observations and Remote Sensing, 15*, 261–269.

Induced Impact of El Niño-Southern Oscillation and Haze Events on Aerosol Optical Depth (AOD) in the Tropical Climate of Borneo Island

Carolyn Payus, Siti Irbah Anuar, Fazlina Nurashilah and Justin Sentian

Abstract

Previous studies demonstrated that the El Niño–Southern Oscillation (ENSO) could modulate regional climate, thus influencing air quality in tropical latitude regions like South-east Asia. However, such influence has never been analysed in Borneo Island as it is always strongly affected by the El Niño and haze events due to the complications of the Walker and Hadley circulations that severely impact the weather in this region. To fill the gap, this study investigated aerosol concentration by aerosol optical depth (AOD) and particulate matter (PM_{10}) in Borneo Island during the whole dynamic development of ENSO phases. The findings from this study showcased that the AOD values were usually highest in September months, mostly affecting Central Kalimantan, Indonesia, as compared to other regions in Borneo. The results suggest strong positive correlations between the aerosol optical depth (AOD) and El Niño phases, as high AOD occurred during the El Niño event. Such correlations are mainly attributed to atmospheric circulation and precipitation variation during the ENSO phase. This study further quantified the AOD variation from 2007 to 2019 attributed to changes in ENSO phases and PM_{10} contributions by > 50% due to anthropogenic emissions. The result also shows the importance of El Niño analysis by varying aerosol concentration (PM_{10}) in a short-term period and a more targeted evaluation of the detected events and the specific El Niño and haze episodes.

Keywords

El Niño · Southern oscillation · Haze · Aerosol · Aerosol optical depth (AOD) · Tropical · Borneo

1 Introduction

The measurement of aerosol is known as aerosol optical depth (AOD), which is known as the extinction coefficient of sunlight in the surrounding air because of aerosol (Omari et al., 2019). Retrieval of AOD includes ground-based and satellite-based methods. However, ground-based monitoring stations have limited availability and spatial coverage, making remote sensing the best substitute for monitoring AOD in larger spatial coverage (Yogesh et al., 2019). Particulate matter is the composition of suspended liquid and solid particles in the atmosphere, whilst aerosol is the composition of suspended particles mixed with the surrounding gases. However, both matters influence the air quality in the surrounding. In addition, El Nino phenomena can be linked with AOD and particulate matter. El Niño could be defined as the unusual increase of temperature at the ocean surface, which occurs irregularly at two- to seven-year intervals (Mendi & Dwa-rakish, 2015). The occurrence of El Niño event in the period was related to the fluctuations of the AOD data (Mielonen et al., 2018); Therefore, the main purpose of this study is to employ the remote sensing technique of AOD concentrations in Borneo to determine the magnitude of interference or impact of El Niño event onto the air pollution, particularly on haze event that is causing higher particulate matter.

2 Materials and Methods

The study was conducted at Borneo Island (0.9619° N, 114.5548° E) in South-east Asia (SEA). Borneo Island was administered by three different countries: Malaysia

C. Payus (✉) · J. Sentian
Natural Disasters Research Centres, Universiti Malaysia Sabah, 88400 Kota Kinabalu, Sabah, Malaysia
e-mail: cpayus@gmail.com

C. Payus · S. I. Anuar · F. Nurashilah · J. Sentian
Faculty of Science and Natural Resources, Universiti Malaysia Sabah, 88400 Kota Kinabalu, Sabah, Malaysia

H. Chenchouni et al. (eds.), *Recent Advancements from Aquifers to Skies in Hydrogeology, Geoecology, and Atmospheric Sciences*, Advances in Science, Technology & Innovation, https://doi.org/10.1007/978-3-031-47079-0_38

(Sabah, Sarawak, Federal Territory of Labuan), Brunei, and Kalimantan, Indonesia. The period of study is 2007–2019. AOD data were collected from January 2007 to October 2019 from NASA Giovanni (https://giovanni.gsfc.nasa.gov). El Niño data from 2007 to 2019 were obtained from NOAA (https://www.climate.gov). PM_{10} data for Borneo were obtained from the Department of Environment (DOE) and collected only for two years (2009 and 2015). Specifically, 2009 data are only at Palangkaraya, Indonesia, and Sibu, Sarawak, and for 2015 Kota Samarahan, Sarawak, and Tawau, Sabah. For fire distribution in this study, it was only retrieved for the year 2009 and obtained from the Fire Information for Resources Management System (FIRMS) (https://firms.modaps.eosdis.nasa.gov). Correlations were also made to analyse the relationship between AOD–ONI/El Niño and AOD–PM_{10}.

3 Results and Discussion

Over the study period, 2009, 2015, and 2019 showed the highest values of AOD recorded in September, as shown in Figs. 1, 2, and 3, respectively. For 2009, the highest AOD value was above 0.907 (September), and the lowest was 0.045 (March). For 2015, the highest was 1.765 (October), and the lowest was 0.040 (January–July and December). In 2019, the highest was 1.243 (September), and the lowest was 0.040. During the three years, the most affected region with high AOD was Central Kalimantan, and AOD was seen to be higher on land than the ocean. For yearly averaged comparison, the highest mean was on 2015 (0.222), and the lowest was on 2010 (0.107). For the monthly averaged AOD from 2007 to 2019, as illustrated in Fig. 4, the highest AOD value was 0.480, and the lowest was 0.050. Again, September and October showed the highest values, and the affected region is mostly Kalimantan.

From the 13-year study period, it was observed that there were three El Niño occurrences, including 2009–2010 (moderate El Niño), 2015 (strong El Niño), and 2018–2019 (weak El Niño). For 2009–2010, the AOD–ONI relationship showcases a moderate negative relationship ($R = -0.547$). On the other hand, AOD–ONI for 2015 illustrates a moderate positive relationship ($R = +0.547$), and for 2018–2019, it showcases a moderate positive relationship ($R = +0.394$). PM_{10} monitoring in 2009 showed that in January–July, Palangkaraya is lower than in Sibu. Concentration increases drastically in August with high peaks in September, which decrease afterwards. The amount of PM_{10} release corresponds with the intense forest fire occurrence in Borneo in 2009. This can be proven as Palangkaraya, located in South Borneo, has a huge forest fire in September, which correlates with having the highest PM_{10} in September 2009. The AOD–PM_{10} correlation results showed that for 2009, the

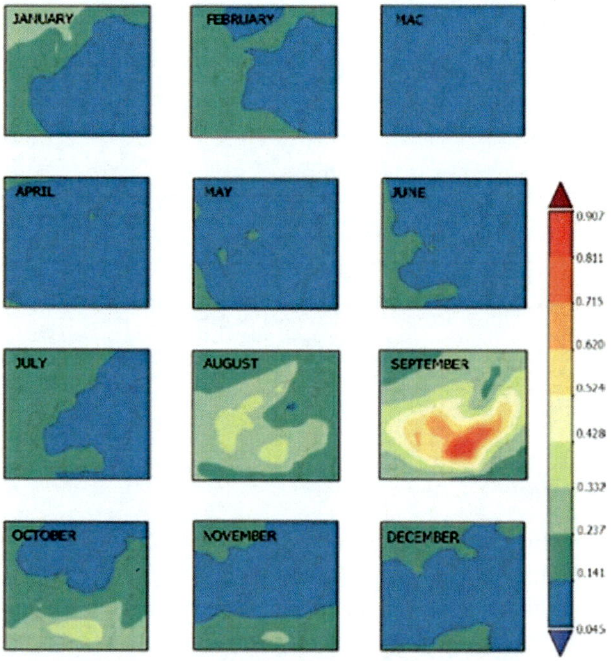

Fig. 1 Borneo monthly average aerosol optical depth (AOD) in 2009

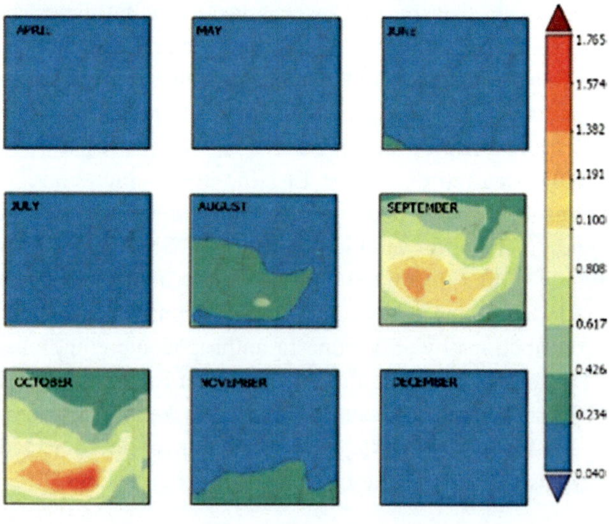

Fig. 2 Borneo monthly average aerosol optical depth (AOD) in 2015

AOD–PM_{10} indicates a moderate positive relationship for both Sibu ($R = +0.526$) and Palangkaraya ($R = +0.439$). For 2015, the correlation coefficient in Kota Samarahan ($R = +0.874$) and Tawau ($R = 0.980$) indicates a strong positive relationship. The moderate to a strong positive relationship suggests that as AOD increases, PM_{10} also increases in the atmosphere.

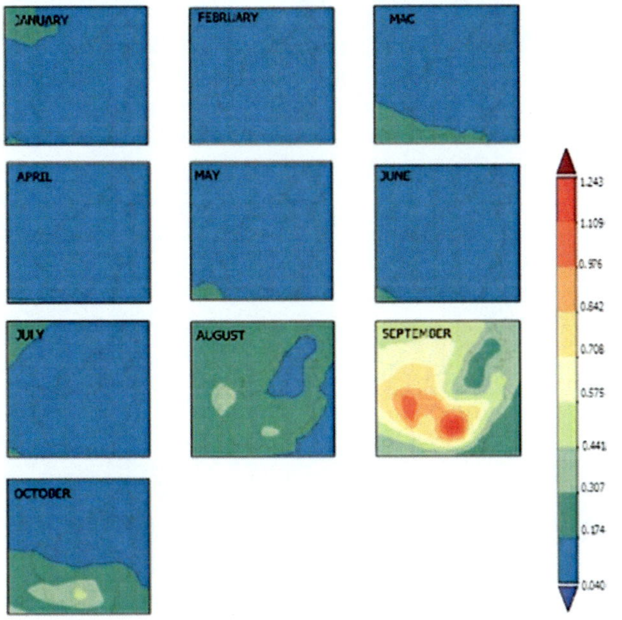

Fig. 3 Borneo monthly average aerosol optical depth (AOD) in 2019 for January until October

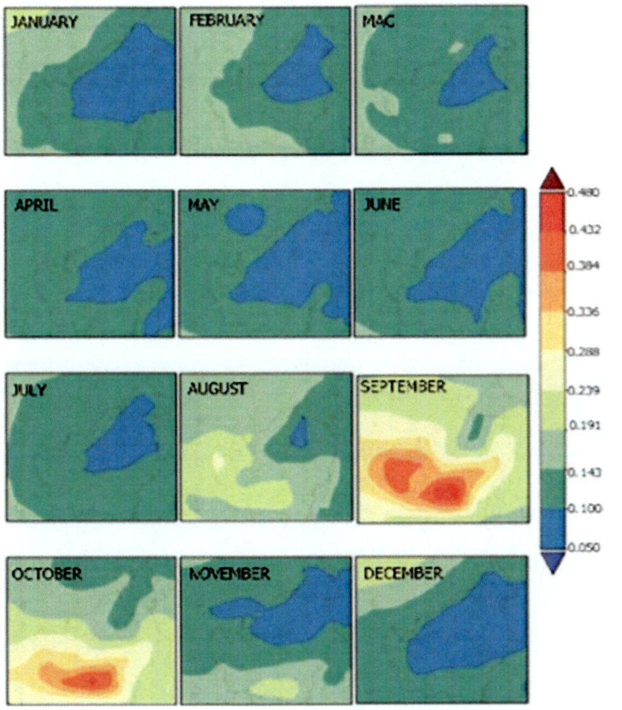

Fig. 4 Borneo monthly average aerosol optical depth (AOD) (2007–2019) for January until December

4 Conclusion

Due to the dry season association, the highest AOD reading is usually in September. The most affected region was typically Central Kalimantan in Indonesia, compared to other countries (Malaysia and Brunei) in Borneo, which usually showcased lower AOD readings. The years with the highest AOD values throughout the study were 2009, 2015, and 2019. This study suggests that there are correlations between AOD and El Niño as high AOD occurred during the El Niño event. Such correlations are mainly attributed to atmospheric circulation and surface temperature and precipitation variations during the corresponding El Niño events (ENSO phase). This study further concluded and quantified that the AOD variation from 2007 to 2019 attributed to > 30% of the changes in ENSO phases (El Niño events), and contributions from PM_{10} by > 50% influenced the AOD. The result also shows the importance of El Niño analysis by varying aerosol concentrations (PM_{10}) in a short-term period analysis and a more targeted evaluation of the detected events and the specific El Niño and haze episodes. Hence, this information will be useful in air quality management in the region regarding health protection, particularly when dealing with haze (PM) during El Niño.

References

Mendi, V., & Dwarakish, G. S. (2015). El Niño: A review. *International Journal of Earth Sciences and Engineering, 8*(2), 130–137.

Mielonen, T., Hienola, A., Kühn, T., et al. (2018). Summertime aerosol radiative effects and their dependence on temperature over the Southeastern USA. *Atmosphere, 9*(5), 180.

Omari, K., Abuelgasim, A., & Alhebsi, K. (2019). Aerosol optical depth retrieval over the city of Abu Dhabi, United Arab EMirates (UAE) using Landsat-8 OLI images. *Atmospheric Pollution Research, 10*(4), 1075–1083.

Yogesh, S., Kulkarni, S., Gupta, P., Kaginalkar, A., Islam, S., & Gargava, P. (2019). Application of moderate resolution imaging spectroradiometer (MODIS) Aerosol Optical Depth (AOD) and Weather Research Forecasting (WRF) model meteorological data for assessment of fine particulate matter (PM2.5) over India. *Atmospheric Pollution Research, 10*(2), 418–434.

Accurate Monitoring and Timely Prediction of Ionospheric Scintillation Using Support Vector Machine

Hina Magsi, Arif Hussain, Syed Hadi Hussain Shah, Madad Ali Shah, Safdar Ali Abro and Junaid Ahmed

Abstract

The monitoring and prediction of the ionospheric scintillation effect which is characterized by rapid fluctuations in the phase and/or amplitude of the satellite signals is considered to be a very important aspect of space weather monitoring and remote sensing. In addition, the ionospheric scintillation poses a significant threat to satellite-based navigation services. Typically, severe scintillation can result in the occurrence of cycle slip at the tracking stage or loss of lock events which can lead to degraded navigation performance. Thus, monitoring and timely prediction of ionospheric scintillation can help in improving the accuracy and reliability of navigation services, space weather monitoring, and remote sensing. Most of the traditional methods for scintillation monitoring are based on very basic event triggers which compare the scintillation values with fixed thresholds and then identify the scintillation event. However, these conventional methods having fixed thresholds may not be effective because the occurrence and severity of ionospheric scintillation is highly dependent on the geographical location of the receiver (i.e. latitude), time of the day, and geomagnetic and solar activity. In this paper, an efficient method for accurate, timely, and automatic prediction of amplitude scintillation events is proposed. Support vector machine (SVM) is the most efficient supervised machine learning method used for classification and regression, and it classifies the data based on kernel functions. In this paper, the proposed method applies the SVM on raw global navigation satellite system (GNSS) data to detect and classify the amplitude scintillation into three main categories, i.e. no Scintillation, moderate, and severe scintillation. For this purpose, raw GNSS data were acquired by deploying a Septentrio multi-frequency GNSS receiver at Sukkur, Pakistan. The results show that the machine learning algorithms can help in developing efficient scintillation prediction models that can accurately monitor scintillation occurrence patterns and predict the scintillation events. The proposed SVM-based model achieved an accuracy of 98% which is very close to manual classification driven by human. Moreover, the prediction responsiveness is also improved, allowing early scintillation alarms.

Keywords

GNSS · Ionospheric scintillation · SVM · Prediction models

1 Introduction

Satellite signals propagation through earth atmosphere has a disruptive impact on Earth or ground GNSS receiver. Irregularities and disturbances in the most ionized layer in the atmosphere ionosphere that is comprised of high amount of free electrons can produced the rapid deviations in the signal magnitude and phase that are referred as ionospheric scintillations. Under highly disturbed scintillations conditions, GNSS receivers are more prone to loss of lock, phase errors, deviated CNR which gives rise to the uncertainties in the position estimation (Kintner et al., 2007). Scintillation is a major hazard for GNSS; hence, they have

H. Magsi · M. A. Shah
GNSS and Space Weather Lab, Sukkur IBA University, Sukkur, Pakistan

S. H. H. Shah
Muhammad Ali Jinnah University, Karachi, Pakistan

M. A. Shah · J. Ahmed
Department of Electrical Engineering, Sukkur IBA University, Sukkur, Pakistan

A. Hussain (✉) · S. A. Abro
The Benazir Bhutto Shaheed University of Technology and Skill Development Khairpur Mirs, Khairpur, Pakistan
e-mail: arif.hussain@iba-suk.edu.pk; arif@bbsutsd.edu.pk

H. Chenchouni et al. (eds.), *Recent Advancements from Aquifers to Skies in Hydrogeology, Geoecology, and Atmospheric Sciences*, Advances in Science, Technology & Innovation, https://doi.org/10.1007/978-3-031-47079-0_39

significant impact on the navigation performance especially when reliable and accurate positioning services are required as in safety and critical applications (Magsi et al., 2019; Seo et al., 2011; Todd et al., 2021).

Accurate and timely detection of scintillation events becomes significant for space weather-related domains (Lee et al., 2017). However, there are limited studies on scintillation detection techniques presented in the literature. The majority of the work is related to the simple triggering of scintillation events, that majorly dependent on the scintillation indices given by ground receivers with certain values (Taylor et al., 2012). However, these techniques require more signal characteristics and detrending actions. Some more other algorithms were proposed in comparison with conventional scintillation approaches, i.e. wavelet decomposition techniques (Parmar et al., 2017) and adaptive time–frequency approaches (Savas & Dovis, 2019). There 2017are some machine learning algorithms exploited for scintillation detection like data mining approaches that provide online forecasting services (Jiao et al., ; Linty et al., 2019). However, these techniques mainly depend on the external data sources that are always available and rely on the complicated and complex computational operations on receiver architectures. The paper proposed the machine learning approach support vector machine (SVM)-based amplitude scintillation prediction method that takes the GNSS raw data to detect and classify the amplitude scintillation into three main categories, i.e. no scintillation, moderate, and severe scintillation.

2 Methodology

The most important and basic stage of machine learning models is the relevant and accurate data acquisition that is to be used to make useful predictions. In order to predict the amplitude scintillation, raw GNSS data were acquired by deploying Septentrio PolarRx5S receiver at Sukkur, Pakistan.

The next stage is the feature engineering in which input and output features are selected. In this paper, we have chosen S_4 parameters as input feature and classes as target variable to support vector machine (SVM) model to classify the amplitude scintillation. Based on the fixed threshold, S_4 is labelled as shown in Table 1.

The final stage is the model selection, and this paper employed SVM algorithm that is mostly configured for multi-class classification. SVM is the most popular classification algorithm and has been used as recognition, sorting, and other domains of machine learning. SVM draws the hyperplane to classify the target variable into various classes.

3 Results

Figure 1 shows the data feature selected S_4 values that is plotted on x-axis and y-axis and illustrates the classes categorized based on Table 1. As it can be observed that each class has its own proportion, blue coloured represents the Class 0, red colour is for Class 1, and Class 2 is represented by green colour. SVM model is trained on these data to accurately predict the scintillation classes in the test data. Confusion matrix is mostly used to quantify the right and wrong predictions and evaluate the performance of any classification models based on test data.

Figure 2a and b shows the confusion matrix of SVM model in which Class 2 was predicted right, whereas we found some wrong predictions in Class 0 and Class 1. Due to the overlapping data that lead to cause some noise in data and result in low predictions in Class 0 and Class 1, Fig. 2b gives the predictions in terms of accuracy percentages. Class 2 has the highest accuracy amongst the other two; it means severe scintillation events can be predicted through SVM and can provide the scintillation alerts based on these predictions. Figure 3 represents the actual classes versus SVM predictions. It is noted that there is dispersion in predictions of classes due to the misclassification in the Class 0 and Class 1. Classification performance proved to be best due to no misclassification and all classes are correctly predicted.

Table 1 List of different classes

S_4 Threshold	Classes
$0 < S_4 < 0.3$	No scintillation (Class 0)
$0.3 < S_4 < 0.6$	Moderate scintillation (Class 1)
$S_4 > 0.6$	Severe scintillation (Class 2)

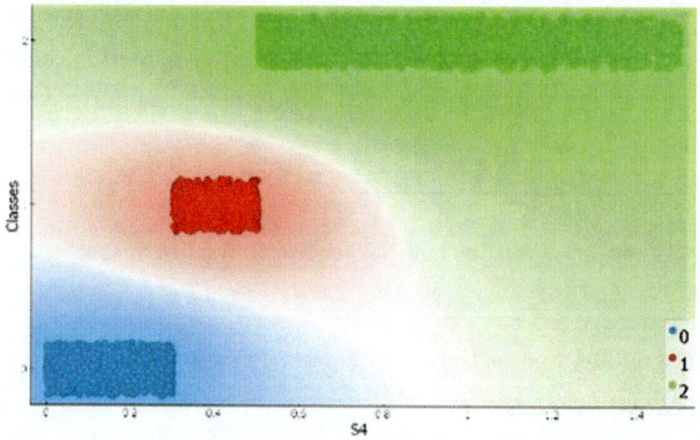

Fig. 1 S_4 values vs actual classes based on classification given in Table 1

(a)

Actual \ Predicted	0	1	2	Σ
0	2006	0	4	2010
1	5	334	4	343
2	0	0	293	293
Σ	2011	334	301	2646

(b)

Actual \ Predicted	0	1	2	Σ
0	99.8 %	0.0 %	0.2 %	2010
1	1.5 %	97.4 %	1.2 %	343
2	0.0 %	0.0 %	100.0 %	293
Σ	2011	334	301	2646

Fig. 2 **a** Confusion matrix in total number of right and wrong predictions, **b** confusion matrix in percentage of predictions for SVM model trained and test on scintillation data

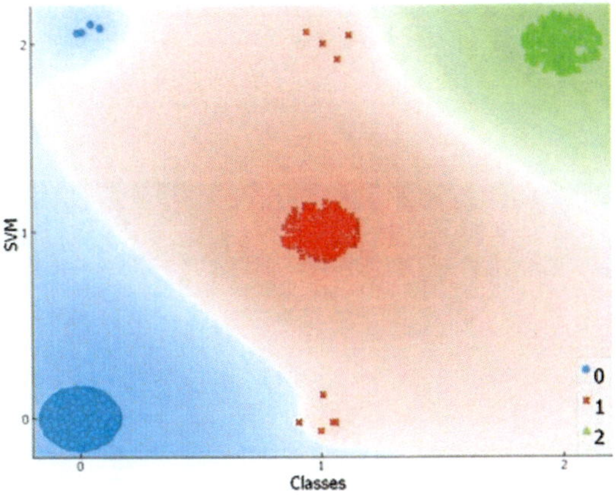

Fig. 3 Total number of predictions and actual predictions based on SVM classification model

Table 2 Performance metrics of SVM

Model	AUC	CA	F1-score	Precision	Recall
SVM	0.998	0.989	0.989	0.989	0.989

To quantify the performance of machine learning models, there are various metrics used like AUC (area under curve), CA (classification accuracy), precision, F1-score, and recall. The detailed performance metrics of SVM model is listed in Table 2 in terms of AUC, CA, F1, precision, and recall. It is seen that all five metrics having value of 98% that is due to the SVM predictions performance. There are various SVM models already proposed for scintillation prediction having low classification accuracy as in Savas and Dovis (2019) and Jiao et al., (2017). SVM model provides the classification accuracy of 91% and 95%. From the results presented in this paper, we observed the best accuracy performance scintillation predictions based on SVM. Hence, it is established that SVM based scintillation predictor is the best model for prediction of scintillation events due to its accuracy and less misclassifications.

4 Conclusion

This paper presents an efficient algorithm for accurate and timely predictions of amplitude scintillation events. The proposed method is based on SVM algorithm to detect and classify the amplitude scintillation into three main classes, i.e. no scintillation, moderate, and severe scintillation. The SVM model obtained the accuracy of 98%, and it can help in enabling early scintillation alerts.

References

Jiao, Y., Hall, J. J., & Morton, Y. T. (2017). Automatic equatorial GPS amplitude scintillation detection using a machine learning algorithm. *IEEE Transactions on Aerospace and Electronic Systems, 53*(1), 405–418. https://doi.org/10.1109/TAES.2017.2650758

Kintner, P. M., Ledvina, B. M., & De Paula, E. R. (2007). GPS and ionospheric scintillations. *Space Weather, 5*(9), 9003. https://doi.org/10.1029/2006SW000260

Lee, J., Morton, Y. T. J., Lee, J., Moon, H. S., & Seo, J. (2017). Monitoring and mitigation of Ionospheric anomalies for GNSS-based safety critical systems: A review of up-to-date signal processing techniques. *IEEE Signal Processing Magazine, 34*(5), 96–110. https://doi.org/10.1109/MSP.2017.2716406

Linty, N., Farasin, A., Favenza, A., & Dovis, F. (2019). Detection of GNSS ionospheric scintillations based on machine learning decision tree. *IEEE Transactions on Aerospace and Electronic Systems, 55*(1), 303–317. https://doi.org/10.1109/TAES.2018.2850385

Magsi, H., Ahmed, A., & Hussain, A. (2019). Real-time monitoring and logging of ionospheric scintillation and total electron content. In Sixth International Conference on Aerospace Science & Engineering ICASE 2019, Nov. 2019. https://doi.org/10.1109/ICASE48783.2019.9059198

Parmar, S. N., Dalal, U. D., & Pathak, K. N. (2017). Predicting and analyzing TEC Variations using wavelet and Hilbert Huang decomposition for IRNSS signals at low latitude location, Surat, India. *Procedia Computer Science, 115*, 39–46. https://doi.org/10.1016/J.PROCS.2017.09.074

Savas, C., Dovis, F. (2019). The impact of different kernel functions on the performance of scintillation detection based on support vector machines. Sensors, *19*(23), 5219. https://doi.org/10.3390/s19235219.

Seo, J., Walter, T., & Enge, P. (2011). Availability impact on GPS aviation due to strong ionospheric scintillation. *IEEE Transactions on Aerospace and Electronic Systems, 47*(3), 1963–1973. https://doi.org/10.1109/TAES.2011.5937276

Taylor, S., Morton, Y., Jiao, Y., Triplett, J., & Pelgrum, W. (2012). An improved ionosphere scintillation event detection and automatic trigger for GNSS data collection systems. (pp. 1563–1569), Feb 01, 2012. Accessed: Oct 01, 2022. [Online]. Available: http://www.ion.org/publications/abstract.cfm?jp=p&articleID=10034

Todd, P. M. K., Humphreys, E., & Psiaki, M. L. (2021). GPS carrier tracking loop performance in the presence of ionospheric scintillations. https://repositories.lib.utexas.edu/handle/2152/63325 (Accessed Jul 03, 2021).

Observations of Fireballs with the UAE Meteor Monitoring Network

Maryam Essa Sharif, Aisha Al-Owais, Ilias Fernini, Masa Al-Naser and Hamid Al-Naimiy

Abstract

Space debris is often perceived as a severe threat to planet Earth. These can be natural ones such as meteors and fireballs or artificial ones like rocket boosters. The UAE Meteor Monitoring Network (UAEMMN) was established to monitor space debris in the UAE sky. The UAEMMN is hosted by the Sharjah Academy for Astronomy, Space Sciences, and Technology (SAASST) and is sponsored by the UAE Space Agency (UAESA) and the University of Sharjah (UoS). The network consists of three towers distributed over three different UAE locations. The top of each tower is equipped with 17 cameras placed around a ring-like structure directed toward the sky. The system automatically captures any movement from sunset to sunrise. This paper showcases the fireballs detected by the UAEMMN system over the UAE. The UFO Analyzer and Orbit programs were utilized to define and analyze fireballs based on their apparent magnitude and visual assessment by the observer. From the date of operation of the system (September 2018) until December 2020, the network was able to classify 223 fireballs. The number of fireball detections varied over the period. However, the study reports a consistent pattern, showing that the number of fireballs tends to increase by the end of each year in November and December, coinciding with the Leonids and Geminids showers. Our UAE fireballs observations help the International Meteor Organization build a worldwide database to compensate for the region's lack of reported observations.

M. E. Sharif · A. Al-Owais · I. Fernini (✉) · M. Al-Naser · H. Al-Naimiy
Sharjah Academy for Astronomy, Space Sciences, and Technology, University of Sharjah, POB 27272, Sharjah, United Arab Emirates
e-mail: ifernini@sharjah.ac.ae

I. Fernini · H. Al-Naimiy
Department of Applied Physics and Astronomy, University of Sharjah, POB 27272, Sharjah, United Arab Emirates

Keywords

Meteoroids · Meteorites · Fireball · Bolide · UAEMMN

1 Introduction

Earth encounters meteoroids as it orbits the Sun. Higher-velocity fireballs are more likely to drop meteorites due to mass loss, or ablation. Observing meteors and fireballs helps retrieve meteorites (Hawkins, 1959; Ceplecha et al., 1998). Therefore, multiple monitoring systems were established all around the world. The Sharjah Academy for Astronomy, Space Sciences, and Technology (SAASST) and the UAE Space Agency launched the UAE Meteor Monitoring Network (MMN) in September 2018 to observe fireballs and meteors in the MENA region, specifically the gulf. This study presents UAE MMN fireball counts, contributing to the IMO's worldwide database and accounting for the region's lack of recorded sightings.

2 Materials and Methods

The materials used for this study were obtained from the UAEMMN data. The UAEMMN consists of three stations placed at different spots over the UAE. To maximize the network coverage, each station has multiple charge-coupled device (CCD) cameras with different lens sizes. The cameras are directed toward the sky to capture moving objects every day, from sunset till sunrise since the network was established back in September 2018 (Fernini et al., 2020). The detected objects can be natural such as meteors and fireballs or man-made like rocket boosters and airplanes.

To capture and analyze the data, UFO series package was utilized. The series comprises three software packages: UFO

H. Chenchouni et al. (eds.), *Recent Advancements from Aquifers to Skies in Hydrogeology, Geoecology, and Atmospheric Sciences*, Advances in Science, Technology & Innovation, https://doi.org/10.1007/978-3-031-47079-0_40

capture, UFO Analyzer, and UFO Orbit. The first software is used to automatically capture moving objects, while the second and the third are used to analyze the detected objects. Using UFO analyzer and UFO Orbit software, the magnitude and orbital elements are calculated, and the observations are classified as a possible shower and meteor stream.

Several factors were considered to distinguish a meteor from a fireball, including magnitude, shape, duration, and the observer's overall visual assessment. The observation's magnitude was determined using UFO analyzer. The minimum accepted value of apparent magnitude for a meteor to be classified as fireball was set to −3. According to Hawkins (1959), the apparent magnitude of a fireball ranges from −3 to −10, and in extreme situations might exceed −10 (Madiedo et al., 2014) that can be separated into two types: fireballs and bolides (Ceplecha et al., 1998).

3 Results

3.1 Fireball Statistics

Visual classification and magnitude value were used to determine whether a meteor was a fireball. This was applied to count fireballs. The network recorded 223 fireballs and bolides from September 2018 to December 2020. Twenty-one of these detections were paired, indicating that the fireball was detected by two sites. The resulted fireball statistics are reflected in Fig. 1.

3.2 Analysis Result

A full UFO software series analysis requires paired fireball detections to establish the object's orbit, radiant point, and trail across the ground and star map. Multiple detections help provide diverse viewpoints on the captured fireball. To generate appropriate analysis results, a substantial number of visible stars must also be recognized. Figure 2 depicts an example of a fireball detected by two stations, as well as the masked image with the scintillation masks to the reference stars were linked to determine the fireball's magnitude. The four maps in Fig. 3 display the analysis result of the mentioned fireball in order to demonstrate the software's capability.

4 Discussion

Although the meteors detected by the UAEMMN towers are often faint, certain fireballs coincided with the peak period of specific showers.

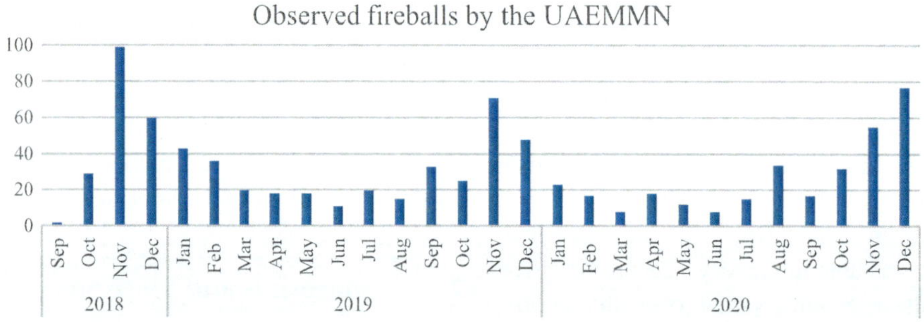

Fig. 1 Number of observed fireballs per month by the network over the study period from September 2018 to December 2020

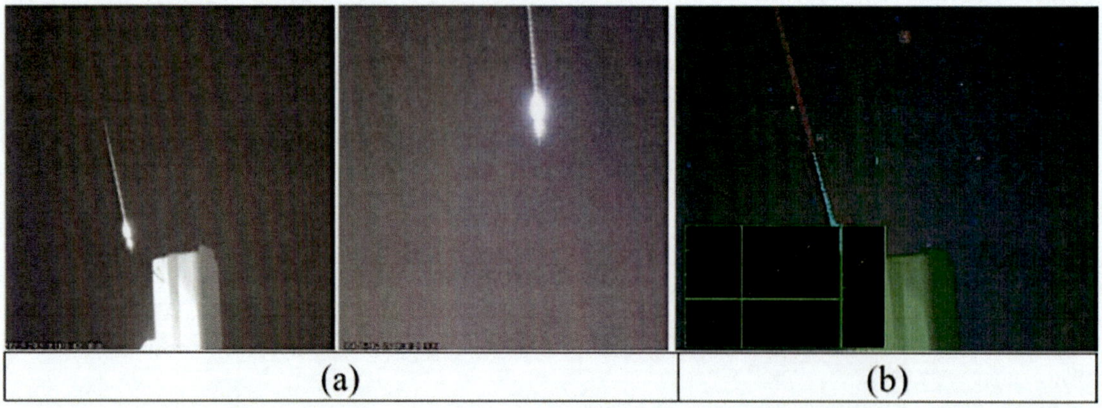

Fig. 2 **a** Captured fireball by two UAEMMN stations on September 2, 2020. **b** The fireball masked image with the magnitude result

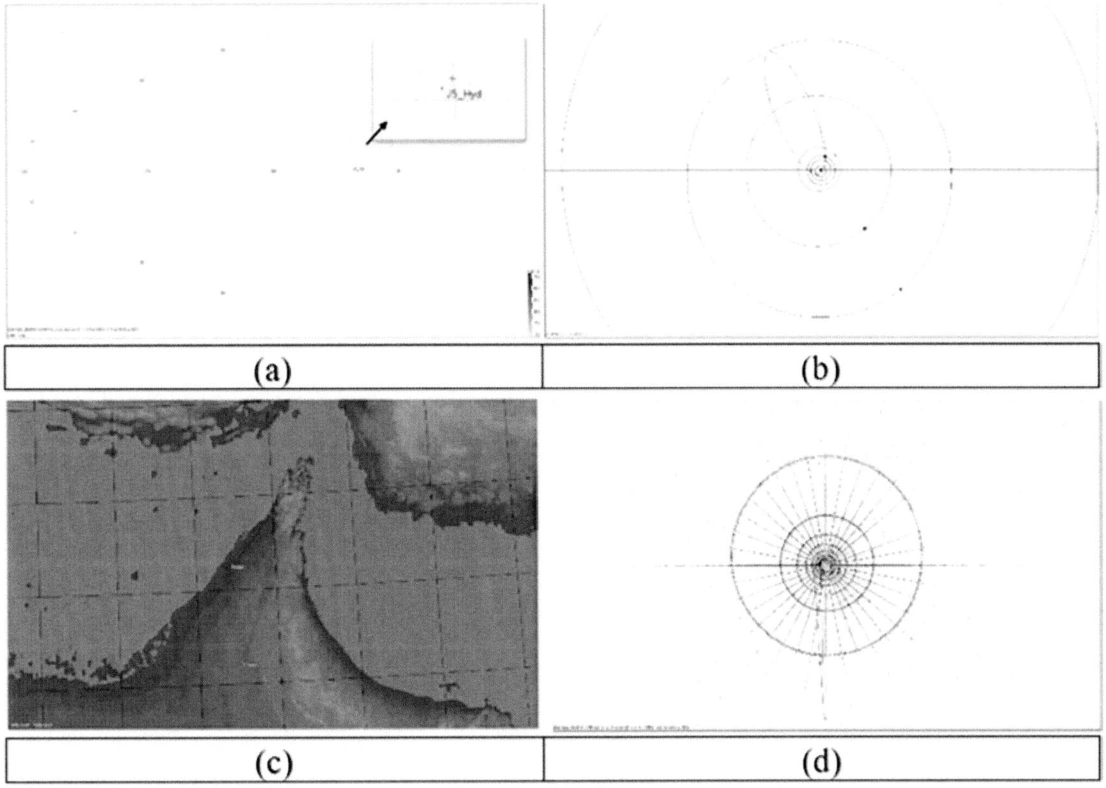

Fig. 3 The four maps resulted from UFO Orbit: **a** the fireball's radiant point, **b** fireball's orbit, **c** the fireball trail on ground map, **d** the fireball trail over the star map

Most fireballs were typically detected in November and December. For example, in 2018 and 2019, we observed 25 and 22 fireballs in November, respectively, outnumbering the 10 and 12 detections in December. However, in 2020, there were 25 fireballs in December, compared to 18 in November. The Leonids and Geminids are well-known showers of the indicated period, and they are known to be exceedingly numerous, especially when observed from this region. As a result, it is projected that fireball detections will rise during those showers. Except for November and December, August, which is known for its Perseid meteor shower, had a high quantity of fireballs. Despite this, the majority of the fireballs were sporadic and appeared at random times. Although there is no clear reason for the rise in fireball detection other than the showers, perhaps more meteoroids were in the area at the time. Furthermore, those fireballs could have come from the same meteoroid but impacted the Earth's atmosphere in chunks at different periods of the month.

Our observations of fireballs in the UAE would contribute to the global database of meteors by the IMO, which helps in compensating the lack of observations reported in the region.

5 Conclusion

The UAEMMN is a capable and essential system in this region. During nearly three years of observation, 223 fireballs were detected. The last two months of each year seemed to have the most fireball detections, due to the Leonids and Geminids showers. The UFO Series is astonishingly useful for examining meteors and fireballs. This will allow us to construct a database for fireball counts, records, and a system with a lot of information on such occurrences. By having this network, the UAE complements similar projects worldwide and invites neighboring countries to do the same to increase regional sky coverage and reported observations.

References

Ceplecha, Z., et al. (1998). Meteor phenomena and bodies. *Space Science Reviews, 84*, 327–471.

Fernini, I., et al. (2020). The UAE meteor monitoring network. *Journal of Instrumentation, 15*, T06007.

Hawkins, G. (1959). The relation between asteroids, fireballs and meteorites. *The Astronomical Journal, 64*, 450.

Madiedo, J., et al. (2014). Trajectory, orbit, and spectroscopic analysis of a bright fireball observed over Spain on April 13, 2013. *Astronomy & Astrophysics, 569*, A104.

Accurate, Stable, and SI-Traceable Reference Gas Mixtures of VOCs Relevant for Climate at Atmospheric Levels

Maitane Iturrate-Garcia, Annarita Baldan, Jianrong Li, Christophe Sutour, Tatiana Macé, Tobias Bühlmann and Céline Pascale

Abstract

Volatile organic compounds (VOCs) have effects on the oxidative capacity of the lower atmosphere, air quality and radiative forcing, and are thereby contributing to climate change. To identify atmospheric trends in VOC abundance and composition, comparable datasets of atmospheric measurements at different spatial and temporal scales are needed. Traceability to the international system of units (SI) is key to achieve comparability. However, in many cases, this is hardly accomplished due to the lack of SI-traceable reference gas mixtures (RGMs) to calibrate the measuring instruments. Within the framework of the EMPIR project "Metrology for Climate Relevant Volatile Organic Compounds", new SI-traceable RGMs of selected VOCs were prepared. The RGMs were produced at atmospheric amount-of-substance fractions (<100 nmol/mol) following different dynamic and static methods. For most of the selected compounds (halogenated VOCs and oxygenated VOCs, such as methanol, ethanol, and acetone), relative expanded uncertainties (U) lower than 5% were achieved fulfilling stakeholders' needs. This was not the case for specific oxygenated VOCs (acetaldehyde; U < 10%) and terpenes (α-pinene, β-pinene, myrcene, and β-caryophyllene, U = 7–13%) because of surface reactions (e.g., adsorption) and impurities. Further development of passivation treatments, as well as closer collaboration among manufacturers, metrologists and atmospheric researchers are needed to facilitate the production of accurate, stable, and SI-traceable RGMs and to enhance, in turn, data comparability and climate trend identification.

Keywords

Atmospheric VOCs · Reference gas mixtures · SI-traceability · Uncertainty · Climate change

1 Introduction

In the atmosphere, volatile organic compounds (VOCs) are present at trace levels (nmol/mol or pmol/mol). Many VOCs, except for most halogenated ones, are reactive compounds with atmospheric short lifetimes (minutes to days). VOCs play an important role in the oxidative capacity of the lower atmosphere (Atkinson & Arey, 2003) and have direct and indirect effects on climate (Taipale et al., 2008). In addition, VOCs have a potential to be used as tracers of atmospheric processes such as source emissions, transport, deposition, particle formation, and scavenging. Therefore, several atmospheric monitoring networks (e.g., ACTRIS, AGAGE, and WMO-GAW) include VOCs as part of their measurement strategy to enhance our understanding of the physics and chemistry of the atmosphere.

Reliable and comparable VOC measurements are essential to identify atmospheric trends of VOC composition and amount-of-substance fractions. However, VOC monitoring is challenging due to VOC reactivity and low atmospheric abundance. Despite significant progress made during the past years, stable and accurate gas references traceable to the international system of units (SI), along with suitable measuring methods and well-defined calibration procedures—all needed for data comparability—are lacking for most VOCs.

M. Iturrate-Garcia (✉) · T. Bühlmann · C. Pascale
Department of Chemical and Biological Metrology, Federal Institute of Metrology METAS, 3003 Bern-Wabern, Switzerland
e-mail: maitane.iturrate@metas.ch

A. Baldan · J. Li
Chemistry Group, VSL, National Metrology Institute, 2629 JA Delft, Netherlands

C. Sutour · T. Macé
Department of Gas Metrology, Laboratoire National de Métrologie Et d'Essais LNE, Paris CEDEX 15, 75724 Paris, France

H. Chenchouni et al. (eds.), *Recent Advancements from Aquifers to Skies in Hydrogeology, Geoecology, and Atmospheric Sciences*, Advances in Science, Technology & Innovation, https://doi.org/10.1007/978-3-031-47079-0_41

The project "Metrology for Climate Relevant Volatile Organic Compounds" (MetClimVOC, 2020–2023, https://www.metclimvoc.eu)—funded by the European Metrology Programme for Innovation and Research (EMPIR)—pursued to minimize these limitations. For that purpose: (i) new accurate, SI-traceable, and stable VOC reference gas mixtures (RGMs) of VOCs at atmospheric levels were produced; (ii) sampling and analytical methods used in monitoring stations were evaluated; and (iii) fit-for-purpose traceable working standards were developed. Here, we present the efforts done within the project to produce RGMs at atmospheric amount-of-substance fraction fulfilling the data quality objectives (DQOs) of monitoring networks, such as WMO-GAW and ACTRIS, in terms of accuracy and temporal stability.

2 Materials and Methods

2.1 Gravimetric method—Oxygenated VOC (Oxy-VOC) RGMs

We prepared two sets of oxy-VOCs (Table 1) in high-pressure cylinders following the gravimetric method according to ISO 6141-1 (ISO6142-1, 2015). Both sets were prepared at 100 nmol/mol level in nitrogen (BIP+, air products, purity > 99.99990%). Binary mixtures (individual oxy-VOC at 100 µmol/mol in nitrogen) were prepared by liquid injection (ISO6145-4, 2004) and were combined into two mother multi-component mixtures at 5 µmol/mol. We further combined and diluted both mother mixtures to obtain the RGMs at 100 nmol/mol level. In both sets, we added n-hexane as internal standard.

2.2 Dynamic method—Terpene RGMs

We generated dynamically terpene RGMs, based on the permeation method according to ISO 6145-10 (ISO 6145-1, 2002), using a dilution system coupled to a magnetic suspension balance (MSB; TA Instruments GmbH, Germany). We placed the permeation units containing individual liquid pure compounds of interest—α-pinene (VICI Metronics, WA, USA) and β-pinene, myrcene, and β-caryophyllene (Fine Metrologly, Italy)—in the MSB chamber, which was pressure (1.08 bar) and temperature (40 °C) regulated. Each permeation unit remained in the MSB chamber for several days (5–7 days) to ensure the stability of the membrane. Their permeation rates were estimated by monitoring their mass loss with time. In addition to these RGMs, we used a portable generator (Pascale et al., 2017) to generate terpene RGMs onsite. All the stainless steel surfaces in contact with the VOCs were coated with SilcoNert® 2000 to minimize adsorption. RGMs at different amount-of-substance fractions were generated by changing the flow rates of the dilution gas (nitrogen obtained from a liquid nitrogen distillation; purity > 99.9990%).

2.3 Cryo-Filling method—Halogenated VOC RGMs

We generated RGMs of the selected halogenated VOCs (CH_2Cl_2, CCl_4, 1,2-dichloroethane, HFC-32, HFC-365mfc, and HFO-1336mzzZ) at atmospheric levels by using an improved cryo-filling system based on previous work (Guillevic et al., 2018). RGMs were generated using the permeation system described above (MSB). We used synthetic air (99.9996% purity; PanGas, Switzerland) as matrix gas. The diluted RGMs exiting the MSB were subsequently filled into two 34-L Essex Industries and six 2.25-L FITOK cylinders via the cryo-filling system. The cylinders were cleaned prior the filling by evacuating and filling them with nitrogen three times and evacuating them again to < 0.1 mbar. After that, we prefilled the cylinders with wet synthetic air to further minimize adsorption of the halogenated VOC on the surface of the SilcoNert® 2000 treated cylinders.

3 Results

The amount-of-substance fractions of the VSL oxy-VOC RGMs were verified against dynamically prepared RGMs. The values were confirmed at around 100 nmol/mol with a relative expanded uncertainty in the range 3–6% for all

Table 1 Sets of gravimetric RGMs generated for the priority oxy-VOCs

Set	Compounds	Cylinder type	Number cylinders	Pressure (bar)
VSL	Methanol, ethanol, acetone, acetaldehyde, methyl vinyl ketone (MVK), methacrolein, methyl ethyl ketone (MEK)	10-L aluminum with a gas specialty improved type B proprietary treatment	6	130
LNE	Methanol, ethanol, acetone	10-L treated (gas specialty proprietary) aluminum + 10-L coated stainless steel (SilcoNert® 2000)	3	130

compounds, except for acetaldehyde (< 10%). A coverage factor $k = 2$ was used for all the uncertainties. The choice of material and handling at the RGM preparation was crucial, considering the interaction of these compounds with contact surfaces. Instability of the prepared RGMs as well as their analytical verification was additional sources of uncertainty. The LNE set of oxy-VOC RGMs was produced at amount-of-substance fractions between 90 and 100 nmol/mol with a relative expanded uncertainty lower than 5%. Similar uncertainty sources as for VSL were identified.

Dynamic RGMs of terpenes were generated at amount-of-substance fractions ranging between 5 and 50 nmol/mol. Their relative expanded uncertainty (>7%) was greater than the targeted 5%. The main source of uncertainty of the primary RGMs was the permeation unit purity; temporal stability and purity of the permeation units were the main sources for the RGMs generated using the portable generator.

The amount-of-substance fraction values of the halogenated VOCs in the prepared RGMs were in the low pmol/mol range. Their relative expanded uncertainty was lower than 3%. Main sources of uncertainty were the permeation rate stability during the filling of each compound and the purity of the permeation unit.

4 Discussion

Instrument calibration using accurate and SI-traceable RGMs is essential for high quality measurements of atmospheric VOCs. However, producing RGMs at atmospheric amount-of-substance fraction levels remain challenging for most VOCs.

Here, we used different methods to prepare RGMs of VOCs, which stakeholders considered priority, and explored their suitability. Uncertainty results showed that each method had its particular difficulties to achieve the targeted accuracy. Identification of sources contributing to the overall uncertainty of the RGMs gave insight into the modifications of the preparation procedure needed to improve RGM accuracy. For example, the use of treated surfaces (e.g., SilcoNert® 2000) and highly pure matrix gases will contribute greatly to reduce RGM uncertainty. In other cases, improvements are not possible without the collaboration of manufactures as for terpene RGMs, where improvement on the purity and temporal stability of the permeation units is needed before considering the permeation method suitable for the generation of terpene RGMs.

Besides fulfilling DQOs defined by the monitoring networks, metrology can help to achieve data comparability at local and spatial scales. Even if SI-traceability and complete and well-defined uncertainty budgets are essential for that purpose, the complexity of these metrological aspects may limit their systematic use at monitoring station. Because of that, collaborative research activities between metrologists and atmospheric researchers, such as the MetClimVOC project, are paramount.

5 Conclusion

Further development of passivation treatments, as well as a closer collaboration among manufacturers, metrologists, and atmospheric researchers are needed to facilitate the production of accurate, stable, and SI-traceable RGMs and to enhance, in turn, data comparability and climate trend identification.

Acknowledgements This project 19ENV06 MetClimVOC has received funding from the EMPIR programme co-financed by the Participating States and from the European Union's Horizon 2020 research and innovation programme.

References

Atkinson, R., & Arey, J. (2003). Atmospheric degradation of volatile organic compounds. *Chemical Reviews, 103*, 4605–5638.

Guillevic, M., Vollmer, M. K., Wyss, S. A., Leuenberger, D., Ackermann, A., Pascale, C., Niederhauser, B., & Reimann, S. (2018). Dynamic-gravimetric preparation of metrologically traceable primary calibration standards for halogenated greenhouse gases. *Atmospheric Measurement Techniques, 11*(6), 3351–3372.

ISO 6142-1. (2015). Gas analysis—Preparation of calibration gas mixtures—Part 1: Gravimetric method for Class I mixtures.

ISO 6145-1. (2002) Gas analysis—Preparation of calibration gas mixtures using dynamic volumetric methods—Part 10: Permeation method.

ISO 6145-4. (2004). Gas analysis—Preparation of calibration gas mixtures using dynamic volumetric methods—Part 4: Continuous syringe injection method.

Pascale, C., Guillevic, M., Ackermann, A., Leuenberger, D., & Niederhauser, B. (2017). Two generators to produce SI-traceable reference gas mixtures for reactive compounds at atmospheric levels. *Measurement Science and Technology, 28*, 124002.

Taipale, R., Ruuskanen, T. M., Rinne, J., Kajos, M. K., Hakola, H., Pohja, T., & Kulmala, M. (2008). Technical note: Quantitative long-term measurements of VOC concentrations by PTR-MS—measurement, calibration, and volume mixing ratio calculation methods. *Atmospheric Chemistry and Physics, 8*, 6681–6698.

Statistical Evaluation of the Distribution of PM_{10} and NO_2 in the Ambient Air Due to Urban Forms: A Case Study in the Agglomeration of Cergy-Pontoise

Souad Lagmiri and Salem Dahech

Abstract

This work investigates the spatio-temporal trends of air quality and its relationship with urban form in the Cergy-Pontoise conurbation from April 8 to June 6, 2022. Air quality data were obtained from a network of 16 PM_{10} and NO_2 sensors, installed specifically to fill the information gap regarding pollution levels achieved in its different areas. The evaluation of the relationship between urban form and pollution identified different sources of pollution between PM_{10} and NO_2. Morphological parameters such as building density, degree of openness to the sky, nature of the substrate, distances to roads, and traffic flow/speed were shown to influence the concentrations of each pollutant. Concerning the temporal variability, the daily amplitudes are well-marked. NO_2 concentrations are higher during peak hours and lower in the middle of the day. In the evening, the concentrations slowly decrease until the early hours of the morning. The diurnal cycle of PM_{10} shows a high morning peak and a lighter one in the late afternoon. Unlike NO_2, PM_{10} concentrations are higher during the day due to turbulence and pollen activity.

Keywords

Air pollution · Urban forms · Spatial clustering · Temporal variability

1 Introduction

In Europe, the Ambient Air Directive 2008/50/EC has defined the minimum number of fixed air quality monitoring stations (AQMS) according to human density. Typically, an official AQMS covers, on average, about 100,000 people in cities in developed countries. However, this distribution is insufficient to obtain accurate information on emissions' spatial distribution and identify their sources. Several researchers have implemented experiments to compensate for the inadequacy of ground measurements. Ung (2004) defined the concept of "virtual stations" as a means of artificially densifying the measurement network. Other studies have explored the potential of simulation software to predict pollution under different scenarios (Jonhson et al., 2010; Liu et al., 2012; Misra et al., 2019; Naughton et al., 2018; Pepe et al., 2016; Xing et al., 2015).

Only one measuring instrument is installed in "Cergy-Pontoise" (CACP). It is dedicated to permanent measurements of PM_{10}; this represents a resolution of 84 km^2 (> 200,000 inhabitants). It gives very accurate information but at a very low spatial resolution (Lagmiri & Dahech, 2023). Consequently, the maps published each day assess global air quality, i.e., an average over the territory. The monitoring of NO_2 concentrations was carried out only once and for a short period (from 13 November to 11 December 2018). It was carried out using micro-sensors mounted on a set of lampposts. They recorded serious exceedances of the limit value set by the WHO. This measurement campaign was carried out at a single sampling site on the edge of a busy road, which implies that the observed NO_2 levels were strictly site-specific and did not characterize the spatial distribution in the agglomeration. The present work using a network of 16 outdoor Ecosmart pollution sensors is the first attempt at parallel multi-site measurement of NO_2 and PM_{10} pollutants.

S. Lagmiri (✉) · S. Dahech
UMR PRODIG, Paris Cité University, 75006 Paris, France
e-mail: souad.lagmiri@u-paris.fr

H. Chenchouni et al. (eds.), *Recent Advancements from Aquifers to Skies in Hydrogeology, Geoecology, and Atmospheric Sciences*, Advances in Science, Technology & Innovation, https://doi.org/10.1007/978-3-031-47079-0_42

2 Materials and Methods

2.1 Monitoring Network and Site

A network of 16 pollution measurement stations is installed on the territory of the CACP by electrical connection on the lampposts immediately along the roads except for two stations; A16 installed in the middle of a park and A1 installed 77 m from the road. The network constantly monitors the gas and particle concentrations (NO_2, PM_{10}) via a chemiluminescence component for the former and an optical component for the latter. Measurements are taken every minute and transmitted to a Web server. The network is distributed to cover a diversity of urban landscapes. Representative urban sites were selected: *seven typical downtown sites, *seven suburban residences, *one urban park, and one rural area. These sites are characterized by differences in building type, density of use, percentage of greenery, street layout, traffic density, and speed.

2.2 Statistical Processing of Data

Pollutants are continuously monitored from April 8 to June 6, 2022, at a time resolution of one minute. A preprocessing based on the calculation of the median value of each hour is carried out to eliminate the very low values and the instantaneous peaks linked to the passage of vehicles. The data from the filter was used as input for the agglomerative hierarchical classification (AHC). Sites with a similar pollution pattern are grouped using Ward's method. The Euclidean distance is the algorithm chosen in this classification to maximize the homogeneity of attributes within groups and the heterogeneity between groups. The results of the spatial clustering are used as input for the analysis of the spatio-temporal variation of pollution.

3 Results and Discussion

3.1 Spatial Classification of the Air Quality Monitoring Site

The ascending hierarchical classification (AHC) following Ward's method was applied to a database composed of 16 columns (16 distinct morphological zones) and 1417 rows (1417 hourly concentration series) for the period from April to June and for each pollutant. This process resulted in the identification of five groups for NO_2 and four for PM_{10} with a high level of spatial homogeneity. The composition of the groups by site category shows that NO_2 and PM_{10} do not essentially come from the same source or the dispersion conditions act differently (Fig. 1 and Fig. 2).

3.2 Spatial and Temporal Variability of NO_2 Pollution

The weekly and diurnal cycles of the groups identified by the CAH make it possible to identify where the pollution comes from (Fig. 3). The results isolate three zones of concentration: high, moderate, and low. Group D4 is classified as a poor air quality site. It corresponds to a suburban sampling site near the agglomeration's busiest motorway (A15). During the morning and evening rush hours, traffic is around 1088 car/h, and the average speed drops from

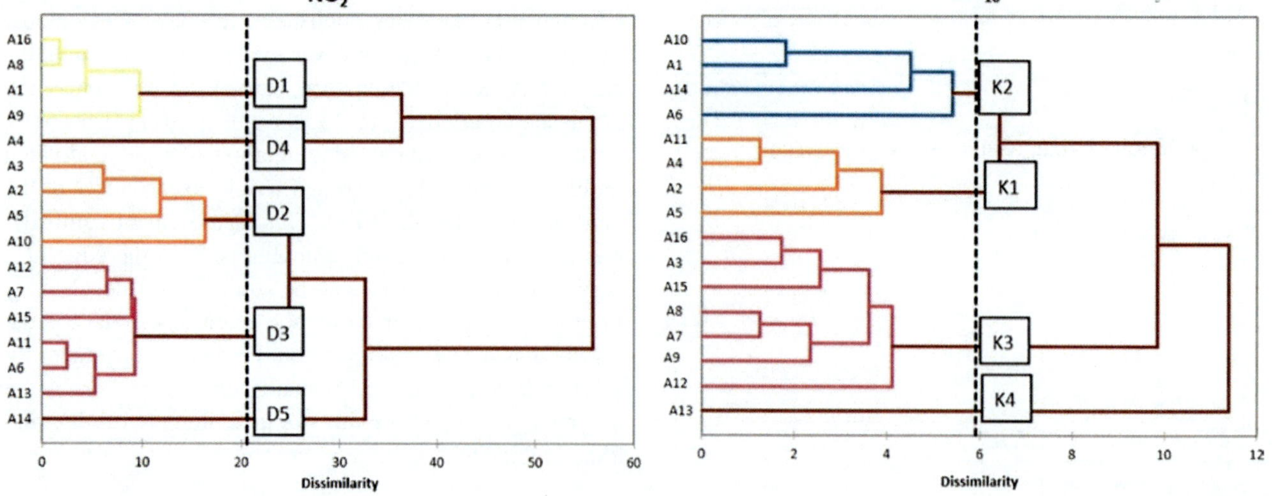

Fig. 1 Dendrogram showing the spatial classification of air quality monitoring sites (NO_2 on the right and PM10 on the left; A_i is the code sites)

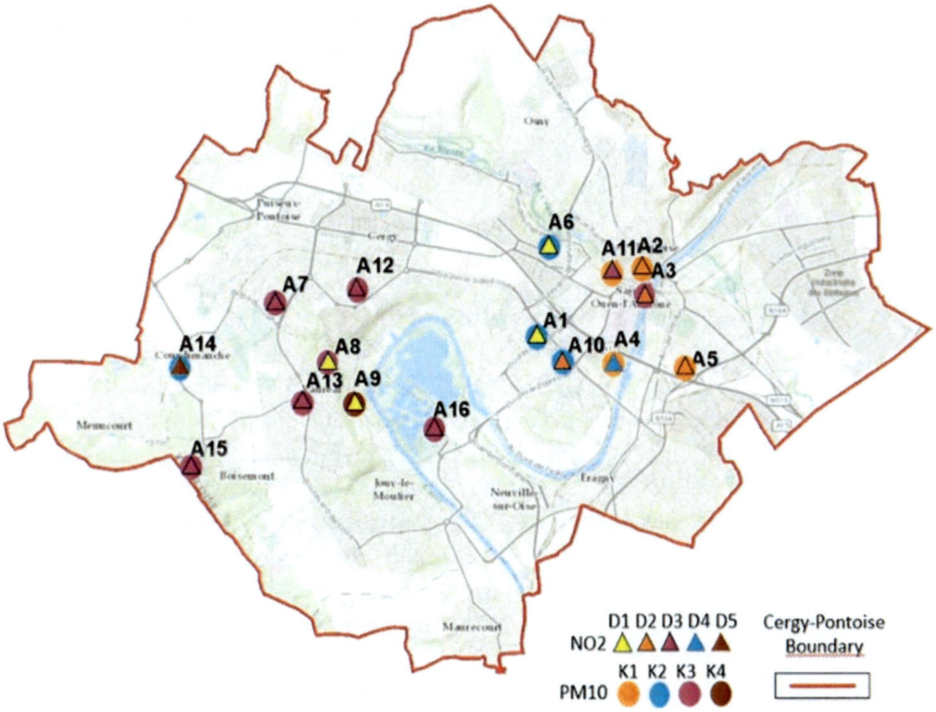

Fig. 2 Location of the 16 pollution sensors in the Cergy-Pontoise conurbation (circles indicate the NO_2 groups; triangles indicate the PM_{10} groups)

110 to 40 km/h. During this period, concentrations are highest (70–115 ppb). In the middle of the day, concentrations decrease due to a decrease in emissions compared to the peak hours combined with increased turbulence in the boundary layer due to solar heating and chemical reactions (a decrease of up to 40 ppb). After the evening peak hour, concentrations slowly decrease until the early morning hours. Overall, they remain higher than during the day due to a weak lower boundary layer. The same hourly pattern is observed for the rest of the groups from Monday to Friday. The morning peak is reduced on Saturday and absent on Sunday during the weekend.

Group D1 is the healthiest zone from 10 p.m. to 8 a.m. It is associated with the lowest concentrations (56–42, 25 ppb). It includes a site in the city center of the university campus 77 m from the road, a site in the middle of the largest urban park in the agglomeration, and two peripheral residential sites. In the latter, traffic is much lower than in the built-up areas. Traffic in the park (25 ha) is only recorded at the entrance to a car park some 189 m from the sample site. The city center site (university campus) corresponds to a sector with a lot of car traffic that the sampling carried out at 77 m from the road places this site in the low-emission category.

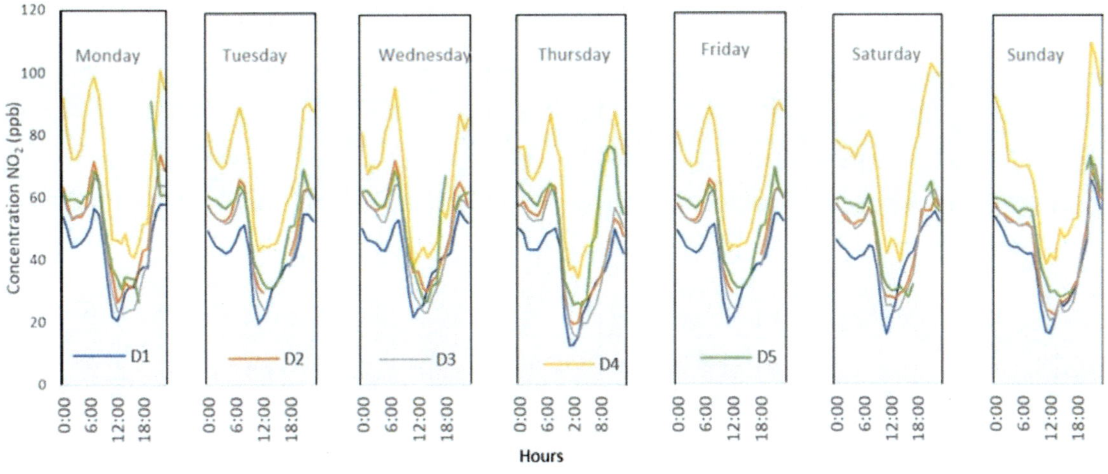

Fig. 3 Diurnal/weekly cycle of hourly NO_2 averages (in ppb for the five groups from the CAH)

Groups D2, D3, and D5 are classified as intermediate quality from 10 p.m. to 8 a.m., i.e., 72.06–53.14 ppb, 68.13–53 ppb, and 71.53–59.05 ppb, respectively. The urban forms in these sites range from suburban (four sites) to city center with a lot of shops (six sites) and a rural site. Outside this period, the spatial distribution of pollution is fairly homogeneous in all groups except D4.

3.3 Spatial and Temporal Variability of PM_{10} Pollution

For particulate pollution, the patterns are very different in time and space. Except for a few point exclusions, the highest concentrations are noted in the K2 group; composed of two sites in the urban center, a rural site, and a peri-urban site at the bottom of a vegetated valley (Fig. 4). The urban sites are subject to the influence of traffic due to the proximity of numerous avenues with heavy vehicle traffic and a large bus terminal with 19 lines. In the other two sites, pollution is mainly due to the presence of non-concrete surfaces. There is less car traffic, but each pass can easily raise dust from the ground. Particles also come from biogenic sources (Monnier et al., 2015; Chenchouni et al., 2022). In sites with a rural and wooded character, pollen activity increases the level of PM_{10} pollution.

Group K3 includes four residential sites with mainly local traffic, two sites in the urban center, and one site in the middle of the park. Group K4 corresponds to a site in the central square of a town on the outskirts of the conurbation, interfacing with the rural part. These two groups show intermediate levels of concentration throughout the week, except Thursday. On this day, group K4 shows fairly similar concentrations to group K2 and a peak between 3 and 6 pm corresponding to the weekly market time. The levels recorded in the morning are probably related to the influx of market traders' trucks. The peaks observed at this site on Friday evening and Saturday morning correspond to the patterns of visits to the city center supermarkets for the weekly food supply.

The lowest concentration levels are found in group K1; consisting of two sites in the urban center of the city of Pontoise (low car traffic), one site in a sports complex near the A15 motorway, and one site in a collective residence near the A86/A15 interchange. The barriers (hedge line and wall line) set up between these two heavy traffic routes and the houses prevent the propagation of particles to these sites. We have shown above that the sports complex near the A15 is the most polluted with NO_2. It can be concluded that the hedge line, which is a porous barrier, is only effective in case of particulate pollution. Furthermore, the low concentrations reached in group K1 are also justified by the distance from the pollen sources.

For all sites, the diurnal cycle on Sunday shows a concentration peak that occurs later in the morning. This is associated with leisure and sports activities.

Like NO_2, the daily amplitudes are well-marked, showing that PM_{10} concentrations are also influenced by environmental conditions. The diurnal cycle shows a high morning peak and a lower peak in the late afternoon. In addition to transport, this may be related to the transition rate between the nocturnal and convective layers. According to Menut (1997), the morning transition between these layers at mid and low latitudes is rapid and occurs shortly after sunrise, in contrast to the slow late-day transition. PM_{10} trapped during the night is rapidly reinjected at the surface

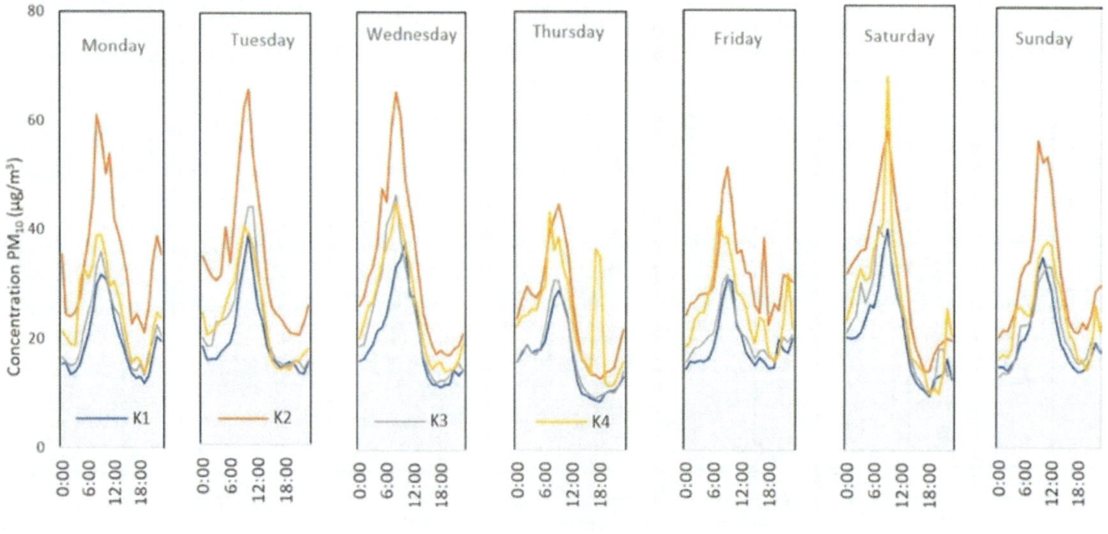

Fig. 4 Diurnal/weekly cycle of hourly PM_{10} averages (in µg/m³) for the four groups from the CAH

in the morning (Marley et al., 2021). In addition, pollen emission is particularly high in the morning which justifies the PM_{10} peak (Burki et al., 2019).

4 Conclusions

This work focused on the analysis of the distribution of pollution in urban forms. The results reveal that the morphological characteristics of the measurement site have a strong impact on pollutant concentrations and their spatial variability. They also show that the concentration levels recorded depend on the time of day and the weekly rhythm of anthropogenic activity. The 16 sensors are used to allow the concentrations of pollutants to be quantified at each moment of the day. However, their use at specific locations does not contribute to drawing generalizable conclusions for the entire conurbation. In this respect, the proposed results will be supplemented by those established from itinerant measurements using mobile sensors. Temporal differences in pollution recorded according to the type of weather will also be taken into account in the analysis.

References

Burki, C., Sikoparija, B., Thibaudon, M., & Oliver, G. (2019). Artificial neural networks can be used for Ambrosia pollen emission parameterization in COSMO-ART. *Atmospheric Environment, 218*, 116969. https://doi.org/10.1016/j.atmosenv.2019.116969

Chenchouni, H., Chaminé, H.I., Khan, M.F., Merkel, B.J., Zhang, Z., Li, P., Kallel, A., & Khélifi, N. (2022). New Prospects in Environmental Geosciences and Hydrogeosciences. Cham, Springer. https://doi.org/10.1007/978-3-030-72543-3

Lagmiri, S., & Dahech, S. (2023). Weather types and theirinfluence on PM_{10} and O_3 urban concentrations in the Cergy-Pontoise conurbation. *Journal of AppliedMeteorology and Climatology, 62*, 549–561. https://doi.org/10.1175/JAMC-D-22-0161.1

Jonhson, M., Isakov, V., Touma, J. S., Mukerjee, S., & Ozkaynak, H. (2010). Evaluation of land-use regression models used to predict air quality concentrations in an urban area. *Atmospheric Environment, 44*(30), 3660–3668. https://doi.org/10.1016/j.atmosenv.2010.06.041

Liu, Y., He, K. B., Li, S. S., Wang, Z. X., Christiani, D. C., & Koutrakis, P. (2012). A statistical model to evaluate the effectiveness of $PM_{2.5}$ emissions control during the Beijing 2008 Olympic Games. *Environment International*, 44, 100-105. https://doi.org/10.1016/j.envint.2012.02.003

Marley, H. G., Dirks, K. N., Neverman, A. J., McKendry, I., & Salmon, J. A. (2021). The relationship between Brown haze, atmospheric boundary layer structure, and air pollution in an urban area of complex coastal terrain. *Atmospheric Pollution Research, 12*(5), 101057. https://doi.org/10.1016/j.apr.2021.101057

Menut, L. (1997). *Etude expérimental et théorique de la couche limite atmosphérique en agglomération parisienne*. Doctoral thesis, Univ. Pierre et Marie Curie.

Monnier, S., Thibaudon, M., Besancenot, J.-P., & Michelot, N. (2015). Pollution, pollen et pollinoses: Retour sur l'épisode de pollution de mars 2014 en France. *Pollution. Atmosphérique, 25*, 157–164.

Misra, P., Imasu, R., & Takeuchi, W. (2019). Impact of urban growth on air quality in Indian cities using hierarchical Bayesian approach. *Atmosphere, 10*, 517. https://doi.org/10.3390/atmos10090517

Naughton, O., Donnelly, A., Nolan, P., Pilla, F., Misstear, B. D., & Broderick, B. (2018). A land use regression model for explaining spatial variation in air pollution levels using a wind sector-based approach. *Science of the Total Environment, 630*, 1324–1334. https://doi.org/10.1016/j.scitotenv.2018.02.317

Pepe, N., Pirovano, G., Lonati, G., Balzarini, A., Toppetti, A., Riva, G. M., & Bedogni, M. (2016). Development and application of a high resolution hybrid modelling system for the evaluation of urban air quality. *Atmospheric Environment*, 141, 297-311. https://doi.org/10.1016/j.atmosenv.2016.06.071

Ung, A. (2004). *Cartographie de la pollution atmosphérique en milieu urbain a l'aide de données multisources*. Doctoral thesis, Univ. Paris 7.

Xing, J., Mathur, R., Pleim, J., Hogrefe, C., Gan, C. M., Wong, D. C., Wei, C., Gilliam, R., & Pouliot, G. (2015). Observations and modeling of air quality trends over 1990–2010 across the Northern Hemisphere: China, the United States, and Europe. *Atmospheric Chemistry and Physics, 15*(5), 2723–2747. https://doi.org/10.5194/acp-15-2723-2015

Heat Content from Gliders and Satellites: Eastern Mediterranean Case Study

Yael Amitai, Hezi Gildor and Aldo Shemesh

Abstract

Heat content measured directly by ocean gliders in the Eastern Mediterranean (EM) Sea is compared to heat content inferred from remotely sensed sea surface properties, assuming a simple reduced-gravity model. Since the Middle East is one of the most responsive regions to climate changes, expected to be warmer and dryer, monitoring and studying the sea heat capacity and the effects on the local weather regime is of great importance. For example, EM heat content during fall was found to be significant in predicting the amount of precipitation over Israel in the following winter. Heat content is a function of both the mixed layer thickness and its mean temperature. Using temperature profiles observed by gliders, we can verify heat content inferred from remotely sensed sea surface temperature and sea level anomalies under reduced-gravity approximation and study its spatial and temporal variability. Sea gliders missions, conducted in high temporal resolution, provide detailed information on the water column thermal structure along its path. The comparison between the heat content obtained from gliders and the heat content calculated from remotely sensed properties shows that there are places and periods where a distinct separation between the surface and subsurface is found in the EM. This separation occurs when a subsurface dynamical eddy, e.g., is observed in the profiles collected by the glider but has no surface signature, hence is not detected by satellites. Furthermore, satellite products seem to poorly represent the heat content of coastal water which exhibit a different thermal structure than an open-water thermal structure. Another finding from the heat content comparison, obtained by two different methods, is that the subsurface is better represented by the surface during summer than during winter. This indicates that the strength of the stratification is also crucial when remotely monitoring heat content. In summary, heat content obtained directly by gliders, remotely by satellites, and the comparison between the two provides several important insights about water column thermal structure variability in one of the most sensitive regions to climate change. Ocean gliders mapping is needed when accurate heat content estimation is required.

Keywords

Heat content · Glider · Satellite · Sea surface temperature · Sea surface anomaly

1 Introduction

Eastern Mediterranean Sea water temperature, hence its heat content, is expected to significantly rise in the next decades. Additionally, Eastern Mediterranean heat content at fall may be used to predict the winter precipitation over Israel (Amitai & Gildor, 2017). Prediction of perception in a region that suffers from a shortage in natural water resources, due to the combination of population growth and droughts, is of great importance. Therefore, a continuous and accurate monitoring of heat content in the Eastern Mediterranean (EM) is needed, and the challenges in acquiring a reliable product from remote sensing are being addressed here.

Heat content (HC) of the water column can be directly measured by in situ measurements and also can be inferred from remote sensed parameters under a reduced-gravity

Y. Amitai (✉)
Israel Oceanographic and Limnological Research, Migdal, Israel
e-mail: yael.amitai@ocean.org.il

H. Gildor
The Hebrew University of Jerusalem, Jerusalem, Israel

A. Shemesh
Weizmann Institute of Science, Rehovot, Israel

H. Chenchouni et al. (eds.), *Recent Advancements from Aquifers to Skies in Hydrogeology, Geoecology, and Atmospheric Sciences*,
Advances in Science, Technology & Innovation, https://doi.org/10.1007/978-3-031-47079-0_43

model. Measurements are, in its nature, sporadic in time and space, while merged satellites products are continuous and lately become more available. Here, we compared HC measured by sea gliders with HC that is calculated from remote sensed sea surface temperature and elevation under a two-layer assumption.

2 Methods

The upper layer HC is defined as the heat stored in the upper mixed layer and is calculated as follows:

$$HC(t) = \rho_0 C_p \int_{D_{16}}^{\eta} T(z,t)\mathrm{d}z \quad (1)$$

where ρ_0 and C_p are the reference density and the heat capacity of seawater, respectively. The thickness of the upper layer is defined from the depth of isotherm 16 °C, which is D_{16}, until the surface, where η is the sea level anomaly. The value of the integral in Eq. 1 is acquired by multiplying the upper layer thickness and its mean temperature. Directly observed HC was measured by SeaExplorer gliders of ALSEASMAR (https://www.alseamar-alcen.com) equipped with a Sea-Bird pumped payload CTD. The gliders are buoyancy-controlled underwater vehicles that move through the water and collect depth profiles along their route. In this work, eight glider's missions piloted during the years 2016–2019 in the coastal shelf of Israel were used. General information and routes maps of the missions are presented in Table 1 and Fig. 1, respectively. The gliders measured a depth profile of temperature to provide the depth of the isotherm 16 °C and the mean temperature of the layer above the chosen isotherm.

Remote-sensed HC is calculated under the assumption that the thermocline variation is depicted in sea level anomalies (SLA; https://doi.org/10.48670/moi-00141), in what called a two-layers reduced-gravity model. We compute the depth (thickness) of the upper layer from SLA variations and obtain its average temperature from a linear relationship, previously obtained in Amitai and Gildor (2017), between remotely sensed sea surface temperature (SST; https://doi.org/10.48670/moi-00173) and in situ temperature observations.

Table 1 Information of each glider mission

Glider ID	Mission ID	Start	End	Dives
SEA012	M103	20-May-16	31-May-16	157
SEA013	M108	3-Jul-16	12-Jul-16	115
SEA012	M125	23-Feb-17	7-Mar-17	118
SEA012	M138	10-May-18	22-May-18	189
SEA013	M115	10-May-18	3-Jun-18	248
SEA013	M117	10-Jun-18	4-Jul-18	213
SEA013	M119	11-Oct-18	5-Nov-18	237
SEA013	M124	20-Feb-19	7-Mar-19	195

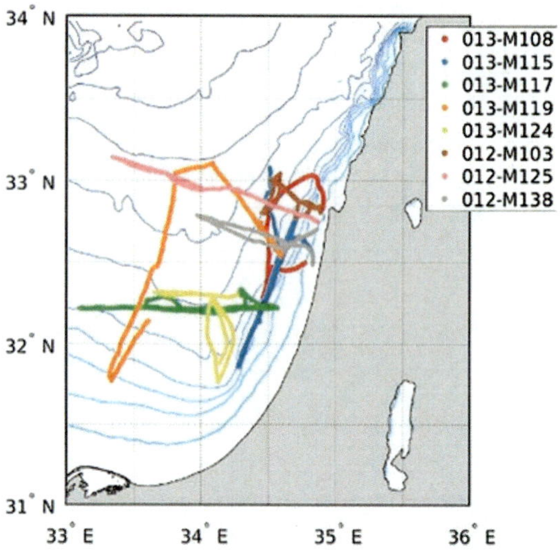

Fig. 1 Gliders routes, in the coastal water of Israel, used in this study

3 Results

Comparing merged satellite product of SST from Copernicus (https://doi.org/10.48670/moi-00173) with the upper 2 m from the Gliders shows that although there is an overall agreement, there is also large variability between missions (Fig. 2). It seems that missions took place during winter months exhibit less correlation than those took place during other seasons. High correlation between the remote sensed and directly measured SST is also more robust in long missions that capture a seasonal warming or cooling, such as in mission 115 of glider 13.

However, when comparing calculated HC from SLA and SST with measured by gliders HC (Fig. 3), the correlations drops and vertical spreads seen in some cases (e.g., M125 and M117). We also note that in most missions, there are two separate clouds of HC values that might represent the coastal and open-water difference.

4 Discussion

Vertical spreads in the satellite's HC means that the satellite products do not represent the glider variability, probably due to resolutions differences. The satellites products horizontal resolution is ~ 4.5 km while the glider samples a full profile every ~ 400 m. When examining every case separately (not shown here), we see that when the surface and the subsurface are disconnected, by subsurface eddy for

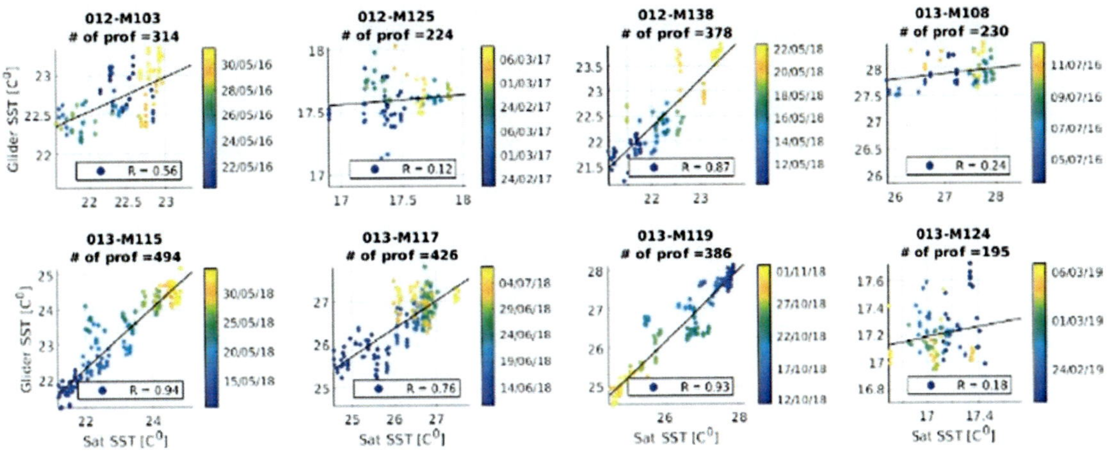

Fig. 2 Satellites SST versus in situ measured SST based on the upper 2 m of gliders profiles

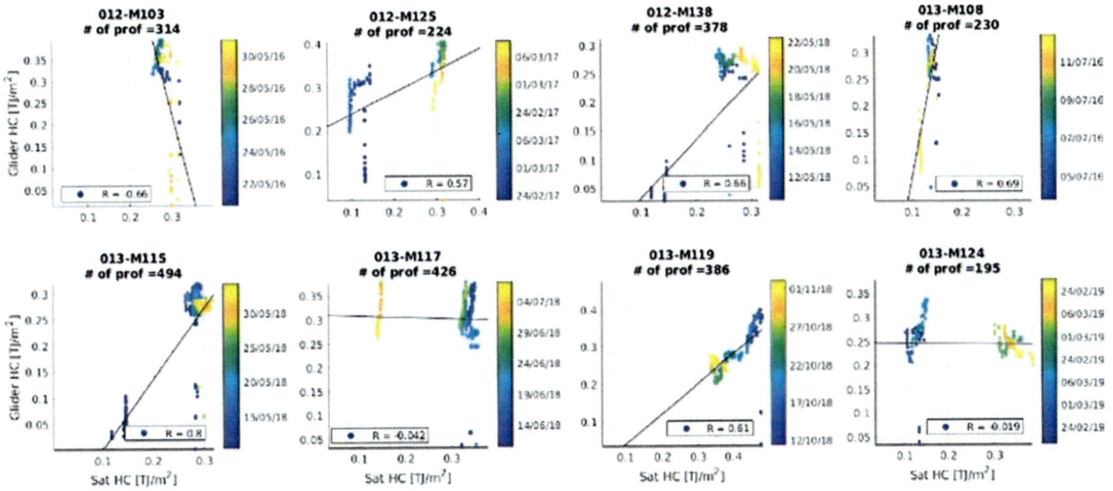

Fig. 3 Calculated HC based on satellites versus in situ measured HC from gliders profiles

example, the agreement is weaker. We also note from the case studies analysis that when the gliders measure strong shallow stratification, the satellite product calculation overestimate the HC since it deduces warmer upper layer temperature than the actual temperature.

5 Conclusion

This study aimed to understand heat content variability in the eastern Mediterranean as inferred from remote-sensed parameters and observed by sea gliders. The comparison shows periods of a distinct separation between the surface and subsurface. Results show that satellite resolution is not enough to properly represent differences in the thermal structure of the water column across coastal and open water in the Israeli shelf. Therefore, sea gliders monitoring is needed when accurate HC estimation is required.

References

ALSEAMAR homepage: https://www.alseamar-alcen.com

Amitai, Y., & Gildor, H. (2017). Can precipitation over Israel be predicted from Eastern Mediterranean heat content? *International Journal of Climatology, 37*(5), 2492–2501.

Copernicus SLA homepage. https://doi.org/10.48670/moi-00141

Copernicus SST homepage. https://doi.org/10.48670/moi-00173

Non-stationary Similarity in Trends of Seasonal and Monthly Rainfall in the Tuscan Apennine Alps (Middle Italy)

Fabrizio D'Aprile, Matteo Gentilucci and Gilberto Pambianchi

Abstract

Climate change raises several questions about its effects on a wide range of environmental factors. Forests are not immune to these effects, as the variability of rainfall over time affects several aspects, such as growth rate, water stress, regeneration, species composition and, possibly, population displacement. Assuming that the response of a tree species to climatic variability is territorially homogeneous within a common general trend may be erroneous. In fact, although the agglomerative hierarchical clustering technique revealed a certain similarity between the four weather stations analyzed, always greater than 0.62, both the amount of precipitation, and the similarity in the various months between them change, creating problems for the adaptation of silver fir especially in the summer season. With this in mind, it was also very important to evaluate through Fourier analysis possible cycles within the time series, in order to understand possible fluctuations that may influence the adaptation of the tree species analyzed. The annual precipitation required for a silver fir is over 1500 mm, and only one of the four rain gauges, the one with the highest altitude (Abetone, 2449 mm), is well above this threshold, while two others are well below (La Verna and Vallombrosa). Moreover, all the rain gauges investigated due to climate change show a significant trend (alpha level 0.05) of decreasing rainfall, which could lead to limitations in the growth of silver fir. The greatest problems occur in summer where the season is often dry and rainfall averages less than 200 mm at La Verna and Vallombrosa. This analysis is innovative, firstly because it identifies the recurring climatic and growth periods for the forest species in question, as well as understanding the influence that climate change has on vegetation in neighboring weather stations and assessing possible adaptation problems. This research can facilitate the development of strategies that favor the maintenance of environments. Therefore, effective adaptive forest management should take into account the variability of climatic factors at the local forest scale, as general models could be misleading.

Keywords

Forests · Precipitation · Temperature · Climate change

1 Introduction

Rainfall is an important factor in the distribution and growth of plants; warming of seasons and months, especially summer (Bartolini et al., 2008), is expected to increase the intensity and/or frequency of periods of water stress for trees in Italy. Many climatic analyses have highlighted important increases in average temperatures both in the Mediterranean area (Hertig & Jacobeit, 2008) and in Italy (Toreti & Desiato, 2008), including the Apennines (Gentilucci et al., 2020). Temperature is influential for trees in many respects; for example, there are important correlations with the phenological phases, growth, and water demand (Allen et al., 2010; Chmielewski & Rötzer, 2001). Therefore, it is important to verify whether and how precipitation varies over time and among sites to identify potential impacts on trees under changing climate conditions.

Silver fir (*Abies alba* Mill.) is a moisture demanding tree species susceptible to water stress and damage due to

F. D'Aprile
Monash University, Affiliate, Australia-Italy, Clayton, Australia

M. Gentilucci (✉) · G. Pambianchi
University of Camerino, 62032 Camerino, Italy
e-mail: matteo.gentilucci@unicam.it

H. Chenchouni et al. (eds.), *Recent Advancements from Aquifers to Skies in Hydrogeology, Geoecology, and Atmospheric Sciences*, Advances in Science, Technology & Innovation, https://doi.org/10.1007/978-3-031-47079-0_44

drought or insufficient moisture availability (Camarero et al., 2011); besides, rainfall is influential on growth (Macias et al., 2006). In this study, modifications in the regime and/or amount of rainfall have been investigated in the Tuscan Apennine Alps to understand if any impact on silver fir growth may occur. As this species is sensitive to water stress, aridity, and increase of temperature and it is the main conifer species of permanent montane forests in the Italian Peninsula, it can also be somewhat considered as a forest environmental indicator (Dobrowolska et al., 2017; Luszczyńska et al., 2018). Given the high environmental variability in the Italian range of silver fir, the response to growth, which is also a complex quality index, is likely to vary between sites even at relatively short distances, depending on possible differences in rainfall patterns over time. Therefore, the variability of seasonal and monthly rainfall in the twentieth century in the study area and the similarity of trends between sites were analyzed, respectively, to identify the type of changes in rainfall patterns that can potentially affect silver fir growth and to verify whether the influence of potential alterations in rainfall patterns varies by site.

2 Materials and Methods

2.1 Geography of the Study Area

The study area lies predominantly in the Apennine zone of the Tuscan region, and this area is home to the largest silver fir forests in Apennine Italy. The forest nuclei are located in the areas of the stations we have reported, namely Abetone, La Verna, Vallombrosa, Camaldoli. The region of Tuscany overlooks the Tyrrhenian Sea and has an extension of almost 23,000 km^2, and the landscape is mostly hilly, 67% of the total, 25% mountainous, and the remaining part flat.

2.2 Data Collection and Analysis Procedures

Four weather stations were considered: Abetone (ABE), La Verna (LAV), Camaldoli (CAM), and Vallombrosa (VAL). Data of these weather stations have been further validated and homogenized to make them even more reliable (Gentilucci et al., 2018a, b). First, annual precipitation trends at all sites were assessed to identify potential long-term alterations in precipitation patterns in the study area at both monthly and seasonal levels. Variations in rainfall above and below the long-term mean have been identified at all sites to highlight peaks and troughs in trends; moving averages have been used to smooth trends in rainfall series and enhance medium-long term variability. In addition, the Mann–Kendall trend test has been applied to detect the presence of trends.

$$\text{sign}(T_i - T_j) = \begin{Bmatrix} 1 & \text{if } T_i - T_j > 0 \\ 0 & \text{if } T_i - T_j = 0 \\ -1 & \text{if } T_i - T_j < 0 \end{Bmatrix} \quad (1)$$

$$S = \sum_{k=1}^{n-1} \sum_{j=k+1}^{n} \text{sign}(T_i - T_j) \quad (2)$$

The sum of these results determines the S value, and this value can be entered into the Z test.

$$Z = \begin{matrix} \frac{S-1}{\sqrt{\text{VAR}(S)}} & \text{if } S > 0 \\ 0 & \text{if } S = 0 \\ \frac{S-1}{\sqrt{\text{VAR}(S)}} & \text{if } S < 0 \end{matrix} \quad (3)$$

The variance of S is calculated using the following equation:

$$\text{VAR}(S) = \frac{1}{18}\left[n(n-1)(2n+5) - \sum_{i=1}^{g} t_p(t_p - 1)(2t_p + 5)\right] \quad (4)$$

g = number of tides groups.
t_p = number of observations in the pth group.

Sen's test assesses the true slope of a trend line only if the trend can be assumed to be linear of the following form:

$$f(t) = Qt + B \quad (5)$$

Q = slope of the line,
B = constant.

$$Q_i = \frac{x_j - x_k}{j - k} \quad (6)$$

The median of the N pairs of Q_i values (slope of the line), which are obtained from the differences between the previous and the next values ($xj - xk$), is necessary for the estimation of the angular coefficient according to the method of the Theil–Sen estimator. Then, from the obtained values, after being arranged in increasing order, the angular coefficient was calculated as follows:

$$Q = Q_{\left[\frac{(N+1)}{2}\right]} \quad \text{if } N \text{ is odd} \quad (7)$$

$$Q = \frac{1}{2}\left(Q_{(N/2)} + Q_{[(N+2)/2]}\right) \quad \text{if } N \text{ is even} \quad (8)$$

The confidence interval is calculated from the following formula:

$$C_\alpha = Z_{1-\alpha/2}\sqrt{\text{VAR}(S)} \quad (9)$$

Thus, the upper threshold is identified as follows:

$$M_1 = (N - C_\alpha)/2 \quad (10)$$

with Q_{min} represented by the (M_1)th largest in the N estimates of ordered slope Q_i.

While the lower threshold is calculated as follows:

$$M_2 = (N + C_\alpha)/2 \quad (11)$$

with Q_{max} represented by the (M_2+1)th largest in the N estimates of the ordered slope Q_i.

In order to obtain the value of the constant B, the n values of the differences xi-Qti were subtracted by calculating the confidence intervals with a similar procedure as described above. All these procedures were performed using Microsoft Excel software, through XLSTAT, which is a complete statistical add-in for Microsoft Excel (Gentilucci et al., 2021).

Matrix correlation has been used to test the level of association in both the seasonal and monthly rainfall series among sites. To identify the hierarchy in the associations of rainfall series among sites, both the Pearson and Kendall coefficients have been applied to verify if results are similar. The agglomerative hierarchical clustering (AHC) with the Pearson's correlation coefficient has been applied under both the forms "unweighted pair-group average" and "complete linkage" to better visualize the "proximity" in similarity of the seasonal and monthly rainfall series among sites. Although matrix correlation and AHC can estimate the presence and level of association in rainfall among sites, they do not show if similarity in trends of rainfall is stationary over time. The presence of periodicity or cycles in seasonal rainfall series has been verified by applied spectral (Fourier) analysis for all sites in the study area. Results have shown a series of intervals that are multiples of a basic period of 6 years. Thus, the Pearson's r correlation of 7-year moving average has been applied to test whether similarity in rainfall series among sites is non-stationary during the twentieth century in the study area (Fig. 1).

3 Results

3.1 Descriptive Analysis and Association of Annual Rainfall Series

In Italy, the optimal annual rainfall for silver fir is greater than 1.500 mm, but rainfall during the growing season and especially in summer is more important than annual rainfall for silver fir growth (Table 1).

The level of association in annual rainfall series among sites has been tested by correlation matrix tests (Pearson, Kendall) (Table 2). Results show that the association levels vary from low-moderate to moderate.

Fig. 1 Methodological flowchart

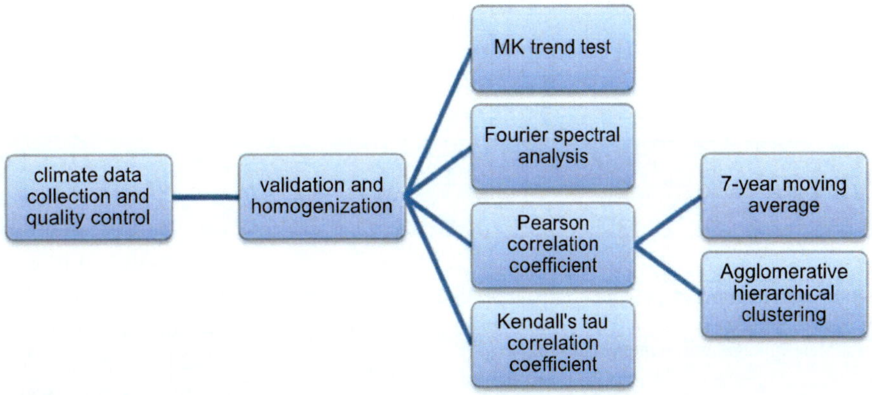

Table 1 Descriptive statistics of the weather stations analyzed

Period	CAM 1885–2005	VAL 1872–2006	ABE 1931–2000	CAM 1931–2000	LAV 1931–2000	VAL 1931–2000
Mean	1721.5	1401.8	2449.1	1649.9	1235.8	1340.9
St. dev	393.1	286.4	508.4	306.6	355.2	262.9
Max	3131.7	2148.9	3958.7	2451.2	2843.4	2148.9
Min	857.5	664.0	1382.4	857.5	435.2	664.0

Table 2 Pearson's and Kendall's matrix correlations of the annual rainfall series in the study area

	Pearson's *r*				Kendall's *τ*			
	ABE	CAM	LAV	VAL	ABE	CAM	LAV	VAL
ABE	1	0.57	0.38	0.57	1	0.39	0.34	0.35
CAM	0.57	1	0.56	0.73	0.39	1	0.42	0.52
LAV	0.38	0.56	1	0.47	0.34	0.42	1	0.40
VAL	0.57	0.73	0.47	1	0.35	0.52	0.40	1
	p-value Pearson				*p*-value Kendall's *τ*			
	ABE	CAM	LAV	VAL	ABE	CAM	LAV	VAL
ABE		<0.01	0.02	<0.01		0.02	0.03	0.03
CAM	<0.01		<0.01	<0.01	0.02		0.02	<0.01
LAV	0.02	<0.01		<0.01	0.03	0.02		0.02
VAL	<0.01	<0.01	<0.01		0.03	<0.01	0.02	

All the coefficients are statistically significant at the alpha level 0.05, and the bottom of the table also shows the *p*-value

Table 3 Hierarchy of clusters in monthly rainfall in the study area

Months	1st order	2nd order	3rd order
Mar, Apr, Jun, Aug, Sep, Oct	CAM-VAL	LAV	ABE
Dec, Feb	CAM-VAL	ABE	LAV
Jan	CAM-LAV	ABE	VAL
Jul	CAM-LAV	VAL	ABE
May, Nov	LAV-VAL	CAM	ABE

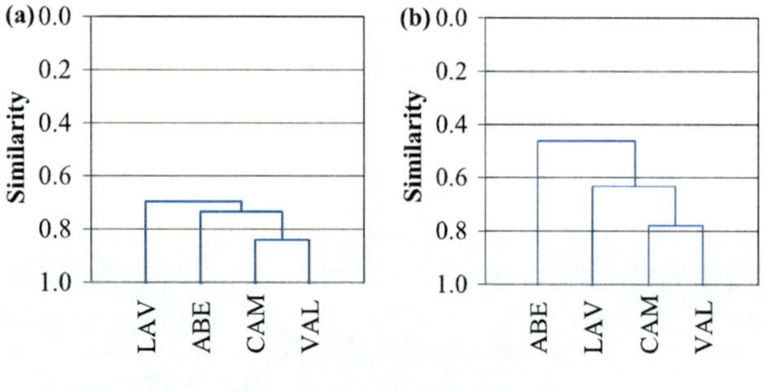

Fig. 2 Dendrogram of proximity rainfall series at the study sites as shown by AHC based on the Pearson's correlation coefficient, method 'complete linkage' **a** in winter, **b** in summer

3.2 Grouping of Seasonal and Monthly Rainfall in the Study Area: Agglomerative Hierarchical Clustering

The similarity in monthly rainfall series has been analyzed using the same AHC method, which has shown that similarity varies on a monthly level (Table 3). Normally, it tends to decrease from the CAM-VAL cluster to LAV and then ABE. CAM and VAL form the closest similar monthly precipitation series in all months except January and July, when CAM and LAV become the closest series. Similarity in monthly rainfall among sites differs the most in January with respect to the other months. The presence of significant correlation in each month between sites and the absence of correlation between months within sites help to understand that seasonal rainfall may not detect changes in rainfall that can affect tree growth on a monthly basis.

The AHC is also reported seasonally, with two examples in winter and summer, and it is evident that CAM and VAL are always the most similar rain gauges, while especially in summer, the ABE coefficient drops a lot (Fig. 2).

3.3 Spectral Fourier Analysis

The analysis identified the following periods of symmetrical variation for all analyzed rain gauges of (2–4), 6–7, 9–11, 14–16, 18–19, 22–23, 27–28, 34–36, 45, 55, 67–70, and 110–112 years. In particular, the periods (2–3), 6–7, 22–23, 34–36, 67–70, and 110 years are the most frequent in the seasonal series of mean precipitation. It is also noted that the harmonic mean of all periodogram values varies little between sites in all seasons (Fig. 3), and the 7-year period, which is derived from a harmonic mean ranging between 5.5 and 7.5, is the most frequent interval of all. In light of these results, the 7-year interval was obtained, which approximates the harmonic mean of the peak periodicities over the entire series of mean seasonal precipitation in the study area, which is a subdivision of the longer periodicities that occur with regularity.

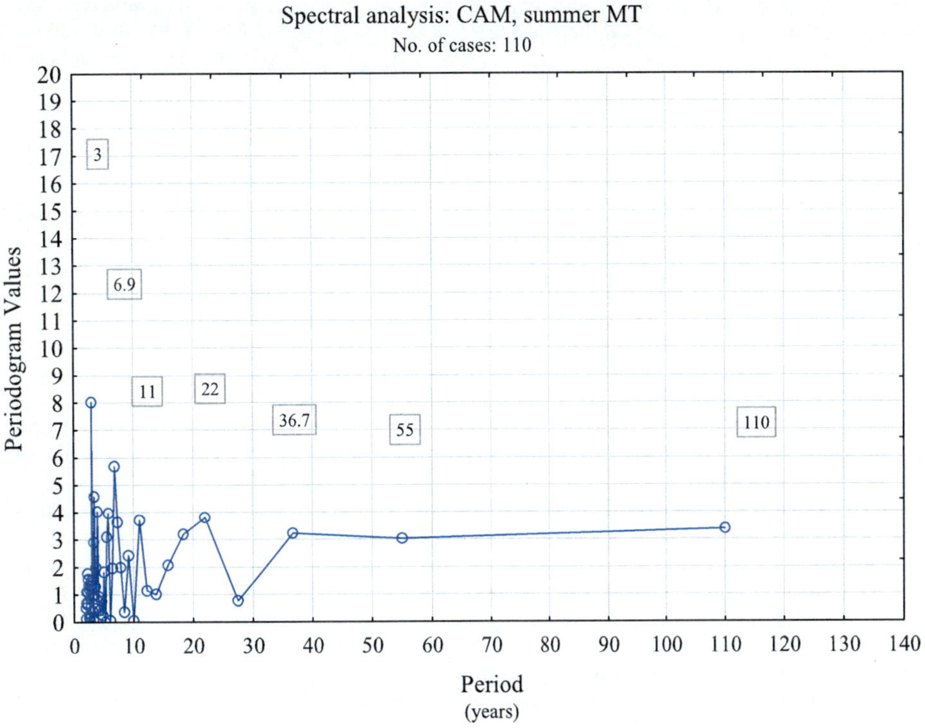

Fig. 3 Examples of periodogram for the weather station CAM in summer, the values in the squares show the prevailing period length

3.4 Non-stationary Association of Monthly Rainfall Series Among the Study Sites

Similarity in trends of monthly rainfall among sites has been tested using Pearson's r 7-years moving averages between pairs of sites (Fig. 4). Results show that in the period analyzed (1931–2004) there are periods when similarity in monthly rainfall varies highly among sites. The occurrence and extent of these periods vary by month and site, and the variations in the association level of trends in monthly rainfall are so great that the similarity can range from $r>0.80$ to $0.0<r<0.1$–0.2 and even negative ($r<-0.30$). These results show that similarity in trends of monthly rainfall among sites is non-stationary during the twentieth century in the study area.

4 Discussion

The low temperature and short growing season are not limiting factors for silver fir growth in most of the southern native area as they are in the central and northern European regions (Camarero et al., 2011; Macias et al., 2006). The low temperature and short growing season are not limiting factors for silver fir growth in most of the southern native area as they are in the central and northern European regions. In summer, monthly rainfall can be very low in the study area. Low monthly rainfall in summer can have an impact on silver fir growth and health conditions (i.e., water stress and its consequences), especially if combined with high average monthly temperatures. In the study area, the similarity in monthly rainfall series between sites is

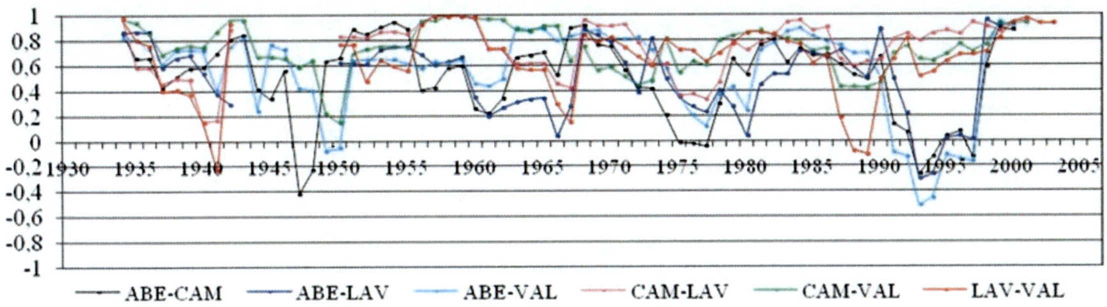

Fig. 4 Examples of variability in Pearson's r of 7-years moving averages between pairs of monthly rainfall series at all sites in the study area. August is represented

non-stationary during the twentieth century even at short distances (CAM, LAV, and secondarily VAL) and/or similar elevations (CAM and LAV). This variability is expected to be reflected in silver fir growth between sites, as monthly rainfall variability is normally high in the study area. In other words, the similarity in monthly rainfall trends varies unevenly by month and site in the study area and, as a result, the influence on silver fir growth is unlikely to be homogeneous over time and between the sites.

5 Conclusion

Inhomogeneous trends in monthly rainfall can differentiate the response of silver fir growth at the forest landscape level even within a common regional pattern of changing climate conditions. This raises the question of whether the silver fir growth response to both variability and non-stationary similarity in monthly rainfall trends during the twentieth century is common to all study sites or differs across sites. Answering this question is very important in order to identify the growth response behavior of silver fir, which can also be used as a health and/or environmental indicator, in the first decades of the twenty-first century; but also to check whether potential impacts differ within and/or between sites and finally to help develop adaptive management and mitigation strategies for the impacts of climate variability on silver fir forests. Indeed, the Tuscan Apennines is a unique environment where changing climatic conditions and their potential effects on silver fir forests are not yet fully known.

References

Allen, C. D., Macalady, A. K., Chenchouni, H., Bachelet, D., McDowell, N., Vennetier, M., et al. (2010). A global overview of drought and heat-induced tree mortality reveals emerging climate change risks for forests. *Forest Ecology and Management, 259*(4), 660–684.

Bartolini, G., Morabito, M., Crisci, A., Grifoni, D., Torrigiani, T., Petralli, M., et al. (2008). Recent trends in Tuscany (Italy) summer temperature and indices of extremes. *International Journal of Climatology, 28*(13), 1751–1760.

Camarero, J. J., Bigler, C., Linares, J. C., & Gil-Pelegrín, E. (2011). Synergistic effects of past historical logging and drought on the decline of Pyrenean silver fir forests. *Forest Ecology and Management, 262*(5), 759–769.

Chmielewski, F. M., & Rötzer, T. (2001). Response of tree phenology to climate change across Europe. *Agricultural and Forest Meteorology, 108*(2), 101–112.

Dobrowolska, D., Bončina, A., & Klumpp, R. (2017). Ecology and silviculture of silver fir (*Abies alba* Mill.): A review. *Journal of Forest Research, 22*(6), 326–335.

Gentilucci, M., Barbieri, M., & Burt, P. (2018b). Climatic variations in Macerata province (Central Italy). *Water, 10*(8), 1104.

Gentilucci, M., Materazzi, M., Pambianchi, G., Burt, P., & Guerriero, G. (2020). Temperature variations in Central Italy (Marche region) and effects on wine grape production. *Theoretical and Applied Climatology, 140*(1), 303–312.

Gentilucci, M., Moustafa, A. A., Abdel-Gawad, F. K., Mansour, S. R., Coppola, M. R., Caserta, L., Inglese, S., Pambianchi, G., & Guerriero, G. (2021). Advances in Egyptian Mediterranean coast climate change monitoring. *Water, 13*(13), 1870.

Gentilucci, M., Barbieri, M., & Burt, P. (2018). Climate and territorial suitability for the Vineyards developed using GIS techniques. In *Conference of the Arabian Journal of Geosciences* (pp. 11–13). Springer.

Hertig, E., & Jacobeit, J. (2008). Downscaling future climate change: Temperature scenarios for the Mediterranean area. *Global and Planetary Change, 63*(2–3), 127–131.

Luszczyńska, K., Wistuba, M., & Malik, I. (2018). Reductions in tree-ring widths of silver fir (*Abies alba* Mill.) as an indicator of air pollution in southern Poland. *Environmental and Socio-Economic Studies, 6*(3), 44–51.

Macias, M., Andreu, L., Bosch, O., Camarero, J. J., & Gutiérrez, E. (2006). Increasing aridity is enhancing silver fir *Abies alba* mill.) water stress in its south-western distribution limit. *Climatic Change, 79*(3–4), 289–313.

Toreti, A., & Desiato, F. (2008). Temperature trend over Italy from 1961 to 2004. *Theoretical and Applied Climatology, 91*(1–4), 51–58.

Evaluation of Statistical and Deep Learning Methods for Short-Term Weather Forecasting in Semi-arid Regions

Chouaib El Hachimi, Salwa Belaqziz, Saïd Khabba and Abdelghani Chehbouni

Abstract

Numerical methods represent a powerful tool for weather forecasting. However, they still face various limitations related to energy consumption and the time it takes to run simulations. To overcome these weaknesses, various statistical and deep learning models were developed to combine precision, time, and energy efficiency criteria. In this paper, we evaluated models of both classes for the task of short-term weather forecasting (one day, three days, and one-week forecasts), namely, the autoregressive integrated moving average, the theta method, and fast Fourier transform as statistical models, versus a long short-term memory, neural basis expansion analysis for interpretable time series, and temporal convolutional neural network as deep learning architectures. The dataset used in this study is sourced from the automatic meteorological station installed in the Marrakech region (center of Morocco) covering the period from January 3, 2013, to December 31, 2020, on a half-hour scale. These include air temperature (T_a), air relative humidity (H_r), and global solar radiation (R_g). Before feeding the data to our models, we first used the ERA5-Land Reanalysis data to impute missing values found in our time series. Results show that the TCNN model outperforms the others in terms of the coefficient of determination (R^2) and the root mean square error (RMSE).

Keywords

Weather time series forecasting · Statistical models · Deep learning models · LSTM · Temporal convolutional neural network · N-BEATS

1 Introduction

Weather time series forecasting is the process of making projections of the future state of the atmosphere for short-term periods. It is a subfield of climate modeling and is a challenging topic for researchers, managers, and end-users. Mastering climate dynamics is the key to making informed decisions in various domains such as agriculture (El Hachimi et al., 2023), transportation, and emergency management. Theory-driven or numerical methods for weather prediction (NWP) have been used for a long time and have shown great success in this field. They are based on solving mathematical equations that characterize the heat and matter interactions between Earth and the atmosphere. However, these methods have several drawbacks; they are time-consuming, energy-intensive processes and require supercomputers to achieve precise results (Carreño et al., 2015), not to mention the chaotic nature of the atmosphere which makes the approximation of physics' laws governing the complex processes which occur at very small scales a very difficult task (Kalnay et al., 1998).

With the emergence of artificial intelligence and data-driven approaches, new statistical and machine learning-based methods have shown promising results in solving regression and time series problems (Schultz et al., 2021).

This study aims to evaluate the performance of these state-of-the-art models from both classes (statistical and

C. El Hachimi (✉) · S. Belaqziz · S. Khabba · A. Chehbouni
Center for Remote Sensing Applications (CRSA), Mohammed VI Polytechnic University (UM6P), Benguerir, Morocco
e-mail: chouaib.elhachimi@um6p.ma

S. Belaqziz
LabSIV Laboratory, Department of Computer Science, Faculty of Science, Ibn Zohr University, Agadir, Morocco

S. Khabba
Department of Physics, Faculty of Sciences Semlalia, LMFE, Cadi Ayyad University, Marrakesh, Morocco

A. Chehbouni
Centre d'Etudes Spatiales de La Biosphère (CESBIO), Université de Toulouse, Toulouse, France

H. Chenchouni et al. (eds.), *Recent Advancements from Aquifers to Skies in Hydrogeology, Geoecology, and Atmospheric Sciences*, Advances in Science, Technology & Innovation, https://doi.org/10.1007/978-3-031-47079-0_45

machine learning models) and whether or not they can achieve a compromise of precision, time, and energy consumption for weather data in a semi-arid region.

2 Materials and Methods

2.1 Study Area and Data

Located 40 km east of Marrakesh in Morocco, the R3 district, our study area, has a semi-arid climate, typically a South-Mediterranean with an average annual precipitation of about 250 mm. The automatic weather station (AWS) installed there enabled us to collect a dataset of half-hourly measurements of weather from January 3, 2013, to December 31, 2020. This data includes air temperature (T_a), air relative humidity (H_r), and global solar radiation (R_g). A full description of this data is given in Table 1 which also includes statistics for missing values.

2.2 Statistical Models

This study evaluates three different statistical models. The first being ARIMA, which is an auto-regressive-based model that takes advantage of the inherent serial correlation between observations. Next is an exponential smoothing-based method (4theta) which is a modification of the original theta method (Assimakopoulos & Nikolopoulos, 2000). Finally, the fast Fourier transform method which uses the discrete Fourier transform formulas (Brigham & Morrow, 1967).

2.3 Deep Learning Models

The study also evaluates three state-of-the-art neural network architectures in the field of time series forecasting. The first is a long short-term memory, which is a recurrent neural network variant whose main objective is to overcome the shortcomings of the vanishing gradient and exploding gradient problem. The second model is the neural basis expansion analysis for interpretable time series forecasting or N-BEATS (Oreshkin et al., 2019) for short. It makes use only of residual blocks and fully connected layers to perform accurate forecasts. Lastly, the temporal convolutional neural network (TCNN) (Chen et al., 2020) is an architecture based on the encoder–decoder principle where the encoder is composed of stacked causal convolutional layers that captures the spatio-temporal dependencies in data, while the decoder component, in turn, contains a variant of residual neural networks (ResNet).

Table 1 Description of the meteorological station data

Variables	Description	Unit	Missing values
T_a	Air temperature at 2 m	°C	2631
R_g	Global solar radiation	W m^{-2}	5169
H_r	Air relative humidity	No unit	2626

2.4 Baseline Approach

The study followed the steps shown in Fig. 1. First, the ERA5-Land Reanalysis data was used to impute missing values in the three time series. The choice of using this remote data source was motivated by the fact that numerous studies have demonstrated its reliability in different climates and environmental conditions (Sebbar et al., 2023; Vanella et al., 2022). Missing data imputation consists of filling the gap found in the three time series, in this step, T_a was filled by converting ERA5-Land-t2m to Celsius, H_r was estimated using ERA5-Land-d2m and ERA5-Land-t2m based on the Magnus formulas, and R_g was filled with ERA5-Land-ssrd after unit conversion and disaggregation. Next is modeling, where the clean data is fed to the six models to train and perform forecasts (1 day, 3 days, and 1 week).

To evaluate the models, the root mean squared error (Eq. 1), and R^2, the coefficient of determination (Eq. 2), were used.

$$\text{RMSE} = \sqrt{\frac{\sum_{i=1}^{N}\left(y_i - \widehat{y}_i\right)^2}{N}} \quad (1)$$

$$R^2 = 1 - \frac{\text{SS}_{\text{res}}}{\text{SS}_{\text{tot}}} \quad (2)$$

where SS_{res} is the sum of squares of residuals and SS_{tot} is the total sum of squares.

The implementation of all parts of the flowchart were done using mainly the libraries: Darts, data science toolkit, and ClimatFiller (El Hachimi et al., 2022; Oreshkin et al., 2019; El Hachimi et al., 2023). Also, the UM6P high-performance computing platform (HPC) was used to run the simulations.

3 Results and Discussions

Table 2 presents the performance evaluation of the six compared models in terms of the RMSE. This latter has the same unit of the corresponding meteorological variables (Celsius, percentage, and W m^{-1} for T_a, H_r, and R_g, respectively). Lower values of RMSE indicate a better fit of the

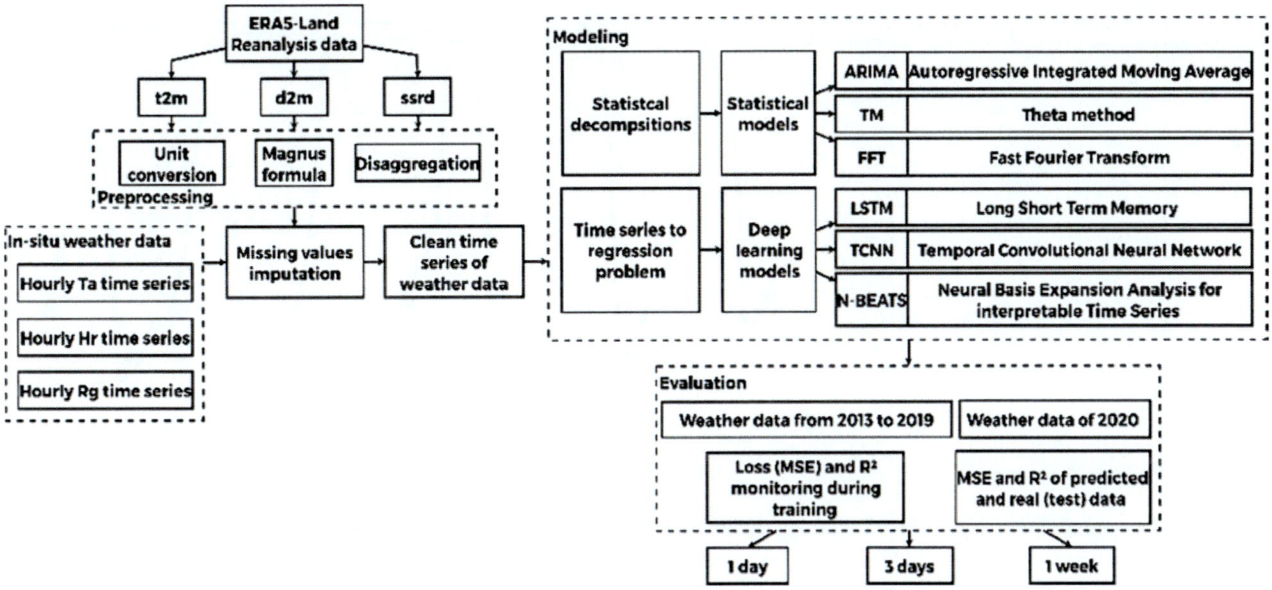

Fig. 1 Flowchart of the implemented approach

model to the data while higher values indicate greater error between the forecasts and real values. We also calculated R^2 for best-performing models to measure the explainability capability of the model, with 0 indicating that the model does not explain any of the variations in the dependent variable, and 1 indicating that all variations in the corresponding meteorological variable can be explained by the model.

As reported in Table 2, statistical models show acceptable performance for T_a, especially for the FFT model which has an RMSE lower than 1.6 °C and an R^2 greater than 0.9 in all periods. This is due to the nature of this time series (seasonality, cyclical movements, less unexpected variations, etc.) which makes it easier to be approximated by mathematical equations used in statistical models. In contrast, for the challenging variables such as H_r, deep learning models were the best. For example, the RMSE is greater than 11.6% for all models in all periods (Table 2).

For deep learning models, and despite the fact that dealing with time series and sequential data has been largely associated with recurrent neural network architectures such as the LSTM, this latter architecture was unable to capture the pattern of the R_g time series for all periods expressed by a high value of RMSE that is in the order of the measurement itself (~ 356.49 W m^{-2}). However, TCNN was the best for a 1-day forecast (R^2>0.9) and had similar results to N-BEATS for 3 days and 1-week forecasts (the difference between R^2 did not exceed ~ 0.1).

Finally, when it comes to the trade-off between capturing time series complexity versus precision versus the forecast period, the TCNN model outperforms all the others.

Table 2 RMSE of the compared models

		Deep learning models			Statistical models		
		LSTM	TCNN	N-BEATS	ARIMA	Fast Fourier transform	Theta method
1 day	T_a	1.30	0.93	1.64	5.08	1.13	6.69
	H_r	6.69	4.05	4.74	12.31	13.61	17.16
	R_g	322.14	30.38	57.27	107.05	128.93	128.76
3 days	T_a	2.24	2.01	1.14	6.46	1.37	6.60
	H_r	9.25	7.12	6.69	16.60	11.67	15.65
	R_g	348.16	26.60	38.05	139.33	108.42	160.45
1 week	T_a	4.94	2.88	2.42	6.64	1.54	5.95
	H_r	14.96	13.73	15.02	23.17	14.30	16.88
	R_g	356.49	26.08	23.82	141.47	104.64	162.72

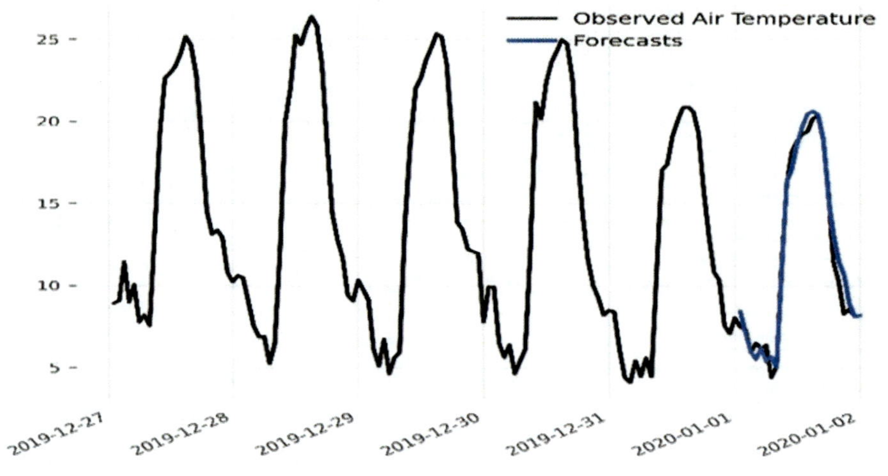

Fig. 2 Visual results of a 1-day air temperature forecast

For the best model (TCNN), Fig. 2 shows the visual results of the real and estimated air temperature values for a forecast period of 1 day (24 h). There is a clear agreement between the TCNN model forecasts and reality.

4 Conclusion

In this study, we evaluated statistical and deep learning models to test their effectiveness in short-term weather forecasting. The TCNN model performed better than the others for the three used weather time series. This model can also be deployed to make inferences instantly once trained, which represents a good compromise: time of execution, energy consumption, and most importantly precision.

Acknowledgements This study was supported by and conducted within the Center for Remote Sensing Applications (CRSA) (https://crsa.um6p.ma), at the Mohammed VI Polytechnic University (UM6P) in Morocco. Additionally, we would like to thank the Moroccan Ministry of Higher Education, Scientific Research and Innovation and the OCP Foundation who funded this work through the APRD research program (GEANTech). And also, we would like to recognize the contributions of the AGREEMed, PRIMA-S2-ALTOS-2018 and ASSIWAT projects in directing the outcomes of this research.

References

Assimakopoulos, V., & Nikolopoulos, K. (2000). The theta model: A decomposition approach to forecasting. *International Journal of Forecasting, 16*(4), 521–530. https://doi.org/10.1016/S0169-2070(00)00066-2

Brigham, E. O., & Morrow, R. E. (1967). The fast Fourier transform. *IEEE Spectrum, 4*(12), 63–70. https://doi.org/10.1109/MSPEC.1967.5217220

Carreño, E. D., Roloff, E., & Navaux, P. O. A. (2015). Challenges and solutions in executing numerical weather prediction in a cloud infrastructure. *Procedia Computer Science, 51*, 2832–2837. https://doi.org/10.1016/J.PROCS.2015.05.443

Chen, Y., Kang, Y., Chen, Y., & Wang, Z. (2020). Probabilistic forecasting with temporal convolutional neural network. *Neurocomputing, 399*, 491–501. https://doi.org/10.1016/J.NEUCOM.2020.03.011

El Hachimi, C., Belaqziz, S., Khabba, S., & Chehbouni, A. (2022). Data Science Toolkit: An all-in-one python library to help researchers and practitioners in implementing data science-related algorithms with less effort. *Software Impacts, 12*, 100240. https://doi.org/10.1016/J.SIMPA.2022.100240

El Hachimi, C., Belaqziz, S., Khabba, S., Ousanouan, Y., Sebbar, B., Kharrou, M.H., Chehbouni, A. (2023). ClimateFiller: A python framework for climate time series gap-filling and diagnosis based on artificial intelligence and multi-source reanalysis data. *Software Impacts, 18*, 100575. https://doi.org/10.1016/j.simpa.2023.100575

El Hachimi, C., Belaqziz, S., Khabba, S., Sebbar, B., Dhiba, D., Chehbouni, A. (2022). Smart weather data management based on artificial intelligence and big data analytics for precision agriculture. *Agriculture, 13*, 95. https://doi.org/10.3390/AGRICULTURE13010095

Kalnay, E., Lord, S. J., & McPherson, R. D. (1998). Maturity of operational numerical weather prediction: Medium range. *Bulletin of the American Meteorological Society, 79*(12), 2753–2770. https://doi.org/10.1175/1520-0477(1998)079<2753:MOONWP>2.0.CO;2

Oreshkin, B. N., Carpov, D., Chapados, N., & Bengio Mila, Y. (2019). N-BEATS: Neural basis expansion analysis for interpretable time series forecasting. https://doi.org/10.48550/arxiv.1905.10437

Schultz, M. G., et al. (2021). Can deep learning beat numerical weather prediction? *Philosophical Transactions of the Royal Society, 379*(2194). https://doi.org/10.1098/RSTA.2020.0097

Sebbar, B., et al. (2023). Machine-learning-based downscaling of hourly ERA5-land air temperature over mountainous regions. *Atmosphere, 14*(4), 610. https://doi.org/10.3390/ATMOS14040610

Vanella, D., et al. (2022). Comparing the use of ERA5 reanalysis dataset and ground-based agrometeorological data under different climates and topography in Italy. *Journal of Hydrology: Regional Studies, 42*, 101182. https://doi.org/10.1016/J.EJRH.2022.101182

Impact of Air Pollution on the Incidence and Health Risk of Adolescents

Emiliya Valeeva, Natalya Stepanova, Gulgena Ismagilova, Elena Minakova, Amr S. Elbahnasawy and Galia Skvortsova

Abstract

Atmospheric air pollution is one of the most critical environmental factors affecting the child, adolescent, and adult populations. Our research aimed to study the atmospheric air pollution effect on adolescents' health risk. For that purpose, correlation analysis was used to reveal the relationship between atmospheric air pollution and adolescents' health indicators. The Mann–Whitney test was applied for the comparison of the two samples' distribution. We identified four districts in the city of Kazan for research: 1—Vakhitovsky, 2—Sovetsky, 3—Kirovsky, and 4—Privolzhsky districts. The target group of people was adolescents aged 15–17 years old, study period (2008–2019). Comparative analysis of the zones in the level of harmful substances in the air was performed. We presented the major air pollutants with significant differences ($p < 0.05$) in zoning (carbon (soot), carbon oxide, nitrogen dioxide, suspended particles PM 2.5 and PM 10, formaldehyde, benzol, and benzene). "From the health issues considered in this study—respiratory diseases (RD), allergic manifestations, anemias, congenital malformations (CM), and malignant neoplasms (NP)—RDs are at the top of the list: their proportion in primary morbidity structure varied insignificantly in all zones, the highest levels being found in the second zone. Analysis of chemicals' content in atmospheric air showed that the third zone was in a worse position than the first and fourth zones. The second zone had the worst indices in significant evidence-based criteria. The current ecological situation in Kazan reflects the level of primary morbidity and prevalence of RD, NP, CM, and other adolescent diseases in the most neglected second and third city zones. Elevated concentration of formaldehyde, nitrogen dioxide, suspended particles PM 2.5, and benzol was observed in the second zone. Negative correlations between air pollutants and different human diseases, which can be explained by delayed body response to the accumulation of harmful substances, were also revealed in this zone. Suspended particles PM 2.5 and PM 10 have the strongest accumulative effect in this zone.

Keywords

Atmospheric air · Morbidity · Health risk · Suspended particles

1 Introduction

Air pollution is one of the most important environmental factors affecting the health of children, adolescents, and adults. Diseases of the respiratory organs (RO), endocrine system (ES), congenital malformations (CM), and malignant neoplasms (NO) are currently classified as pathologies with high growth rates but insufficiently studied in terms of the role of atmospheric air pollution in their formation in assessing the well-being of territories. The state of the atmospheric air, which is associated with 3411.9–3979.4 thousand additional cases of morbidity in the Russian population (Valeeva et al., 2019), is relatively widely represented both by indicators of socio-hygienic and environmental monitoring and by scientific research. The issues of long-term exposure to low concentrations of chemicals, including low-dose exposure, in terms of possibly hidden, remote time changes in the body are becoming more relevant and more often attract the attention of researchers (Maludze et al., 2020; Stepanova et al., 2022).

The purpose of the study was to assess the impact of atmospheric air quality on the health status of adolescents

E. Valeeva · N. Stepanova (✉) · G. Ismagilova · E. Minakova · G. Skvortsova
Kazan Federal University, Republic of Tatarstan, 420044 Kazan, Russia
e-mail: Stepmed@Mail.Ru

A. S. Elbahnasawy
National Research Centre, Dokki, Cairo 12622, Egypt

H. Chenchouni et al. (eds.), *Recent Advancements from Aquifers to Skies in Hydrogeology, Geoecology, and Atmospheric Sciences*, Advances in Science, Technology & Innovation, https://doi.org/10.1007/978-3-031-47079-0_46

(in the example of the city of Kazan) and to calculate the risk of developing non-carcinogenic effects.

The relevance of the topic is given by the dynamism of the development of the region, the qualitative and quantitative indicators of the technogenic load on the atmospheric air of the Republic of Tatarstan and in particular the city of Kazan, which has a highly developed sector of industry, oil production, petrochemistry, and mechanical engineering.

The work aims to study the impact of atmospheric air pollution on the incidence and health risks of adolescents.

2 Methods

Correlation analysis was used to determine the relationship between air pollution and adolescent health indicators. To compare air quality indicators and the incidence rate in different zones, the following was used: firstly, the classical *T*-test with a preliminary check for the normality of distributions and the nonparametric Mann–Whitney test were used. The work revealed differences in the averages, reliably identified by at least one of the criteria. The achieved levels of significance (*p*-value) are also given. The assessment of non-carcinogenic risk was carried out based on the results of studies performed based on an accredited laboratory of the FBUZ "Center for Hygiene and Epidemiology in the Republic of Tatarstan" for risk assessment for public health (R 2.1.10.1920-04). The risk of developing non-carcinogenic effects in adolescents from exposure to atmospheric air pollution was assessed according to R 2.1.10.1920-04 (Sabirzyanov & Khamitova, 2017) by hazard coefficients.

The target of the study was adolescents aged 15–17 years old, study period (2008–2019), attached to the corresponding children's polyclinic (CP) in the city and corresponding to the studied areas. We have identified four districts of the city of Kazan for research: first zone Vakhitovsky (CP No. 2), second zone Kirovsky (CP No. 6), third zone Sovetsky (CP No. 4), and fourth zone Privolzhsky district (CP No. 10).

3 Results

Every year, about 360 types of chemical pollutants with a total mass of about 104–107 thousand tons are emitted into the atmospheric air of the city, where the share of industrial emissions in total pollution is 29.4–32 thousand tons. Among the impurities that worsen the air quality in the city are toxic substances of the first and second hazard classes (chromium, benzene, phenol, acrolein, and formaldehyde) and substances with a low irritant threshold (hydrogen sulfide, ammonia, sodium hydroxide, etc.). Information on gross emissions from stationary sources in Kazan showed that most of the emissions are accounted for by motor vehicles (71.4–72.8%), and the rest is accounted for by industry. During the studied years in Kazan, cases of exceeding the maximum permissible concentrations (MPC) were recorded: for suspended solids, nitrogen dioxide, ammonia, formaldehyde, ethylbenzene, xylene, acetone, hydrogen sulfide, and phenol. Average annual concentrations of all measured harmful substances did not exceed MPC.

A sample of data on the medical care sought by adolescents in terms of morbidity was carried out according to environmentally determined groups of diseases and nosologies: respiratory diseases (RD), allergic manifestations, anemia, congenital malformations (CM), and malignant neoplasms (NP). ORD is the leading class: their share in the structure of primary morbidity in all zones varied slightly: (18.1–39.58) %, having determined in zone 2 – (31.40–39.58) % (95% CI 714, 69–1078.15) as the highest levels. As is known, respiratory diseases are largely etiologically associated with technogenic atmospheric air pollution, since the mucous membranes of the upper respiratory tract are the first target for harmful environmental substances.

During the studied years, with maximum atmospheric emissions, there was a high incidence of diseases in respiratory organs in adolescents in children's city clinics in Kazan.

The statistical analysis carried out allows us to assert the following with the achieved level of significance p-value < 0.0001: the prevalence of respiratory diseases in zones 1, 2, and 3 is higher than in zone 4; the prevalence of RD in zone 2 is higher than in zone 1, the prevalence of BHO in zone 2 is higher than in zones 1 and 3, the prevalence of NS in zone 3 is higher than in zones 1 and 2, the prevalence of MS in zones 2 and 3 is higher compared to studied zones, and the prevalence of CA 3 zone has maximum values (Table 1).

In addition, a comparative analysis was made between the zones in terms of the level of harmful substances in the air. We present the main air pollutants that have significant differences (p-value < 0.05) in terms of distribution by zones (Table 2).

The results of the nitrogen dioxide analysis showed that its highest values were obtained in the second and third zones under study. The indicators of carbon dioxide were distributed as follows: the highest in zones 2 and 3, and the lowest in zone 4. Suspended PM 10 is highest in zone 4 and lowest in zone 3. The highest concentration of soot was determined in zones 2 and 3, lower in zone 1, and the lowest in zone 4.

The main atmospheric air pollutants are presented with significant differences ($p < 0.05$) in zoning (carbon (soot), carbon monoxide, nitrogen dioxide, and suspended particles), PM 2.5 and PM 10, formaldehyde, and benzene.

Table 1 Average values of the prevalence of diseases by zones in Kazan for 2008–2019 among adolescents

Zone	1	2	3	4
Respiratory diseases (RD)	802.50	925.81	602.23	237.30
Malignant neoplasms (NO)	1.59	4.75	2.95	2.62
Diseases of the blood and hematopoietic org. (BHO)	8.47	25.84	7.59	14.79
Diseases of the nerve system (NS)	31.59	41.84	73.87	73.81
Diseases of the musculo-skeletal system (MS)	46.24	110.39	115.70	102.73
Diseases of congenital anomalies (CA)	0.20	0.73	2.57	1.49

Table 2 Average values of levels of harmful substances by zones in Kazan for 2008–2019

Zone	1	2	3	4
Nitrogen dioxide	0.088	0.106	0.101	0.092
Carbon oxide	3.447	3.777	3.863	2.941
Carbon (soot)	0.126	0.165	0.145	0.085
Suspended solids	0.140	0.156	0.152	0.157
Suspended particles PM 10	0.070	0.0849	0.063	0.097
Suspended particles PM 2.5	0.030	0.037	0.028	0.040
Petrol	0.003	0.006	0.003	0.0008
Formaldehyde	0.001	0.001	0.001	0.001

4 Discussion

Respiratory diseases occupy the first place in the list: their share in the structure of primary morbidity varied slightly in all zones, and the highest levels were noted in the second zone. The dependence of the incidence rate on the level of harmful substances in the air in each zone was studied separately. Analysis of the content of chemicals in the atmospheric air showed that the second zone had the worst performance, and the third zone was in a worse position than the first and fourth. Thus, in the second zone, an increased concentration of formaldehyde, nitrogen dioxide, suspended particles of PM 2.5 and PM 10, gasoline, and carbon (soot) was observed, and in the third zone, the concentration of carbon monoxide was increased.

A similar picture was revealed by a comparative analysis of the incidence rate between zones. Thus, the level of primary morbidity and prevalence of NO, BHO, and other diseases in the second zone is the highest, and the level of MS is the highest in zone 3. Thus, the current environmental situation in Kazan is reflected in the incidence of adolescents in the most neglected second and third city zones. Negative relationships between atmospheric air pollutants and various human diseases were also found in these zones, which can be explained by a slow reaction of the body to the accumulation of harmful substances. Suspended particles PM 2.5 and PM 10, as well as carbon oxide and formaldehyde, have the strongest accumulative effect in these zones.

5 Conclusion

Our study showed that zones 2 and 3 are in a worse position compared to zones 1 and 4 since the indicators of pollutant concentrations and the dynamics of morbidity are higher in them. The current ecological situation in the city of Kazan is reflected in the level of morbidity and prevalence of diseases of AML, OI, congenital malformations, and other diseases of adolescents in the most disadvantaged zones 2 and 3 of the city. The share contribution of chemicals showed that the first ranking place of influence on the occurrence of negative effects on the health of adolescents is occupied by nitrogen dioxide, which amounted to 21.45% in zone 1, while in zones 2 and 3 (soot), respectively, 24.56% and 23.3%. Third place is PM 2.5 suspended particles, and the spread of values by districts was (15.03–18.68%).

Acknowledgements This paper has been supported by the Kazan Federal University Strategic Academic Leadership Program (PRIORITY-2030).

References

Maludze, G., Stepanova, N., Zakirova, O., et al. (2020). Risk assessment of the incidence of cardiovascular diseases based on evolutionary model. *European Journal of Clinical Investigation, 52*, 20. https://doi.org/10.1111/eci.13796

Sabirzyanov, A. R., & Khamitova, R. Ya. (2017). Dependence of indicators of endocrinological morbidity of the adult population on the use of chemicals in agriculture and air pollution. *Public Health and Habitat, 10*(295), 24–26.

Stepanova, N., Ilyasova, A., Yusupova, N., Khairullina, L., Fomina, S., & Saifullin, R. (2022). Assessment of risk from atmospheric air pollution and traffic load intensity in the city of Kazan (Republic of Tatarstan). *Advances in science, technology, and innovation* (pp. 247–249).

Valeeva, E. R., Ismagilova, C. A., Ziyatdinova, A. I., Elagina, D. S., & Arkhipova, N. S. (2019). Inhalative impact of chemical pollutants of atmosphere on the organism of teenagers and risks for their health. *Journal of International Pharmaceutical Research, 46*(4), 529–534.

Bromide Variability in Wet Atmospheric Precipitation at a Coastal Location in Southwestern Europe

João Moreno, Filipa Moreno, Francisco Fatela, Ana Russo, Alexandre Ramos and Orquídia Neves

Abstract

During the last decades, scientific interest in bromine (Br) has been fueled by its influence on the climate, namely via the depletion of ozone and elemental mercury. Today, there is compelling evidence that the heterogeneous/multiphase Br-reactive chemistry in the troposphere, extremely important in polar regions, is also abundant in other areas of the globe, including tidal coastal areas. This likely impacts the composition of wet precipitation for bromide (Br^-) and sustains Br transport over relatively large distances inland, enriching continental depositional environments. Thus, contributing to the quantification of Br^- in rainwater concerning its main natural source (i.e., the ocean), and possible influential meteorological factors (e.g., wind speed/direction), seems important to better understand the role of the oceanogenic Br in paleoclimatology as a proxy for storminess. In this study, we used Br^- concentrations ([Br^-]) in a set of twenty-one rainfall samples collected, on an event basis, between June 2018 and November 2019, at three sites in a coastal area of NW Portugal (Iberian Peninsula, IP; SW Europe), to investigate source–receptor relationships. The sampling sites were selected to represent a sea-to-inland gradient (~ 30 km long). Chemical analyses were performed by ion chromatography (anions) and inductively coupled plasma atomic emission spectrometry (cations). [Br^-] data, though without statistical outliers, showed high variability (12–375 μg/L), the median of the period being 155 μg/L, while the average of the recorded values was 161 ± 100 μg/L. In the sampling period, two [Br^-] peaks in rainfall were observed: in winter (January 30, 2019; 375 μg/L) and in summer (June 21, 2018; 329 μg/L), both at the innermost site. HYSPLIT model analysis suggests that the first event was related to air masses passing directly over the Atlantic before reaching the sampling area, while in the second, air masses arrived at the target region from the NE-to-E sector, after traveling through the IP landmass. PCA analysis (after *z*-score conversion of the values) allowed us to recognize three main groups of rainwater samples following similar tendencies on Br^-. It is expected that our results will be useful for ongoing and future research on (paleo)precipitation and (paleo)cyclogenesis.

Keywords

Bromide · Rainwater · Mid-latitude westerly wind belt · NE Atlantic coastland · NW Portugal

1 Introduction

Systematic research on the chemistry of atmospheric bromine (Br) has made clear its complexity, also shedding new light on Br activation hotspots (Simpson et al., 2015). Bromine is released from the open ocean via sea spray aerosol (SSA), containing sea salt bromide (Br^-), or by emissions of a wide range of volatile organobromines, synthesized by ocean biota (Simpson et al., 2015). Once in the atmosphere, it undergoes heterogeneous/multiphase (re) cycling (Simpson et al., 2015), able to maintain the transference of oceanogenic Br from the marine to the terrestrial

J. Moreno (✉) · F. Fatela · A. Russo · A. Ramos
Faculdade de Ciências, IDL—Instituto Dom Luiz, Universidade de Lisboa (FCUL), 1749-016 Lisboa, Portugal
e-mail: jcmoreno@fc.ul.pt

F. Moreno
Caminho da Portela 97, 4940-061 Bico PCR, Portugal

F. Fatela
Departamento de Geologia, FCUL, Universidade de Lisboa, 1749-016 Lisboa, Portugal

O. Neves
CERENA, Instituto Superior Técnico, Universidade de Lisboa, Av. Rovisco Pais 1, 1049-001 Lisboa, Portugal

H. Chenchouni et al. (eds.), *Recent Advancements from Aquifers to Skies in Hydrogeology, Geoecology, and Atmospheric Sciences*, Advances in Science, Technology & Innovation, https://doi.org/10.1007/978-3-031-47079-0_47

milieu, and likely impacting the chemical composition (as regards Br^-) of rainfall.

This work was conducted on the NW façade of Portugal (SW Europe), an Atlantic location with high average annual rainfall (1200–4200 mm) and strongly influenced by the moisture brought by the Northern Hemisphere mid-latitude westerly wind belt. The main goal was to examine the rainwater Br^- variability in relation to the transport of air masses from the North Atlantic Ocean, helping to better constrain the significance of Br while proxy for storminess (and windiness).

2 Materials and Methods

Rainfall samples were collected using a glass funnel connected to a 1-L polyethylene bottle, acid-cleaned and pre-rinsed with Milli-Q water, at three locations: Nearshore (NS), Arga (A), and Corno do Bico (CB), defining a roughly W–E (coast/inland) gradient. The sampling took place between June 2018 and November 2019 during precipitation events associated with westerlies varying in magnitude. In the laboratory, the pH and conductivity values of the samples were measured; after filtration through 0.45 µm membrane filters, Br− and other anions (Cl^-, $SO_4^{2}-$, NO_3^-, $PO_4^{3}-$, and F^-) concentrations were measured using ion chromatography, while cations (Na^+, K^+, Ca^{2+}, and Mg^{2+}) were determined by ICP-OES. The analytical methods ensured a precision better than 5% for anions and uncertainties between 11 and 17% for cations ($p<0.05$). Chemical data were tested by ion balancing methods; then principal component analysis (PCA) was applied (after *z*-score conversion), using the PAST software v. 4.03. Backward trajectories were derived by the NOAA hybrid single-particle Lagrangian integrated trajectory model 4 (HYSPLIT4), which was run based on the dataset of the National Centers for Environmental Prediction (NCEP).

3 Results

3.1 Bromide Variations in the Wet Precipitation Influenced by the Westerlies

During the sampling period, [Br−] in rainfall coupled to westerly winds were highly variable, though without any detected statistical outlier, ranging from 12 to 375 µg/L. The median was 155 µg/L, while the average (± 1SD) was 161 ± 100 µg/L. The two highest [Br^-] (> 300 µg/L) events occurred on January 30, 2019 (375 µg/L) and on June 21, 2018 (329 µg/L), both at the CB site (527 m a.m.s.l.; 30 km from coast).

From PCA, three principal components, accounting for 94.6% of the total variance in the data, were retained. Principal component 3 (PC3), explaining 8.5% of the variance, is almost entirely related to Br−, with high positive loading (0.89), so variations in PC3 mainly reflect changes in the Br^- content of rainwater. Three groups of samples may be distinguished (Fig. 1), with group 2 roughly expressing the average [Br−] in our dataset, and groups 1 and 3 representing high and low [Br^-], respectively. The sample NS2 ([Br−] = 277 µg/L), with 219 mg/L of Cl^-, 177 mg/L of Na^+, and 39 mg/L of SO_4^{2-}, is visibly "contaminated" by direct SSA and washout of giant sea salt particles.

3.2 Influence of Atmospheric Trajectories on Major [Br−] Peaks

The HYSPLIT results showed air masses reaching the three sampling sites along (northern and southern) westerly transport trajectories, having passed directly over the Atlantic before arrival. The highest [Br^-] (sample CB4) was associated with the North Atlantic storm Gabriel, which formed on January 28, 2019, and underwent explosive cyclogenesis during the following 30 h (Fig. 2, left; http://www.aemet.es/en/conocermas/borrascas/2018-2019/estudios_e_impactos/gabriel). The second highest value (sample CB1) was related to the passage of weakly active frontal waves, with air motion arriving at the CB site from the NE-to-E sector, after traveling over land, throughout the interior of the IP, for ~ 75 h (Fig. 2, right). As can be seen, at 1500 m, the air inflow direction was from the south, and airstreams at that height derived from North Africa.

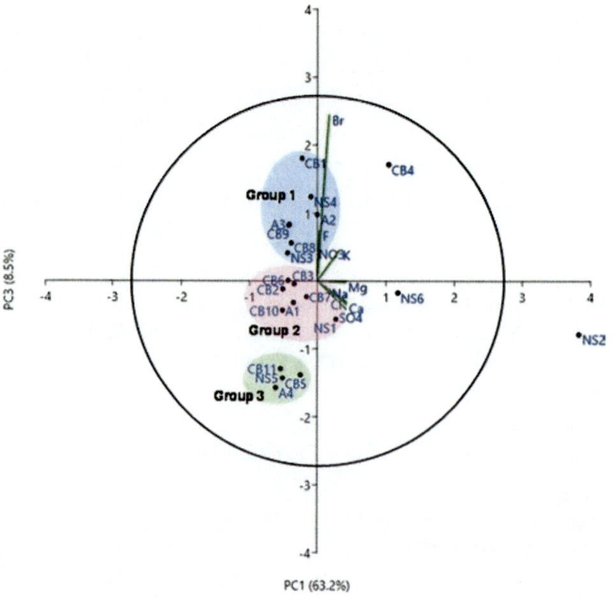

Fig. 1 Biplot of the PC1 x PC3 obtained by PCA on chemical data

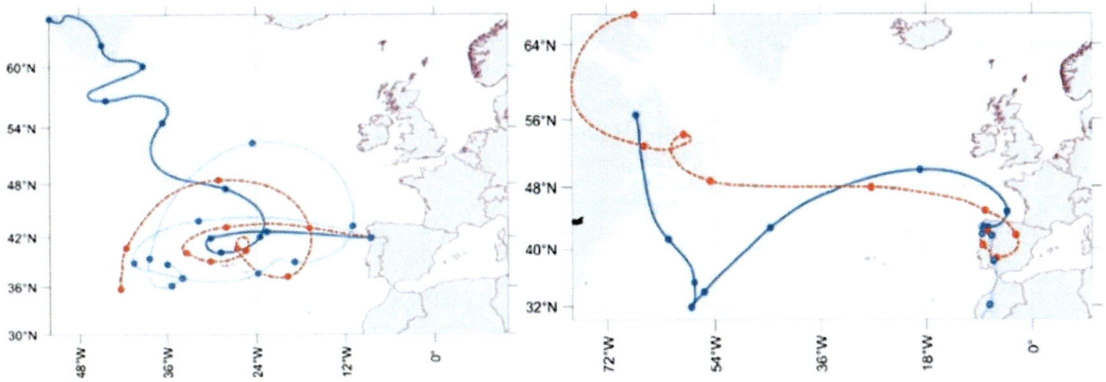

Fig. 2 Backward trajectories calculated by HYSPLIT (based on NCEP reanalysis data) at three starting heights: 500 m (red), 1000 m (dark blue), and 1500 m (dotted light blue) above sea level

4 Discussion

Although sea salt aerosols, produced as bubbles burst at the oceanic surface, are the dominant source for ions such as Cl−, Na^+, SO_4^{2-}, Mg^{2+}, Ca^{2+}, K^+, and Br− in the marine boundary layer and rainwater in coastal regions, the PCA performed on our dataset disclosed a different behavior for Br^-, which does not group with the sea salt indicators (associated with PC1; Fig. 1). This shows that [Br^-] in the wet precipitation at the three sites is not a simple product of SSA, suggesting chemical modifications in the atmosphere (Simpson et al., 2015). Based on the known affinity of Br for organic matter, we hypothesize that Br^- could be somehow related to dissolved organic carbon, a key component of marine rainfall.

One explanation for the highest [Br−] during the development of storm Gabriel is that extratropical cyclones can reach higher altitudes in the troposphere than frontal waves, thereby wet scavenging more of the solutes and gaseous species present there. The strength of the jet stream winds might be said to have a great potential for high concentrations and deposition of Br^- in our study area. The second highest [Br^-] could be affected by the input of continental (natural and anthropogenic) sources during air mass transport over IP, such as the emission of methyl bromide (CH_3Br) from biomass burning (e.g., forest fires) (Manö & Andreae, 1994), practically constant in this region along summertime.

5 Conclusion

In this study, we focused on the Br^- content in rainfall linked to the westerlies in NW Portugal for one and a half years. The highest [Br−] in rainwater was detected in connection with an extratropical cyclone, which seems a promising result concerning the proposed use of Br as a proxy for storminess. However, another rainfall event showing similar [Br^-] might have been induced by continental sources (natural/anthropogenic). Thus, further efforts are needed to provide a more accurate picture of the most likely source-to-sink modification processes that can shape the [Br−] of rainwater in our study area, including the contribution (if any) of local moisture recycling.

Acknowledgements This work was funded by the Fundação para a Ciência e a Tecnologia, I.P./MCTES through national funds (PIDDAC)—UIDB/50019/2020.

References

http://www.aemet.es/en/conocermas/borrascas/2018-2019/estudios_e_impactos/gabriel. Accessed in August 31, 2022.

Manö, S., & Andreae, M. O. (1994). Emission of methyl bromide from biomass burning. *Science, 263*(5151), 1255–1257.

Simpson, et al. (2015). Tropospheric halogen chemistry: Sources, cycling, and impacts. *Chemical Reviews, 115*, 4035–4062.

Using the Support Vector Machine Classification to Identify Water Masses in the Western Alboran Sea

Ayoub Belattmania, Abdelkrim El Arrim, Adam Ayouche, Karim Hilmi and Bouchta EL Moumni

Abstract

Quasi-synoptic hydrological data collected in the Western Alboran Sea, during a glider cruise, dedicated to sample the Western Alboran Gyre (WAG) in late fall 2020 have been used to examine the distribution of the water masses in the region. The spatial variability has been addressed by means of a Support Vector Machine (SVM) classification applied on TS diagram through a training set of more than 3000 vertical profiles of temperature and salinity from the World Ocean Database (WOD). The training process is based on a manual labelling data in which the frontiers between the water masses have been defined on the basis of the values of temperature, salinity, and density, inferred from traditional expert analysis of the TS diagram in the area. Compared to conventional clustering analysis methods, this approach is appropriate to distinguish water masses in regions where intense mixing occurs such as the Alboran Sea. This new method highlights the WAG's characterization by recent AW advected from the Strait of Gibraltar, the presence of the Levantine Intermediate Water (LIW) in the northern two-third of the basin and the uplift of the dense Mediterranean Waters in the south, in the sill surroundings, roughly parallel to the Moroccan continental slope where they mix directly with AW.

A. Belattmania (✉) · A. El Arrim
Faculté des Sciences et Techniques de Tanger, Tangier, Morocco
e-mail: ayoub-bel17@hotmail.com

A. Ayouche
Laboratory for Ocean Physics and Satellite Remote Sensing (LOPS), UMR6523, Ifremer, Univ. Brest, CNRS, IRD, Brest, France

K. Hilmi
Institut National de Recherche Halieutique, Casablanca, Morocco

B. EL Moumni
Université Abdelmalek Essaâdi, Tétouan, Morocco

Keywords

Alboran Sea · Western Alboran Gyre · Water masses · Support vector machine classification

1 Introduction

The use of the TS diagram has been almost universal in physical oceanography since Helland-Hansen introduced it in his monograph (Swift, 1986). Since then, several methods have been proposed in the field of water mass classification, notably the volumetric T-S diagram analysis (Watelet et al., 2020). Apart from these traditional TS diagrams analysis, several methods have been proposed in the field of water mass classification. The most common and conventional one is the cluster analysis (Naranjo et al., 2015).

In this study, on the basis of World Ocean Database profiles and TS data observed in the Western Alboran Sea in late fall 2020, we propose a new approach based on Support Vector Machine classification to identify the boundaries between the water masses in the area.

2 Data and Methods

In this study, we distinguish two types of data, as shown in Fig. 1. The first concerns TS observations from the World Ocean Database (https://www.ncei.noaa.gov/products/world-ocean-database) in the Western Alboran Sea, used as a training dataset for the classification model. The second is related to TS data from a glider mission performed in the same region in late fall 2020. In order to keep the length of the manuscript within a reasonable limit, only first transect results (from November 11 to November 19, 2020) are included in the text.

To retain the physical meanings in the WOD labeling process, boundaries between water masses have

H. Chenchouni et al. (eds.), *Recent Advancements from Aquifers to Skies in Hydrogeology, Geoecology, and Atmospheric Sciences*, Advances in Science, Technology & Innovation, https://doi.org/10.1007/978-3-031-47079-0_48

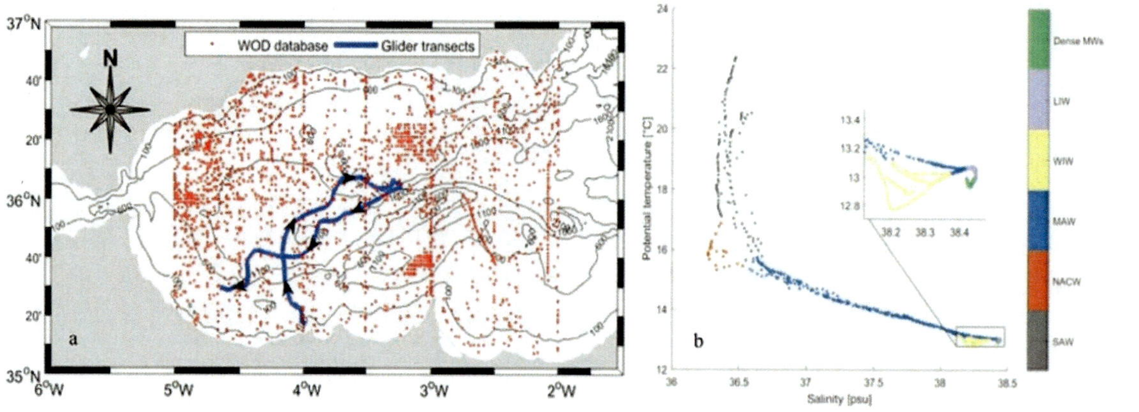

Fig. 1 **a** Map of the western Alboran Sea showing the position of the vertical profiles of WOD18 (red dots) and the glider transects (blue line). **b** Example of a TS labeled diagram

been defined manually in the TS diagram. In this process, Atlantic Waters (AW) were separated into Surface Atlantic Waters (SAW), North Atlantic Central Waters (NACW), and Modified Atlantic Waters (MAW), whereas Mediterranean Waters were separated into Winter Intermediate Waters (WIW), Levantine Intermediate waters (LIW), and Dense Mediterranean Waters (DMW) (see Fig. 1).

After preparing the training dataset, we applied a classical soft margin SVM model, formulated as follows:

$$\mathrm{Min}_{w,b}\frac{1}{2}||w||^2 + C\cdot\sum_{i=1}^{l}\xi_i \quad \boldsymbol{s.t} \quad y_i\left(\sum_{k=1}^{K} w_k x_{i,k} + b\right) \geq 1 - \xi_i \quad (\xi_i \geq 0\forall_i)$$

In this formulation, w and b variables are the coefficients of the optimal separator hyperplane and ξ_i variable represents the deviation associated with misclassified samples. C is a penalty parameter, regulating the balance between the maximization of the margin and the minimization of the deviation.

Also, a cluster analysis was applied to the TS data to classify the water mass in the glider transect with the aim of making a comparison with the SVM classification results.

3 Results

3.1 Glider Transect Data

TS vertical profiles acquired during the second transect are represented in Fig. 2. Warm (17–20 °C) and fresh waters of Atlantic origin are found in the surface. The dominant feature is a layer of low salinity ($S<36.60$) almost homogenous with a vertical gradient of temperature that forms the WAG, characterized by the deepening of isothermal and isohaline layers of several ten meters up to 180 m.

3.2 Glider Transect Classification

Based on k-fold cross-validation, the SVM classification method applied on glider transect shows an accuracy greater than 99%. The WAG core is mainly formed by SAW, while its lower vertical extension is composed of the MAW (see Fig. 3). Beyond the isopycnal 28.9 kg m^{-3}, we found the MW, containing the LIW and Dense MW. NACW and WIW were not detected by the algorithm. The

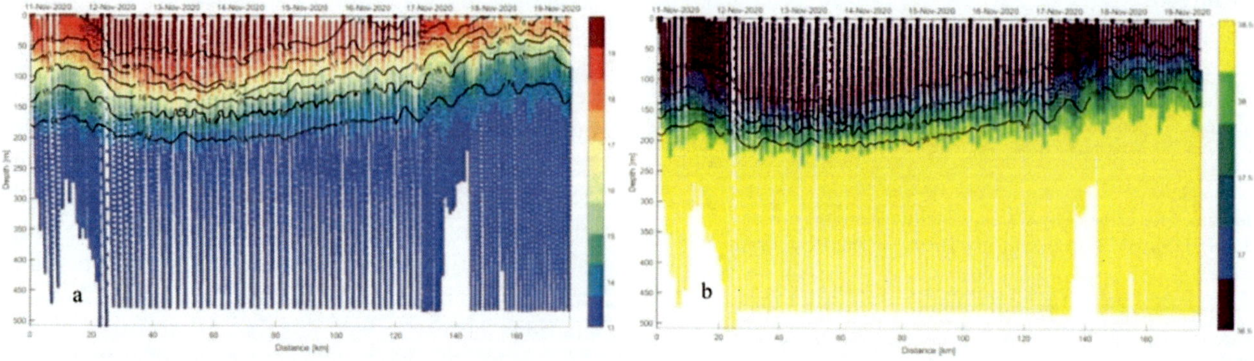

Fig. 2 Temperature (**a**) and salinity (**b**) along the first glider transect

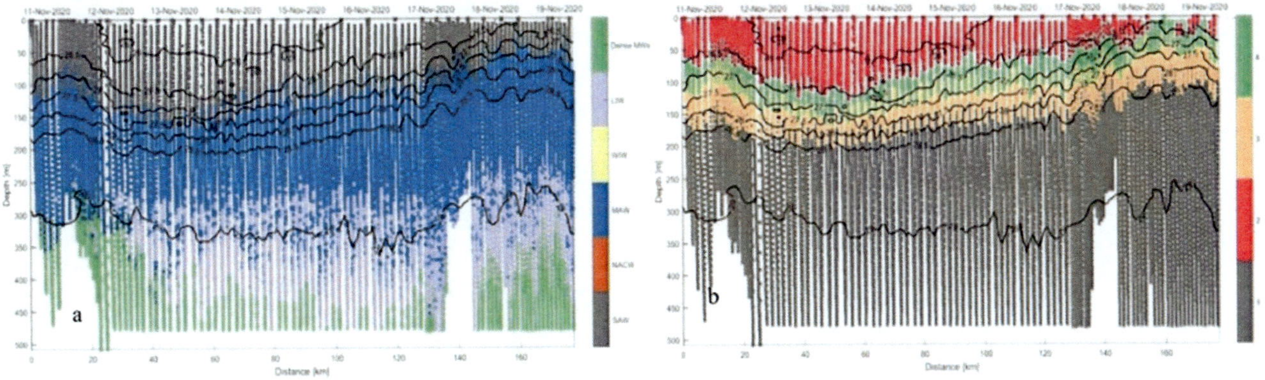

Fig. 3 SVM classification results (**a**) and the cluster analysis results (**b**)

classification results using clustering analysis show a good performance in the detection of SAW particularly in the core of the WAG (see Fig. 3). However, clustering analysis cannot separate between MAW, LIW, and Dense MW.

4 Discussion

Our results show the efficiency of a SVM method to do water mass classification. On the contrary of the conventional clustering analysis, SVM classification method can automatically distinguish water masses when several AW and MW are considered. Thus, clustering analysis are not always appropriate to discriminate water masses in regions where intense mixing occurs such as the Western Alboran Sea.

The SVM classification results are in good agreement with the circulation schemes established in previous studies in Western Alboran Sea and deduced from the traditional method based on TS diagram expert analysis. The homogeneous salinity layer ($S<36.6$) was classified as a SAW in agreement with (Vélez-Belchí et al., 2005). The LIW layer was not detected near the Moroccan coasts, and it is mainly concentrated in the central and Northern Alboran Sea; this spatial distribution agrees with (Millot, 2014). The dense MW are uplifted and directly linked to AW, near the Moroccan coasts in concordance with (Millot, 2014). The seasonal and spatial variability of Atlantic surface waters and Mediterranean waters, respectively, are taken into account in the supervised learning process. Thanks to this subjective labeling approach, our methodology retains the physical meaning of defining the boundaries between water masses.

5 Conclusion

Using the SVM classification, the boundaries between water mass can be accurately defined. The water masses in the Western Alboran Sea in fall season are classified into four categories: SAW, MAW, LIW, and dense MW. Compared with clustering analysis, we can conclude that classifying water masses using this new methodology is efficient and robust.

References

Millot, C. (2014). Heterogeneities of in- and out-flows in the Mediterranean sea. *Progress in Oceanography, 120*, 254–278.

Naranjo, C., Sammartino, S., Garcıa-Lafuente, J., Bellanco, M. J., & Taupier-Letage, I. (2015). Mediterranean waters along and across the strait of Gibraltar, characterization and zonal modification. *Deep Sea Research Part I: Oceanographic Research Papers, 105*, 41–52.

Swift, J. H. (1986). The Arctic waters. In B. G. Hurdle (Ed.), *The Nordic seas*. Springer.

Vélez-Belchí, P., Vargas-Yáñez, M., Tintoré, J. (2005). Observation of a western Alborán gyre migration event. *Progress in Oceanography, 66*(2–4), 190–210.

Watelet, S., Skagseth, Ø., Lien, V., Sagen, H., Østensen, Ø., Ivshin, V., & Beckers, J.-M. (2020). A volumetric census of the Barents Sea in a changing climate. *Earth System Science Data, 12*, 2447–2457. https://doi.org/10.5194/essd-12-2447-2020

WOD18 Homepage. https://www.ncei.noaa.gov/products/world-ocean-database

Comparative Study of Boussinesq Wave Module and Potential Swell Theory in Calculating the Port Agitation: Case of Moroccan Atlantic Façade

Oumaima Gharnate, Mohktar Abdenour, Laila Mouakkir, Mohamed Chagdali and Dalila Loudyi

Abstract

Along the Moroccan coast, swell is generally unrelated to the weather on the coast. It is, most often, the consequence of barometric depressions passing between the Azores Archipelago and Iceland. North Atlantic swells are the most frequent, and they are characterized by fairly long wavelengths (period of 10–18 s) and are refracted on all or part of the continental shelf. In this context, the aim of the study is to compare Boussinesq and Birkhoff theories for the calculation of port agitation in two harbors: Casablanca and Jorf Lasfar. The comparison is done based on two models for the calculation of port agitation: The first model is based on the Birkhoff equation (MIKE 21 EMS) obtained from the potential swell theory, and the second is based on the Boussinesq equation (MIKE 21 BW). First, an analysis of agitation limit conditions for vessels in the ports is performed, then time series of offshore sea states are analyzed, and the propagation of the reconstructed sea states is performed, after which simulation scenarios are chosen. The research shows that Boussinesq waves are adequate to be used with long wavelengths swells and that Birkhoff theory MIKE 21 EMS is more pertinent to be employed in smaller coastal locations, where diffraction and wave breaking are critical, and the forcing wave conditions can be represented by a monochromatic and unidirectional wave. This phenomenon is important enough to be studied as it leads in guaranteeing good circulation conditions for ships which are very sensitive and important ships, and its simulation using pertinent theory helps provide optimal solutions leading to allow good harbor conditions.

1 Introduction

Diffraction is the phenomenon by which a swell suddenly changes shape and orientation, without there being necessarily a variation in depth or presence of marine currents. This situation occurs when the swell meets a natural or artificial barrier, such as a maritime pier (Holthuijsen et al., 2003): thus, in Jorf Lasfar.

Diffraction is essentially the bypassing of the obstacle encountered; as a result, in the case of port enclosures, part of the swell energy is diffused in the protected area. The sheltered water area is then the seat of a problematic agitation for the cargoes, the ships, and the quayside installations.

The port of Jorf Lasfar is a port with a mineral and conventional vocation. It has a favorable geographical location in a region in full economic, social, industrial, agricultural, cultural, and tourist development. It is characterized by the presence of important seabeds (5–14 m depth of draught). It is located at 33°07′ North and 8°38′ West on the Atlantic coast, 17 km from El Jadida, and 120 km southwest of Casablanca. It is halfway between Casablanca and Safi.

For the Port of Jorf Lasfar, it was a great need to extend the quays of phosphate traffic due to their increase. Following this fact, the agitation in some critical areas of the port has undergone a substantial change. It was therefore considered necessary to reconsider this phenomenon within the general framework of improving the protection of the sheltered water body.

In this work, the agitation in the port will be studied using MIKE 21 Boussinesq waves (Lynett, 2006; Madsen et al., 1991; Madsen & Sorensen, 1992; Panigrahi et al., 2015; Shi et al., 2012; Woo & Liu, 2004) and an analysis of the potential use of MIKE 21 EMS (Behrendt & Jonsson, 1984) is clarified.

O. Gharnate (✉) · M. Abdenour · L. Mouakkir · M. Chagdali
Polymer Physics Mechanical Sciences and Materials Laboratory, Faculty of Sciences Ben M'sik, University Hassan II, PO BOX 7955, Casablanca, Morocco
e-mail: oumaima.gharnate@gmail.com

D. Loudyi
Faculty of Sciences and Technics, Hassan II University of Casablanca, Mohammedia, Morocco

H. Chenchouni et al. (eds.), *Recent Advancements from Aquifers to Skies in Hydrogeology, Geoecology, and Atmospheric Sciences*, Advances in Science, Technology & Innovation, https://doi.org/10.1007/978-3-031-47079-0_49

For the calculations, only MIKE 21 BW was used. This choice is based on the fact that MIKE 21 EMS is suitable for small coastal areas where diffraction and wave breaking are important and where the wave conditions could be represented by monochromatic and unidirectional wave. In the case of our study, the extent of the model along the *x*-axis is approximately 5060 and 2200 m along *y*-axis where the maximum distance of wave propagation is defined from the offshore model boundary to the furthest point in the harbor is about 2200 m. Also, we note that the directions 290°, 300°, 310°, 320°, and 330° are representative of the majority of the incident swells. The use of MIKE 21 BW is more adequate for the study.

The study confirmed the presence of the port agitation, and a discussion for potential solutions is provided.

The critical areas where the agitation is the most critical are:

- Zone 1: Portion of the channel adjacent to the turning circle totally sheltered by the two dikes;
- Zone 2: Part of the channel sheltered by the main dike only;
- Zone 3: Unsheltered portion of the channel.

2 Methods

2.1 Analysis of the Wave Signal and Choice of Simulation Scenarios

The first part of the study concerns the processing and analysis of the swell signal in the port of Jorf Lasfar. The applied approach consists of analyzing three-hour time series of offshore sea states spanning over 22 years (1992–2014). First, the offshore sea states were determined, and it is established by the combined use of WaveWatch III for all years. In a second phase, the propagation of the reconstructed sea states was performed to the study points, using the SWAN propagation software. The probability of occurrence of swell height by direction and the directional distribution of the peak period Tp were analyzed, and the distribution of Hm0 occurrences relative to Tp was used in the choice of the simulation scenarios.

The strongest storms were first selected at P1 located in the immediate vicinity of the port: P1: $X = 195{,}500$ m/$Y = 286{,}000$ m by a depth of ~ 27 m/ZH.

2.2 Hydrodynamic Model and Model Setup

In this study, the two-dimensional hydrodynamic model MIKE 21 Boussinesq waves is used, and it solves the Boussinesq equations improved by an implicit finite difference technique with variables defined on a scaled rectangular grid which can reproduce the combined effects of wave phenomena in harbor and coastal engineering.

MIKE 21 BW also includes "Porosity" for the simulation of partial reflection and transmission through piles and breakers. Sponge layers are applied when full absorption of wave energy is required. Finally, MIKE 21 BW also includes internal wave generation.

The bathymetry for the study area was first generated using MIKE Zero bathymetry editor and MIKE 21 setup planner, and the SHOM 2015 Map was used. The model has a special resolution of 3.5 m and a temporal resolution of 0.1 s.

Sponge layers has been used to provide radiation boundary conditions adsorbing the energy of waves propagating out of the model area, thus ensuring stability of the numerical solution.

Also, a porosity layer map is used to model partial reflection and/or transmission through different types of structures.

3 Results

3.1 Comparison of MIKE 21 BW and MIKE 21 EMS

The elliptic mild-slope module MIKE 21 EMS is based on "mild-slope" wave equation, which describes the motion of time harmonic water waves of infinitesimal height (linear waves) on a gently sloping bathymetry with arbitrary water depth.

MIKE 21 EMS is employed in smaller coastal locations, where diffraction and wave breaking are critical, and the forcing wave conditions can be represented by a monochromatic and unidirectional wave. The model is applicable to any water depth, and the only restriction is that nonlinear effects such as amplitude-dispersion and wave-wave interaction are not included (Booij, 1983). The model can also be applied to the study of short period wave disturbance in harbors, but since the model operates with monochromatic and linear waves the MIKE 21 BW is used for this purpose (DHI, 2017).

3.2 Agitation Simulation

The selection of the agitation conditions to be simulated was made based on the results of the wave study at the site. This selection took into account the frequency of sea states, but also their potential effect in the basin, favoring the most unfavorable ones (either in terms of direction or swell heights).

As a results; the directions 290°, 300°, 310°, 320°, and 330° are representative of the majority of incident swells.

Statistical results, such as the agitation coefficient or the significant wave height, are of great practical importance. In most cases, the most interesting result is the agitation coefficient (or relative wave height).

The simulation outputs signals have been extracted in 10 points with different depths varying from − 5 to − 20.

We can say that all points are characterized by agitation coefficients which are higher than the thresholds set for ships in the port of Jorf Lasfar as specified in Sect. 2.1, especially for point 1 which belongs to the area 3 of the channel which is not sheltered. The situation is less critical for the point 4 and 2, but the admissible limit heights will be exceeded for swell heights > 2 m; in fact by returning to the climate of the swell, we find: Swell height exceeds 2.5 m, 3 m, and 3.5 m heights: 12.2% (44.5 d/yr), 6.6% (24 d/yr), and 3.5% (13 d/yr) of the time, respectively. Thus, access and navigation problems are then justifiable.

The agitation coefficients are generally acceptable for the other points inside the harbor docks.

3.3 Investigating a Solution

One of the possible solutions for the agitation problem is to extend the dike in order to limit different directions of swells entering the outer harbor. The extension of the main breakwater will also allow a better dissipation of the wave energy.

In order to choose the optimal extension scenario, several attempts were made for different extension lengths.

To observe the impact of the extension, one of the directions that are representative of the majority of incident swells is simulated, the one with a 23% probability of occurrence is chosen: Tp = 16 s, Dir = 310°.

Many extension scenarios have been simulated, and the results have been extracted for the same reference points where agitation was calculated. It was therefore observed that the best extension scenario is an extension of 300 m with a 20-degree inclination. With this extension, the agitation coefficient went from 1.059 to 0.495 m, so we have reduced the swell by 53% at the entrance of the harbor.

To observe the impact of the extension, we simulate one of the directions that are representative of the majority of incident swells, and we choose the one with a probability of occurrence of 23%: Tp = 16 s, Dir = 310°.

The best scenario of extension is the one of 300 m with an inclination of 20°. With the extension, the agitation coefficient went from 1.059 to 0.495 m, thus reducing the swell by 53% at the entrance of the harbor.

4 Discussion

In this study, we focused on evaluation the agitation in the Port of Jorf Lasfar, investigating the port's ships handling and station keeping thresholds.

First the signal is analyzed, and projections are made in order to extract the most accurate simulations scenarios. Throughout the study, a lot of highlights were highlighted. Digitalizing the correct bathymetry with the exact depths, the bathymetry was created from scratch, the SHOM map was used to manually draw the contours to cover the maximum details. Moreover, dealing with the three open boundaries using sponge layers and figuring the right number of layers was also resolved.

The used software presented a challenge as well, dealing with very long computational time and the complexity of interface as well.

As per the analysis performed, a comparison of the two modules, MIKE 21 EMS and MIKE 21 BW, is made in order to choose the most adequate one. The results clearly confirm the agitation issue in the port, and therefore, investigating solutions such as extending the principal dike with the right orientation was done.

Further, studies can be performed in order to investigate the mud transport in the port.

5 Conclusions

In shallow water such as near Jorf Lasfar's port, the small-scale presentation of the bathymetry is crucially important in estimating the agitation in the port. This phenomenon is important to be studied in order to guarantee good conditions for its very sensitive ships.

Regarding our research question, we tried to select the model where all relevant processes are considered.

In order to obtain the wave agitation, it is important to consider all the free parameters that control the physical processes related to the energetic transformation of the wave within the harbor which include, but not limited to, the partial wave reflection coefficients, bathymetry detail, and the geometry establishment. This aspect was one of the most critical ones to realize during the presented work.

References

Behrendt, L., & Jonsson, I. G. (1984). The physical basis of the mild-slope wave equation. In *Proceedings of the 19th Conference on Coastal Engineering*, Houston, 1984. ASCE.

Booij, N. (1983). A note on the accuracy of the mild-slope equation. *Coastal Engineering, 7*, 191–203.

DHI. (2017). *MIKE 21 EMS, Elliptic mild-slope wave module, user guide*. Hørsholm, Denmark.

Holthuijsen, L. H., Herman, A., & Booij, N. (2003). Phase-decoupled refraction-diffraction for spectral wavemodels. *Coastal Engineering, 49*, 291–305.

Lynett, P. J. (2006). Nearshore wave modeling with high-order Boussinesq-type equations. *Journal of Waterway, Port, Coastal and Ocean Engineering, 132*(5), 348–357.

Madsen, P. A., Murray, R., & Sorensen, O. R. (1991). A new form of the Boussinesq equations with improved linear dispersion characteristics. *Coastal Engineering, 15*(4), 371–388. https://doi.org/10.1016/0378-3839(91)90017-B

Madsen, P. A., & Sorensen, O. R. (1992). A new form of the Boussinesq equations with improved linear dispersion characteristics. Part 2. A slowly-varying bathymetry. *Coastal Engineering, 18*(3), 183–204.

MIKE 21, Boussinesq waves module, step-by-step training guide

Panigrahi, J. K., Padhy, C., & Murty, A. (2015). Inner harbour wave agitation using Boussinesq wave model. *International Journal of Naval Architecture and Ocean Engineering, 7*(1), 70–86.

Shi, F., Kirby, J. T., Harris, J. C., Geiman, J. D., & Grilli, S. T. (2012). A high-order adaptive time-stepping TVD solver for Boussinesq modeling of breaking waves and coastal inundation. *Ocean Modelling, 43–44*, 36–51.

Woo, S.-B., & Liu, P.L.-F. (2004). Finite-element model for modified Boussinesq equations. II: Applications to nonlinear harbor oscillations. *Journal of Waterway, Port, Coastal and Ocean Engineering, 130*(1), 17–28.

Ocean Observing Systems on Board Fishing and Cargo Vessels

A. Miguel Piecho-Santos, Anabela J. Carvalho and Teresa L. Rosa

Abstract

The Portuguese Ocean Observing Program (OBSERVA.PT) under development at IPMA is presented. The activities started to be funded under two main projects: OBSERVA.PT that aims to implement appropriate technologies for monitoring the marine environment for the operational production of marine meteorological and oceanographic information. Meteorological and oceanographical equipment are being installed on cargo vessels, *en route* between Portugal Mainland and Madeira, Azores, and Cape Verde (NE Atlantic), as well as in fishing vessel; and OBSERVA.FISH that aims to develop a fully autonomous system to be installed on all types of fishing vessels, integrating various meteorological, oceanographical, and vessels behavior parameters. We will show the first results of the installation of a "ferrybox"-type (Undersee_water) equipment in a cod trawler vessel during her route from Portugal Mainland to the Grand Banks of Newfoundland. The "ferrybox" measured continuously temperature, salinity, chlorophyll, dissolved oxygen, pH, and turbidity from water pumped from the surface. The installation in the cargo and fishing fleet will allow the collection of data with great spatial coverage and high temporal resolution, difficult to obtain with other observation platforms.

Keywords

Ocean observing systems · Fishing vessels · Cargo ships · Ferrybox · Autonomous systems

A. Miguel Piecho-Santos (✉) · A. J. Carvalho · T. L. Rosa
IPMA-Portuguese Institute for the Sea and the Atmosphere, Lisboa, Portugal
e-mail: amsantos@ipma.pt

A. Miguel Piecho-Santos · T. L. Rosa
CCMAR-Centre of Marine Sciences, University of Algarve, Faro, Portugal

1 Introduction

The operation of autonomous observing systems (AOS) on board Voluntary Observing Ships (VOS) for collecting in situ atmospheric, oceanic, and biogeochemical data had been one of the many ways to observe the ocean and coastal seas, and probably the older one for long-term sustainable ocean measurements (Rosa et al., 2021). Although, ferries and cargo ships have been used more frequently, the use of fishing boats as ocean observing platforms has not been fully explored (Patti et al., 2016; Vranken et al., 2020).

The best known and older of these programs is the World Meteorological Organization's (WMO) Voluntary Observing Ships (VOS) Program, dating back to 1853, proposed by Matthew F. Maury to establish a uniform ocean collection system of meteorological and oceanographical data that in return could benefit maritime navigation (Corbin, 1888; Grady, 2015). Presently, the longer data set of ocean surface measurements is the International Comprehensive Ocean–Atmosphere Data Set (ICOADS) that archives surface global marine data, since 1750 to present days (Freeman et al., 2017). The surface information include air temperature (AT), sea surface temperature (SST), wind speed and direction, humidity, sea-level pressure (SLP), cloud cover, sea state, and sea ice that are reported from a given ship, buoy, coastal platform, or oceanographic instrument (Freeman et al., 2017). However, ocean observations obtained from fishing vessels are very rare and under experimental and/or demonstrate programs (Patti et al., 2016; Rosa et al., 2021; Vranken et al., 2020), but fishing vessels have enormous potential to be used as data collection platforms because they ply coastal seas at almost all times and in almost all weather conditions. At the same time, commercial cargo vessels steam over the ocean along traditional routes, thus having limited spatial resolution but with a high repetition of observations in certain areas of the ocean allowing to built long time series along their routes.

H. Chenchouni et al. (eds.), *Recent Advancements from Aquifers to Skies in Hydrogeology, Geoecology, and Atmospheric Sciences*, Advances in Science, Technology & Innovation, https://doi.org/10.1007/978-3-031-47079-0_50

In this study, it is described the Portuguese autonomous ocean observing systems that are being installed by IPMA (the Portuguese Institute for the Sea and the Atmosphere) on fishing and cargo vessels operating in the north Atlantic Ocean.

2 The Portuguese Ocean Observing System (OBSERVA.PT)

This is the main ocean observation program under development and ongoing at IPMA, in which the main objective is the development of methodologies for automatic ocean and meteorological observations, especially on board national vessels of opportunity (e.g., cargo, fishing, and cruise vessels). The motivation is because "Portugal is Sea", since about 98% of its territory is in the ocean, which corresponds to an area of jurisdiction of about 3.8 million km^2 of immersed territory (Table 1; Fig. 1). Thus, there are many opportunities for the exploitation of marine resources and to increase sea-related economic activities, but also major challenges for the management, monitoring, and sustainable use of these marine resources and ecosystems. However, this must be supported by a deep knowledge of the entire "Portuguese Sea", but this knowledge will only be possible with long-term sustainable ocean observing systems, which will require large operational capacity and financial effort.

The OBSERVA.PT program aims to obtain information necessary to support the management of marine ecosystems and maritime meteorology (mainly) in this vast area of the northeastern Atlantic Ocean. The program is divided in four main activities: (i) surface marine observations based on "ferrybox"-type equipment and meteorological stations, being the latter ones under EUMETNET E-SURFMAR initiative; (ii) subsurface observations based on sensors installed on fishing gears; (iii) biogeochemical indices from data obtained from surface water collected in the ferrybox water circulation system, for nutrient measurement and identification of planktonic communities, including non-indigenous species, as well as cetacean observations obtained by trained observers; and (iv) subsurface observations using Argo floats (https://argo.ucsd.edu/). These data serve to create indicators for Marine Strategy Framework Directive (MSFD) descriptors and contribute to the evaluation of the Good Environment Status of European waters, such as for biodiversity (D1), exotic species (D2), trophic chains (D4), and eutrophication (D5). The start of this program was funded by two main projects: (i) "OBSERVA.PT—Observations on board national commercial vessels to support the protection and restoration of biodiversity in marine ecosystems of the Portuguese Sea (16-01-04-FMP-002)", funded by the EU and the Portuguese Government-Mar2020 Program (OBSERVA.PT, 2022; Santos et al., 2018); and (ii) "OBSERVA:FISH-Autonomous Observation Systems aboard fishing vessels to support the management of marine ecosystems (PTDC/CTA-AMB/31141/2017)", funded by Portugal-Lisboa-Algarve2020 (Piecho-Santos et al., 2021).

Table 1 Territory of Portugal

Territory	Area (km^2)	Percentage (%)
Land (Mainland and Islands)	92,000	2.4
Economic Exclusive Zone (EEZ)	1,600,000	41.6
Outer limits of the continental shelf (as submitted by Portugal to the United Nations in 2009)	2,150,000	56.0
Total (Portugal territory)	3,842,000	100

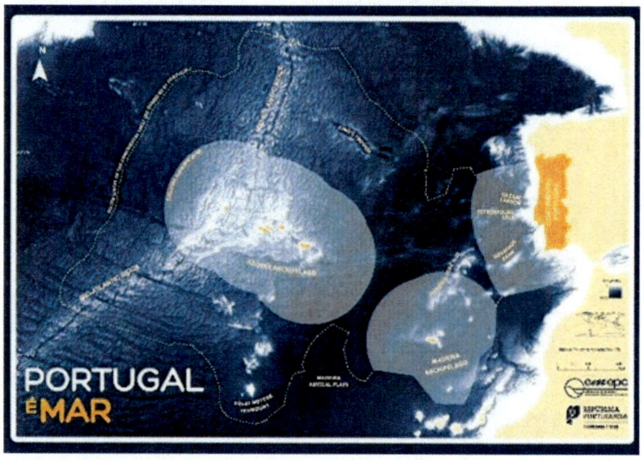

Fig. 1 Portuguese Sea. In this map, the Portuguese land territory (including the Azores and Madeira Islands) is represented in orange, the Economic Exclusive Zone (EEZ) in gray, and the dashed line represents the proposed outer limits of the continental shelf as submitted by Portugal to the United Nations in 2009. Translation: "Portugal é Mar" means Portugal is Sea. (*source* EMEPC https://en.emepc.pt/, last accessed 2022/09/02)

3 Materials and Methods

The ferrybox-type tested and used is the Undersee_water device (Fig. 2a), developed by a Portuguese SME (Materospace, Ltd.) that measures continuously surface temperature, salinity, chlorophyll, dissolved oxygen, pH, and turbidity (Piecho-Santos et al., 2021; Rosa et al., 2021). The vessel motion sensor unit is a gyrocompass manufactured by IXSEA model OCTANS (Fig. 2b) that is both a fiber-optic survey graded IMO-certified and a motion reference unit, which provides true-heading, roll, pitch, yaw, heave, surge, sway, rates of turn, and accelerations (Piecho-Santos et al., 2021), and should ideally be installed as near as possible to ship's gravity center. The main meteorological station is the standard European Common Shipborne

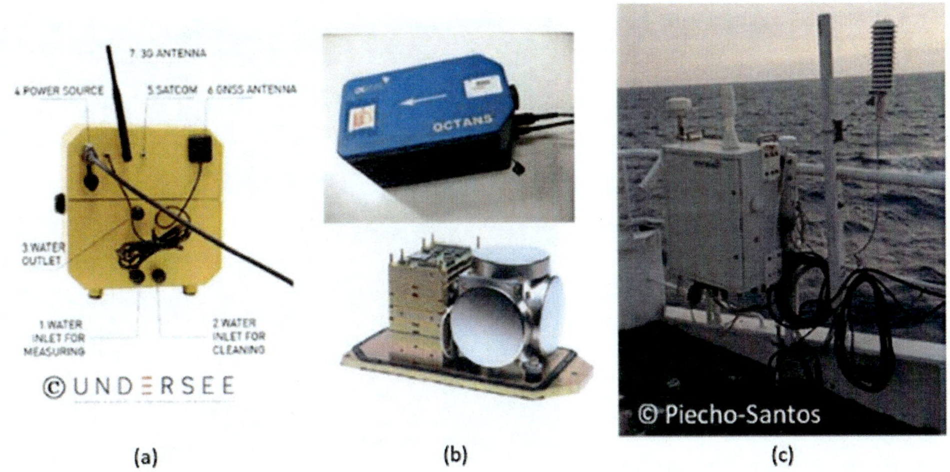

Fig. 2 Autonomous integrated ocean observing system: **a** ferrybox; **b** gyrocompass; and **c** meteorological station. Note that the wind sensor is installed in the ship must

Fig. 3 Example of sea surface temperature obtained during the tests on board R. V. Diplodus (inset photography)

Automatic Weather Station (AWS) Neptune (Fig. 2c), developed as part of the EUCAWS project in cooperation with EUMETNET (Cohuet et al., 2016), with sensors for measurements of sea-level atmospheric pressure, air temperature and humidity, and wind speed and direction. This autonomous integrated ocean observing system integrate all the measurements with date, time, and GPS position to produce single data points.

4 Results and Discussion

The autonomous integrated ocean observing system was tested (Fig. 3) in several IPMA's surveys using research vessels as case studies that simulated the several types of fishing vessels (Rosa et al., 2021): R. V. Diplodus[1] (e.g., artisanal fishing and longline vessels) and R. V. Mário Ruivo[2] (trawler). These tests allow to detected and fixed problems, and at the end of the campaigns, the system was fully operational. Automatic Weather Stations are being installed in several ships of the company Transinsular and in R. V. Mário Ruivo in the frame of E-SURFMAR. Finally, the ferrybox was installed in a cod trawler (M. V. Coimbra) that fish in the Grand Banks of Newfoundland and the Barente Sea. These data are being sent to the EMODnet portal (Fig. 4). COVID-19 pandemic is delaying the installation of the systems, and it was decided to make the installations as possible, which means that the instruments are being installed independently from each other and not in an integrated way, because of the different needs of each other (e.g., ships docking).

5 Conclusions

The ocean observing system presented is now fully tested and prepared to be installed in several fishing and cargo vessels attaining a Technological Readiness Level (TRL) between 7 and 8. The next activities will be to install the meteorological station in MV Coimbra and the ferrybox in

[1] https://www.ipma.pt/en/navios/diplodus/.

[2] https://www.ipma.pt/en/navios/mario-ruivo/.

Fig. 4 First data acquired by the trawler MV Coimbra during the steaming for the Grand Banks of Newfoundland. The insets are the photography of the trawler and the ferrybox Undersee_water installed in that vessel

Transinsular's cargo vessels. However, all these equipment only measured surface parameters and the next developments will be toward methodologies for the installation of sensors on fishing gears (e.g., bottom and subsurface longlines, trawl, and purse seine nets) that allow water column measurement, and video cameras for estimation of fishing catches.

Finally, the autonomous integrated ocean observing system on board fishing and cargo vessels presented here is a component of a more complete observing system—the Portuguese Ocean Observing System (OBSERVA.PT). OBSERVA.PT had also components related with the use of Argo floats under the ARGO.PT program that is the Portuguese contribution to the International Argo Program (Freeman et al., 2017) that are deploying Argo floats, including BGC-Argo floats, in the Portuguese Seas; and with the European Multidisciplinary Seafloor and water-column Observatory (EMSO) ERIC under EMSO.PT, which keep the Iberian Margin Observatory in the Gulf of Cadiz (http://emso.eu/observatories-node/iberian-margin/).

Acknowledgements This study was conducted in the frame of the projects OBSERVA.PT (16-01-04-FMP-002) and OBSERVA.FISH (PTDC/CTA-AMB/31141/2017) funded by the EU, the Portuguese Government, FCT, MAR2020, Portugal2020, Lisboa2020, and Algarve2020. This is also a contribution to projects UIDB/04326/2020, UIDP/04326/2020 and LA/P/0101/2020 funded by FCT.

References

Cohuet, J. B., Kleta, H., Rozema, J., Blouch, P., Peterman, E., & Poli, P. (2016). The Shipborne European Common Automatic Weather Station (EUCAWS). EUMETNET, Brussels. https://www.wmocimo.net/eventpapers/session2/O2(5)_Cohuet_The%20Shipborne%20European%20Common%20Automatic%20Weather%20Station%20(EUCAWS).pdf. Last accessed September 02, 2022.

Corbin, D. F. M. (1888). Life of Matthew Fontaine Maury, USN and CSN. Author of 'physical geography of the sea and its meteorology'. Compile by his daughter Diana F. Maury Corbin. Sampson Low, Marston, Seatle and Rivington, Limited.

Freeman, E., Woodruff, S. D., Worley, S. J., Lubker, S. J., Kent, E. C., Angel, W. E., Berry, D. I., Brohan, P., Eastman, R., Gates, L. Gloeden, W., & Smith, S. R. (2017). ICOADS Release 3.0: A major update to the historical marine climate record. *International Journal of Climatology, 37*(5), 2211–2232.

Grady, J. (2015). Matthew Fontaine Maury, Father of Oceanography. A Biography, 1806–1873. McFarland & Company, Inc.

http://emso.eu/observatories-node/iberian-margin/. Last accessed September 02, 2022.

https://argo.ucsd.edu/. Last accessed September 02, 2022.

OBSERVA.PT. (2022). https://www.researchgate.net/project/OBSERVAPT-Obser-vations-on-board-national-commercial-ships-to-support-the-conservation-of-marine-bio-diversity-in-the-Portuguese-Seas. Last accessed September 02, 2022.

Patti, B., Martinelli, M., Aronica, S., Belardinelli, A., Penna, P., Bonanno, A., Basilone, G., Fontana, I., Giacalone, G., Gabriele Gallì, N., Sorgente, R., & Mazzola, S. (2016). The fishery and oceanography observing system (FOOS): A tool for oceanography and fisheries science. *Journal of Operational Oceanography, 9*(sup1), s99–s118.

Piecho-Santos, A. M. P., Hinostroza, M. A., Rosa, T., & Soares, C. G. (2021). Autonomous observing systems in fishing vessels. In C. Guedes Soares & T. A. Santos (Eds.), *Developments in maritime technology and engineering* (Vol. 2, pp. 805–808). CRC Press.

Rosa, T. L., Piecho-Santos, A. M., Vettor, R., & Guedes Soares, C. (2021). Review and prospects for autonomous observing systems in vessels of opportunity. *Journal of Marine Science and Engineering, 9*(4), 366.

Santos, A. M. P., Lourenço, N., Carvalho, A., Sousa Pinto, I., & Oliveira, A. (2018). OBSERVA.PT—Observações a bordo de navios comerciais nacionais nos ecossistemas do Mar Português. In *Actas das 5.as Jornadas de Engenharia Hidrográfica* (pp. 113–116). Instituto Hidrográfico.

Van Vranken, C., Vastenhoud, B. M., Manning, J. P., Plet-Hansen, K. S., Jakoboski, J., Gorringe, P., & Martinelli, M. (2020). Fishing gear as a data collection platform: Opportunities to fill spatial and temporal gaps in operational sub-surface observation networks. *Frontiers in Marine Science, 7*, 485512.

The Risk of Marine Submersion Along the Ain Sbâa Coastline with a Maximum Tide and Pessimiste Scenario of Sea Level Rise (Atlantic, Morocco)

Taoufiq Chtioui, Mounir Hakkou, Abdelhaq Aangri, El Mostapha Zakariya and Aicha Benmohammadi

Abstract

This work employs a technique based on spatiotemporal projection onto a digital elevation model (DEM) to establish the geographical extent of exceptional sea levels. This makes it easier to study their influence. Using IPCC forecasts, maximum and extreme sea levels were computed for present and future (2100) time circumstances (RCP8.5). The characterization of the tidal components (astronomical amplitude and surge), the computation of the runup, and their statistical modeling according to the generalized extreme value theory (GEV) are all used in the calculation of these levels. The physical qualities of the shoreline are also taken into account while estimating the runup parameter (in particular the variation of beach slopes). The findings reveal that Ain Sbâa's shoreline is poorly protected from coastal flooding during powerful storms. The vulnerable areas are located near the industrial sector. The extension of potentially floodable areas would include areas of 10.9 km^2 and 12 km^2, respectively, with a maximum continental width of submersion between 1114 and 1225 m, according to projections of sea levels for present temporal circumstances and at the horizon of 2100. The static flood mapping methodology utilized in this "bathtub" has a number of flaws, including failing to account for the flows and time required to overflow the zones identified here as sensitive to flooding danger, but is widely used for coastal hazard maps. The highest values of the runup, the astrometric tide, the storm surge, and the sea level rise (SLR) are used in this approach's computations, with the chance of their coincidences remaining extremely low. This coincidence, though, is still imaginable.

Keywords

Sea level · Submerged · Ain Sbâa · IPCC (RCP 8.5) · GEV

1 Introduction

According to the IPCC's Sixth Assessment Report 2021, by 2100, sea levels might have risen by up to one meter, and severe sea level events that happen once every 100 years could happen every year by the end of the century. Flooding is one of the most common and destructive natural disasters, with devastating consequences for infrastructure and human societies (Salman & Yue, 2018). Low-lying coastal settings are estimated to account for 2% of the global land area and house more than 10% of the world's population (Pollard et al., 2019) and are among the most developed lands in the world (Aangri et al., 2022).

The major marine and atmospheric variables causing coastal flooding include changes in mean sea level, tides, storm surges, waves, river flow, and precipitation, which can be exacerbated when two or more of these processes occur concurrently (Marcos et al., 2019). During these events, an increase in water levels occurs as a result of ocean energy interacting with the coastline, the magnitude of which is determined by nearshore bathymetry, coast orientation with respect to the storm, beach profile, presence of sea defense structures, dune morphology, or vertical ground motion (Hakkou et al., 2019). Assessing and mapping this risk is critical for coastal

T. Chtioui (✉) · A. Aangri · A. Benmohammadi
Faculty of Science, Ibn Tofail University, Kenitra, Morocco
e-mail: taoufiq.chtioui@uit.ac.ma

M. Hakkou
Scientific Institute, Mohammed V University, Rabat, Morocco

E. M. Zakariya
Royal Naval School, Casablanca, Morocco

H. Chenchouni et al. (eds.), *Recent Advancements from Aquifers to Skies in Hydrogeology, Geoecology, and Atmospheric Sciences*, Advances in Science, Technology & Innovation, https://doi.org/10.1007/978-3-031-47079-0_51

management and engineering activities such as set-back line design and delimitation (Hakkou et al., 2019). Static or numerical modeling can be used to make this assessment. Storms and sea level rise impacts are widely studied using static models (Heberger et al., 2009). This model utilizes hydrodynamic data from a regional spectral wave and hydrodynamic model to drive on N coastal profiles using a computationally efficient set of empirical shoaling, breaking, runup, and overtopping equations (Stokes et al., 2021). It is based on a basic comparison of total water levels (TWL) and backshore heights. TWL denotes the mean water of the measured tide plus storm wave runup (Ruggiero et al., 2001). The formulation of Stockdon et al. (2006) is frequently used to estimate the runup (Vousdoukas et al., 2016).

After the end of the twentieth century, the Ain Sbâa coast experienced rapid development, which led to the growth of urbanized areas. High-energy waves with significant heights, sometimes surpassing 8 m, are an increasing threat to urban, industrial, and tourist region. A study of the risks of marine flooding in this area is inevitable to help coast managers and maybe improve coastal community preparation for floods induced by storm surges and expected sea level rise.

The approach used to evaluate the vulnerability to coastal flooding hazard in this case study of the Ain Sbâa coast is simple to implement in geographic information system (GIS) environments, and it consists of projecting TWL for today and by 2100 onto a digital terrain model to outline the inundation contour and its inland movement. This method has been widely utilized for studies of different scales because of its simplicity (Vousdoukas et al., 2016).

2 Study Area

The coastal section of Ain Sbâa, extending over 12 km, is located north of Casablanca in the northeastern part of the Moroccan Atlantic Ocean between the latitudes of 33° 41′ 00 and 33° 37′ 00, and the longitudes of 007° 26′ 22 and 007° 32′ 15. This section is the most important industrial region in Morocco. Morphologically, it is made up of sandy beaches and sandy rocky coasts. The tidal range varies from 0.9 to 4.09 m. The prevailing winds are from this direction. During the winter and spring, the waves are significantly larger, and major wave heights approach 8 m offshore during severe storms (Hakkou et al., 2011). According to NOAA WaveWatch III (NWW3) wind statistics, winds from the NW–SE sectors have dominated in this region between 1979 and 2018.

3 Material and Methods

The delimitation of flood-prone zones along the Ain Sbâa coast in this study is based on the calculation of the extreme sea level (ESL) and its cartographic projection. The greatest shoreward elevation of water is referred to as the ESL (Ruggiero et al. 2001). It is caused by the interaction of the predicted astronomical tide (η_a), storm surge (S_s), and the 2% exceedance wave runup height ($R_2\%$).

$$\text{TWL} = \eta_a + S_s + R. \tag{1}$$

We take a maximalist approach, using the extreme values of both (S_s) and (η_a) in all of our calculations.

Future forecasts (by 2100) are created by combining the TWL estimations with the possibilities for the IPCC's global sea level rise forecast (2021), and Eq. 1 becomes

$$\text{TWL} = \eta_a + S_s + R + \Delta\text{SLR}, \tag{2}$$

where (ΔSLR) denotes the change in sea level rise as a result of global warming. The highest emission scenario's maximum value (SSP5-8.5) (IPCC, 2021) will be used.

The coastal flood forecast is produced in this study in four primary stages:

First, the most active yearly sea states from 1979 to 2018 were collected from the NWW3 model along with the appropriate water level and wind conditions. The nearest model grid point to the research region is used in this instance which is about 32 km offshore (34° 00′ 00″ N/007° 30′ 00″ W) and then propagated from the open sea to the coast using the SWAN model (Simulating Waves Nearshore) (Booij et al., 1999).

Secondly, the outputs of this model, H_0 and L_0, were taken at a depth of 20 m opposite each cross-shore transect and used to forecast the runup using parametric Eqs. (3) and (4) and then to determine the yearly TWL maximums Eq. (1) at various locations along the shoreline. The beach's pente and the overall raider of the waves control its size. The topo-bathymetric digital elevation model was used to extact six typical cross-shore profiles, indicating the most at-risk locations. We use the empirical model (Stockdon et al., 2006) expressed in two ways depending on the number of Iribarren ε_0 (Battjes & Stive, 1985) for its estimation:

$$R_{2\%} = 1.1(0.35\tan\beta(H_0L_0)^{1/2} + \left[H_0L_0(0.563\tan\beta^2 + 0.004)\right]^{1/2}/2) \quad \text{for} \geq 0.3, \tag{3}$$

and

$$R_{2\%} = 0.043(H_0L_0), \quad \text{for} < 0.3 \tag{4}$$

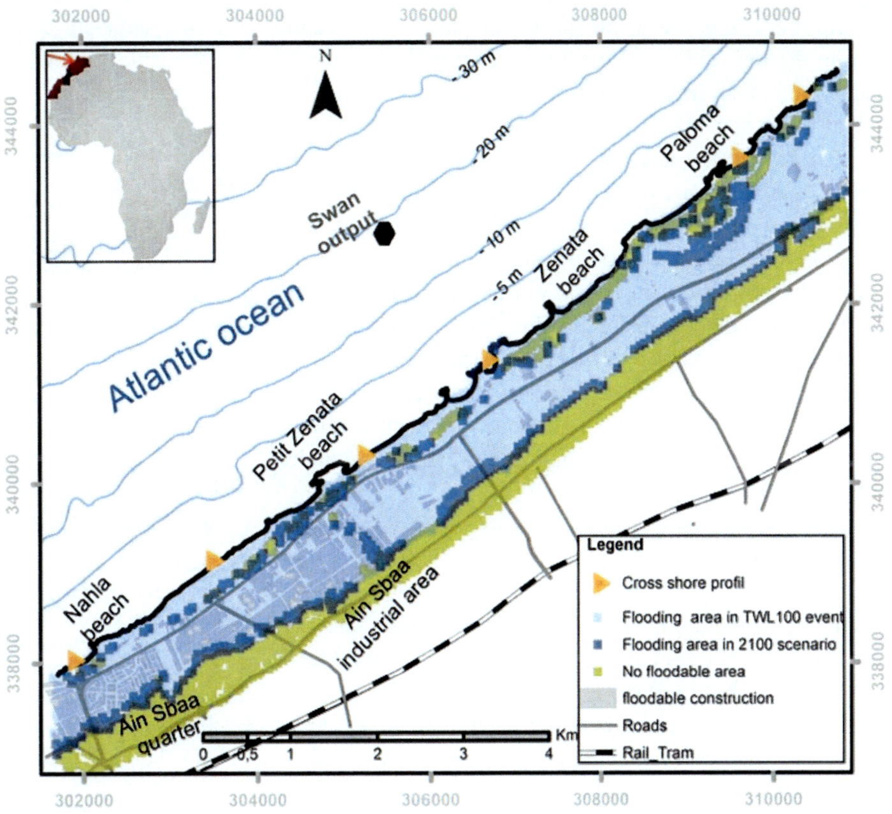

Fig. 1 Flooding extent in Ain Sbâa area

where tanβ is the mean foreshore slope, H_0 is the deep-water significant wave height, L_0 is the deep-water wave length, and ε_0 is the deep-water Iribarren parameter (Battjes & Stive, 1985), calculated as

$$\varepsilon_0 = \frac{\tan\beta}{\left(H_0/L_0\right)^{\frac{1}{2}}}. \quad (5)$$

Thirdly, the yearly TWL maximums used as well as the input to the theory (GEV) to determine the extreme TWL values for various return periods.

Fourth, high resolution topographic data, the 100-year return period TWLs, and (ΔSLR) are used to assess the likely extent of episodic coastal flooding on present and future time scales (2100).

4 Result

The topography of the research area indicates low-lying areas, especially along the coastline. However, the elevation inland varies by location, and they are still vulnerable to coastal flooding.

The most vulnerable places in this area are close to the industrial sector, where maximum flood widths of 1114 and 1225 m at around 2100 from the coastline. The district of Ain Sbâa is likely to be flooded up to 1046 and 1111 m in 2100. The areas of Paloma beach could be flooded from 919 to 992 m in 2100. The total region's area that could flood during the extreme centennial event is 10.9 km^2, compared to 12 km^2 in 2100 (Fig. 1).

5 Discussion

The topography of the research area indicates low-lying areas on several parts of the dune cordon (Fig. 1). Like the entire Moroccan coastline, clandestine sand extraction on these beaches has increased with the proliferation of nearby construction sites (Hakkou et al., 2015); thus, the continuity of the marine dunes that previously provided a buffer against storm waves is currently disrupted by several breaches. Furthermore, the entire back-dune land is low lying and thus unprotected during storms.

To support marine storm mitigation measures and to inform decision makers on areas to be protected, the assessment and mapping of this risk is essential for the management of this coast. Generally, two methods can be used in this type of study: hydrodynamic modeling and static modeling. The available data and the width of the study area, which extends over 12 km, favored the use of a simple static model. In addition, the delimitation of the floodplain is simplified. It is based on the technique of the bathtub wetting the low zone. The credibility of this type of model is approved in several regions of the world.

However, it can cause uncertainty in the reliability of the SLR projections and flow formula, the statistical model (GEV), and data quality. In addition, the bathtub mapping method ignores the process of water circulation behind the shoreline. However, these uncertainties are unavoidable. Future advances in this scientific field should therefore reinforce the conclusions of this study. In addition, our approach remains very pessimistic; the calculations use the largest values of the runup, astrometric tide, storm surge, and SLR, with the possibility that their coincidences remain very weak. This coincidence remains, however, imaginable.

6 Conclusion

According to the study of topographic data, the dune cordon along the coast of Ain Sbâa has several breaks, leaving the back beach directly exposed to marine storms. Therefore, the danger of flooding in the low-lying areas along this coastline is still present and is expected to increase due to climate change.

In order to identify the geographical extent of extreme sea levels in this area, this work adopts the extreme TWL estimate for the 100-year event that was statistically calculated using the predicted values of the annual 40-year TWL maxima as input on the GEV distribution. The annual TWL is the sum of the maximum runup and the extreme tide. The extreme TWL expected for 2100 is estimated here by a simple addition of the maximum SLR of the scenario (RCP8.5). The results are projected onto a digital elevation model (DEM).

The most vulnerable locations are those that are low-lying, including the Ain Sbâa quarter and the industrial sector, while the maximum floodable surface area is between 8721.4 km^2 and 10,284.2 km^2 in 2100, with a maximum continental width of submersion between 1114 and 1225 m at the end of this century.

The methodology employed in this work has some weaknesses, primarily as a result of doubts about the accuracy of the SLR forecasts, runup formula, statistical model (GEV), and data quality. However, these uncertainties are unavoidable. Future scientific progress could resolve some of these limitations.

References

Aangri, A., et al. (2022). Predicting Shoreline change for the Agadir and Taghazout Coasts (Morocco). *Journal of Coastal Research, 38*(5), 937–950.

Battjes, J. A., & Stive, M. J. F. (1985). Calibration and verification of a dissipation model for random breaking waves. *Journal of Geophysical Research, 90*, 9159–9167.

Booij, N., et al. (1999). A third-generation wave model for coastal regions, part I, model description and validation. *Journal of Geophysical Research, 104*(C4), 7649–7666.

Hakkou, M., et al. (2011). Wave climate and morphosedimentary characteristics of the Kenitra–Bouknadel sandy coast, Morocco. *Environmental Earth Sciences, 64*(7), 1729–1739.

Hakkou, M., et al. (2015). Perspectives d'optimisation technique de la gouvernance environnementale des activités du dragage du sable marin au Maroc. *Bulletin de l'Institut Scientifique, Rabat, Section Sciences de la Terre, 37*, 1–11.

Hakkou, M., et al. (2019). Assess and mapping the flooding hazards using geospatial tools and empirical model along Kenitra coast, Morocco. *Ocean and Coastal Management, 169*, 264–272.

Heberger, M., et al. (2009). *The impacts of sea-level rise on the California coast*. California Climate Change Center.

IPCC (2021) *Summary for policymakers. In Climate Change 2021: The physical science basiscontribution of working group i to the sixth assessment report of the intergovernmentalpanel on climate change*. Masson-Delmotte, V., Zhai, P.A., Pirani, S.L., Connors, C., Péan,S., Berger, N., Caud, Y., Chen, L., Goldfarb, M.I., Gomis, M., et al. (Eds.) CambridgeUniversity Press.

Marcos, M., et al. (2019). Increased extreme coastal water levels due to the combined action of storm surges and wind waves. *Geophysical Research Letters, 46*, 4356–4364.

Pollard, J. A., et al. (2019). Harmonising topographic & remotely sensed datasets, a reference dataset for shoreline and beach change analysis. *Scientific Data, 6*, 42.

Ruggiero, P., et al. (2001). Wave runup, extreme water levels, and the erosion of properties backing beaches. *Journal of Coastal Research, 17*(2), 407–419.

Salman, A. M., & Li, Y. (2018). Flood risk assessment, future trend modeling, and risk communication: A review of ongoing research. *Natural Hazards Review, 19*(3), 4018011.

Stockdon, H. F., et al. (2006). Empirical parameterization of setup, swash, and run-up. *Coastal Engineering, 53*, 573–588.

Stokes, K., et al. (2021). Forecasting coastal overtopping at engineered and naturally defended coastlines. *Coastal Engineering, 164*, 103827.

Vousdoukas, M. I., et al. (2016). Developments in large-scale coastal flood hazard mapping. *NHESS, 16*, 1841–1853.

Numerical Simulation of the Nador Lagoon by the Lattice Boltzmann Method

Ali Haddach, Hassan Smaoui and Bouchaib Radi

Abstract

Coastal flows are generally simulated by computer codes based on conventional discretization methods such as finite difference method, finite element method, or finite volume method. They are rarely simulated by the lattice Boltzmann method. The aim of this paper is to present the preliminary results of the numerical resolution of the Saint–Venant equations based on the new lattice Boltzmann method. The code thus developed was successfully applied to the real case of the Nador lagoon (Mediterranean cost of Morocco). It allows to reproduce with great precision the tidal flow in the lagoon generated by the main semi-diurnal wave M2. This method has the advantage of being fast since its implementation is much faster than the conventional methods mentioned above. With its parallel version on graphics processing unit, it can be adopted for simulations of phenomena requiring rapid responses such as flash floods.

Keywords

Lattice Boltzmann · Shallow water equations · Tidal flow · Numerical resolution · Nador lagoon

1 Introduction

Lagoons are fragile and vulnerable aquatic environments when they are subjected to anthropogenic pressures by socioeconomic interests. It is therefore essential that the authorities have action plans to protect them in order to find a compromise between a sustainable ecological balance and these sometimes-uncontrolled human interventions. For this, it is necessary to understand beforehand the often-complex functioning of these lagoons and their interactions to ensure better protection. This understanding goes through experimental scientific studies in situ and/or by numerical modeling.

The aim of this paper is to study by numerical simulations the dynamics of the tide in the Nador lagoon located in the Mediterranean coast in the North of Morocco. This lagoon has been the subject of several numerical studies: hydrodynamic (Hilmi et al., 2015), sedimentological (Boyauzan & Irzi, 2015; Raji, 2014), and planning effects (Boyauzan & Irzi, 2015). The numerical simulations of all these studies were carried out by solving the shallow water equations SWE (these equations are also called the Saint–Venant equations) by the classical discretization methods such as finite differences methods (FDM), finite element methods (FEM), or finite volume methods (FVM).

As far as we are concerned, we present here the simulations of the tidal flow by solving SWE by the lattice Boltzmann method (LBM). This recent method has been widely applied for the resolution of Navier–Stokes equations and tested only on various academic cases solving SWE (Zhou, 2004), but rarely applied on real cases such as coastal areas, estuaries, rivers, and lagoons. LBM is a simple method to implement. Due to its design, it is highly parallelizable and in particular on graphics processing unit (GPU). Even in the serial calculation version, LBM is much faster (in central processing unit (CPU) time) compared to the classic methods, FDM, FEM, and FVM. To check the LBM performance in computation speed (which motivated the implementation of this method), we compare the LBM CPU time with the Model of Estuarine and Coastal Circulation Assessment (MECCA) code CPU time which solves the SWE equations by the FDM (Smaoui & Radi, 2001).

A. Haddach (✉) · B. Radi
Faculty of Sciences and Technics, LIMII, Hassan First University, Settat, Morocco
e-mail: a.haddach@uhp.ac.ma

H. Smaoui
Cerema/DTecREM, 134, Rue de Beauvais, 60280, Margny-Lès-Compiègne, Associate Researcher in Laboratoire Roberval, UTC, Compiègne, France

H. Chenchouni et al. (eds.), *Recent Advancements from Aquifers to Skies in Hydrogeology, Geoecology, and Atmospheric Sciences*, Advances in Science, Technology & Innovation, https://doi.org/10.1007/978-3-031-47079-0_52

2 Brief Presentation of Shallow Water Equation and the LBM

2.1 Governing Equations for Lagoon Flows

Lagoon flows are free surface flows. It can be described using SWE. These equations are obtained by taking the depth-averaged integration of the Navier–Stokes equations. The SWE are represented by the continuity equation:

$$\frac{\partial h}{\partial t} + \frac{\partial (hu)}{\partial x} + \frac{\partial (hv)}{\partial y} = 0, \tag{1}$$

and the 2 equations of momentum are

$$\frac{\partial hu}{\partial t} + \frac{\partial huu}{\partial x} + \frac{\partial huv}{\partial y} = -g\frac{\partial}{\partial x}\left(\frac{h^2}{2}\right) + \vartheta \nabla^2(hu) + F_x, \tag{2}$$

$$\frac{\partial hv}{\partial t} + \frac{\partial huv}{\partial x} + \frac{\partial hvv}{\partial y} = -g\frac{\partial}{\partial y}\left(\frac{h^2}{2}\right) + \vartheta \nabla^2(hv) + F_y, \tag{3}$$

where h is the water depth, u and v are the velocity components in x and y directions, respectively, g is the gravitational acceleration, and ϑ is the kinematic viscosity of the fluid. F_x and F_y are the components of external forces which may include the effects of topography, Coriolis force, friction with the bottom and wind force.

2.2 Lattice Boltzmann Method (LBM)

The LBM is a relatively new numerical method compared to the classical approaches used in numerical simulation of flows. It was presented in the 90s, to process gases on the lattice by cellular. LBM is similar to the formalism of statistical physics to solve numerically the Boltzmann Equation (BE). This equation is interested, not only in macroscopic quantities, but also to the distribution of the different particles constituting a fluid.

Note that in macroscopic description (mechanics of continuums), the variables are obtained by the average on the representative elementary volume (REV). On the other hand, in mesoscopic description (kinetic theory) on which LBM is based, the variables are obtained by calculating the successive moments of the particle distribution function (PDF) defined on the REV and depending on position vector $\boldsymbol{x}$, particle vector velocity $\boldsymbol{c}$ and time t denoted $f = f(\boldsymbol{x}, \boldsymbol{c}, t)$. The PDF is governed by BE expressed as

$$\frac{\partial f}{\partial t} + \nabla .\boldsymbol{c}f + \frac{\boldsymbol{F}}{m}\frac{\partial f}{\partial \boldsymbol{c}} = -\frac{1}{\tau}(f - f^{eq}). \tag{4}$$

$\boldsymbol{F}$ is the external force vector. The right-hand side of the Eq. (4) denotes the collision operator. It was proposed by Bhatnagar et al. (1954) (known as BGK collision operator) to express that the effects of the collision inertia are damped after a certain time called relaxation time and noted (τ). At the end of the collision duration, the PDFs are stabilized around the equilibrium PDF denoted (f^{eq}). The expression of f^{eq} is given by the Maxwell–Boltzmann PDF (Zhou, 2004). Afterward, in order to discretize the BE (4), we use a lattice pattern of nodes and velocities. For instances, in a two-dimensional situation, generally we use the D2Q9 lattice: "D2" for 2 dimensions and "Q9" for 9 possible directions toward the neighboring points while considering the zero direction (Zhou, 2004).

3 Numerical Test for Nador Lagoon

Morocco is bordered by approximately 3500 km of coastline, which extends over the Atlantic coast and the Mediterranean Sea. The Moroccan coasts have several lagoons. Among these lagoons is the Nador lagoon which is the largest one of the Moroccan coastlines located at (02° 45′ W to 02° 55′ W and meridionally from 35° 06′ N to 35° 16′ N). This lagoon is extended on 115 km^2 of area, separated from the Mediterranean Sea by a long spit (25 km^2), with a single inlet that was opened in 2011 with 300 m wide (Fig. 1). The average depth of the lagoon is about 4.8 m and the maximum depth not exceed 8 m (Haddach et al., 2023) (Fig. 2). This lagoon has been the subject of several research works in order to better understand the functioning of its ecosystem.

The bathymetric values were extracted from the EMODnet bathymetric database (https://www.emodnet-bathymetry.eu/). The model was forced by the superposition of the M_2 and S_2 tidal wave whose values were extracted from the TPXO world tide database (https://www.tpxo.net/global/tpxo9-atlas).

The computational domain has been discretized with a regular spatial step $\Delta x = \Delta y = 78.5$ m, i.e., a lattice of 290×100 points. Turbulence was modeled by the large eddy simulation (LES) approach, which is well suited for LBM (Haddach et al., 2023; Zhou, 2004). The model was forced by $M_2 + S_2$ tidal wave at the inlet of the lagoon.

Since we are limited by the number of pages, we will analyze most of the simulations conducted by LBM. Figure 3 shows the superposition of the free surface and the streamlines at high water − 04 h (HW − 04 h). This figure shows the formation of two recirculation zones (in the opposite direction) on either side of the inlet of the lagoon. The nonlinearity of the SWE is shown in this figure by the non-symmetry of these two vortices. This dynamic

Fig. 1 Overview of the Nador lagoon with the new inlet

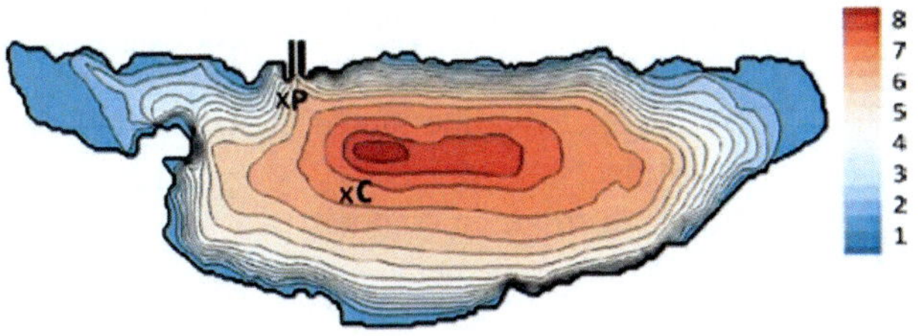

Fig. 2 Bathymetry of Nador lagoon in meter (m) and *P* and *C* locations points

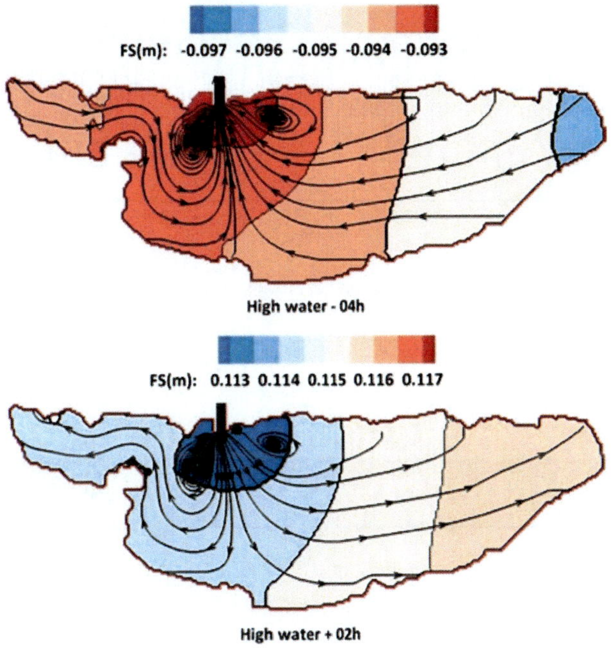

Fig. 3 Free surface level and streamlines during the ebb and the flood (spring tide)

continues until the dissipation of these two zones to reappear 6 h later, i.e., at HW + 02 h (Fig. 3), but in reverse circulation.

The dynamics of these two vortices is undoubtedly due to the configuration of the flow in the vicinity of the inlet. Indeed, during the tidal cycles, the exchange of water masses entering and leaving the lagoon takes place via a channel (the inlet) 14 km long and 300 m wide. Therefore, the velocities are confined in this channel with very high intensities to flow into the lagoon in the form of a powerful jet. This jet meets the southern limit of the domain which forces it to change direction toward the inlet. Finally, note that these results are in perfect agreement with the numerical simulations of other studies carried out on the Nador lagoon (Haddach et al., 2023; Hilmi et al., 2015; Maicu et al., 2021).

Figure 4 shows the free surface at point (*P*) located near the inlet exit (Fig. 2). It is observed that the model is able to reproduce the modulation of the tidal spring–neap cycle. This result is not a surprise, since it is well known that in wave theory that the superposition of two signals of close periods generates a new signal of much larger periods of the two signals. This is exactly the case here, since the model generated the spring/neap tide signal with a period of 14.5 days (half lunar month) from the two tidal signals M_2 (with a period of 12.42 h) and S_2 (with a period of 12 h). From Fig. 4, we can then read that the amplitude of spring tide is of the order of 13 cm, while this amplitude decreases to the value of 6 cm during neap tide. These values are in perfect agreement with the values of the literature (Haddach et al., 2023; Hilmi et al., 2015; Maicu et al., 2021).

For velocities, we present in Figs. 5 and 6 the two components $(\overline{u}, \overline{v})$ of the depth-averaged velocity at point (*P*) and at point (*C*) during a spring/neap tidal cycle. For the point (*P*), we can read an amplitude of 37 cm/s for $\overline{u}$ during spring tide and 20 cm/s during neap tide. On the other hand, for point (*C*), these values are respectively reduced to 3 cm/s for spring tide and 2 cm/s at neap tide. This sudden drop in the values of the velocities amplitudes between point (*P*) and point (*C*) (the center of the lagoon) has also been highlighted by previous studies on the lagoon (Haddach et al., 2023; Hilmi et al., 2015; Maicu et al., 2021).

For the two points, we notice a negligible component $\overline{v}$ compared to $\overline{u}$ with asymmetry in the two velocity components (but very accentuated for the component $\overline{v}$). This is due to the nonlinearity of the terms of friction and the gradient of the bathymetry.

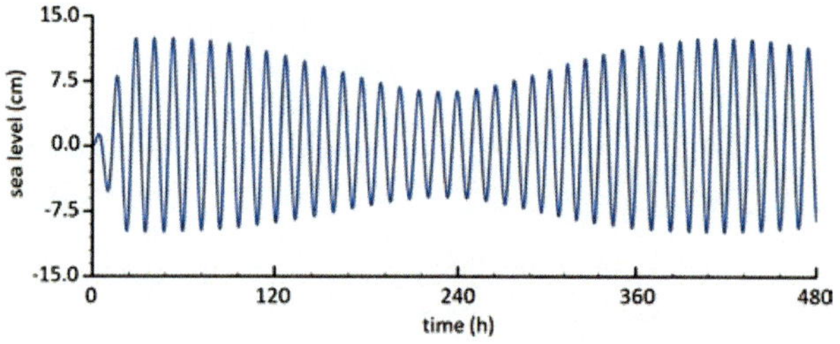

Fig. 4 Free surface level at point *P*, during spring–neap tidal cycle

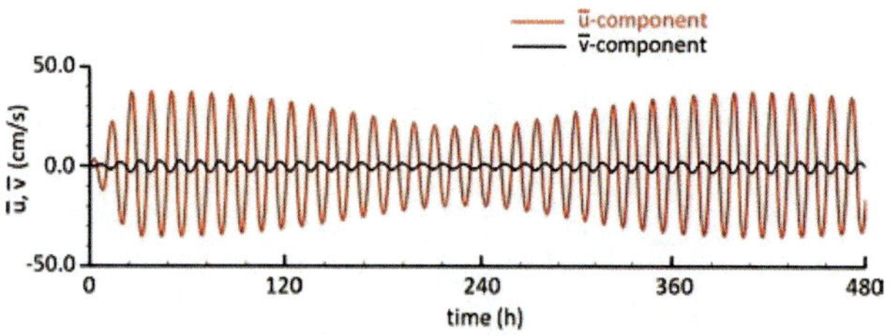

Fig. 5 Velocity components ($\overline{u}, \overline{v}$) at point *P*, during spring–neap tidal cycle

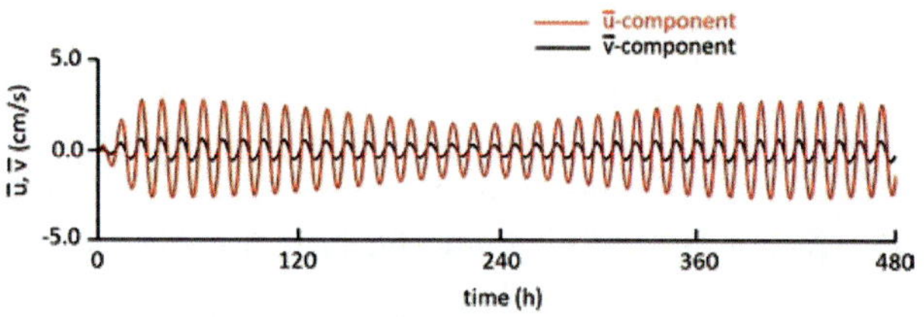

Fig. 6 Velocity components ($\overline{u}, \overline{v}$) at point *C*, during spring–neap tidal cycle

4 Conclusion

We have developed a numerical model for the resolution of the SWE (which models free surface flow) by LBM. For the velocity scheme, we have adopted the D2Q9 model, which is sufficient to cover the SWE from the BE. In our code, we have integrated the effects of Coriolis forces, the turbulence, bottom friction forces as well as the effect of bathymetry. The developed model was applied to study the tidal flow in the lagoon of Nador. The analysis of the simulations showed that the LBM model is capable of reproduce with good precision all the characteristics of the tidal flow in the lagoon (reputed to be quite complex). The comparison with other simulations resulting from classical discretization schemes has also shown that LBM is a serious alternative for the modeling of flows governed by SWE, at least by its speed in computing time. To verify this, the MECCA code was applied to the same area to simulate 480 h of tide (i.e., 20 days). The CPU time recorded for this simulation is about 1836s on a mono-processor machine with a 4-core processor (Intel Xeon E552, @2.27 GHz), while for the LBM model, we recorded 1120 s. In perspective, we intend to develop the OpenMP parallel LBM version and the GPU version which will allow us to save considerable computation time. With the parallel versions, we will develop the 3D version of the LBM code to include the effects of wind, temperature, and salinity which seem to strongly influence the hydrodynamics of the Nador lagoon.

Nomenclature

BE	Boltzmann Equation
BGK	Bhatnagar-Gross-Krook
CPU	Central processing unit
FDM	Finite difference method
FEM	Finite element method
FVM	Finite volume method
GPU	Graphics processing unit
LBM	Lattice Boltzmann method
LES	Large eddy simulation
MECCA	Model of Estuarine and Coastal Circulation Assessment
PDF	Particle distribution function
REV	Representative elementary volume

SWE Shallow water equationReferences

Bhatnagar, P., Gross, E., & Krook, M. (1954). A model for collision process in gases. I.: Small amplitude process in charged and neutral one-component systems. *Physics Review, 94*(3), 511–525.

Boyauzan, A., & Irzi, Z. (2016). Effets des aménagements sur les dynamiques morpho-sédimentaires de l'île barrière de la lagune Nador (Nord-Est du Maroc, région de l'oriental). *Méditerranée, 125, 2015*, 85–94.

Haddach, A., Smaoui, H., & Radi, B. (2023). 2D *Numerical simulation of the tidal flow: Application to the Nador lagoon, Paper in revision for the Advanced In Applied Mathematics and Mechanics* (39 p.).

Hilmi, K. (2005). *Numerical modeling of the circulation of two paralic environments of the Moroccan coast: the lagoon of Oualidia (Atlantic) and the lagoon of Nador (Mediterranean)* (186 p.). Ph.D. thesis (in French), Faculty of Sciences, Hassan II University, Mohammedia.

Hilmi, K., Makaoui, A., Idrissi, M., Abdellaoui, B., & El Ouehabi, Z. (2015). Circulation marine de la lagune de Nador (Morocco) par modélisation hydrodynamique. *European Scientific Journal, 11*(32), 418–428.

Maicu, F., Abdellaoui, B., Bajo, M., Chair, A., Hilmi, K., & Umgiesser, G. (2021). Modelling the water dynamics of a tidal lagoon: The impact of human intervention in the Nador Lagoon (Morocco). *Continental Shelf Research, 228*, 104535.

Raji, O. (2014) *Évènement extrêmes du passé et paléoenvironnements : reconstitution à partir des archives sédimentaires de la lagune de Nador* (199 p.). Ph.D. thesis, Faculty of Sciences, University of Rabat.

Smaoui, H., & Radi, B. (2001). Comparative study of different advectives schemes: Application to the MECCA model. *Environmental Fluid Mechanics, 1*, 361–381.

Zhou, J. G. (2004). *Lattice Boltzmann methods for shallow water flows*. Springer.

Causes and Impacts of Global Change

Aerosol Monitoring and Radiative Forcing Assessment in Southwestern Iberia

Maria João Costa, Vanda Salgueiro, Daniele Bortoli and Miguel Potes

Abstract

Aerosols are short-lived atmospheric constituents, acting as climate forcers and accountable for major uncertainties in the total radiative forcing estimates, as reported by the Intergovernmental Panel on Climate Change (IPCC). Several aerosol microphysical, optical, and geometrical properties are measured at Évora ground-based station since 2002, using in situ instrumentation, as well as remote sensing techniques. Data obtained with sun photometers included in AErosol RObotic NETwork (AERONET) are analyzed here. The intra-annual variability of aerosol parameters is examined on a monthly basis, and the inter-annual trends are evaluated. Results show that the region is frequently affected by long-range transport of atmospheric aerosols from different origins as desert dust or forest fire, with higher aerosol optical depth (AOD) and lower Ångström exponent (AE) values in Spring and Summer with respect to Autumn and Winter. Trend analysis point to decreasing trends of AOD and AE during the 20-year period. However, when only extreme events are considered (AOD > 0.4), the AOD trend inverts and shows an increasing tendency at a rate of 1% per year, even more pronounced in Springtime at a rate of 2.9% per year, both trends statistically significant at the 95% confidence level. Aerosol radiative forcing (ARF) presents the lowest (more negative) values at the surface, and moderately low (negative) values at the top of the atmosphere, indicating a cooling effect at both levels, whereas a warming effect is obtained for the atmosphere. ARF values are consistent with aerosol properties measured and present significant trends indicating more pronounced effects of extreme events at all levels, especially during Winter and Spring.

Keywords

Aerosols · Remote sensing · Radiative forcing · Iberia

1 Introduction

Aerosols play a fundamental role in many atmospheric processes over a broad range of temporal and spatial scales, and consequently in climate. These particles are responsible for the largest uncertainty in the total radiative forcing estimates, according to the IPCC, partly due to their spatial and temporal variability and also owed to their interactions with clouds. Monitoring and characterization of atmospheric aerosols is thus critical, and AERONET ground-based network is contributing to aerosol research for more than 25 years by providing long-term open-access standardized and detailed aerosol properties worldwide (Giles et al., 2019).

Évora, a town with approximately 60 thousand inhabitants, is located inland in southwestern Iberian Peninsula, distancing about 120 km from the Atlantic coast, in a rural area with no local sources of pollution, apart from urban traffic and, in the cold season, biomass burning from domestic heating (Salgueiro et al., 2021). The site is chosen for the study since it is on the pathway of frequent long-distance aerosol transport from different origins, with little influence from local sources. Passive remote sensing sun photometer spectral measurements from Évora are included in AERONET since 2003, yielding a reasonably long time series of aerosol parameters, which is analyzed here.

M. J. Costa (✉) · V. Salgueiro · D. Bortoli · M. Potes
Earth Remote Sensing Laboratory (EaRSLab), Universidade de Évora, Évora, Portugal
e-mail: mjcosta@uevora.pt

M. J. Costa · V. Salgueiro · D. Bortoli · M. Potes
Institute of Earth Sciences—ICT (Évora Pole), Institute for Advanced Studies and Research (IIFA), Universidade de Évora, Évora, Portugal

M. J. Costa · V. Salgueiro · D. Bortoli · M. Potes
Department of Physics, School of Sciences and Technology, Universidade de Évora, Évora, Portugal

H. Chenchouni et al. (eds.), *Recent Advancements from Aquifers to Skies in Hydrogeology, Geoecology, and Atmospheric Sciences*, Advances in Science, Technology & Innovation, https://doi.org/10.1007/978-3-031-47079-0_53

The study refers to the twenty-year period 2002–2022 and aims at analyzing the variability of climate-relevant aerosol parameters as aerosol optical thickness, Ångström exponent and single scattering albedo, as well as radiative forcings induced in the region, and on the other hand to evaluate the existence of trends in the 20-year data record.

2 Materials and Methods

Data used here is obtained at Évora AERONET site (ICT; 38.5° N, 7.91° W, 293 m asl). Aerosol optical depth (AOD), Ångström exponent (AE), and single scattering albedo (SSA), along with radiative forcing (ARF) at the bottom of the atmosphere (BOA), top of the atmosphere (TOA) and atmosphere (ATM) are obtained from AERONET Version 3 (level 1.5 Direct Sun and Inversion Algorithms) (Giles et al., 2019). AOD is calculated at the reference wavelength of 550 nm, using AOD at 440 nm and the Ångström exponent obtained between the wavelength pair 440, 870 nm. SSA and ARF are only considered when AOD440 nm is larger than 0.4 to ensure that SSA accuracy is within ±0.03 (Dubovik et al., 2000).

The intra-annual variability of the data is examined on a monthly basis. The inter-annual trends are obtained applying the nonparametric Mann–Kendall test to the deseasonalized data, to assess the statistical significance and the magnitude of the trend is calculated using Theil-Sen's method.

3 Results

The intra-annual variabilities are represented in Fig. 1, where the two upper plots presenting AOD and AE, respect the complete dataset. The four lower plots refer to the subset of higher AODs, corresponding to the condition AOD440 nm larger than 0.4, thus guaranteeing SSA and ARF accuracy within acceptable limits (Dubovik et al., 2000). The violin plots are very useful to represent data distributions, complementing the boxplot statistics. The monthly mean values are represented by the turquoise diamonds, and number of data available each month (*N*), for both datasets considered, is indicated in the plots.

The trend analysis is summarized in Table 1, similarly to Fig. 1, for the complete AOD and AE dataset (upper rows) and for the subset of all variables in the case of extreme AOD events (lower rows). Bold numbers signal statistically significant trend values at the 95% confidence level.

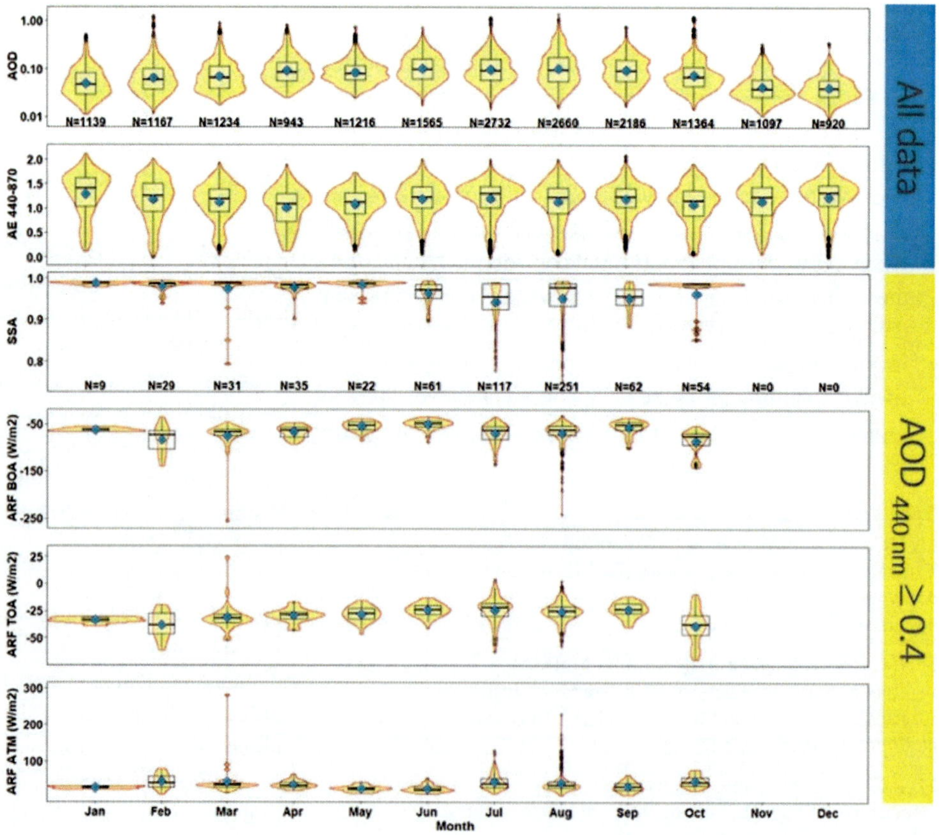

Fig. 1 Boxplots combined with violin plots showing the monthly distributions of AOD, AE, SSA, and ARF at the BOA, TOA, and ATM. Turquoise diamonds represent monthly means

Table 1 Trends obtained for the 20-year period, in units or (%) per year

	Trend (units (%)/year)					
	Variables	MAM	JJA	SON	DJF	Annual
All data	AOD	**– 0.0017 (– 1.4)**	– 0.0006 (– 0.4)	**– 0.0017 (– 1.5)**	**– 0.0012 (– 1.3)**	**– 0.0016 (– 1.4)**
	AE	**– 0.0055 (– 0.5)**	**– 0.0048 (– 0.4)**	**– 0.010 (– 0.9)**	**– 0.0067 (– 0.6)**	**– 0.0082 (– 0.7)**
Extreme events $(AOD_{440 \geq 0})$	AOD	**+ 0.011 (+ 2.9)**	**+ 0.0021 (+ 0.5)**	+ 0.0027 (+ 0.8)	+ 0.0026 (+ 0.4)	**+ 0.0039 (+ 1.0)**
	AE	**– 0.046 (– 6.8)**	– 0.0039 (– 0.4)	+ 0.276 (+ 2.6)	**– 0.11 (– 10.6)**	**– 0.033 (– 2.8)**
	SSA	**– 0.0008 (– 0.08)**	~ 0	– 0.0002 (– 0.02)	– 0.0008 (– 0.08)	~ 0
	ARF BOA (W/m^2)	**– 1.4 (+ 2.7)**	**– 0.5 (+ 0.9)**	**– 0.4 (+ 0.7)**	**– 2.6 (+ 4.5)**	**– 0.8 (+ 1.6)**
	ARF TOA (W/m^2)	**+ 1.4 (– 1.07)**	**+ 0.02 (– 0.1)**	– 0.2 (+ 1.2)	**– 0.8 (+ 3.2)**	**– 0.1 (+ 0.5)**
	ARF ATM (W/m^2)	**+ 1.0 (+ 3.7)**	+0.4 (+1.3)	+ 0.03 (+ 0.07)	**+ 3.0 (+13.2)**	**+ 0.6 (+ 2.2)**

Periods are March, April and May (MAM), June, July and August (JJA), September, October and November (SON), December, January and February (DJF)

Bold numbers indicate statistically significant trend values at the 95% confidence level.

4 Discussion

Figure 1 shows that AOD presents the lowest values in Nov. and Dec., with values steadily increasing from Jan. to Aug. The opposite is registered for AE, with values slightly decreasing from Winter to Spring. SSA presents a few extreme low values in Mar., and a distribution showing a persistence of lower values (absorbing aerosols) from Jun. to Oct., related with wildfires in the country (Salgueiro et al., 2021). The BOA ARF shows negative values, indicating a cooling effect at the surface; the TOA ARF presents the same tendency, yet presenting a few positive values. The ATM ARF indicates a warming of the atmosphere. A few extreme values are obtained in Mar. and Aug., consistent with a few low SSA values observed in the same months.

Table 1 shows decreasing trends both for AOD and AE, when the complete dataset is considered, most of them statistically significant at the 95% confidence level. When the condition AOD_{440} nm ≥ 0.4 is imposed, the AOD trends invert showing an increasing tendency, which is significant on an annual basis and for Spring and Summer, indicating + 2.9%/year in Spring. AE maintains a decreasing trend but in the case of extreme events it is more pronounced. As for ARF at BOA and TOA, decreasing trends are found, inducing a greater cooling at these levels and a consequent increased ATM warming.

5 Conclusion

AOD presents an annual cycle in the region, with higher values in Spring–Summer with respect to Autumn–Winter. The trend analysis revealed an increasing AOD trend for extreme aerosol events, inducing more pronounced radiative forcing at all levels, reinforcing surface and TOA cooling effects and atmospheric warming. This trend analysis should be carried out also for other sites, checking its spatial significance.

Acknowledgements The work was co-funded by Portuguese funds through FCT—Fundação para a Ciência e Tecnologia, I.P. (projects UIDB/04683/2020 and UIDP/04683/2020), as well as ATMO-ACCESS (Grant Agreement ID: 101008004) and ACTRIS-2 Research Infrastructure Project and Implementation Project of the European Union's Horizon 2020 Research and Innovation program (grants nº 654109 and 871115). The authors are thankful to AERONET/PHOTONS and RIMA networks for the scientific and technical support. This work is related to activities within the COST Actions CA18235 PROBE and CA21119 HARMONIA.

References

Dubovik, O., Smirnov, A., Holben, B., et al. (2000). Accuracy assessments of aerosol optical properties retrieved from aerosol robotic network (AERONET) sun and sky radiance measurements. *Journal of the Geological Society Australia, 105*, 9791–9806.

Giles, D. M., Sinyuk, A., Sorokin, M. G., et al. (2019). Advancements in the aerosol robotic network (AERONET) Version 3 database—Automated near-real-time quality control algorithm with improved cloud screening for Sun photometer aerosol optical depth (AOD) measurements. *Atmospheric Measurement Techniques, 12*, 169–209.

Salgueiro, V., Costa, M. J., Guerrero-Rascado, J. L., et al. (2021). Characterization of forest fire and Saharan desert dust aerosols over south-western Europe using a multiwavelength Raman Lidar and sun-photometer. *Atmospheric Environment, 252*, 118346.

What is and What Will Be? The Future of Climate Through Climate Modelling: A Study from Sub-tropical Region (Pakistan)

Muhammad Tahir Waseem, Muhammad Imran Shehzad, Abdul Majid Khan, Abdul Ghaffar and Jay Quade

Abstract

The regional climate models are developed to assess the future trends of the climatic regimes and provide fine-scale information which is applied to check the vulnerability, impacts and adaptation in regard to climate change. The sub-tropical regions worldwide and specifically in Pakistan house a considerable diversity of faunal elements which is subjected to adverse impacts of climate change. Thus, we aim to provide here the details of upcoming climate change in sub-tropical regions of Pakistan (Chakwal and Jhelum) and its possible impacts on the recent mammalian communities of the area. For this purpose, different climatic models including RegCM4.4, CMIP5, and GCMs under the Coordinated Regional Climate Down-scaling Experiment (CORDEX) framework for the South Asian (SA) domain were used. We found that over the upcoming years (by 2099), the precipitation will slightly increase for a shorter Monsoon timespan, which may support a C3 ecosystem where browsers will be at an advantage (future Mean Annual Precipitation (MAP) > 900 mm/year). However, the current MAP is below 900 mm/year which supports a diversity of vegetation and available habitats to the mammals. Furthermore, a steady increase in the Mean Annual Temperature (MAT) was noted which may impact the faunal diversity of the region. The bovids, suids and rodents can bear the temperature elevation as they witness higher temperatures in other parts of the country as well. However, cervids in the region are already "nearly threatened" and may further be at a disadvantage due to climatic change and temperature elevation. The current MAT of the region is slightly below 22 °C; however, after three decades, it may rise above 25 °C ultimately threatening the extant diversity of the mammals of the region.

Keywords

Chakwal · Jhelum · Climate change · Biodiversity · Conservation

1 Introduction

Many investigations have illustrated the robust changes in past climate of Potwar Plateau of Pakistan (Farooqi et al., 2005). As a result of these environmental conditions, Siwalik Formations witnessed a quantitative change in the mammalian fauna (Barry et al., 2002). The period of transition began at ~ 20 Ma and continued at ~ 2.5 with a broad progression of the biome change from forest ecosystem to savanna to grassland leading to the immigration or extinction of Siwaliks fauna (Webb & Opdyke, 1995). Nowadays, many scientists have accelerated the assessment of past environmental states and climatic trends through the construction of numerical models including RegCM4.4, CMIP5, and GCMs under the Coordinated Regional Climate Down-scaling Experiment (CORDEX) framework for the South Asian (SA) domain. These investigations

M. T. Waseem (✉)
Zoological Science Division, Pakistan Museum of Natural History, Shakarparian, Islamabad 44000, Pakistan
e-mail: tahirmuhammad1213@gmail.com

M. I. Shehzad
Department of Meteorology, COMSATS University, Islamabad 45550, Pakistan

A. M. Khan
Institute of Zoology, University of the Punjab, Lahore 54590, Pakistan

A. Ghaffar
Department of Zoology, Cholistan University of Vetrinary and Animal Sciences, Bhawalpur, Punjab, Pakistan

J. Quade
Department of Geosciences, University of Arizona, Tucson, AZ 85721, USA

H. Chenchouni et al. (eds.), *Recent Advancements from Aquifers to Skies in Hydrogeology, Geoecology, and Atmospheric Sciences*, Advances in Science, Technology & Innovation, https://doi.org/10.1007/978-3-031-47079-0_54

enabled us to assimilate observational datasets and understand the Earth's evolution. The regional climate models are developed to assess the future trends climatic regime which may affect the region in many ways and provide fine-scale information which is applied to check the vulnerability, impacts and adaption in regard of climate change (Ali et al., 2016). So far, in this dissertation, we have evaluated the impacts of past climate change on the mammalian communities. Average annual temperature of Pakistan has been increased by 0.6 °C during the last century (Giorgi et al., 2019). Meteorological Department analysed the time series of area-weighted MAT and showed the warming trend with temperature increasing at 0.6 °C from the year 1901 to 2007 (Aslam et al., 2019). The Himalayan region of Potwar Plateau experienced the increase in precipitation during second half of the previous century (Aslam et al., 2019). These increasing trends in MAT and MAP can impose a threat of current mammalian diversity of Potwar Plateau (UNEP, 2022).

2 Materials and Methods

All the Regional Climate Model RegCM version 4 (RegCM4.4) of the International Centre for Theoretical Physics (ICTP), has down-scaled six different Coupled Model Inter-comparison Project 5 (CMIP5) Global Climate Models (GCMs), i.e. CCCMA-CanESM2, CNRM-CM5, CSIRO-MK3-6-0, IPSL-CM5A-LR, MPI-ESM-MR and GFDL-ESM2M, under the Coordinated Regional Climate Down-scaling Experiment (CORDEX) framework for South Asian (SA) domain. The fourth version of the RegCM (Afzaal et al., 2009) driven by CCMA-CanESM2 GCM boundary conditions, model with horizontal grid resolution of 0.44 degrees is used to study the projected (2006–2099) inter-annual variability of temperature and precipitation over the two stations, viz. Chakwal (32.93° N and 72.88° E) and Jhelum (32.82° N and 73.35° E) located in the Potwar Plateau of Pakistan, under two emission scenarios, i.e. RCP 4.5 and RCP 8.5. A recent study conducted by Ain et al. (2020) showed that the RegCM4.4 simulations forced with CCMA-CanESM2 GCM boundary conditions well approximate the observations, therefore, have been selected for this analysis.

3 Results and Discussion

The Potwar Plateau of Pakistan receives abundant rainfall not only in summer months but in winter months as well (under southwest monsoon) however, the rainfall in pronouncedly higher in summer months creating a seasonal climate of the area (Giorgi & Anyah, 2012; Sajjad & Ghaffar, 2018). The rainfall can vary from less than 50 mm in winters to more than 200 mm in summers (Fig. 1), whereas temperature also ranges from very cold winters (~<5 °C) to very hot summers (~>40 °C). However, Mean Annual temperature is increasing steadily (Fig. 2) which may impose a threat to the mammalian diversity of the area.

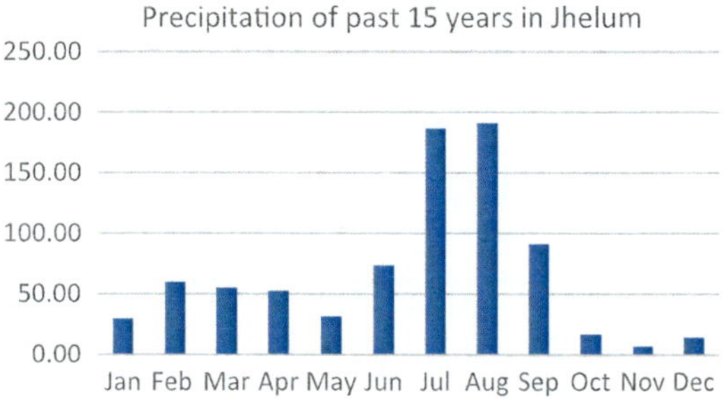

Fig. 1 Seasonal average monthly precipitation for Jhelum station for past 15 years

At present, the mammals of Siwalik area of Pakistan include suids, cervids, bovids, equids, felids and a variety of rodents (Latif et al., 2017). Although, many studies have assessed that climatic extremes in MAT and MAP in Pakistan (Latif et al., 2017; UNEP, 2022), but the long-term impact of such changes on mammalian diversity is yet to be established which seeks its answer in animal's adaptability according to the climatic regime. Thus, we applied climatic models on the available data to assess that how the seasonal changes are going to happen and we have drawn some conclusion on these models regarding the mammalian diversity and impact of climate on extant mammals. We have found the cyclic interval of precipitation change in the studied region which follows overall trends of monsoon in Pakistan in which 10–11 comparatively direr years are followed by more humid years with the interval of 10–11 years as reported by Ain et al. (2020). The same trend can be seen in our simulation of future climate (Fig. 2).

According to RCP 8.5, the precipitation over the years will slightly increase which will ultimately increase the MAP, which may support a C_3 ecosystem where browsers will be at an edge (future MAP>900 mm/year). However, the current MAP is below 900 mm/year which supports a diversity of vegetation and a diversity of available habitats to the mammals. Extant suids and bovids are found to be the main grazers in both the studied areas, whereas cervids showed a shy behaviour and preferred comparatively closer areas which were close to water streams and ponds. The current precipitation trends possess no threats to the diversity of the extant mammals of Potwar Plateau of Pakistan (PPP) and indicate a strengthening of monsoon in upcoming years as well which may support of a diverse vegetational system from forests to woodlands and grasslands. All of these vegetation types can be observed in present

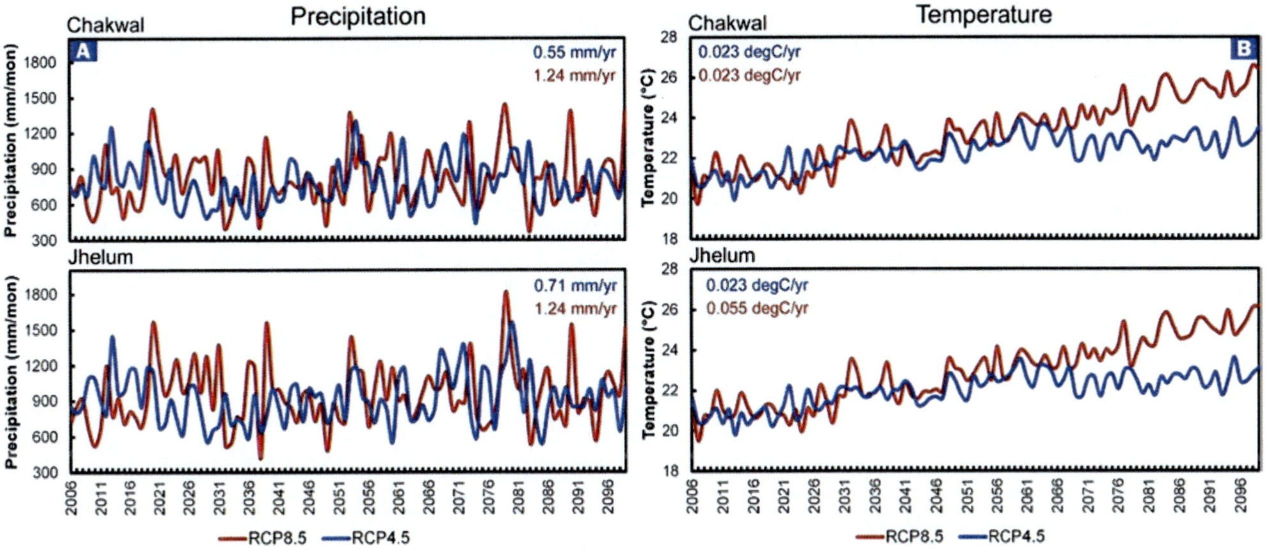

Fig. 2 **a** Climatic models of precipitation for the Chakwal and Jhelum stations respectively indicating the yearly fluctuations in rainfall. The values in blue (red) colour indicate the slope of the trend line (mm/year) for RCP4.5 (RCP 8.5) scenario. The units of precipitation are mm/month. **b** The climatic models of temperature for the Chakwal and Jhelum stations respectively indicating the yearly fluctuations in MAT. The values in blue (red) colour indicate the slope of the trend line for RCP4.5 (RCP 8.5) scenario. The units of precipitation are mm/month

day Siwaliks of Pakistan. However, grasslands comparatively dominate the plains (Quade et al., 1995). The data simulations for Mean Annual Temperature (MAT) possess a threat towards faunal diversity of the region overall. Figure 2 explains the temperature trends from 2006 to 2096 at RCP8.5 and RCP4.5. Both the models indicate a steady increase in the MAT over the upcoming years which may impact the faunal diversity of the region. The bovids, suids and rodents can bear the temperature elevation as they witness higher temperature in other parts of the country as well (Rashid & Rasul, 2011). However, cervids in the region are already "nearly threatened" and may further be threatened due to climatic change and temperature elevation. However, the climatic models developed the states the region as humid were precipitation is somewhat more than evapotranspiration (5%).

One of the pronounced impact of increased temperature is the disturbance in relative proportion of the available diet to the herbivores. Currently, cervids browse in a woodland habitat (field observation); however, with the increasing temperature, C_4 grasses will be at edge, whereas C_3 vegetation, which already not very abundant, will decrease. Such change in vegetation will pronouncedly impact the extant mammals of the region (Ahmed et al., 2019; Roberts, 1977). The current MAT of the region is slightly below 22 °C for Jhelum and Chakwal; however, after three decades, it may rise above 25 °C ultimately threatening the extant diversity of the mammals of the region (Fig. 2).

4 Conclusion

Sub-tropical regions of Pakistan are vulnerable to robust climatic conditions as RCP 8.5 predicted the increase in precipitation over the coming years which will ultimately increase MAP. These environmental extremes will be less favourable for C_4 grasses and may lead to the onset of C_3 ecosystem. MAT data simulations by RCP4.5 and RCP8.5 predict the rise in temperature in upcoming years which may cause ecological disruptions resulting the extinction or immigration of the mammalian fauna. Bovids, suids and cervids are grazers inhabiting the habitats of sub-tropical region which may be threatened due to this temperature elevation and climatic changes.

References

Afzaal, M., Haroon, M. A., & Zaman, Q. (2009). Interdecadal oscillations and the warming trend in the area-weighted annual mean temperature of Pakistan. *Pakistan Journal of Meteorology, 6*(11), 13–19.

Ahmed, K., Shahid, S., Wang, X., Nawaz, N., & Najeebullah, K. (2019) Evaluation of gridded precipitation datasets over arid regions of Pakistan. *Water, 11*(2), 210.

Ain, N. U., Latif, M., Ullah, K., Adnan, S., Ahmed, R., Umar, M., & Azam, M. (2020). Investigation of seasonal droughts and related large-scale atmospheric dynamics over the Potwar Plateau of Pakistan. *Theoretical and Applied Climatology, 140*(1), 69–89.

Ali, S., Khattak, M. S., Khan, D., Sharif, M., Khan, H., Ullah, A., & Malik, A. (2016). Predicting future temperature and precipitation over Pakistan in the 21st century. *Journal of Engineering and Applied Sciences, 35*(2), 61–76.

Aslam, S., Khan, A. M., Ahmad, R. M., Iqbal, A., & Akhtar, M. (2019). Systematic study of the new remains of *Propotamochoerus hysundricus* (Suidae, Mammalia) from the Late Miocene-Early Pliocene Middle Siwaliks (Pakistan). In: Boughdiri, M. (eds) Paleobiodiversity and Tectono-Sedimentary Records in the Mediterranean Tethys and Related Eastern Areas. CAJG 2018. (pp. 61–63). Springer, Cham. https://doi.org/10.1007/978-3-030-01452-0_15

Barry, J. C., Morgan, M. E., Flynn, L. J., et al. (2002). Faunal and environmental change in the late Miocene Siwaliks of northern Pakistan. *Paleobiology, 28*(S2), 1–71.

Farooqi, A. B., Khan, A. H., & Mir, H. (2005). Climate change perspective in Pakistan. *Pakistan Journal of Meteorology, 2*(3).

Giorgi, F., & Anyah, R. O. (2012). The road towards RegCM4. *Climate Research, 52*, 3–6.

Giorgi, F., Raffaele, F., & Coppola, E. (2019). The response of precipitation characteristics to global warming from climate projections. *Earth System Dynamics, 10*(1), 73–89.

UNEP. http://www.unep.org/resources/report/environment-and-climate-change-outlook-pakistan. Last accessed July 15, 2022.

Latif, M., Syed, F. S., & Hannachi, A. (2017). Rainfall trends in the South Asian summer monsoon and its related large-scale dynamics with focus over Pakistan. *Climate Dynamics, 48*(11–12), 3565–3581.

Quade, J., Cerling, T. E., Andrews, P., & Alpagut, B. (1995). Paleodietary reconstruction of Miocene faunas from Paşalar, Turkey using stable carbon and oxygen isotopes of fossil tooth enamel. *Journal of Human Evolution, 28*(4), 373–384.

Rashid, K., & Rasul, G. (2011). Rainfall variability and maize production over the Potohar Plateau of Pakistan. *Pakistan Journal of Meteorology, 8*(15), 63–74.

Roberts, T. J., Bernhard. (1977). Principe d'Olanda. The mammals of Pakistan.

Sajjad, H., & Ghaffar, A. (2018). Observed, simulated and projected extreme climate indices over Pakistan in changing climate. *Theoretical and Applied Climatology, 137*(1–2), 255–281.

Webb, S. D., & Opdyke, N. D. (1995). Global climatic influence on Cenozoic land mammal faunas. Effects of past global change on life, 184–208.

Aerosol Forcing Dominating Late-Summer Precipitation Change Over East Asia's Transitional Climatic Zone in CMIP6 Model Simulation

Paul Adigun, Koji Dairaku and Precious Ebiendele

Abstract

Utilizing the historical simulations from the Coupled Model inter-comparison Project Phase 6 (CMIP6), the contributions of greenhouse gasses, anthropogenic aerosols, and natural forcing to the declining trend in late summer precipitation over East Asia's transitional climatic zone (TCZ) were evaluated. Late summer precipitation across the TCZ has decreased significantly between the period of 1951 and 2013, as agreed upon by the gridded dataset utilized. CESM2, ACCESS-ESM1-5, GFDL-ESM4, and HadGEM2-GC31 out of the 13 estimated models capture the amplitude of the observed trend. Our results show that the decrease in precipitation across the TCZ is mainly attributable to the influence of aerosol forcing, which has a relative contribution of 139%. In comparison, greenhouse gases show an increasing trend in precipitation with a relative contribution of − 77%. We further examine the long-term trend in consecutive dry days during JAS and find that anthropogenic aerosol emissions significantly influence their occurrence. Our findings suggest that aerosols play a significant role in the observed changes in precipitation over the TCZ.

Keywords

Attributable · Consecutive · Anthropogenic · Relative contribution · Aerosols

P. Adigun (✉) · K. Dairaku · P. Ebiendele
Department of Engineering Mechanics and Energy, University of Tsukuba, Tsukuba, Japan
e-mail: Pauladigun7@gmail.com

1 Introduction

The East Asian transitional climatic zone (TCZ) is a region of complex weather patterns characterized by a monsoonal climate. It is located between the subtropical and tropical monsoon regions (arid and humid regions) and experiences a pronounced seasonal precipitation cycle. Recent decades have shown that the region's precipitation is unstable (Wang et al., 2017), increasing worries regarding its fundamental causes. The observed fluctuation is likely due to forced changes encompassing natural and anthropogenic-induced change (Ayodele et al., 2022). Anthropogenic aerosols can suppress light precipitation and increase the likelihood of precipitation events (Wang et al., 2011). The direct impact of aerosols on precipitation is primarily linked to their ability to serve as cloud condensation nuclei (CCN) or ice nuclei (IN); consequently, aerosols can affect cloud optical thickness, longevity, and the spatial distribution of clouds, thereby modulating the amount and distribution of precipitation (Lal et al., 2013). The aerosol indirect effect can suppress or enhance precipitation depending on the environmental conditions and aerosol properties. Extensive efforts have been made to study the aerosol effect on precipitation over monsoon-affected areas, especially for the causes of the variations (Tian et al., 2018). Zhao et al. (2021) examine the role of anthropogenic forcing in decreasing the trend of August precipitation over the TCZ; their result shows that aerosols have suppressed precipitation by about 10% since the 1950s. To conserve and manage these regions, it is necessary to investigate anthropogenic contribution to their possible dry land expansion (Li et al., 2015). The current drying trend in northern China is influenced by internal climate variability and anthropogenic forcing (Chen et al., 2017). Previous modeling studies have shown that tropospheric aerosols and human-induced land cover changes play key roles in the variations in summer rainfall across eastern China. Increasing aerosols have a direct cooling impact, which contributes to the weakening of East Asian

H. Chenchouni et al. (eds.), *Recent Advancements from Aquifers to Skies in Hydrogeology, Geoecology, and Atmospheric Sciences*, Advances in Science, Technology & Innovation, https://doi.org/10.1007/978-3-031-47079-0_55

summer monsoon (EASM) circulation, both of which contribute to the increasing frequency of drought across North China (Zhang et al., 2017). Aerosol properties can vary seasonally, making the late summer period of particular interest due to its potential influence on regional water resources and ecosystems. These studies aim to quantify the possible contribution of anthropogenic forcing to the observed precipitation decline in late summer precipitation over the last decades.

2 Data and Methods

The REGEN daily precipitation datasets from 1951 to 2013 were used in this study (Steefan et al., 2019). (ii) Climate Research Unit gridded Time Series (CRUTS) Version 4.0.3 from the University of East Anglia (Harris et al., 2020), and University of Delaware (UDEL) precipitation datasets version 3.01 for 1951–2013 on a 0.5° × 0.5° grid (Menne et al., 2012). We used 13 historical simulations from 13 CMIP6 models from Detection and Attribution Model Intercomparison Project (DAMIP), from 1951–2013 July-August-September (JAS). DAMIP estimates natural and anthropogenic forcing on climate variables at multiple spatial scales. This study used the DAMIP historical ALL forcing run (hist-ALL), which is forced with WMGHGs, BC, OC, SO_2, SO_4, NO_x, NH_3, solar irradiance, stratospheric aerosols, land use, etc. The historical aerosols-only run (hist-AER) uses SO_2, BC, NO_x, OC, SO_4, NH_3, CO, and NMVOC. The historical well-mixed GHG-only run (hist-GHG) is forced with WMGHGs, and historical natural forcing (hist-NAT) is forced with solar irradiance and stratospheric aerosols. We calculate the relative contribution percentage RCP as the ratio of the trend in all forcing runs to the trend in individual external forcing runs which is given as in Eq. (1).

$$\mathrm{RCP} = \frac{\mathrm{trend}_i}{\mathrm{trend}_{\mathrm{hist}}} \times 100 \tag{1}$$

where the subscript i refers to AER, GHG, NAT, while RCP denotes the relative contribution that is brought about by the forcing i. Signal strength of individual forcing component (aerosols, greenhouse gas, and natural aerosols) is then obtained using the multi-linear regression method (MLRM) which is written as

$$\Delta\mathrm{OBS} = \beta 1 \Delta\mathrm{NAT} + \beta 2 \Delta\mathrm{GHG} + \beta 3 \Delta\mathrm{AER} + \varepsilon \tag{2}$$

where "Δ" denotes in annual mean change of the variable from 1951 to 2013 and ε represents the small residual terms. The regression coefficients ($\beta 1$, $\beta 2$, and $\beta 3$) were estimated by NAT, GHGs, AER, respectively, to fit the observation (OBS) at each grid point during 1951–2013. In this study, all the data were normalized by using the value of the five-year moving average.

3 Results

Figure 1 shows spatial distributions of annual and JAS mean aerosol optical depth distribution at 550 nm distribution across the East Asia region from 1951 to 2013. The annual mean value shows a high AOD greater than 0.4 in southern regions. This is consistent across GFDL-ESM4, MIROC6, and HadGEM3-GC31LL, where the mean annual values throughout the Tibetan plateau and northern regions are about 0.08–0.24. GFDL-ESM4 and HadGEM3-GC31LL show high mean AOD values during the JAS months. GFDL-ESM4 demonstrates high AOD values greater than 0.48 across the South China region, which expands to the transactional climatic zone.

A significant decrease in late summer precipitation (JAS) over the East Asian transitional climatic zone (TCZ) since the 1960s is shown in Fig. 2a. This decrease is evident in all three gridded observation datasets and in the CESM2, ACCESS-ESM1-5, GFDL-ESM4, and HadGEM3-GC31 as shown in Fig. 2b. There are several possible explanations for this decrease. One possibility is that it is due to anthropogenic forcing, such as aerosols emissions or changes in land use. Another possibility is that it is due to internal climate variability. However, the observed decrease has essential implications for the region. The TCZ is a significant source of precipitation for East Asia, and the decrease in precipitation could lead to water shortages, crop failures, and other problems (Ayodele et al., 2019). Therefore, it is essential to understand the possible anthropogenic contribution to this decrease and take steps to mitigate its effects.

Figure 3 depicts spatial trends of JAS precipitation over the TCZ in observations and external forcing. All forcing runs shows a decreasing trend similar to that seen in observation as shown in Fig. 3b. The NAT forcing runs show a decreasing trend across the TCZ with a relative contribution percentage of 97%. However, they cannot exceed the 95% confidence level, implying that natural forcing has a minor contribution to the recent decline in TCZ precipitation in late summer. The AER forcing run indicates a total contribution ratio of 139%, suggesting that AER forcing is the main cause of the observed decreasing trend. The GHG forcing run shows increasing precipitation over TCZ with a relative contribution percentage of − 77%, indicating that increased GHG forcing could lead to a wetter condition.

Area-averaged consecutive dry days (CDD) anomaly relative to 1960–1990, and over TCZ is depicted in Fig. 4. The ALL forcing simulations (black lines) reveal a discernible pattern in the occurrence of CDDs across TCZ, which suggests that CDDs have significantly increased. Comparison with the AER forcing run (yellow line) indicates a similar significant increasing trend in CDD with a higher

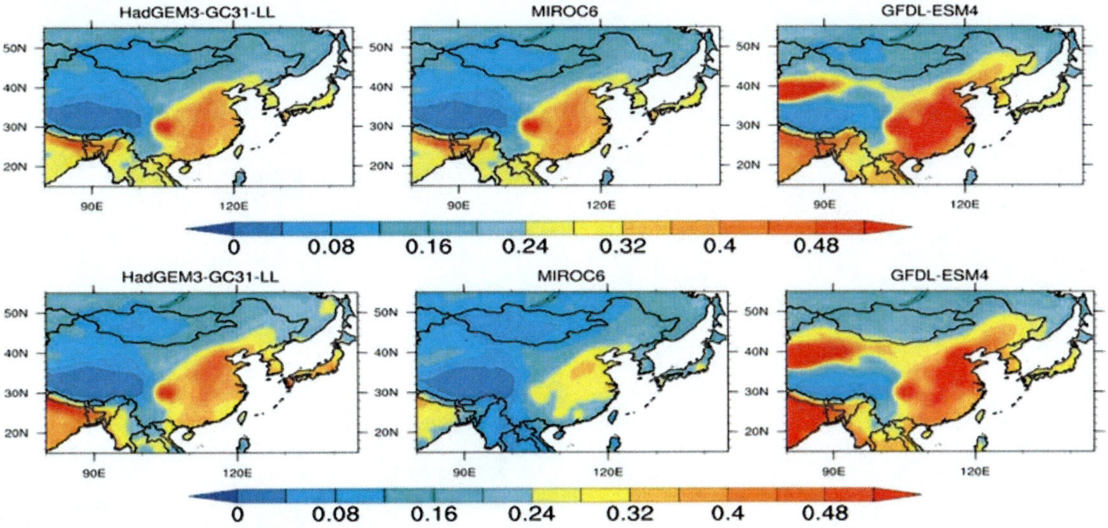

Fig. 1 Spatial distribution of aerosol optical depth over East Asia in HadGEM3-GC31LL, MIROC6, and GFDL-ESM4 during 1951–2013. Upper panel annual mean value and lower panel JAS mean value

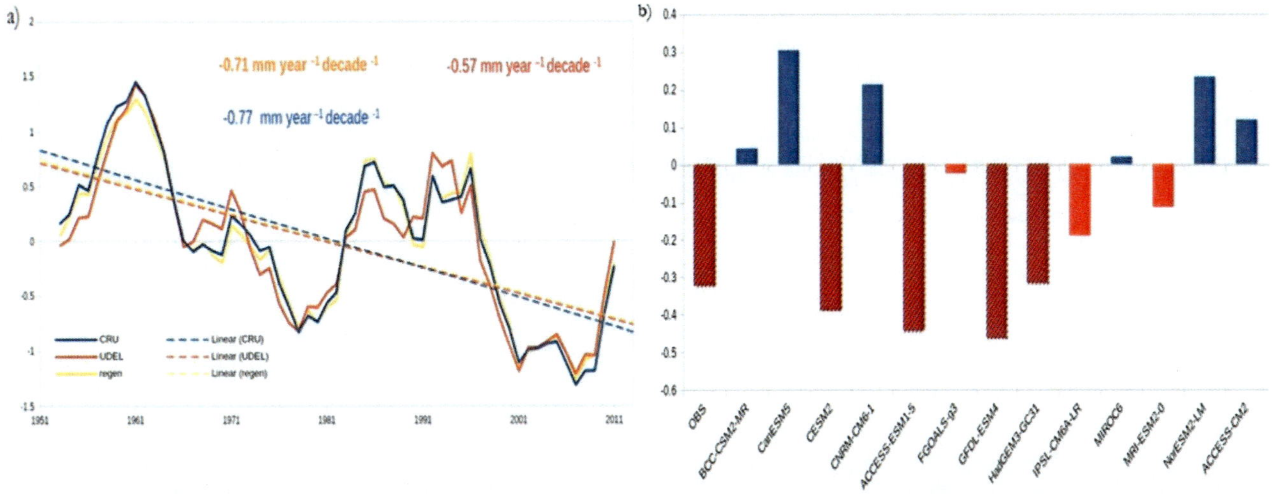

Fig. 2 Linear trend of standardized JAS precipitation anomalies (mm year−[1]) over transitional climatic zone **a** 5-year running average, **b** 13-CMIP6 Model. All trends are statistically significant at the 95% level

magnitude over the region. This suggests that an increase in anthropogenic aerosol emission across this region has enhanced the occurrence of CDD events. A decreasing trend is observed in the GHG forcing run; this indicates that CDD events will reduce under the GHG forcing.

Return event for consecutive dry days across the TCZ region was evaluated. Using the hist-NAT ensemble, we determined the recurrence event probability of exceeding the NAT forcing run. The estimated increase was evaluated for ALL and AER forcing runs exceeding the NAT threshold for 5, 20, and 50 return events. AER forcing demonstrate a robust increase in the likelihood of CDD (Fig. 5) in both 5, 20, and 50-year event, which provide further evidence that the increasing trend in CDD in the present climate over the region is primarily due to anthropogenic aerosols.

4 Discussion and Conclusion

Observation evidence shows that precipitation across the transitional climatic zone during JAS (July–august-September) has rapidly declined since the 1960s. CESM2, ACCESS-ESM1-5, GFDL-ESM4, and HadGEM3-GC31

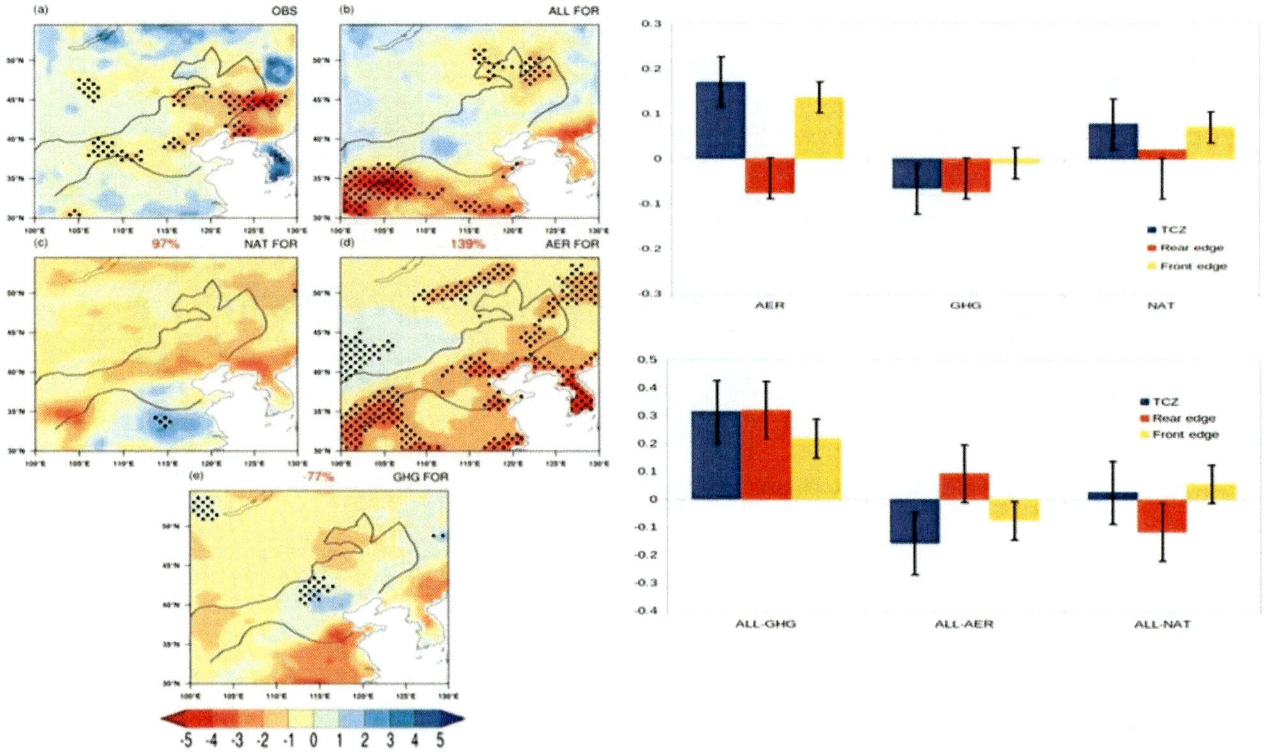

Fig. 3 Spatial trends late-summer precipitation (mm month^{-1} 62 year$-^{1}$) over TCZ from 1951 to 2013 for. **a** Observe, **b** multi-model ensemble (MME) simulations of historical run, **c** natural forcing run, **d** aerosol forcing run, and **e** GHG forcing run. The black curves indicate the boundary of the TCZ domain (**f**, **g**) signal strength of external forcing over the rear edge, front edge, and TCZ, for (aerosols, greenhouse gases, and natural forcing). Error bars denote inter-model standard error

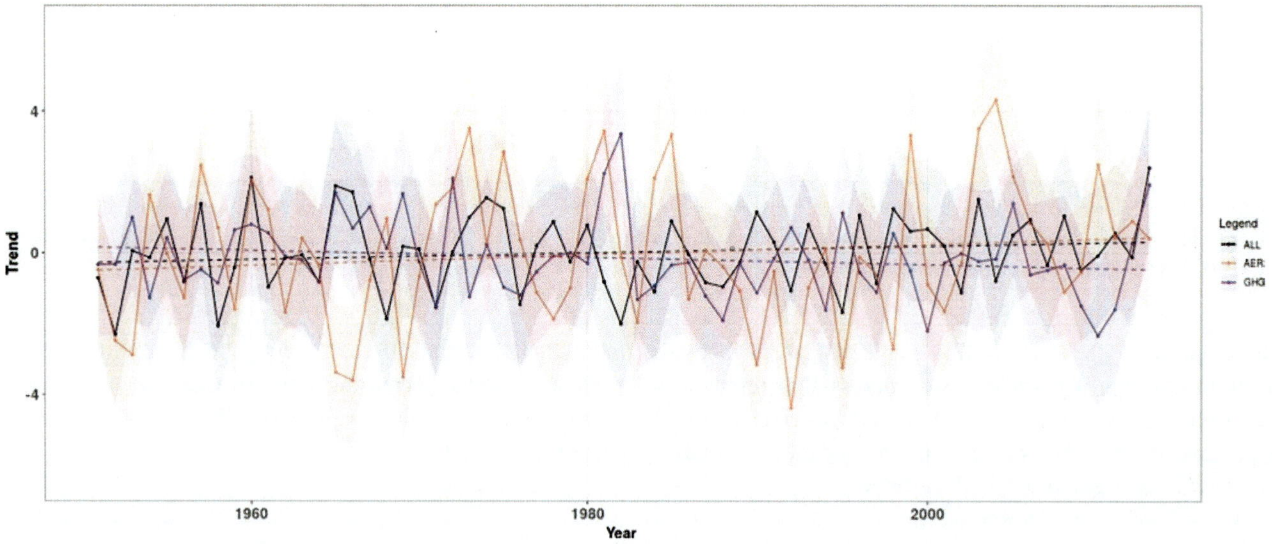

Fig. 4 Area-averaged consecutives dry days (CDD) anomaly during JAS relative to (1960–1990) over TCZ. Black line for ALL, red line for aerosol, and purple for GHG

captured the magnitude of the observed downward trend in late summer. The contributions of greenhouse gases, anthropogenic aerosols, and natural forcing on decreased late summer precipitation across the East Asian TCZ were further investigated. GHG forcing runs show an increasing trend in precipitation with a relative contribution percentage of – 77%, indicating that increased GHG forcing could lead to wetter conditions. On the other hand, natural

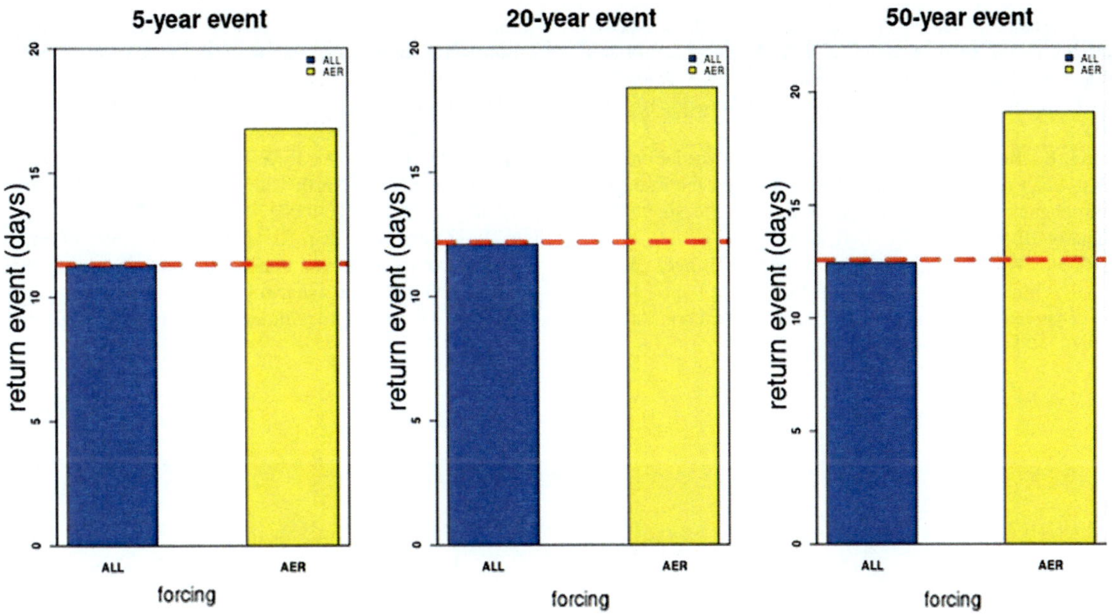

Fig. 5 Return event for CDD events with hist-AER and hist-ALL for 5, 20 years, and 50 return events

forcing reveals a very moderate contribution of 97% to the observed decrease. While the aerosol forcing runs contribute 139% to the decreasing trend in precipitation observed over the TCZ, it suggests that aerosol forcing significantly affected the overall observed decrease in JAS precipitation over the region. We further evaluate the long-term trend in consecutive dry day events over the region, which shows a significant increasing trend during the investigative period. The increase in anthropogenic aerosol emission across this region has enhanced the occurrence of CDD events. AER forcing increases the likelihood of consecutive dry day events, which gives additional evidence that anthropogenic aerosols are the primary contributor to the observed increasing trend in CDD in the present climate throughout the region. TCZ is an industrialized and one of the most populous regions in Asia. Therefore, the ever-increasing human population and activities have resulted in a rapid and continual increase in anthropogenic aerosol emissions (Chenchouni et al., 2022). Studies have shown that aerosols can reduce the rain produced by warm, shallow clouds. Human-made aerosols, including those in air pollution, are known to affect precipitation via radiative and microphysical mechanisms (Rosenfeld et al., 2008; Lal et al., 2013).). Using the multilinear regression approach, uncertainties remain about how much individual external forcing is responsible for the observed decrease trend in precipitation. Consequently, it will be crucial to determine the role of internal climate variability, which will provide a solid understanding of regional climate change responses to external drivers.

References

Ayodele, A. P., Precious, E. E., Brhane, E. S., et al. (2022). CMIP6 multi-model evaluation of summer extreme precipitation over East Asia. *Modelling Earth Systems and Environment*. https://doi.org/10.1007/s40808-022-01433-3

Ayodele, A. P., & Precious, E. E. (2019). Seasonal rainfall prediction in Lagos, Nigeria using artificial neural network. *Asian Journal of Research in Computer Science, 3*(4), 1–10. https://doi.org/10.9734/ajrcos/2019/v3i430100

Chenchouni, H., Chaminé, H.I., Khan, M.F., et al. (2022). *New Prospects in Environmental Geosciences and Hydrogeosciences*. Cham, Springer. https://doi.org/10.1007/978-3-030-72543-3

Chen, H., & Sun, J. (2017). Anthropogenic warming has caused hot droughts more frequently in China. *Journal of Hydrology, 544*, 306–318. https://doi.org/10.1016/J.JHYDROL.2016.11.044

Harris, I., Osborn, T. J., Jones, P., et al. (2020). Version 4 of the CRU TS monthly high-resolution gridded multivariate climate dataset. *Science Data, 7*, 109. https://doi.org/10.1038/s41597-020-0453-3

Lal, D., Patil, S., Singh, H., Ghude, S., Tiwari, S., & Srivastava, M. (2013). Influence of aerosol on clouds over the Indo-Gangetic Plain, India. *Climate Dynamics, 41*, 601–612. https://doi.org/10.1007/s00382-013-1775-z

Li, Y., Huang, J. P., Ji, M. X., & Ran, J. J. (2015). Dryland expansion in northern China from 1948 to 2008. *Advances in Atmospheric Sciences, 32*, 870–876. https://doi.org/10.1007/s00376-014-4106-3

Menne, M. J., Durre, I., Vose, R. S., Gleason, B. E., & Houston, T. G. (2012). An overview of the global historical climatology network-daily database. *Journal of Atmospheric and Oceanic Technology, 29*, 897–910. https://doi.org/10.1175/JTECH-D-11-00103.1

Rosenfeld, D., Lohmann, U., Raga, G. B., O'Dowd, C. D., Kulmala, M., Fuzzi, S., Reissell, A., & Andreae, M. O. (2008). Flood, or drought: How do aerosols affect precipitation? *Science, 321*(5894), 1309–1313. https://doi.org/10.1126/science.1160606. PMID: 18772428.

Steefan, C., Markus, G., Lisa, V., Markus, Z., et al. (2019). Rainfall estimates on a gridded network (REGEN): A global land-based gridded dataset of daily precipitation from 1950–2013. *Hydrology and Earth System Science Discussion*. https://doi.org/10.5194/hess-2018-595

Tian, F., Dong, B., Robson, J., et al. (2018). Forced decadal changes in the East Asian summer monsoon: The roles of greenhouse gases and anthropogenic aerosols. *Climate Dynamics, 51*, 3699–3715. https://doi.org/10.1007/s00382-018-4105-7

Wang, L., Chen, W., Huang, G., et al. (2017). Changes of the transitional climate zone in East Asia: Past and future. *Climate Dynamics, 49*, 1463–1477. https://doi.org/10.1007/s00382-016-3400-4

Wang, Y., Wan, Q., Meng, W., Liao, F., Tan, H., & Zhang, R. (2011). Long-term impacts of aerosols on precipitation and lightning over the Pearl River Delta megacity area in China. *Atmospheric Chemistry and Physics, 11*, 12421–12436. https://doi.org/10.5194/ACP-11-12421-2011

Zhang, L., Wu, P., & Zhou, T. (2017). Aerosol forcing of extreme summer drought over North China. *Environmental Research Letters, 12*. https://doi.org/10.1088/1748-9326/aa5fb3

Zhao, W., Chen, W., Chen, S., et al. (2021). Roles of anthropogenic forcings in the observed trend of decreasing late-summer precipitation over the East Asian transitional climate zone. *Science and Reports, 11*, 4935. https://doi.org/10.1038/s41598-021-84470-9

The Chemistry of Precipitation in Forest Regions Under Anthropogenic Pressure

Justyna Likus-Cieślik, Bartłomiej Woś, Marek Pająk, Piotr Gruba and Marcin Pietrzykowski

Abstract

Atmospheric precipitation (rainwater, wet deposition) affected by industry is one of the sources of pollution that migrate to the ecosystem, especially by the uppermost soil layer and ground water. The chemistry of atmospheric precipitation around the 'Miasteczko Śląskie' zinc smelting plant (HCM) in the Upper Silesia region (southern Poland) was studied. The pH and chemistry of rainwater were analyzed with connection to the meteorological and climatic factors in the research area of the surrounded forest ecosystem area. Water samples were collected once per month (12 months in total) in 2021–2022 season at 20 monitoring points distributed in the grid (5 replications on each) located according to five modeled arches at the following distances from the main emission source: 1640, 2140, 2740, 3340, 3940 m. The control points were located opposite to the dominant wind direction (NW), and with wind directions in the distance above 6 km from the emitter. Pb, Zn, and Cd concentrations were found to exceed the Polish Environmental Protection Inspectorate reference data. Extremely high concentrations of pollutants were observed up to 3 km distance from the smelter: Cd (max. 1.43 mg $L-^1$, mean 0.1) was higher, even than in the most polluted areas in the world (e.g., Mexico City), whereas zinc (max. 21.42 mg L^{-1}, mean 1.71) and the Pb concentration (max. 0.44 mg $L-^1$, mean 0.06) were similar to the highly polluted areas in the world. High concentrations of Cd, Zn, and Pb in rainwater indicate a strong industrial pressure in the adjacent forest ecosystem.

Keywords

Wet deposition · Chemistry of precipitation · Air pollution · Trace elements · Miasteczko Śląskie (HCM)

1 Introduction

Research on the chemical composition of rainwater, which depends on the degree of air pollution, and meteorological and climatic factors, contributes to research on atmospheric area pollution (Guzel, 2021). The highest air pollution existed especially in heavily urbanized, industrialized areas of countries that base their energy on fossil fuels (European Environment Agency, 2022a; Likus-Cieślik et al., 2020; Zeng et al., 2022). Industrial processes determine the impact on the natural environment, including forest ecosystems, and human health (European Environment Agency, 2022b; Manisalidis et al., 2020). The 'Miasteczko Śląskie' zinc smelter (HCM), located in Upper Silesia region is currently the only one in Europe that uses the ISP (Imperial Smelting Process) process. This ISP process enables the simultaneous recovery of zinc and leads from raw materials with a very different chemical composition (HCM, 2022). These processes generate various types of pollutants, including heavy metals and sulfur oxide forms. As a result of many years of pollutants emission into the atmosphere, the areas around the plant have been polluted with Pb, Zn, and Cd. The research investigated the chemistry of precipitation and the concentration of trace elements in wet deposition depending on the distance from the polluting point.

2 Materials and Methods

The research was carried out around the 'Miasteczko Śląskie' zinc smelting plant (HCM, Upper Silesia, southern Poland). The plants are associated with processing zinc-lead ores and recycled waste materials containing these metals.

J. Likus-Cieślik (✉) · B. Woś · M. Pająk · M. Pietrzykowski
Department of Ecological Engineering and Forest Hydrology, Faculty of Forestry, University of Agriculture in Krakow, Al. 29-Listopada 46, 31-425 Kraków, Poland
e-mail: justyna.likus@urk.edu.pl

P. Gruba
Department of Ecology and Silviculture, Faculty of Forestry, University of Agriculture in Krakow, Al. 29-Listopada 46, 31-425 Kraków, Poland

H. Chenchouni et al. (eds.), *Recent Advancements from Aquifers to Skies in Hydrogeology, Geoecology, and Atmospheric Sciences*, Advances in Science, Technology & Innovation, https://doi.org/10.1007/978-3-031-47079-0_56

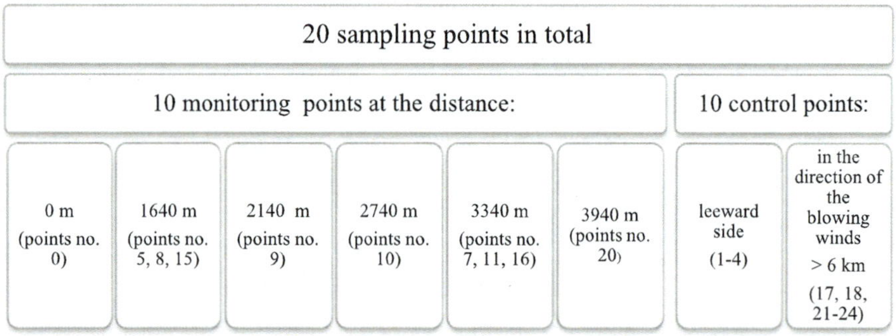

Fig. 1 Sampling points flowchart

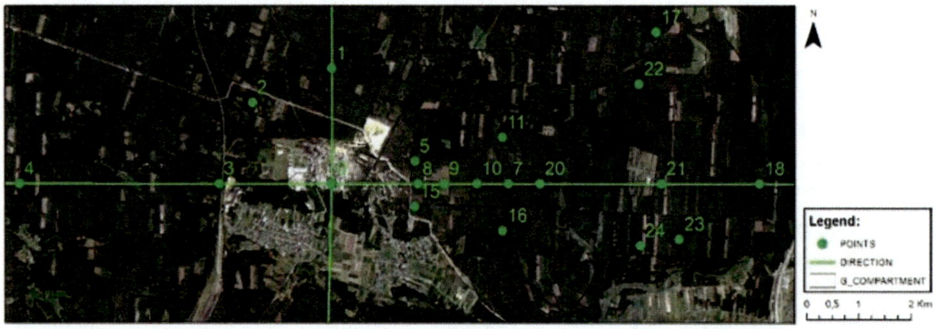

Fig. 2 Localization of sampling points

Wet deposition samples were collected at 20 monitoring points (Fig. 1) located concerning the main emission source (zinc smelter), and five modeled arcs at the distances: 1640 (points 5, 8, 15), 2140 (point 9), 2740 (point 10), 3340 (points 7, 11, 16) and 3940 m (points 20, Fig. 2). The '0' point is located next to the plant in the eastern direction. Control points (1–4) were located on the leeward side, and more than 6 km from the main emission source (17, 18, 21–24). On each sampling point, wet deposition (precipitation) was collected to PVC grabs (5 replications in each monitoring point). Samples were collected once a month during 2021/22 season. The pH and electrical conductivity (EC) were measured potentiometrically (maintaining a ratio of soil: solution 1:5) using Multifunction Meter CX-705. Zn, Cd, and Pb concentrations were determined on the ICP OES ICAP 6000 Series spectrophotometer.

3 Results

Atmospheric precipitation pH values ranged from 4.5 to 6.2 (mean 4.9, Table 1). The highest pH of rainwater was recorded at point 0, located the closest to the zinc smelter. The lowest pH value was observed in February and January (4.6; 4.7, respectively), and the highest was in April (mean 5.3). The mean electric conductivity (EC) in precipitation was 50.04 µS cm^{-1}. The highest value was 147.4 µS cm$-^{1}$ at point 0, while at other points it ranged from 15 to 100 µS cm^{-1}. The highest concentrations of Cd and Zn were found at point 0 (Cd 1.43 mg L$-^{1}$, Zn 21.42 mg L^{-1}). High concentrations of these elements were found in the zone up to 2140 m from the smelter (points 5, 8, 15, 9). The concentration of Cd and Zn decreased with the distance from the smelter. Low values were also observed in the western part (control points 1–4), where the mean concentration of Cd in precipitation was 0.01, and Zn 0.02 mg L$-^{1}$. The mean Pb concentration in precipitation was 0.06 mg L^{-1} (Table 1). The highest Pb concentration was found in point 9 located within 2 km from the smelter. A negative correlation was observed between the pH and Cd, Zn, and Pb

Table 1 Mean pH, EC, Cd, Zn, Pb in precipitation considering the distance from the smelter

The distance	pH	EC	Cd	Pb	Zn
[m]		[µS cm$-^{1}$]	[mg L^{-1}]		
0	6.2	147.40	1.43	0.18	21.42
1640	4.8	47.8	0.07	0.10	1.64
2140	4.5	86.20	0.16	0.44	3.20
2740	4.7	31.60	0.02	0.02	0.44
3340	4.8	53.8	0.02	0.03	0.57
3940	5.2	24.80	0.01	0.00	0.23
>6000[a]	4.8	37.9	0.01	0.01	0.19
>2000[a,b]	5.1	44.7	0.01	0.02	0.29
Total mean	4.9	50.0	0.10	0.06	1.71

[a] control points
[b] opposite to the prevailing wind directions

concentrations. A positive correlation occurred between EC and Cd, Zn, and Pb. The highest concentration of trace elements occurred in autumn and winter, and the lowest was in summer.

4 Discussion

The 'Miasteczko Śląskie' zinc smelter contributes to the contamination of the monitored area, especially in the zone directly adjacent to the smelter. The analysis showed that the concentration of Zn, Cd, and Pb in precipitation in the zone directly adjacent to the smelter (point 0) was from about 500 (Pb, Zn) to 9500 (Cd) times higher than in the background according to Environmental Protection Inspectorate of Poland (background for Poland: 0.036 Zn; 0.00015 Cd; 0.0008 mg $L-^{1}$ Pb) (Liana, 2020). The concentrations of Zn, Cd, and Pb in the precipitation decrease with the distance from the smelter. The most increased values occurred on the leeward side up to about 2 km from the smelter. Spot anomalies at greater distances may indicate the share of other emission sources.

Cd concentration in precipitation (mean 0.1 mg L^{-1}, maks. 1.43 mg $L-^{1}$) exceed the values for highly polluted areas in the world (Mexico City: 0.005 mg L^{-1}), whereas the Pb concentration was similar to highly polluted regions (Báez et al., 2007). Zn (mean 1.71 mg $L-^{1}$, maks. 21.42 mg L^{-1}) in precipitation was also higher than, e.g., Candiota region, in South Brazil, where a high impact of industry (with coal mining, limestone mining, coal industries, and coal-fired power plant) on pollution was found (0.035 mg $L-^{1}$) (Migliavacca et al., 2004). The concentration of the researched trace elements in the rainwater was still higher than that recorded in Bellsund (SW Spitsbergen, Svalbard), where low temperatures and mountainous terrain favor the deposition of pollutants (Lehmann-Konera et al., 2023).

5 Conclusion

High values of Cd, Zn, and Pb in rainwater indicate the influence of the smelter on the closest environment. The values obtained in the immediate vicinity are very high and exceed the values achieved so far as the most polluted places in the world. However, with the distance from the source, the pollution decreases. Despite it, the area must be under permanent monitoring.

References

Báez, A., Belmont, R., García, R., Padilla, H., & Torres, M. C. (2007). Chemical composition of rainwater collected at a southwest site of Mexico City, Mexico. *Atmospheric Research, 86*, 61–75. https://doi.org/10.1016/j.atmosres.2007.03.005

European Environment Agency. (2022a). Air quality in Europe 2022. Online available https://www.eea.europa.eu/publications/air-quality-in-europe-2022/air-quality-in-europe-2022

European Environment Agency. (2022b). EEA Signals 2022—Staying on course for a sustainable Europe (2022). Online available https://www.eea.europa.eu/publications/signals-2022/eea-signals-2022

Guzel, B. (2021). Monitoring of the chemical composition of rainwater in a semi-urban area in the Northern West of Turkey. *Journal of Science, 34*(1), 1–17.

HCM. (2022). http://hcm.com.pl. Last accessed September 06, 2022.

Lehmann-Konera, S., Ruman, M., Frankowski, M., Małarzewski, Ł., Raczyński, K., Pawlak, F., Kozioł, K., & Polkowska, Ż. (2023). Rainwater chemistry composition in Bellsund: Sources of elements and deposition discrepancies in the coastal area (SW Spitsbergen, Svalbard). *Chemosphere, 313*. https://doi.org/10.1016/j.chemosphere.2022.137281

Liana, E. (2020). Monitoring chemizmu opadów atmosferycznych i ocena depozycji zanieczyszczeń do podłoża w latach 2021–2022. Raport Roczny z badań monitoringowych w 2020 roku. Inspekcja Ochrony Środowiska. https://powietrze.gios.gov.pl/pjp/maps/chemistry/stations. Last accessed October 07, 2022. (in polish)

Likus-Cieślik, J., Socha, J., Gruba, P., & Pietrzykowski, M. (2020). The current state of environmental pollution with sulfur dioxide (SO_2) in Poland based on sulfur concentration in Scots pine needles. *Environmental Pollution, 258*, 113559. https://doi.org/10.1016/j.envpol.2019.113559

Manisalidis, I., Stavropoulou, E., Stavropoulos, A., & Bezirtzoglou, E. (2020). Environmental and health impacts of air pollution: A review. *Front Public Health, 20*, 8–14. https://doi.org/10.3389/fpubh.2020.00014

Migliavacca, D., Teixera, E. C., Pires, M., & Fachel, J. (2004). Study of chemical elements in atmospheric precipitation in South, Brazil. *Atmospheric Environment, 38*. https://doi.org/10.1016/j.atmosenv.2003.11.040

Zeng, J., Han, G., Zhang, S., Xiao, X., Li, Y., Gao, X., Wang, D., & Qu, R. (2022). Rainwater chemical evolution driven by extreme rainfall in megacity: Implication for the urban air pollution source identification. *Journal of Cleaner Production, 372*, 20. https://doi.org/10.1016/j.jclepro.2022.133732

Investigating the Relationship Between Water Vapor and Precipitation in Northern Africa

Hoyoung Cha, Jongjin Baik, Hyeon-Joon Kim, Jinwook Lee, Jongyun Byun, Kihong Park, Jeemi Sung, Seunghyun Hwang and Changhyun Jun

Abstract

In this study, we investigated a relationship between integrated water vapor (IWV) and standardized precipitation index (SPI), focusing on historical drought events over the northern part of Africa from 1850 to 2014. Here, monthly data was obtained from two climate models of CMCC-ESM2 and CESM2, representative models in coupled model intercomparison project phase 6 (CMIP6). The IWV was standardized with the gamma distribution for various time intervals from 2 to 12 months, and the standardized IWV (IWVs) was compared with two established drought indicators, namely SPI-6 and SPI-12. We considered Pearson correlation coefficients for evaluating their relationship. As results of correlation analysis with SPI-6, Pearson correlation coefficients were larger than 0.74 from 2 to 4 months and lower than 0.41 from 9 to 12 months. However, the coefficients for SPI-12 with the time interval of 12 months were 0.82 and 0.71 for CMCC-ESM2 and CESM2, respectively, while showing negligible relationships with lower values than 0.21 from 2 to 4 months. It shows there are intra-seasonal and annual patterns in the relationship between IWV and SPIs. Further analysis of temporal patterns in IWV may lead to the potential for improved drought prediction.

Keywords

IWV · SPI · CMIP6 · Drought · Northern part of Africa

H. Cha · C. Jun (✉)
Department of Smart Cities, Chung-Ang University, 06974 Seoul, South Korea
e-mail: cjun@cau.ac.kr

J. Baik · H.-J. Kim · J. Lee · J. Byun · K. Park · J. Sung · S. Hwang · C. Jun
Department of Civil Engineering, Chung-Ang University, 06974 Seoul, South Korea

1 Introduction

Drought frequency is currently on the rise globally, primarily due to the effects of global warming (Alexander, 2006). Drought is divided into several types: meteorological drought, agricultural drought, hydrological drought, etc. (Negm et al., 2020), and affects differently ecosystems and environment depending on geographic region (Allen et al., 2010; Bezzalla et al., 2018). The onset of meteorological drought is one of the causes of other droughts. Patel et al. (2007) used standardized precipitation index (SPI) to analysis spatial patterns of meteorological drought. It showed the relationship between precipitation and vertical water vapor through daily and monthly special sensor microwave/imager (Bretherton et al., 2004). Atmospheric river (AR) is similar in concept to the vertical water vapor. In brief, AR represent a flow of water vapor in the atmosphere and can be categorized into two main components: Integrated Water Vapor (IWV) and Integrated Vapor Transport (IVT). In previous studies, research on the relationship with precipitation has been conducted. Kim et al. (2013) showed the relationship between the AR and precipitation in California from 2001 to 2010. These result indicated that SPI varies depending on the precipitation pattern due to AR. Dettinger (2013) also reported that SPI-6 was profound affected on AR over US West Coast.

In this study, we investigated the correlation between SPIs (SPI-6 and SPI-12) and IWV on the northern part of Africa. North Africa is generally a dry region due to climate and geographical location (Chenchouni & Neffar, 2022). In addition, the continuous drought in the Sahel region is caused by global warming and aerosol emissions from human activities. This vicious circle in North Africa threatens agriculture, water supply, environment, and human health (Ayugi et al., 2022). So, we focused on historical drought events over the northern part of Africa from 1850 to 2014.

H. Chenchouni et al. (eds.), *Recent Advancements from Aquifers to Skies in Hydrogeology, Geoecology, and Atmospheric Sciences*, Advances in Science, Technology & Innovation, https://doi.org/10.1007/978-3-031-47079-0_57

2 Methodology

The study area was defined as North Africa, with a north latitude of 13–35° and a longitude of 345–360° and 0–30°. The process flowchart is depicted in Fig. 1. We utilized specific humidity of each pressure level and precipitation to analyze SPI and IWV through a formula using two products (i.e., Centro Euro-Mediterraneo sui Cambiamenti Climatici-Earth System Model Version 2 (CMCC-ESM2) and the Community Earth System Model Version 2 (CESM2)) of coupled model intercomparison project phase 6 (CMIP6) from 1850 to 2014. We constructed the IWV data through formula (Zhu & Newell, 1998). g, q, and dp represent acceleration of gravity, specific humidity, and pressure difference, respectively (Eq. 1).

$$\mathrm{IWV} = \frac{1}{g}\int_{1000\,\mathrm{hPa}}^{300\,\mathrm{hPa}} q\mathrm{dp}. \tag{1}$$

In general, SPI values are computed using the gamma distribution method (Guttman, 1999). To utilize Standardized IWV (IWVs) values, we estimated the IWV distribution using the gamma distribution for periods ranging from 2 to 12 months. For the analysis of both indicators, we employed the Pearson correlation coefficient.

3 Results and Discussion

3.1 Correlation About Average IWVs, SPI-6, and SPI-12

We figured out how many months on IWVs were more related to SPI-6 and SPI-12. As results of SPI-6, the Pearson correlation coefficient with IWVs-2 was the largest value to 0.8292 for CESM2. On the other hand, IWVs-10 has a very weak negative relationship with SPI-6 for CESM2. It showed a similar pattern in CMCC-ESM2. However, the coefficients for SPI-12 increased as the number of months with applied IWVs increased for CMCC-ESM2 and CESM2. Nonetheless, this relationship exhibited a weak performance with coefficients consistently below 0.21 for periods ranging from 2 to 4 months (Jacob, 1988). By calculating ρ-values, we established the statistical significance of a strong correlation between variables such as IWVs-2 and SPI-6 (Table 1).

3.2 The Tendency in SPI-6 and SPI-12 on Related IWVs

As illustrated in Fig. 2, time series data were presented for two scenarios: SPI-6 and IWVs-4, and SPI-12 and IWVs-12, spanning from 1900 to 2014. CESM2 exhibited a similar trend in both IWVs-12 and SPI-12. The results for SPI-6 and IWVs-4 indicated lower drought severity performance, despite a strong correlation. It became evident that the reduction in IWVs had a more pronounced impact on drought occurrence, while an increase in IWV played a role in drought recovery.

4 Conclusion

SPI-6 showed the characteristics of distinct seasons. The interval between seasonal changes was almost two to four months. The trend of SPI-6 and IWVs-2, 3, and 4 was similar because of the reason. SPI-12 reflected the long-term precipitation patterns. The cumulative precipitation in the previous year was relatively correlated to IWVs in long term. Their correlation values showed that there are characteristics and pattern within a season.

As result, when the IWV approaches a negative value, the likelihood of drought occurrence increases, coinciding with a decrease in precipitation probability. These

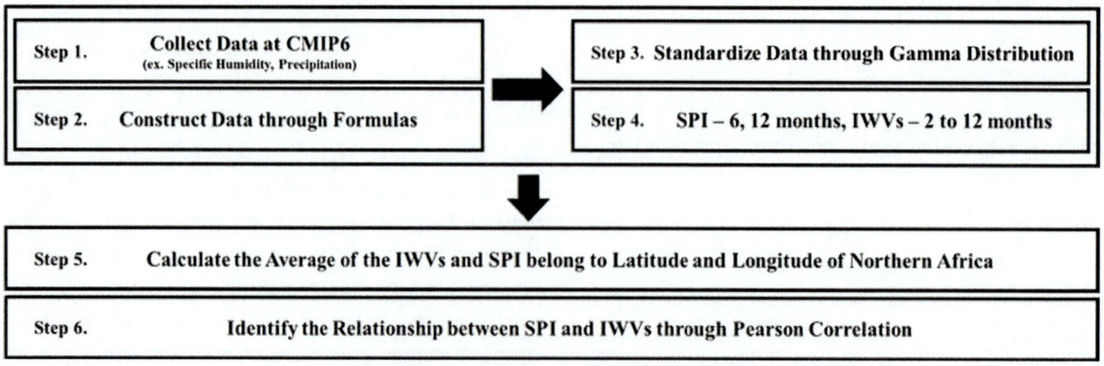

Fig. 1 Flowchart illustrating the historical drought event-based CMIP6 data for assessing the correlation between IWV and SPIs

Table 1 Correlation value on using Pearson about Standardized IWV, SPI-6, and SPI-12

SPI-6	CESM2	CMCC-ESM2	SPI-12	CESM2	CMCC-ESM2
IWVs-2	0.8292***	0.7842***	IWVs-2	0.1154***	0.1685***
IWVs-3	0.8116***	0.8249***	IWVs-3	0.1280***	0.1867***
IWVs-4	0.7465***	0.8202***	IWVs-4	0.1457***	0.2097***
IWVs-5	0.6439***	0.7736***	IWVs-5	0.1686***	0.2390***
IWVs-6	0.5110***	0.6881***	IWVs-6	0.1992***	0.2782***
IWVs-7	0.3421***	0.5615***	IWVs-7	0.2411***	0.3316***
IWVs-8	0.1808***	0.4260***	IWVs-8	0.2995***	0.4039***
IWVs-9	0.0527**	0.3092***	IWVs-9	0.3827***	0.5014***
IWVs-10	-0.0064*	0.2424***	IWVs-10	0.4987***	0.6266***
IWVs-11	0.0709***	0.2687***	IWVs-11	0.6390***	0.7592***
IWVs-12	0.3131***	0.4075***	IWVs-12	0.7122***	0.8220***

p-value—***<0.01, **<0.1, *>0.1

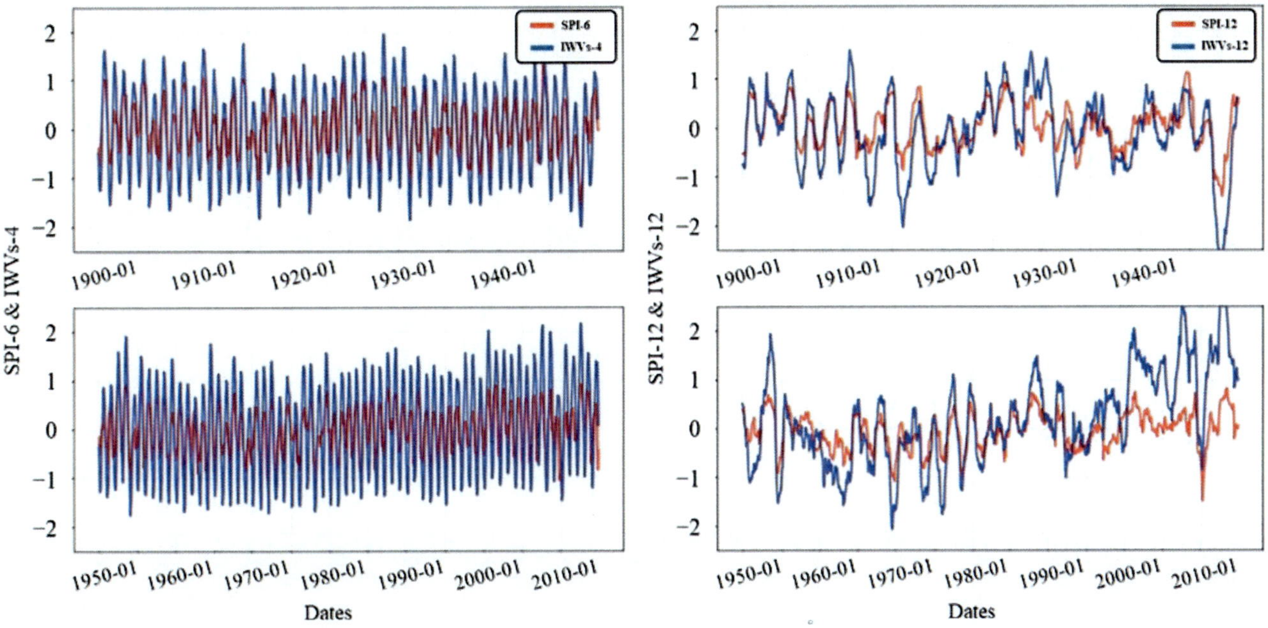

Fig. 2 Time series of IWVs and SPI in 1900–2014 for CESM2

findings suggest that AR are influenced by drought conditions in the northern part of Africa. The drought was resolved over time in statistics when AR was coming. We will use the change of IWVs for predicting the occurrence of drought. In this study, it's a limit to consider the drought index like SPI. In the following study, it is expected that meteorological drought was predicted for using IWVs.

Acknowledgement This research was supported by Korea Environment Industry and Technology Institute (KEITI) through Water Management Innovation Program for Drought (No. 2022003610001) funded by Korea Ministry of Environment and this work was supported by the National Research Foundation of Korea(NRF) grant funded by the Korea government (MSIT) (No. NRF-2022R1A4A3032838).

References

Alexander, D. (2006). Globalization of disaster: Trends, problems and dilemmas. *Journal of International Affairs*, 1–22.

Allen, C. D., Macalady, A. K., Chenchouni, H., Bachelet, D., McDowell, N., Vennetier, M., Kitzberger, T., Rigling, A., Breshears, D., Hogg, E. H., Gonzalez, P., Fensham, R., Zhang, Z., Castro, J., Demidova, N., Lim, J., Allard, G., Running, S., Semerci, A., & Cobb, N. (2010). A global overview of drought and heat-induced tree mortality reveals emerging climate change risks for forests. *Forest ecology and management, 259*(4), 660–684.

Ayugi, B., Eresanya, E. O., Onyango, A. O., Ogou, F. K., Okoro, E. C., Okoye, C. O., Anoruo, C. M., Dike, V. N., Ashiru, O. R., Daramola, M. T., Mumo, R., & Ongoma, V. (2022). Review of meteorological drought in Africa: Historical trends, impacts, mitigation measures, and prospects. *Pure and Applied Geophysics*, 1–22.

Bezzalla, A., Boudjabi, S., & Chenchouni, H. (2018). Seedlings of Argan (*Argania spinosa*) from different geographical provenances reveal variable morphological growth responses to progressive drought stress under nursery conditions. *Agroforestry Systems, 92*(5), 1201–1211.

Bretherton, C. S., Peters, M. E., & Back, L. E. (2004). Relationships between water vapor path and precipitation over the tropical oceans. *Journal of Climate, 17*(7), 1517–1528.

Chenchouni, H., & Neffar, S. (2022). Soil organic carbon stock in arid and semi-arid steppe rangelands of North Africa. *CATENA, 212*(1), 106004.

Dettinger, M. D. (2013). Atmospheric rivers as drought busters on the US West Coast. *Journal of Hydrometeorology, 14*(6), 1721–1732.

Guttman, N. B. (1999). Accepting the standardized precipitation index: A calculation algorithm 1. *JAWRA Journal of the American Water Resources Association, 35*(2), 311–322.

Jacob, C. (1988). *Statistical power analysis for the behavioral sciences*. American Statistical Association.

Kim, J., Waliser, D. E., Neiman, P. J., Guan, B., Ryoo, J. M., & Wick, G. A. (2013). Effects of atmospheric river landfalls on the cold season precipitation in California. *Climate Dynamics, 40*(1), 465–474.

Negm, A., Bouderbala, A., Chenchouni, H., & Barcelo, D. (2020). *Water resources in Algeria—Part I: Assessment of surface and groundwater*. Springer.

Patel, N. R., Chopra, P., & Dadhwal, V. K. (2007). Analyzing spatial patterns of meteorological drought using standardized precipitation index. *Meteorological Applications: A Journal of Forecasting, Practical Applications, Training Techniques and Modelling, 14*(4), 329–336.

Zhu, Y., & Newell, R. E. (1998). A proposed algorithm for moisture fluxes from atmospheric rivers. *Monthly Weather Review, 126*(3), 725–735.

Spatiotemporal Analysis of Extreme Precipitation Characteristics for Prairie Region of Canada

M. Monirul Qader Mirza, William A. Gough and Zahra Noorisameleh

Abstract

The Canadian Prairies are identified as a potential climate change "hotspot" due to their relatively high latitude and central location on the continent. Agriculture is very vital in the southern part of the region, particularly the Prairie's ecozone. Droughts are a frequent visitor in this region and so as floods from convective heavy precipitation events. Over a 24-h period, one single extreme precipitation event could deliver more than 25% of the mean annual precipitation. The behavior of extreme precipitation poses many challenges to flood management, adaptation, and coping mechanisms. The goal of this research is to uncover temporal and spatial aspects of 24-h maximum precipitation in the Canadian Prairie region, including change points, trends, long-term persistence, spatial correlation, upper tail properties, and any clustering to facilitate adaptation and coping mechanisms to extreme floods. 24-h precipitation data collected by tipping bucket rain gauges (TBRG) for 26 stations distributed across Alberta, Manitoba, and Saskatchewan provinces were used. We have investigated abrupt changes in the first two moments of the distribution by applying the Pettit test. The lag-1 (r1) serial correlation coefficient was used to calculate the auto- or serial correlation in the extreme rainfall datasets. The pre-whitening scheme was applied where there was statistically significant serial correlation before trend analysis. Nonparametric Mann–Kendall (MK) test and Theil-Sen's slope estimator were applied for trend detection at 5% significance level. The presence of long-term persistence determined by estimating the Hurst exponent. Statistical modeling was conducted by applying the generalized extreme value (GEV) distribution. The results of the trend analysis suggest that the 24-h extreme precipitation trend is decreasing. Hurst exponent of > 0.5 was found in 65% of the stations demonstrating long-term persistence, while the remaining 35% showed no long-term persistence. In general, 24-h extreme precipitation in the Prairie region is heavy tailed as found in the GEV analysis. All three GEV parameters demonstrate a pronounced decreasing trend from the north to the south while plotted against latitude. Plotted map of 100-year return period values displays three clusters of high precipitation and flood vulnerable areas. One is in the northwest of Alberta and the other two in the South of Saskatchewan and Manitoba.

Keywords

Characteristics · Climate change · Extreme precipitation · Frequency and Prairies

1 Introduction

The Prairie region of Canada, a potential climate change "hotspot" stretches from southern Alberta's Rocky Mountains, through Saskatchewan, and into Manitoba's Red River basin. Internationally, the Canadian Prairies are identified as a potential climate change "hotspot" due to their relatively high latitude and central location on the continent. Agriculture is very vital in the southern part of the region, particularly the Prairie's ecozone (Wheater & Gober, 2013). Precipitation is highly influenced by air circulation and mid-troposphere ridge (Hogg et al., 2000). Droughts are a frequent visitor in this region and so as floods from convective heavy precipitation events. Over a 24-h period, one single extreme precipitation event could deliver more than 25% of the mean annual precipitation. The behavior of extreme precipitation poses many challenges to flood management, adaptation, and coping mechanisms. The goal of

M. M. Q. Mirza (✉) · W. A. Gough · Z. Noorisameleh
University of Toronto Scarborough, Toronto, ON M1C 1A4, Canada
e-mail: monirul.mirza@utoronto.ca

H. Chenchouni et al. (eds.), *Recent Advancements from Aquifers to Skies in Hydrogeology, Geoecology, and Atmospheric Sciences*, Advances in Science, Technology & Innovation, https://doi.org/10.1007/978-3-031-47079-0_58

this research is to uncover temporal and spatial aspects of 24-h maximum precipitation in the Canadian Prairie region, including change points, trends, long-term persistence, spatial correlation, upper tail properties, and any clustering to facilitate adaptation and coping mechanisms.

2 Data and Methods

We have extracted 24-h precipitation data collected by tipping bucket rain gauges (TBRG) for 26 stations (Fig. 1) maintained by Environment and Climate Change Canada (ECCC) distributed across Alberta, Manitoba, and Saskatchewan provinces. We used frequency analysis to find the most common ranges and outliers in the precipitation data. We have investigated abrupt changes in the mean of the distribution by applying the Pettit test (Pettit, 1979). We have applied nonparametric Mann–Kendall (MK) test and Theil-Sen's slope estimator for trend detection. Durbin-Watson test was applied to check significance of autocorrelation. In presence of statistically significant serial correlation, the pre-whitening scheme (Serinaldi & Kilsby, 2016) was applied. Long-term memory or persistence property could lead to prolonged droughts or the temporal clustering of extreme floods (Gil-Alana et al., 2022; Mudelsee, 2006). We have examined the presence of long-term persistence by estimating the Hurst exponent (Ledvinka & Recknagel, 2020). Statistical modeling of 24-h extreme precipitation series is conducted by applying the generalized extreme value (GEV) distribution is defined as

$$F(x) = \exp\left\{-\left(1+\xi\left(\frac{x-\mu}{\sigma}\right)\right)^{-1/\xi}\right\}.$$

The GEV is a three-parameter distribution. They are: μ is the location parameter that represents "center" of the distribution; σ is the scale parameter governs the "size" of the deviations about μ; and ξ is the shape parameter which strictly affects the shape of the distribution and governs the tail of each distribution. For $\xi=0$, $\xi>0$, and $\xi<0$, the GEV becomes a light-tailed Gumbel, a heavy-tailed or Fréchet and a bounded tail or Weibull type distribution, respectively (Gupta & Chavan, 2021; Katz et al., 2005). Moran's I (Fu et al., 2014) was used to examine spatial correlation of the GEV parameters.

3 Results

We conducted a province-wide frequency analysis of 0–25, 25–50, 50–75, 75–100, 100–125, 125–150, and 150–175 mm extreme precipitation. We discovered that precipitation in the 25–50 mm range was dominating in all three provinces. The 50–75 mm range, on the other hand, is most prevalent in Manitoba followed by Saskatchewan and Alberta. Extreme precipitation in the 100–125 mm range was more pronounced in Saskatchewan and Manitoba than in Alberta. Saskatchewan experienced precipitation between 125 and 150 mm and 150–175 mm on one occasion each. The results of the trend analysis suggest that the 24-h extreme precipitation trend is decreasing. A Hurst exponent of>0.5 was found in 65% of the stations that demonstrated long-term persistence, while the remaining 35% did not (Fig. 2). The scale and shape parameters do not display any significant spatial correlation as concluded from the Moran's I values. The analysis of the shape parameters reveals that in 31% of the stations, precipitation distributions are bounded and remaining 69% displays heavy tail behavior.

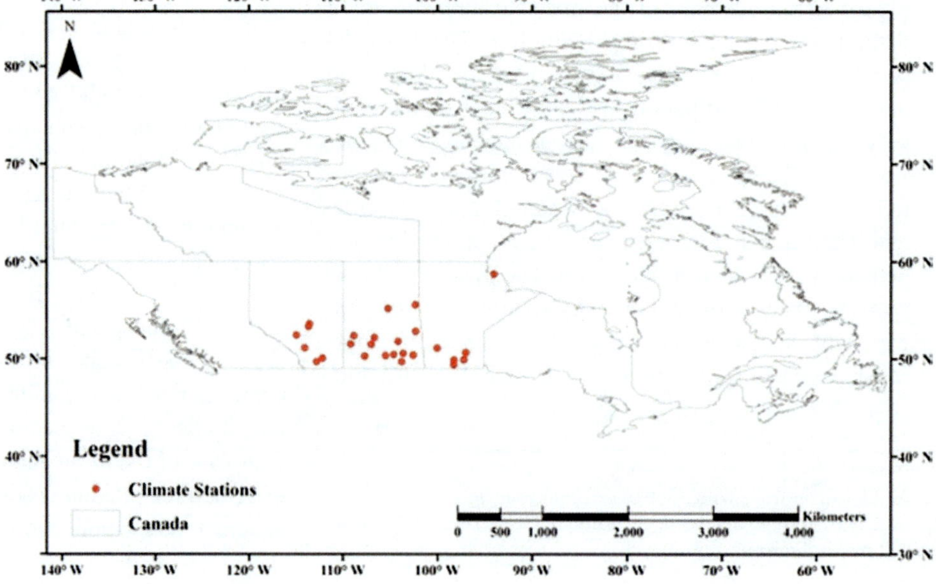

Fig. 1 Selected 26 climatic stations in the Prairie region

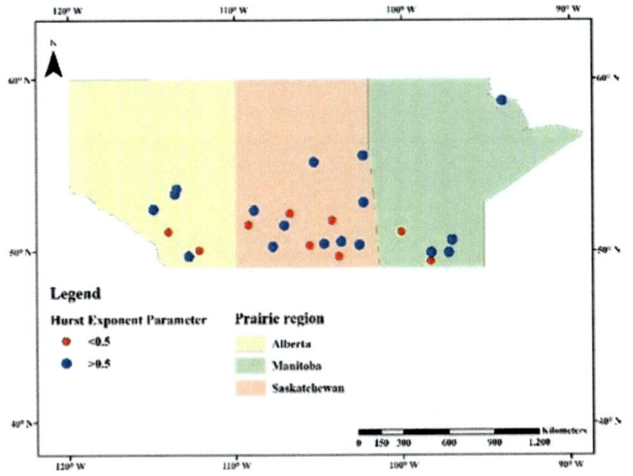

Fig. 2 Spatial distribution of Hurst exponent values

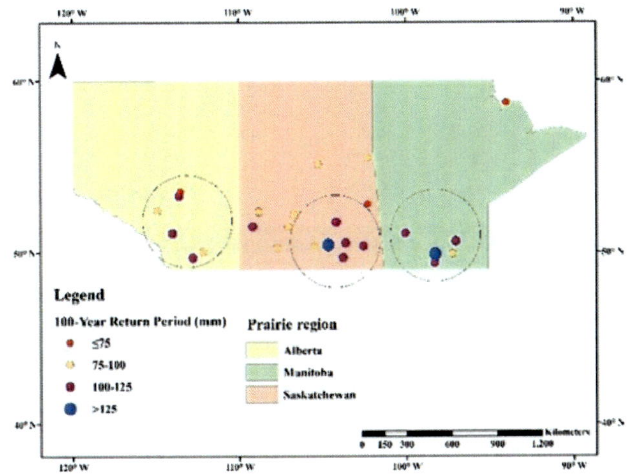

Fig. 3 Spatial distribution of 100-year return period extreme precipitation. High precipitation clusters are displayed with the circles

4 Discussion

Smaller range of extreme precipitation is dominant in the Prairie Region. Overall, the 24-h extreme precipitation trend is decreasing although some stations show increasing trends in the three Prairie Provinces. 24-h extreme precipitation in Alberta and Manitoba are more characterized by long-term persistence than that of Saskatchewan. The possibility of a high value being followed by another high value is one of the implications of long-term persistence. Low values may have a similar pattern of occurrence. In general, 24-h extreme precipitation in the Prairie region is heavy tailed. With altitude, shape parameters display an increasing trend up to certain altitude threshold then show a decreasing trend. Plotted map of 100-year return period values displays three clusters of high precipitation and flood vulnerable areas (Fig. 3).

5 Conclusion

In general, 24-h extreme precipitation in the Prairie region is heavily tailed, implying greater variability. Clustering will help to focus areas for flood management.

References

Fu, W. J., Jiang, P. K., Zhou, G. M., & Zhao, K. L. (2014). Using Moran's I and GIS to study the spatial pattern of forest litter carbon density in a subtropical region of Southeastern China. *Biogeosciences, 11*, 2401–2409.

Gil-Alana, L. A., Gupta, R., Sauci, L., & Carmona-González, N. (2022). Temperature and precipitation in the US states: Long memory, persistence, and time trend. *Theoretical and Applied Climatology, 150*, 1731–1744.

Gupta, N., & Chavan, S. R. (2021). Assessment of temporal change in the tails of probability distribution of daily precipitation over India due to climatic shift in the 1970s. *Journal of Water and Climate Change, 12*(6), 2753–2773.

Hogg et al. (2000). Postulated feedbacks of deciduous forest phenology on seasonal climate patterns in the Western Canadian Interior. *Journal of Climate, 13*, 4229–4243.

Katz et al. (2005). Statistics of extremes: Modeling ecological disturbances. *Ecology, 86*(5), 1124–1134.

Ledvinka, O., & Recknagel, T. (2020). Long-term persistence in discharge time series of mountainous catchments in the Elbe River basin. *Proceedings of IAHS, 383*, 135–140. https://doi.org/10.5194/piahs-383-135-2020

Mudelsee, M. (2006). Long memory of rivers from spatial aggregation. *Water Resources Research, 43*. https://doi.org/10.1029/2006WR005721

Pettitt, A. N. (1979). A non-parametric approach to the change-point problem. *Applied Statistics, 28*(2), 126–135.

Serinaldi, S., & Kilsby, G. S. (2016). The importance of prewhitening in change point analysis under persistence. *Stochastic Environmental Research and Risk Assessment, 14*, 763–777.

Wheater, H. S., & Gober, P. (2013). Water security in the Canadian prairies: Science and management challenges. *Philosophical Transactions of the Royal Society A*. https://doi.org/10.1098/rsta.2012.0409

Woren, R. (2002). Estimating long-range dependence: Finite sample properties and confidence intervals. *Physica A: Statistical Mechanics and Its Applications, 312*(1–2), 285–299.

Predicting Rainfall Onset and Cessation Within the West African Sahel Region Using Echo State Network

Adeyemi Olusola, Samuel Ogunjo and Christiana Olusegun

Abstract

Rainfall onset and cessation are essential factors for the agrarian society in the West African Sahel region. This information determines crops to plant, helps project crop yields, and reduces loss due to climatic factors. Rainfall onset has also been associated with disease outbreaks; adequate knowledge will help in epidemiology and health management schemes. Therefore, predicting the onset will be necessary for the economy of countries in the Sahel region. Historical daily rainfall data from 1901 to 2000 for eight locations within the Sahel region were obtained from the Inter-Sectoral Impact Model Intercomparison Project (ISIMIP) at a $0.5° \times 0.5°$ grid resolution. Rainfall onset and cessation were determined using the Liebmann method based on long-term precipitation in the location of interest. Onset and cessation were predicted using a machine learning approach—Echo State Network. The root mean square error for rainfall onset was 15–24 days, while a range of 9–19 days was obtained for rainfall cessation. Results obtained in this study suggest that it is possible to effectively predict rainfall onset and cessation within the Sahel region. The predictions will enable farmers and government agencies to address agricultural, health, and environmental issues associated with the onset and cessation of rainfall in the region.

A. Olusola (✉)
Faculty of Environmental and Urban Change, York University, Toronto, ON, Canada
e-mail: aolusola@yorku.ca; olusolaadeyemi.ao@gmail.com

S. Ogunjo
Department of Physics, Federal University of Technology, Akure, Nigeria

C. Olusegun
West Africa Science Service Centre On Climate Change and Adapted Land Use, Federal University of Technology Akure, Gaga 340001, Ondo State, Nigeria

Faculty of Physics, Institute of Geophysics, University of Warsaw, 00-927 Warsaw, Poland

Keywords

Echo State Network · Sahel · Rainfall onset · Climate change

1 Introduction

The onset and cessation of rainfall are critical hydro-climatological events within the Sahel region of West Africa. It is crucial for the planning and management of farming practices, the outbreak of diseases, and other environmental challenges. Therefore, precise forecast or prediction of these hydro-climatological events is crucial to human livelihood and a sustainable ecosystem. There have been attempts to model and predict onset and cessation using traditional methods, the Fourier series, and, more recently, various emerging artificial intelligence algorithms (Adeola et al., 2021; Fuwape & Ogunjo, 1811; Olusola et al., 2022). Artificial intelligence and machine learning provide efficient tools for modelling and predicting spatiotemporal and have been successfully applied in various hydro-climatological studies (Fuwape & Ogunjo, 1811; Ogunjo et al., 2022; Ojo & Ogunjo, 2022). This study considered the potential of predicting rainfall onset and cessation using a novel machine learning algorithm. The results will help precision agriculture planning and strategic economic management plans across West Africa.

2 Methodology

The study locations (Fig. 1) were chosen within the Sahel region of West Africa (10°–30° N, 30° W–20° E). Precipitation records between 1901 and 2000 over the

H. Chenchouni et al. (eds.), *Recent Advancements from Aquifers to Skies in Hydrogeology, Geoecology, and Atmospheric Sciences*, Advances in Science, Technology & Innovation, https://doi.org/10.1007/978-3-031-47079-0_59

region were obtained from the archives of the Inter-Sectoral Impact Model Intercomparison Project (ISIMIP) (https://www.isimip.org/gettingstarted/details/4/) (Dirmeyer et al., 2006) at a spatial resolution of $0.5 \times 0.5°$. Long-term annual and decadal trends were estimated using Mann–Kendall and Sen Slope analysis. Onset and cessation were estimated using the approach of Liebmann and Marengo (2001). The Echo State Network (ESN) is a supervised learning algorithm based on recurrent neural networks. Details about ESN can be obtained from Jaeger (2007). The algorithm can be summarized as

$$HC(t) = \rho_0 C_p \int_{D_{16}}^{\eta} T(z,t)\mathrm{d}z \quad (1)$$

Fig. 1 Map of the study region with locations of interest

Echo State Networks are recurrent networks. *fl* is a nonlinear function that makes the current state dependent on the previous state and the current input.

3 Results and Discussion

The yearly rainfall onset and cessation days (Julian) for all the locations under consideration are shown in Figs. 2 and 3, respectively. Variation in rainfall onset across the stations shows the spatial distribution for onset and cessation within the sub-region. For example, Abadji and Abadam stations received the earliest rainfall across the studied within 20 days, followed by Achouadene, while Bakao and Boumboum received rainfall within 50 days or more. For cessation, Bakao and Boumboum, despite having late onset, were also the stations to have the earliest cessation in rains. Observations here align with other West African studies, where reasons for delayed onset and early cessation have been attributed to drought, local anthropogenic activities, and the changing climate (Laux et al., 2008; Omotosho et al., 2000).

The Echo State Network was used to predict the onset and cessation for the last two years and compared with in situ data. The RMSE values for the onset prediction were 15–24 days, while a range of 9–19 days was obtained for

Fig. 2 Rainfall onset across the selected stations

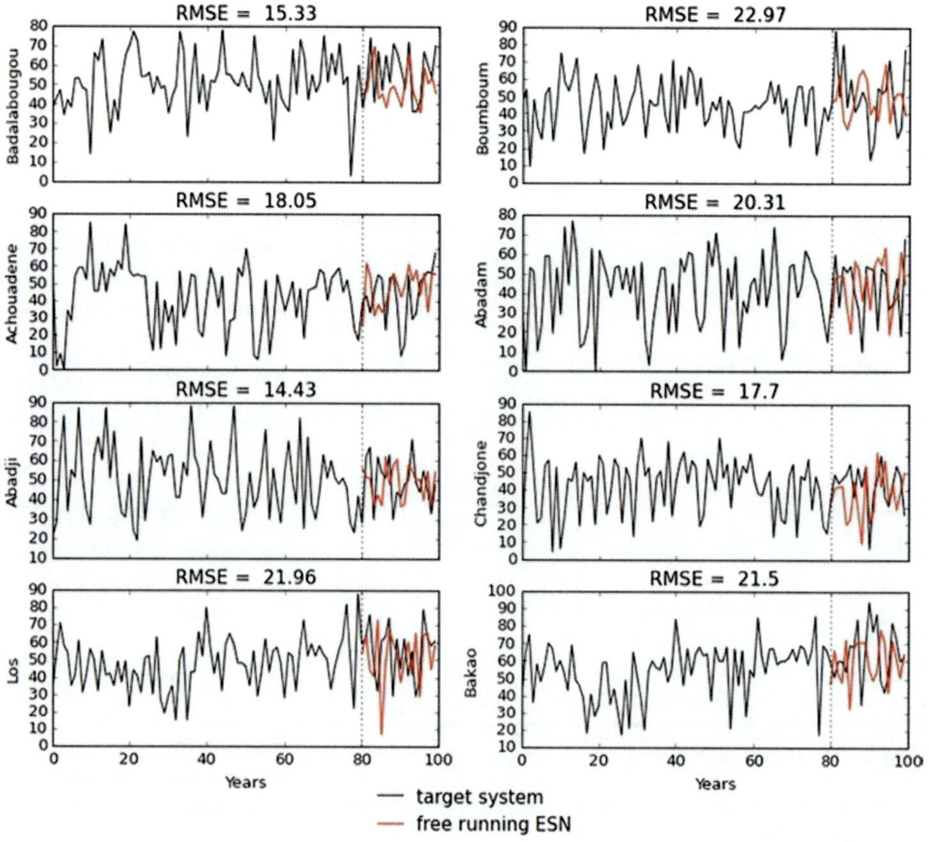

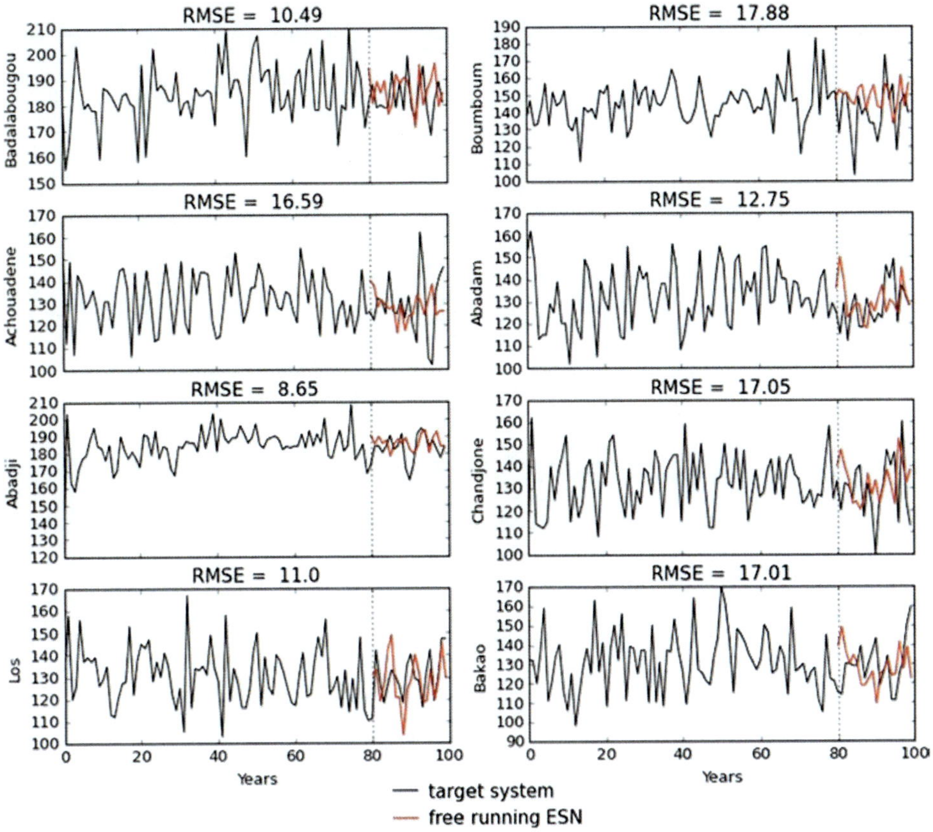

Fig. 3 Rainfall cessation across the selected stations

rainfall cessation. The implication is that the model can predict onset and cessation with errors less than 24 and 19 days, respectively. In most locations, the ESN model captured the general trends of onset and cessation days.

4 Conclusion

In this study, we have demonstrated the capabilities of a machine learning model—ESN in predicting rainfall onset and cessation within the Sahel region of West Africa. The approach will provide information about planting and harvest time well ahead of time with reasonable accuracy. Results from this study can also be used for disease surveillance and control and for designing hydrological budgets for the region.

References

Adeola, O. A., Adeyemi, O., & Onyemaenu, V. (2021). Rainfall–runoff in conterminous tropical river basins of Southwestern Nigeria. *African Geographical Review, 42*, 1–16.

Dirmeyer, P. A., Gao, X., Zhao, M., Guo, Z., Oki, T., & Hanasaki, N. (2006). GSWP-2: Multimodel analysis and implications for our perception of the land surface. *Bulletin of the American Meteorological Society, 87*(10), 1381–1398.

Fuwape, I. A., & Ogunjo, S. T. (2018). Modeling of raining season onset and cessation of tropical rainfall for climate change adaptation in Agriculture. arXiv preprint arXiv:1811.09677

Jaeger, H. (2007). Echo state network. *Scholarpedia, 2*(9), 2330. https://doi.org/10.4249/scholarpedia.2330

Laux, P., Kunstmann, H., & Bárdossy, A. (2008). Predicting the regional onset of the rainy season in West Africa. *International Journal of Climatology: A Journal of the Royal Meteorological Society, 28*(3), 329–342.

Liebmann, B., & Marengo, J. (2001). Interannual variability of the rainy season and rainfall in the Brazilian Amazon Basin. *Journal of Climate, 14*(22), 4308–4318.

Ogunjo, S. T., Dada, J. B., & Ajayi, O. J. (2022). Forecasting tropospheric wet delay using LSTM neural network. In *IOP conference series: Earth and environmental science* (Vol. 993, No. 1, p. 012024). IOP Publishing.

Ojo, O. S., & Ogunjo, S. T. (2022). Machine learning models for prediction of rainfall over Nigeria. *Scientific African, 16*, e01246.

Olusola, A. O, Olumide, O., Fashae, O., & Adelabu, S. (2022). River sensing: The inclusion of red band in predicting reach-scale types using machine learning algorithms. *Hydrological Sciences Journal, 67*, 1740–1754.

Omotosho, J. B., Balogun, A. A., & Ogunjobi, K. (2000). Predicting monthly and seasonal rainfall, onset and cessation of the rainy season in West Africa using only surface data. *International Journal of Climatology: A Journal of the Royal Meteorological Society, 20*(8), 865–880.

Ice Nuclei Activity in Stalagmites: A Case Study of Medieval Climate Optimum and Little Ice Age Periods

F. L. T. Gonçalves, R. I. F. Trindade, A. P. E. Mendes, J. A. Do Carmo, P. F. Jaqueto and C. E. Morris

Abstract

Primary biological aerosol particles (PBAP, also called bioaerosols) are airborne particles from integral or pieces of biological sources. The most abundant PBAP are bacteria, which have been shown in laboratory studies to have shown efficient ice nucleation activity (INA) in the atmosphere. Therefore, a significant amount of INA bacteria was found at ground level and in soil, indicating that those bacteria would percolate the soil layers to the underground and caves. Earth's climate presented a high variability last millennia with two very distinct periods: the optimal medieval, from X to XII century (900–1100 DC), here after called Medieval Climate Optimum (MCO) and the Little Ice Age, from the XVI to XIX centuries (1560–1800), referred to as LIA. The MCO presented warmer climates, with less rainfall in Central Brazil and higher temperatures associated with more savanna types. In turn, LIA presented colder climates, with more rainfall in Central Brazil dominated by C3 plants and trees. Both periods are represented by a stalagmite from the Pau D'Alho cave, located at Central Brazil. The study goal aims to verify IN activity on the different layers of the Pau D'Alho stalagmite, focusing on MCO and LIA periods. This is a first study verifying INA in a stalagmite, for which it was necessary to develop a new methodology to approach it. The methodology was performed through smashing the stalagmite and taking the solution sample to a freezing bath, with and without diluting the preparation. The first results show, for temperatures warmer than − 10 °C, MCO presented 15, frozen tubes against only 6 from LIA, both without dilution. If they are bacterial, they likely were moved from land surfaces via percolation. Tests are colder temperatures might be effective in differentiating the activity in MCO stalagmites from that in LIA stalagmites.

Keywords

Optimal medieval · Little Ice Age · Ice nuclei activity · Stalagmites

1 Introduction

Primary biological aerosol particles (PBAP, also called bioaerosols) are airborne particles from integral or pieces of biological sources (Morris et al., 2011) which can be found worldwide. The most abundant PBAP are bacteria, where have been shown in laboratory studies some of them has shown efficient ice nuclei activity (INA) in the atmosphere which may impact the global distribution of clouds and precipitation, profoundly linked with the latter as their life cycle. A significant amount of INA bacteria is found in the ground and soil. Those bacteria that could percolate into the underground soil layers, reaching caves. In a karstic environment, the percolated water transports trace element that will be incorporated into speleothems when this solution reaches the caves and degassing occurs. Interestingly, studied caves in central Brazil show moisture deficit and wet conditions for the last two millennia through oxygen isotopes, using U-Th method. In these last two millennia, Earth's climate faced a high variability with two very distinct periods: the optimal medieval, from the tenth to twelfth century (900–1100 CE, here after called Medieval Climate Optimum (MCO) and the Little Ice Age, from the sixteenth to nineteenth centuries (1620–1820 CE), also

F. L. T. Gonçalves (✉) · R. I. F. Trindade · A. P. E. Mendes · P. F. Jaqueto
University of São Paulo, São Paulo, Brazil
e-mail: fabio.goncalves@iag.usp.br

J. A. Do Carmo
Institute for Rock Magnetism, University of Minnesota, Minneapolis, USA

C. E. Morris
INRAE, Paris, France

H. Chenchouni et al. (eds.), *Recent Advancements from Aquifers to Skies in Hydrogeology, Geoecology, and Atmospheric Sciences*, Advances in Science, Technology & Innovation, https://doi.org/10.1007/978-3-031-47079-0_60

called the LIA (Vuille et al., 2012). The MCO presented warmer climates, with less rainfall in Central Brazil and higher temperatures associated with more savanna types (Novello et al., 2016). In turn, the LIA presented colder climates, with more rainfall in Central Brazil dominated by C3 plants and trees.

Those differences may lead to the suggestion that INA bacteria could be found in more abundance during the LIA rather than during MCA. Consequently, this study aims to verify possible INA activity in the stalagmites from Pau D'Alho cave in Central Brazil, using a home-made freezer that assured stable temperatures around each aliquot that was tested, as a first approach.

Table 1 Number of frozen samples with two temperature thresholds, totaling 56 for each period

	Cumulative number of frozen tubes at − 8 °C	Cumulative number of frozen tubes at − 9 °C
	− 8 °C	− 9 °C
LIA 1	0	0
LIA 2	4	5
LIA 3	0	1
MCO1	3	8
MCO2	3	7
MCO3	3	7

2 Methods

The stalagmite from Pau D'Alho cave is located in central Brazil where two samples were extracted from the same one. The first one came from the LIA period, corresponding to the seventeenth century (~ 1640 AD), a typical sample of that cold and wet period, keeping it its nomination. The second one came from the MCA, from the tenth century (~ 970 AD), also a typical MCA. From those two samples, a mesh was made with a powder of < 105 μm. After that, the powder was suspended in commercial 0.9% NaCl (saline) to 5 mg/mL. Two first tests were performed with the stalagmite powder: without dilution and with a tenfold dilution. The freezing tests were carried out with volumes of 150 μl for each Eppendorf tube. Those tubes were placed inside a Peltier cooler, with a drilled hole for each tube. The first test focused on about 200 aliquots of each dilution for each time period. [LIA undiluted (WD) and diluted (D), and MCO WD and D]. Then, the tubes were thawed and divided into two groups of equal number that were either boiled or did not receive an additional treatment.

3 Results

The results (Table 1) suggest that the wetter and colder period has less IN activity, corresponding to around 10% of total tubes during LIA and ~ 40% during MCO, warmer and drier.

4 Discussion

4.1 MCO Versus LIA Results

The idea of having more bacteria in the climatic anomaly of the medieval time from the records of speleothems makes sense with the interpretation of less plant cover associated with less rainfall. The smaller total rainfall amounts, though with few intense events, could take all the detritus material from the soil, that would be of sparser quantity, faster to the stalagmites, a sort of "washing" of the soii. This would allow more interaction of the bacteria with the soil, and thus a higher rate of preservation of the bacteria. In the case of LIA, the residence time of rainwater may be higher (greater water availability). The mixing solution with soil pH could destroy and reduce the presence of the bacteria, in this case, with INA.

4.2 Boiled Versus Un-boiled Results

Not shown here, but the results still present a higher INA during MCO (13 frozen tubes) than LIA (2 frozen tubes), from a total of 56. The difference can be attributed to the storage processes. Secondly, the boiled and un-boiled tests have the same results, indicating that there is no biological/organic material involved. The calcification of the cell membranes and all bacteria organic structure (perhaps fossilized) may preserve the ability of INA, though without any biological structure.

5 Conclusion

Therefore, the main conclusion is that during LIA there is lower INA rather than MCO due to the soil pH being affected by higher rainfall amounts at Little Ice Age. This is a very preliminary study, which demands further research and a larger amount of data. It is a new field with a new approach, trying to explain how the INA role in the past since climate can be affected by their distribution, and how climate can also drive different INA concentrations. The next step is to analyze the crystal shapes that may be found in the mesh which cause ice formation. A hypothesis is based on the fact that the hexagonal structure of the bacteria membrane can be fossilized, keeping the shape in the limestone. To answer that, high-resolution microscopy will be performed in the next step.

References

Morris, C. E., Sands, D. C., Bardin, M., et al. (2011). Microbiology e atmospheric processes: Research challenges concerning the impact of airborne micro-organisms on the atmosphere e climate. *Biogeosciences, 8*, 17–25.

Novello, V. F., Vuille, M., Cruz, F. W., et al. (2016). Centennial scale solar forcing of the South American Monsoon System recorded in stalagmites. *Scientific Reports, 6*(1), 1–8.

Vuille, M., Burns, S. J., Taylor, B. L., et al. (2012). A review of the South American Monsoon history as recorded in the stable isotopic proxies over the past two millennia. *Climate of the Past, 8*(4), 1309–1321.

Statistical Analysis of Climate Variability and Prediction of Rainfall in Morocco

Kaoutar Bargach and Soufiana Mekouar

Abstract

The worldwide variations such as climate crisis and global warming are influencing considerably living species and earth due to their implication in the severity and rate of extreme weather. Thus, the seasonal precipitation prediction is of great importance due to the geographical, geomorphological, and seasonal characteristics of Atlantic and Mediterranean positions showing significant events impacting rainfall. In this research, we present a precipitation forecasting model based on multivariate time series where several factors affect the accuracy performance to mitigate extreme events induced by climate change. The purpose is to study the relationship between several climate variables, including surface pressure, wind speed, relative humidity, corrected precipitation, and temperature range, influencing precipitation and the design of an effective multivariate forecasting model showing the temporal evolution of different climate factors. Additionally, the prediction model was designed using a vector autoregression model and tested on monthly data collected from the NASA portal between 1981 and 2021. The introduced model extracts the variations of the time series characteristics and the interactive relationship of several related variables. We performed the Augmented Dickey-Fuller (ADFuller) test for stationarity and the cointegration causality test. The accuracy of the rainfall forecast model was evaluated using regression metrics, including mean absolute error, mean squared error, and root mean squared error, which showed effectiveness in predicting the monthly rainfall of Morocco. Our results showed a decrease in rainfall and a variation in the aridity index in Morocco between 1981 and 2021, reaching a peak in 1996 with a drought index value of 24.25 (semi-humid climate) and a minimum value in 2017 with an aridity index of 5.57 (arid climate). We conclude that during the twentieth and the beginning of the twenty-first centuries (1981–2021), Morocco experienced a change in climate levels from semi-humid to semi-arid and currently knows arid climate. Our model can capture real-world behavior, providing better forecasting performance, while the choice of the best lag value and the split interval of the time series is a challenging task.

Keywords

Climate change · Drought · Precipitation · Aridity index · Variability · Vector autoregression model · Climate levels

1 Introduction

The earth has known along the geological air, an alternation of climatic periods of warming and glaciation. The world is experiencing a climate crisis with considerable influence on all living species and continents, including Morocco. Morocco's climate is of the predominantly arid Mediterranean type. It is relatively contrasting, influenced by the Atlantic Ocean, the Mediterranean Sea, and the Sahara. Morocco is one of the countries most threatened by climate change (Battisti & Naylor, 2009; Hulme et al., 1999; Rifai et al., 2014). Studies on climate change in Morocco have demonstrated the presence of two seasons of equal length, the first wet and slightly cold (November to April) and the second dry and hot (May to October) (Houssa et al., 2017). As well as other works have shown that the cumulative annual rainfall has strongly decreased (Woillez, 2019). Several Moroccan regions have experienced water stress in the last decades, such as

Kaoutar Bargach and Soufiana Mekouar: These authors contributed equally.

K. Bargach (✉) · S. Mekouar
Scientific Institute of Rabat, Mohammed V University in Rabat, Rabat, Morocco
e-mail: kaoutar.bargach@is.um5.ac.ma

H. Chenchouni et al. (eds.), *Recent Advancements from Aquifers to Skies in Hydrogeology, Geoecology, and Atmospheric Sciences*, Advances in Science, Technology & Innovation, https://doi.org/10.1007/978-3-031-47079-0_61

Sous-Massa-Draa, the Saharan, and eastern regions showing a real threat to the agricultural sector (Woillez, 2019). In this study, we used the statistical model of vector autoregression (VAR) to capture the relationship between multiple quantities as they change over time. The VAR is a type of stochastic process model. We focused on this research study on the variation of precipitation in Morocco, taking the whole country for analysis to take the entire picture of the climate events that impact precisely the rainfall.

2 Methodology

We model our forecasting model using the vector autoregression multivariate time series model (Lütkepohl, 2013; Zivot & Wang, 2006) to investigate the dynamic relationships between climate variables impacting each other of order o denoted as VAR (o):

The VAR consists of an ensemble of K endogenous variables $y_t = \{y_t, y_{2t}, \ldots, y_{kt}\}$. . The VAR (o) serves to forecast the current climate variable observations with past observations of itself and past observations of other variables involved in the climate forecasting model. Thus, our model can be written as follows:

$$y_t = \delta + \sum_{i=1}^{\theta} \varphi_i y_{t-i} + \gamma_t$$

with y_t refers $(m \times n)$ endogenous climate variables and y_{t-i} is the i th lag value of y_t .

δ is the intercept $(m \times 1)$ vector of VAR (θ) and φ_i is the time invariant $(m \times n)$ matrix of autoregressive coefficients.

γ_t is the $(m \times 1)$ generalization of white noise vector in which γ_t should satisfies the following constraints: $E[\gamma_t] = 0, E[\gamma_{(t)}^t \gamma_t] = \vartheta, E[\gamma_{(t)}^t \gamma_{t-m}] = 0$, where ϑ is a positive-semidefinite matrix.

We used the de Martonne aridity index, which is computed as follows (Ionac et al., 2015):

$$I = \frac{P}{T + 10},$$

where P refers to total annual precipitation and T is the annual average temperature.

3 Experiments

We use a monthly five-variable VAR collected from the NASA portal https://power.larc.nasa.gov/ (Bosilovich et al., 2015) ranging between January 15, 1981 and October 30, 2021 to investigate the relationships between surface pressure PS, wind speed WS, relative humidity RH2M, corrected precipitation PRECTOTCORR, and temperature at 2-m range T2M_RANGE.

The VAR model reports the dynamic relationships among stationary climate variables. We use the Augmented Dickey-Fuller test ADF (Mushtaq, 2011), which uses a combination of T-statistics and F-statistics to identify the existence of a unit root in time series. Our results show the ADF test on all the statistics is significant at the 5% level. Thus, we can be confident that the climate time series is stationary.

We fit a VAR model with a regression between time series using ordinary least squares (OLS) regression (Craven & Islam, 2011) and select the order of the VAR (o) based on the information criterion of Akaike (AIC) (Sakamoto et al., 1986). So, we split the time series as training and test data and trained the model with a lag of 6 for rainfall prediction. We evaluate our precipitation forecast model with the bias, mean absolute error MAE, root mean square error MSE, root mean square error RMSE, and R2. Next, we evaluate the VAR forecasting model by predicting rainfall using various regression metrics that showed some performance in terms of accuracy in all metrics, as given in Table 2.

Table 1 depicts the granger causality (Diks & Panchenko, 2006) of climate variables and precipitation. Thus, we conclude that all series contribute to causing the rainfall and that the surface pressure showed significance with test statistic 22.29 and p-value inferior to 0.05 in causing precipitation variation.

Figure 1 shows Morocco's historical and future rainfall using the proposed model. We observe a decrease in precipitation over the next ten years from 2021 to 2031, if we do not put in place an environmental strategy that can alleviate the impact of climate change. In 1987, we can observe a peak in precipitation reaching around 10.86 mm with an aridity index of 21.49 (semi-humid climate), compared to previous and upcoming years. However, in 2017,

Table 1 Granger causality test on climate time series

Granger-cause	Test statistic	Critical value	p-value
PS → PRECTOTCORR	22.290	12.592	0.001
WS2M → PRECTOTCORR	10.313	12.592	0.112
RH2M → PRECTOTCORR	7.381	12.592	0.287
T2M_Range → PRECTOTCORR	5.792	12.592	0.447
All → PRECTOTCORR	107.378	36.415	<0.001

Table 2 Evaluation performance metrics on Morocco climate test data

	Bias	MAE	MSE	RMSE	$R2$
Forecasting model	−0.33	0.59	0.46	0.68	0.28

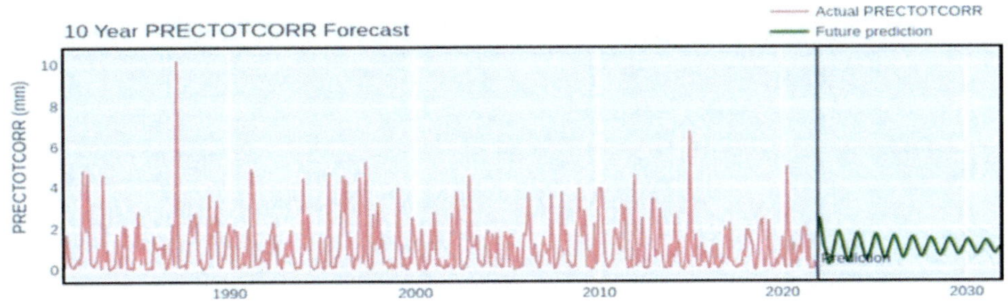

Fig. 1 Future ten-year rainfall forecast using VAR model

Morocco was known for low rainfall with an aridity index of 5.57(arid climate).

4 Conclusion

In this paper, we designed a vector autoregressive model with five-time series and evaluated it to study the relationship between precipitation, relative humidity, surface pressure, and other climate series. The high temperature in the cold season may be the principal cause of the precipitation decrease besides atmospheric conditions and ocean temperature. The granger causality test shows that all series impact precipitation, so each climate series contributes to the rainfall variation. Furthermore, our forecasting methodology can predict the rainfall on test and ten-year future rainfall showing that Morocco will witness a decrease in precipitation. Moreover, Morocco has an arid climate with a variation in the aridity index between 1981 and 2021.

References

Battisti, D. S., & Naylor, R. L. (2009). Historical warnings of future food insecurity with unprecedented seasonal heat. *Science, 323*(5911), 240–244.

Bosilovich, M. G., Lucchesi, R., & Suarez, M. (2015). *MERRA-2: File specification*. No. GSFC-E-DAA-TN27096.

Craven, B. D., & Islam, S. M. (2011). Ordinary least-squares regression. *The SAGE dictionary of quantitative management research*, 224–228.

Diks, C., & Panchenko, V. (2006). A new statistic and practical guidelines for nonparametric Granger causality testing. *Journal of Economic Dynamics and Control, 30*(9–10), 1647–1669.

Houssa, A. A., Drissi, S., Asehraou, A., Asfers, A., Oubaki, L., & Chraibi, H. (2017). Changements climatiques au Maroc: quels systèmes de culture et quelles biotechnologies pour s'y adapter?. *Revue Marocaine des Sciences Agronomiques et Vétérinaires, 5*(3).

Hulme, M., Barrow, E. M., Arnell, N. W., Harrison, P. A., Johns, T. C., & Downing, T. E. (1999). Relative impacts of human-induced climate change and natural climate variability. *Nature, 397*(6721), 688–691.

Ionac, N., Grigore, E., & Constantin, D. (2015). Évaluation des phénomènes de dessèchement et de sécheresse dans la zone continentale du plateau de la Dobroudja du Sud. In XXVIIIe Colloque de l'Association Internationale de Climatologie (pp. 269–274).

Lütkepohl, H. (2013). Vector autoregressive models. *Handbook of Research Methods and Applications in Empirical Macroeconomics*. 139–164.

Mushtaq, R. (2011). Augmented dickey fuller test. SSRN Electron. J. 1911068. https://doi.org/10.2139/ssrn.1911068

Rifai, N., Khattabi, A., & Rhazi, L. (2014). Impact du changement climatique sur les températures journalières minimales et maximales et les précipitations de la région de Tanger (nord-ouest du Maroc) quantifié à partir d'une méthode de descente d'échelle statistique. *VertigO, 14*(3): 15474. https://doi.org/10.4000/vertigo.15474

Sakamoto, Y., Ishiguro, M., & Kitagawa, G. (1986). Akaike information criterion statistics. *Dordrecht, The Netherlands: D. Reidel, 81*(10.5555), 26853.

Woillez, M. N. (2019). Revue de littérature sur le changement climatique au Maroc: observations, projections et impacts. *Papiers de recherche*, 1–33.

Zivot, E., & Wang J. (2006). Vector autoregressive models for multivariate time series. *Modeling financial time series with S-PLUS®* (2006): 385–429.

Spatial Variability of Summer Droughts and Heatwaves in Southern Canada

Zahra Noorisameleh, William A. Gough and M. Monirul Qader Mirza

Abstract

Drought is a short- to long-term risk that challenges access to water. Prolonged droughts could inflict agricultural and economic losses and cause social hardships. Southern Canada is home to the majority of the country's population and serves as the country's primary economic hub. This region has more climatic diversity than the northern part and is prone to drought during the summer months. This study investigates the relationship between the occurrence of daily summer droughts and heat waves in southern Canada. In this regard, from 1985 to 2017, homogenized daily precipitation and maximum temperature data from 26 selected stations were used from May to September. Daily droughts have been calculated using the effective drought index (EDI). The threshold temperature for calculating heat waves was 32 °C. We compared the results from southern Canada to those from the Prairie region. According to the findings, all types of daily droughts occur in southern Canada, but severe and extreme droughts are more common in the southern Prairies. Long-term daily droughts with the highest and lowest mean values were observed at the Toronto and St Jerome stations, respectively. Manitoba had the greatest variability in drought intensity between stations, with 1.96, and Quebec had the least, with 0.76. The increasing trend of daily droughts in Manitoba has been greater than in other southern Canadian provinces. The longest heatwaves occurred in Saskatchewan stations, with daily droughts intensifying in the long run.

Keywords

Maximum temperature · EDI index · Heatwave · Canada · Prairie

Z. Noorisameleh (✉) · W. A. Gough · M. M. Q. Mirza
Department of Physical and Environmental Sciences, University of Toronto Scarborough, Toronto, ON M1C 1A4, Canada
e-mail: z.noorisameleh@mail.utoronto.ca

1 Introduction

Drought is a negative anomaly in rainfall that is not limited to a specific climate (Hateren et al., 1990; Negm et al., 2020). According to the most recent reports, extreme droughts, and heat waves had numerous negative economic consequences, particularly in the agricultural sector of the Prairies and western Canada (News, 2021; Post & New, 2021). Despite increased rainfall, maximum temperatures play a larger role in Canada's droughts (Lu et al., 2019). Human activity has most likely increased the likelihood of compound extreme weather events, such as the frequency of heatwaves and droughts (IPCC: Climate Change, 2021). The occurrence of heatwaves and daily droughts in Canada was investigated in this study to determine their frequency, variability, and relationship.

2 Methodology

The study used daily maximum temperature and precipitation data from 25 weather stations provided by Environment Canada. The geographical distribution of these stations is shown in Fig. 1. Using R software, the daily drought severity of the stations was calculated. Excel and SPSS were used to analyze the frequency and trend of data related to drought, maximum temperatures, and heatwaves. Also, GIS is used for the production of maps in this study.

2.1 The Effective Drought Index (EDI)

The daily drought index of EDI (Noorisameleh et al., 2120) has been selected to monitor the drought because maximum temperatures and heatwaves also occur on a daily and short-term scale. The steps of this index's calculation are introduced in Eqs. 1–3:

H. Chenchouni et al. (eds.), *Recent Advancements from Aquifers to Skies in Hydrogeology, Geoecology, and Atmospheric Sciences*, Advances in Science, Technology & Innovation, https://doi.org/10.1007/978-3-031-47079-0_62

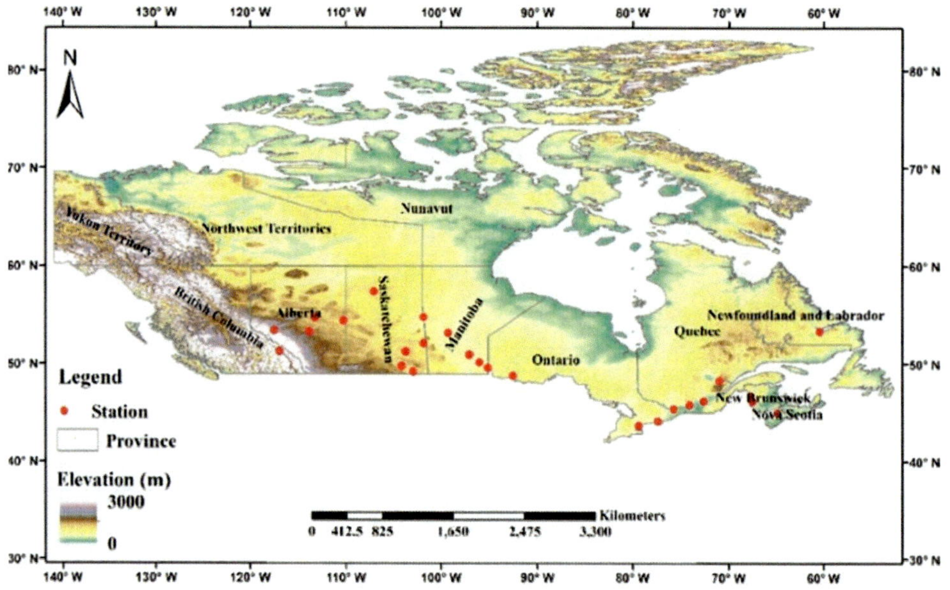

Fig. 1 Geographical distribution of stations and elevation of the study area

$$\mathrm{EDI}_j = \mathrm{PRN}_j/\mathrm{ST(PRN)} \tag{1}$$

$$\mathrm{PRN}_j = \mathrm{DEP}_j \sum_{n=1}^{j} \left(\frac{1}{N}\right) \tag{2}$$

$$\mathrm{DEP} = \mathrm{EP} - \mathrm{MEP} \tag{3}$$

PRN_j indicates precipitation needed for a return to normal. ST refers to the standard deviation of each day's Effective Precipitation (EP). DEP is the deficiency or surplus of water resources. MEP is the mean of EP (Byun & Wilhite, 1999). The classified edi values state that values > 0 denote a surplus of water, values from 7 to 0, ranging from 1.5 to 0.7, and between 2.5 and 1.5 denote dry, moderate, and severe droughts, respectively, and values less than 2.5 refer to extreme drought (Noorisameleh et al., 2120).

3 Results

3.1 Distribution of Daily Droughts in Canada

Figure 2 shows that all types of daily droughts occur in Canada, and dry to moderate intensities were frequent between May to September from 1985 to 2017.

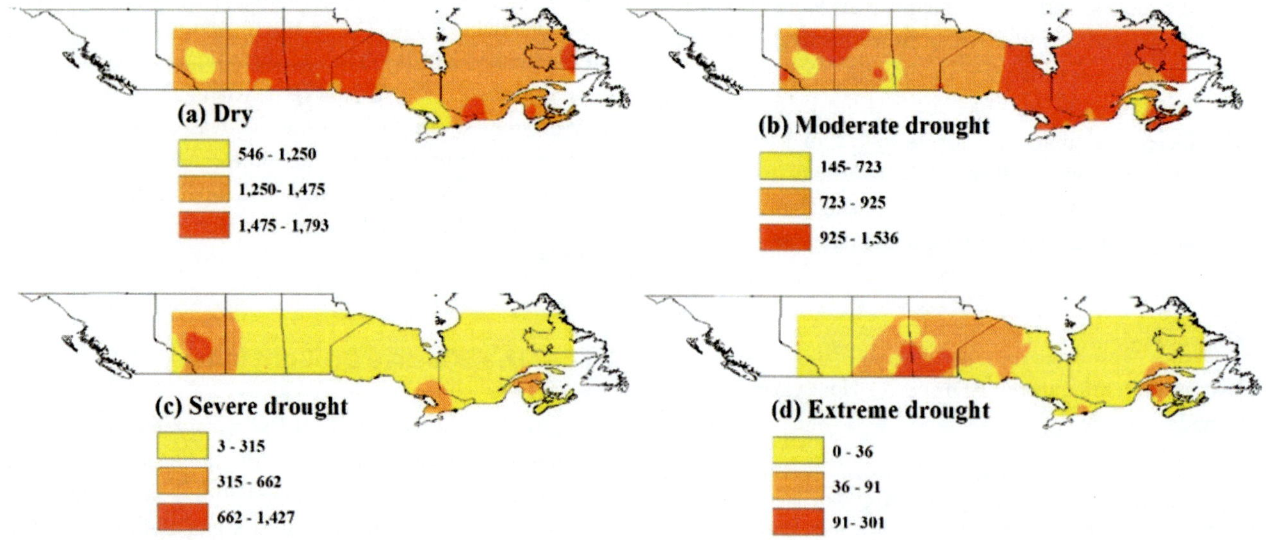

Fig. 2 Daily drought frequency in the study area (1985–2017)

3.2 Maximum Temperature and Heatwaves Frequency

The Prairie region has the highest frequency of maximum temperature (≥ 32) and heatwave length (day). Yellow grass and Estevan stations in southern Saskatchewan province have the highest utilization of these frequencies. As illustrated in Fig. 2, the daily droughts have been more severe in this province. Furthermore, with a maximum wavelength of 5 days after Saskatchewan, Ottawa, and Toronto stations in Ontario have the highest frequency of maximum temperatures. The frequency of maximum temperatures and wavelengths in Canada corresponds to the severity of daily droughts. Although daily droughts are more severe in Saskatchewan, the increasing frequency in Manitoba has been faster than in other southern Canadian provinces (see Fig. 2). Figure 4 shows the 3-year moving average of drought severity based on the EDI index in the study area (1985–2017). It is important to note that contrary to other stations Alberta's daily drought has become more persistent in recent years (Fig. 3).

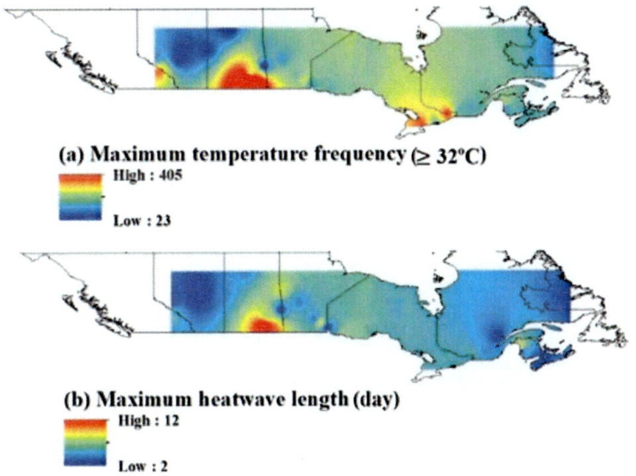

Fig. 3 Frequency of maximum temperature (≥ 32 °C) and heatwave length (day) in the study area

4 Discussion

Canada's warm season is short, but the southern half of the country experiences maximum temperatures, heat waves, and droughts every year. According to data analysis, the province of Saskatchewan is more vulnerable than other provinces to the aforementioned climate risks. Also, based on the 3-year moving average, the trend of increasing daily droughts severity in Alberta has increased significantly (Fig. 4).

5 Conclusion

In southern Canada, heatwaves and drought are climate hazards that could negatively impact Canada's agricultural outputs and public health. Due to the increasing trend of daily droughts and maximum temperatures, comprehensive planning to reduce their effects is necessary.

Fig. 4 The 3-year moving average of drought severity (EDI) in the study area (1985–2017)

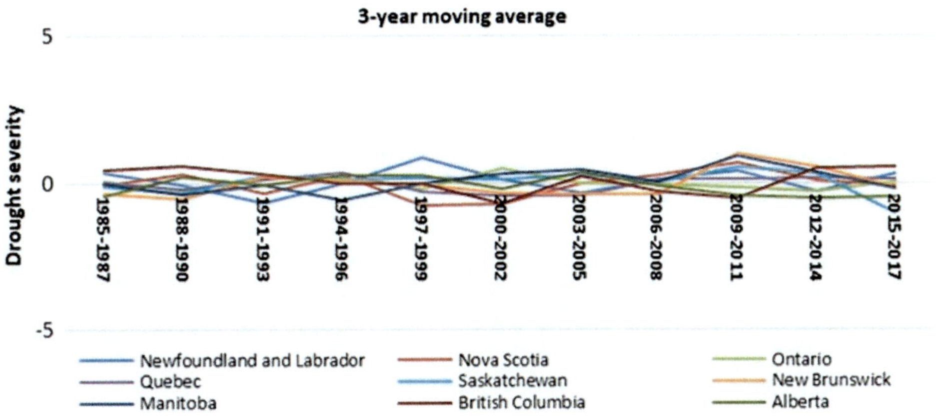

References

Byun, H. R., & Wilhite, D. A. (1999). Objective quantification of drought severity and duration. *Journal of Climate, 12*, 2747–2756.

CBC News (2021). Extreme drought' is threatening parts of the Prairies, says Agriculture Canada. https://www.cbc.ca.

Financial Post New (2021). *Extreme heat and drought in Western Canada wreak havoc on food system*. https://financialpost.com.

IPCC: Climate Change (2021). The physical science basis. Cambridge University Press, Cambridge, United Kingdom and New York, NY, USA. (pp. 2391).

Lu, P., Parker, W. C., Colombo, S. J., & Skeates, D. A. (2019). Temperature-induced growing season drought threatens survival and height growth of white spruce in southern Ontario Canada. *Forest Ecology and Management, 448*, 355–363.

Negm, A., Bouderbala, A., Chenchouni, H., & Barcelo, D. (2020). *Water Resources in Algeria-Part I: Assessment of Surface and Groundwater*. Springer, Cham.

Noorisameleh, Z., Gough, W. A., Mirza, M. M. Q. (2021). Comparison of the temporal distribution pattern of precipitation and the cumulative effect of daily droughts on the local scale in Iran. Arab J Geosci *14*, 1–12.

Van Hateren, T. C., Chini, M., Matgen, P., Teuling, A. J. (1990). Ambiguous agricultural drought: characterising soil moisture and vegetation droughts in Europe from earth observation. *Remote Sensing, 13*(10), 1990.

Fire Weather Anomalies Previous and During Large Forest Fires—A Comparison Between Mt. Carmel and Judean Hills

Haim Kutiel and Lea Wittenberg

Abstract

The Mediterranean region is a highly fire-prone ecosystem with a long fire history, aggravated by the prevailing climate conditions (hot and dry summers) coupled with intensive land use. All forest fires in Israel are manmade, however, the weather conditions after ignition are crucial for its propagation. Given the recent increase in fire activity, the study explores Medium Time Scales (seasonal to annual) climatic variables affecting fire occurrences in Mt. Carmel and the Judean hills region, two 'hotspots' of frequent and devastating forest fires. The main climate variables for rapid forest fire propagation are very low relative humidity and above normal air temperature together with strong winds. Meteorological data for 1988–2019 were collected at two stations representing the weather conditions on Mt. Carmel and the Judean hills. During that period, data for all wildfires larger than 50 ha, on both sites, were obtained from the 'Israel Fire and Rescue Services'. For each fire event, data included burnt area, exact location, date, fire duration and daily maximum temperature, daily minimum relative humidity, daily maximum wind speed and the wind direction during the maximum wind speed. The burnt area size was correlated with of relative humidity and air temperature departures, from the long-term average at the same time and location. Calculations were done for the day of the fire occurrence (day = 0) and for each day during the previous week (day = −7 to day = − 1). In Mt. Carmel, most forest fires (>50 ha) occurred in autumn (September–November), whereas in the Judean hills they occurred in summer (May–October). In summer, temperatures are at their annual maximum, a constant moderate sea breeze blows, and the fuel is relatively dry. However, while in Judean hills, relative humidity is at its annual minimum, on Mt. Carmel, in this period, relative humidity is at its annual maximum. This difference in the relative humidity regime between these two regions, apart from each other only by 120 km, is the leading cause of the difference in the timing of major forest fires. This method is already used for alerting of dangerous weather conditions that may result in a devastating forest fire.

Keywords

Air temperature · Relative humidity · Wildfires · Mt. Carmel · Judean hills

1 Introduction

The Mediterranean basin is a highly fire-prone region with a long fire history. The climatic conditions of wet winters and long, hot and dry summers provoke seasonal fire events, mainly ignited by human activities (Allen et al., 2010; Ganteaume et al., 2021). The probability of wildfire occurrence is associated with the amount, type, arrangement and dryness of the available fuel, ambient weather conditions and presence of ignition sources (Archibald et al., 2009). Fire regimes can often be characterized by variations in these drivers (Clarke et al., 2013).

In the Mediterranean, most of the fires are human caused rather than natural hazards; however, the most critical factors for determining fire regimes are vegetation type and climate/weather patterns. Both the ease of ignition and the parameters affecting fire behavior and spread after ignition are primarily controlled by the amount and connectivity of fuel (fuel structure) and its moisture content (fuel flammability), which in turn depend on weather conditions and climate variability.

H. Kutiel (✉) · L. Wittenberg
University of Haifa, 3190500 Haifa, Israel
e-mail: kutiel@geo.haifa.ac.il

H. Chenchouni et al. (eds.), *Recent Advancements from Aquifers to Skies in Hydrogeology, Geoecology, and Atmospheric Sciences*, Advances in Science, Technology & Innovation, https://doi.org/10.1007/978-3-031-47079-0_63

The impact of weather/climate conditions on fire behavior can be identified at short, medium and long-time scales (Flannigan & Wotton, 2001). Short time scale—STS (minutes to a few hours) refers to rapid processes such as changes in wind speed and/or wind direction, which may alter fire propagation either directly by changing its propagation speed or direction, or indirectly via changes in air temperature (AT) and or relative humidity (RH), which may accelerate, slow or even stop the propagation of the fire. Medium time scale—MTS (seasonal to annual) determines the timing and the duration of the fire seasons when and where climate conditions favor rapid fire propagation, considering a combination of several climatic variables such as AT, RH, wind regime and the time elapsed since the last rain event. Long time scale—LTS (decades) is related to climate change scenarios over decades. Any anticipated change in the AT, RH, wind and/or rainfall regimes may alter the timing and duration of the MTS. LTS impacts should be considered when addressing fire management programs for long-term planning and policy purposes. Given the recent increase in fire activity, the present study explores the MTS climatic variables affecting fire occurrences in both Mt. Carmel and the Judean hills regions.

2 Methods

2.1 Study Area and Data

The Mount Carmel study area (228 km^2) is in Northwestern Israel, south of the city of Haifa. The mountain rises from the Mediterranean coastal plain to 546 m above MSL. The Judean hills study area (313 km^2) is located on the mountainous backbone of Israel at an altitude ranging from 300 to 800 m above sea level.

Meteorological data for the period 1988–2019 were collected at the University of Haifa Met. Station, representing the weather conditions on Mt. Carmel, and Jerusalem Center Meteorological station, representing the weather conditions on the Judean hills. Meteorological data included temperature, relative humidity, wind direction and wind speed. During the above-mentioned period, data for all wildfires larger than 50 ha on both sites were obtained from the 'Israel Fire and Rescue Services'. For each fire event, data included the burnt area, exact location, date and burning duration.

2.2 Data Analysis

For each fire, meteorological data included daily maximum temperature, daily minimum relative humidity, daily maximum wind speed and the wind direction during the maximum wind speed. Daily averages were calculated from 3 h. measurements. The above calculations were done for the day of the fire occurrence (day = 0) and for each day during the previous week (day = −7 to day = −1).

To evaluate to what extent the meteorological conditions before and during the large wildfires were different from normal conditions in the same month, two indices were developed, one for mean DAily maXimum TEmperature Departures (DAXTED) and a second for mean DAily miNimum relative HUmidity Departures (DANHUD).

$$\text{DAXTED}_{(j)} = \frac{\sum_{i=1}^{n} (T\max_j - T\max)A_i}{\sum_{i=1}^{n} A_i}$$

$$\text{DANHUD}_{(j)} = \frac{\sum_{i=1}^{n} (RH\min_j - RH\min)A_i}{\sum_{i=1}^{n} A_i}$$

where $T\max$ is the mean monthly daily maximum temperature; $T\max_i$ is the maximum temperature on the day j, $RH\min$ is the mean monthly daily minimum relative humidity; ($RH\min_i$) is the minimum relative humidity on the day j; Ai is the total burnt area of fire i and n is the total number of analyzed forest fires ($n = 13$ on Mt. Carmel and $n = 27$ on Judean hills). $j = 0$ on the day when the fire started, $j = -1$, on the day before, $j = -2$, two days before and so on. DAXTED and DANHUD were calculated for all days since $j = -7$ until $j = 0$.

3 Results and Discussion

In Mt. Carmel, most of the large wildfires (>50 ha) occurred in autumn (September–November), while in the Judean hills the largest wildfires occurred during the summer months (May–October). These results show a noticeable difference in the timing of large wildfires between Mt. Carmel and Judean hills. This difference in timing should be attributed mainly to differences in the relative humidity between the two sites, as there are only minute differences in other parameters between both locations.

In the period May–August, temperatures are at their annual maximum, there is a constant moderate sea breeze wind regime, and the fuel is relatively dry as normally two to five months have elapsed since the last considerable rain. However, in Judean hills, relative humidity is at its annual minimum, while on Mt. Carmel, relative humidity is at its maximum in this period. The reason for this difference between the two regions is related to the wind and atmospheric pressure regimes in that part of the world which are beyond the scope of the present study.

DAXTED and DANHUD indices for Mt. Carmel and Judean hills show greater departures from average conditions as we approach the day of the fire. On Mt. Carmel, DANHUD constantly decreased from day = −7 to day = 0 from 1% above normal to 27% below normal, respectively. DAXTED shows an increasing trend from 2.0 °C on day = −7 to 4.7 °C on day = 0. In the Judean hills the trends are similar but less significant. DANHUD decreased from about 4% above normal on day = −7 to about 14% below normal on day = 0. Temperature increased to a lesser extent from −0.7 °C on day = −7 to 1.0 °C on day = 0.

4 Conclusion

The main conclusion from the present study is that although both study areas are located relatively close to each other, and the vegetation composition is similar, there are essential differences in the spatio-temporal dynamics of the wildfires, mainly due to differences in the relative humidity regime. Spells of severe drought, heat waves and low humidity (=fire weather) are becoming more extreme. As climate change makes hot and dry conditions prevail, vegetation dries out and landscapes become more flammable, increasing the likelihood of dangerous wildfires.

References

Allen, C.D., Macalady, A.K., Chenchouni, H., et al. (2010). A global overview of drought and heat-induced tree mortality reveals emerging climate change risks for forests. *Forest Ecology and Management*, *259*(4), 660–684.

Archibald, S., Roy, D. P., van Wilgen, B. W., & Scholes, R. J. (2009). What limits fire? An examination of drivers of burnt area in Southern Africa. *Global Change Biology*, *15*(3), 613–630.

Clarke, H., Lucas, C., & Smith, P. (2013). Changes in Australian fire weather between 1973 and 2010. *International Journal of Climatology*, *33*(4), 931–944.

Flannigan, M.D. & Wotton, B.M. (2001). Climate, weather, and area burned. In *Forest fires* (pp. 351–373). Academic.

Ganteaume, A., Barbero, R., Jappiot, M., & Maillé, E. (2021). Understanding future changes to fires in southern Europe and their impacts on the wildland-urban interface. *Journal of Safety Science and Resilience*

Hydrogeological and Climatic Risks: The Emblematic Case of an Exceptional Debris Flow in Central Apennines (Italy)

Domenico Aringoli

Abstract

The present work aims to highlight a particular event where there was a concomitance of particular hydrogeological and climatic conditions, unfortunately in a scenario which could be repeated. In fact, the south-eastern slope of Mount Vermenone (Central Apennines, Italy) was affected by a severe landslide movement of debris flow type. The phenomenon was triggered in the mid-high portion where a vast and thick debris accumulation (coarse, loose, and weakly cemented materials) was present. The aforementioned materials, in addition to being located on a steep slope, were frequently in a condition of water saturation during the winter, given the presence of springs and the characteristic impluvial form in which they were concentrated. In January 1997, a rapid snowmelt increased the water saturations in the thick debris, so that the materials became unstable on the slope and the movement is quickly activated. A fast and dense debris flow was triggered, moving along the steep slope down to the main valley floor: Here, the high energy flow did not stop but continued, spreading its accumulation almost to the built-up area. Only after several days and months the stream was able to resume its course, incising the huge mass of alluvial materials that can still be seen today at the edge of the plain. In addition, the groundwater resource also has been compromised as some springs in the area have changed their regime or even disappeared. The event was of considerable impact, and given its exceptional nature, studies on the dynamics of the landslide phenomenon were carried out. These in particular were conducted on the possibility of reactivations and also toward the monitoring of the adjacent areas. In this regard, a first response from the installed instruments was that no movement was recorded, and therefore, we are facing a very high threshold before triggering, probably due to the high permeability of the materials.

Keywords

Hydrogeological risk · Climatic · Factors · Debris-flow · Landslide hazard · Monitoring · Central Apennines

1 Introduction

A geological and geomorphological study carried out on the south-eastern slope of Mount Vermenone (Central Apennines, Italy), affected by a severe landslide movement, is presented (Fig. 1). The type of landslide phenomenon is debris flow (D'Elia & Tancredi, 1979; Takahashi, 1981; Varnes, 1978), which is moreover recurrent in the area. It has a particularly rapid evolution and very high hazard. Here, it is considered emblematic because several factors, both predisposing and triggering, were involved in its activation (Aringoli et al., 2010; Hutchinson, 1988; Pierson, 1980).

The purpose of this study was to know and analyze the evolutionary dynamics of the phenomenon, also considering the climate and the seismic scenarios (Aringoli et al., 2013, 2015, 2021), with the aim to detect possible similar manifestations along the neighboring slopes, as well as to provide indications for the preparation of the monitoring system, in consideration of the built-up area down the valley floor and functional to respect the high landscape value (Geopark) (Aringoli & Pambianchi, 2009).

An initial zoning was carried out from an early analysis, highlighting the major risk situations; here additional

D. Aringoli (✉)
University of Camerino, 62032 Camerino, Italy
e-mail: domenico.aringoli@unicam.it

H. Chenchouni et al. (eds.), *Recent Advancements from Aquifers to Skies in Hydrogeology, Geoecology, and Atmospheric Sciences*, Advances in Science, Technology & Innovation, https://doi.org/10.1007/978-3-031-47079-0_64

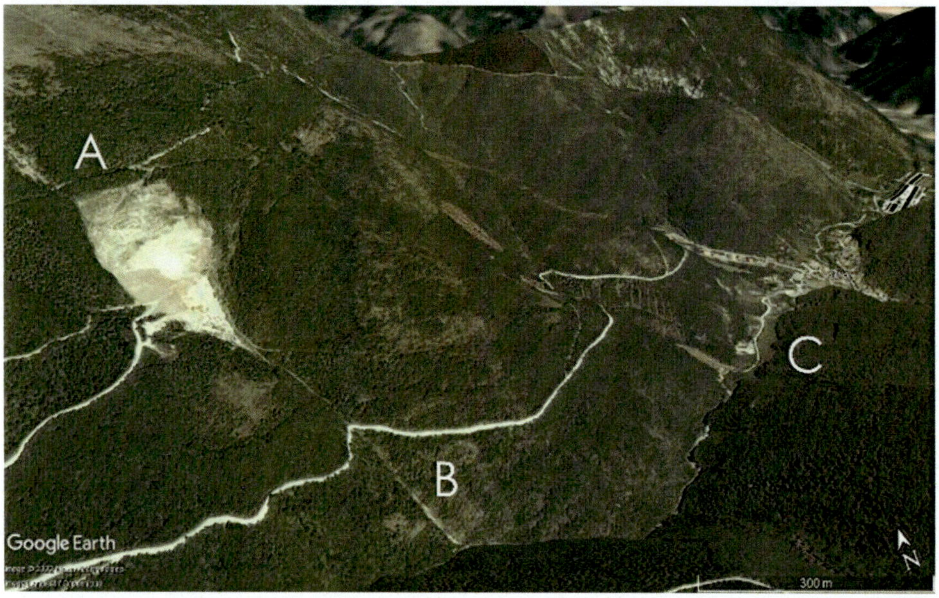

Fig. 1 View of the SE slope of Mount Vermenone in 2004 with the evidence of the 1997 debris flow (clearly recognizable feeding and triggering area (A), flow channel (B) and expansion area (C) along the valley floor down to the village

instruments were installed, providing important data after an initial period (Fig. 2).

2 Materials and Methods

According to the above, the following flowchart was planned and followed (Fig. 3):

The catchment area of a minor stream of the Scarsito River, a tributary of the Potenza River, is analyzed. This basin develops between the altitudes of 505 and 1390 m a.s.l., with an estimated area of 460 ha. Considering the measured elevation differences ($\Delta H = 880$ m and $\Delta h_{thalweg} = 490$ m) and the average slope gradient (> 35°), the energy is considerable. The reliefs consist mainly of calcareous formations, but the lithotypes under observation,

Fig. 2 Synthetic map of Valle dell'Eremita with the areas of potential triggering and/or exposed to debris flows (in yellow) and location of monitoring system points

susceptible to landslides, are essentially debris covers. These are mainly due to weathering, located even above 800 m, somewhere with considerable thicknesses as in the focused area close to the village (Fig. 2). The hydrogeology of the area is strongly influenced by the alternating of limestone and marly lithologies, with high and low permeability properties. Further permeable bodies can act as local aquifers: the thick slope debris accumulations. In these materials, the aquifer is not a resource but a potential geomorphological hazard when it saturates the debris (Aringoli et al., 2010).

The climate data, both rainfall and temperature, were processed and summarized as follows: average annual rainfall 1700 mm/year; monthly rainfall of the highest intensity is around 300–309 mm; recorded rainfall of high intensity is 150–250 mm. These precipitations are very dangerous in triggering mass transport phenomena, especially in the winter months, when rapid melting of the snowpack could occur, with increased meteoric inflow, so several devices were finalized for monitoring (Figs. 2 and 3a, b).

The area was affected in January 1997 by a debris flow-type landslide movement. The phenomenon was triggered in the middle-upper portion of the SE slope of M. Vermenone, where thick debris accumulations, once they reached saturation, were mobilized and, traveling down the steep valley that connects to the Eremita valley floor, acquired enough energy to reach the main watercourse near the built-up area of Sefro. The event was followed by studies of the landslide phenomenon and hydraulic remediation works concentrated along the Valle dell'Eremita stream, and the installation of a monitoring system related to potential unstable areas belonging to the same basin was planned and implemented. Two sites were identified that, due to stratigraphic–geomorphological conditions (i.e., presence of debris accumulations, morphometric parameters of the slope, and particular morphological features), could lead back to a hazard scenario comparable to the one manifested

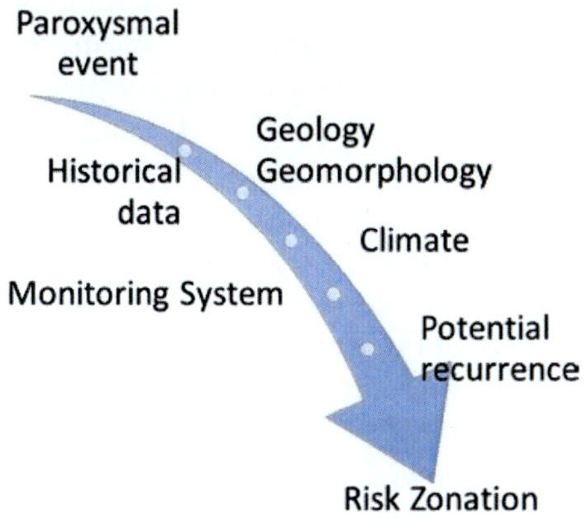

Fig. 3 Flowchart of the main analytical steps

(Fig. 2). At the first site there are debris accumulations of medium-fine grain size, variable thickness, locally and superficially affected by landslide movements: their possible mass mobilization could affect the main valley floor (Valle dell'Eremita), where they would most likely exhaust their energy. Two boreholes of about 12 m were drilled and equipped, hosting a piezometer (p3) and an inclinometer (i3). Within the second site, debris accumulations of considerable thickness and extension, coarse grain size and sometimes cemented, are noted, especially in depth; these deposits are fed upward mainly by the degradation of a conspicuous escarpment (probable crown of an ancient landslide) incised on cemented gravels, while downward they are affected by generally slow surface deformation; important to point out that, just downstream, there are recently built residential settlements.

For the reasons above, instruments were installed both in the upper portion of the site (boreholes up to 27 m with piezometer p1 and inclinometer i1) and in the lower portion of the site (boreholes up to 15 m with piezometer p2 and inclinometer i2); the areas are separated by the roadway. To complete the system, other instruments of a general nature, i.e., concerning the hydropluviometric parameters of the basin, have been installed: just downstream of site 2, near the built-up area (bowling alley), a rain gauge (R) with alarm threshold is placed; in addition, there are also two discharge meters: D1 along the stream (asl elevation 505 m) and D2 at one of the major springs in the basin (850 m). All instruments, starting in January 2003, acquire continuous data that are downloaded and processed at 2–3-month intervals, for more than two years. If the rainfall reaches a certain predetermined alarm level, the control unit placed together with the rain gauge makes an alert call to the competent authorities.

3 Results

3.1 Significant Debris-Flow Phenomena and Zoning

In the study area, several areas of high geomorphologic hazard related to debris flows can be identified, in particular along the aforementioned slope on the left hydrographic side of the Valle dell'Eremita (Fig. 1) facing south-east, which is very sloping and rich in debris covers, even thick ones. The latter slope was affected by a complex and emblematic debris-flow phenomenon, which was triggered between 1000 and 1100 m a.s.l. as a flow, and evolved into a flow along the narrow and steep impluvium below that descends to the ditch (reference height 675 m at the confluence), covering a difference in height of approximately

350 m. Subsequently, given the high energy, the saturated materials reached the first settlements of the built-up area and flowed into the Scarsito River, increasing its liquid and especially solid flow for several days. Along the way, it is still possible to distinguish the maximum height level reached by the chaotic flow of the melt: fine breccias in a sparse sandy matrix are visible at this point, which were also abandoned about 3 m above the present valley floor.

The geomorphological surveys and historical analysis, correlated and compared with studies in areas with similar dynamics, climate and hydrogeological-geomorphological settings (Aringoli et al., 2007, 2008, 2010; Materazzi et al., 2010), made it possible to zonate and localize the areas of potential triggering of the debris-flow phenomenon and obviously those exposed to the flow (Fig. 2). The formers are located in correspondence with the largest debris-flow accumulations (and landslides, even inactive ones) detectable "suspended" on the valley floor, even at altitudes above 800 m; while the areas exposed to the flow, especially the most vulnerable ones, are located at the bottom of the valley and especially near the built-up area, at about 500 m altitude.

Since the materials that can be involved in the potential phenomena are considerable, an adequate monitoring system has been prepared and configured (Figs. 2 and 4a, b). This was possible, considering the geological–geomorphological structure and the potential evolution of the debris flows toward the high-risk areas, also according to comparable studies and analyses (Antonetti et al., 1808; Guzzetti et al., 2008; Minu et al., 2020; Whipple, 1993). By interpreting and processing of the elements resulting from the monitoring system, it was possible to clarify and deepen the evolutionary dynamics of the phenomena and the hazard-risk scenarios.

3.2 Monitoring Data

The system is composed by different instrumentation above acquires data on: rainfall (also extreme rainfall), the saturated zone in the ground, and any movement of the same with location of the depth, so that they can prevent the triggering of a debris flow.

In particular, were installed: 1 rain gauge; 3 sets of inclinometer-piezometer within the largest debris accumulations on the slope; 2 discharge meters, one of which at altitude, in correspondence with a significant spring and one along the valley floor (Fig. 3a). This monitoring system is simultaneously monitoring the amount of rainfall, the level of the water table in the debris (and thus the change in the saturated zone as a function of rainfall) and any, even minimal, movements of the potentially unstable ground (Fig. 3b). In addition, the various data (especially rainfall) are also put in relation to any resulting changes in the discharge of the main stream at the valley floor. This control was also purposed because in the past (1997 event) the same river has not been able to drain excessive water in a timely manner also due to the high solid load received.

The displacements detected by inclinometers, although minimal, are significant, and the vectors are compatible with a slope loosening (Fig. 5). In addition to this outcome, a further result was to attest to the correspondence between the instrumental displacements and the geomorphological evidences detectable on the surface, so they all together help to better define the risk of the area.

4 Discussion and Conclusion

An emblematic phenomenon occurred in the area for several reasons: heavy rainfall (lasting several consecutive days); rapid melting of snowpack due to rising temperatures; rapid saturation of the thick debris, located at high altitudes on steep slopes (>35–40°); presence of a spring. Since similar scenarios are present in the neighboring areas, a monitoring system was set up to predict the triggering of similar phenomena in the debris accumulations that present a potential risk toward the built-up area. Analyzing the data that emerged from the first monitoring period (about 2 years), some considerations were derived based on previous experience in similar situations, according to them (Aringoli et al., 2007, 2008, 2010, 2013, 2015, 2021; Materazzi et al., 2010) and to further studies conducted on triggering thresholds and rain control on debris flows (Antonetti et al., 1808; Guzzetti et al., 2008; Minu et al., 2020; Whipple, 1993).

A good correlation between the instruments was observed (Fig. 4a, b). The increased hourly precipitation is matched by changes in the groundwater level (significant in piezometer 1). These responses are a bit delayed in time, as can also be seen by the changes in spring and stream flow rates, which are, however, aligned with each other. The response of the inclinometers is apparently disconnected from the rainfalls, because there were no particularly long-lasting precipitations and thus no saturation of soils. However, a significant movement is detected at a depth of 4 m, while additional minimal movements occur in the debris at different levels (Fig. 5).

This movement of coverage down to −4 m is located in an area (site 1 in Fig. 2) that does not directly affect the built-up area, as is the case for site 2. In fact, at site 2 we recognize minimal movements at depths of −5, 5–17, and 24 m; because all of these responses indicate potential shear surfaces, there risk is high due to the proximity of housing.

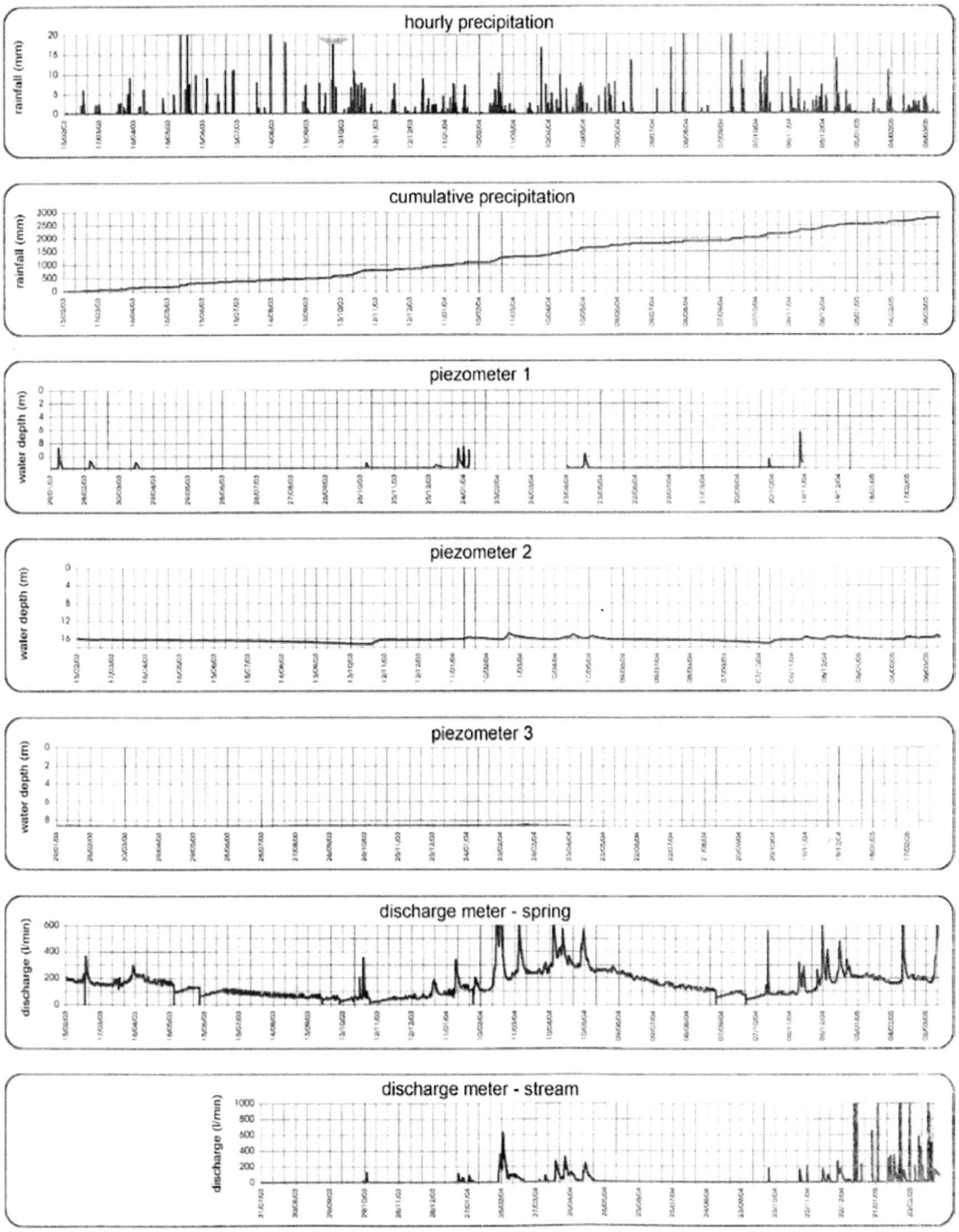

Fig. 4 **a** Data summary after the first monitoring period (rain gauge, piezometers, discharge meters; 2003–2005 every tick is a month). **b** Data summary for the inclinometers i3 (left) and i1 (right) after the first monitoring period; the i2 had a brief period of activity then was broken by agricultural practices

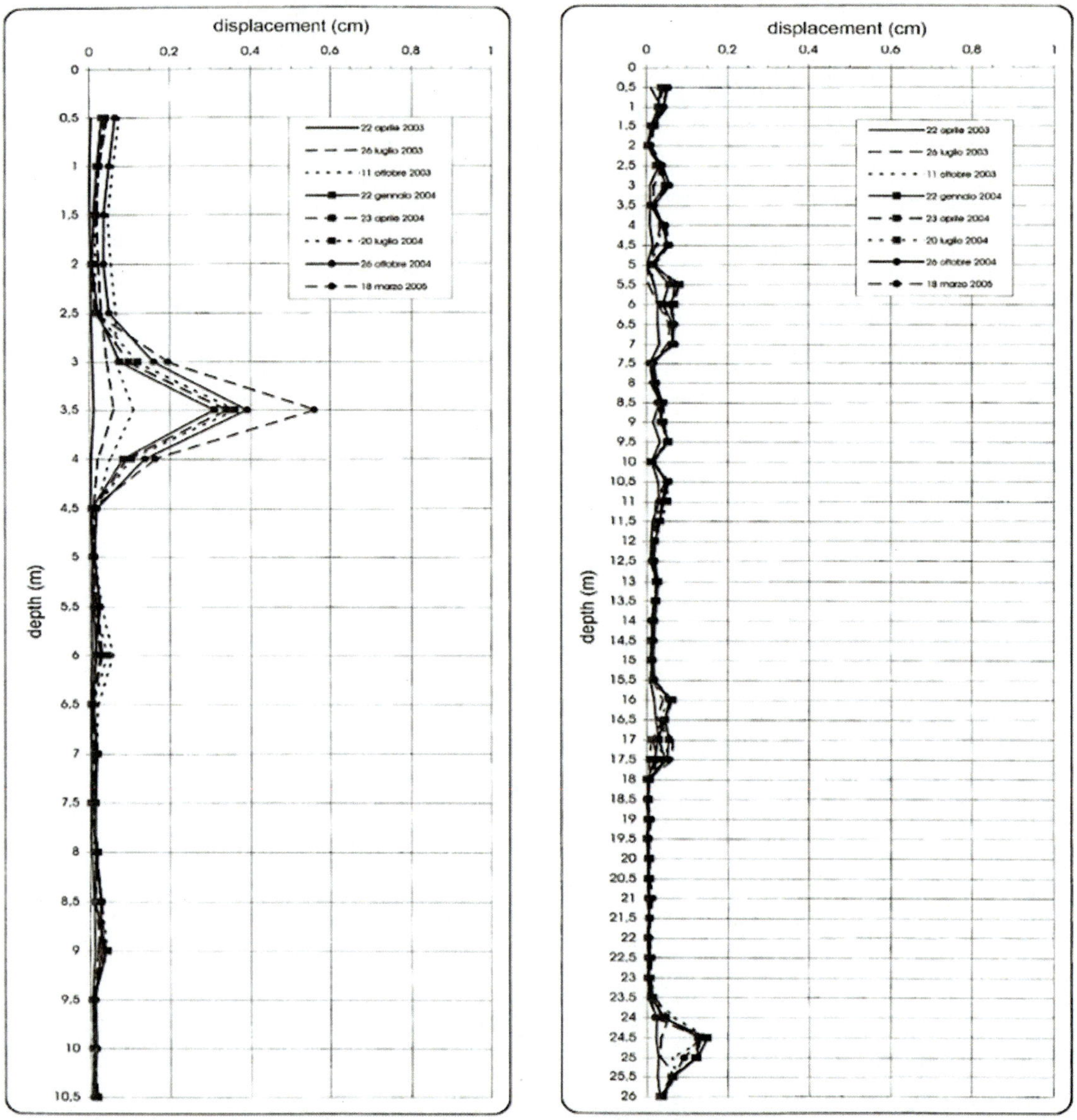

Fig. 4a. (continued)

If the failure were to occur at the greatest depth, the moving mass would be very huge (Fig. 6).

It must be noted that no particularly intense meteorological events were recorded during the observation period, so at the moment the thresholds for triggering significant and hazardous movement in the debris have not yet been established.

In any case, the system proved its reliability, as the responses of the various instruments were concordant and proportioned, and the inclinometers detected the possible failure depth. As a final analysis, the study also led to discover further locations potentially affected by the same type of landslide hazard that obviously should be monitored (yellow areas in Fig. 2).

Fig. 5 Directions of the main displacements detected by inclinometers that, although minimal, are significant (the length of the arrows is not proportional to the measured displacement)

Fig. 6 Thick debris accumulations in the potentially unstable areas located just upslope of the town (site 2)

References

Antonetti, G., Gentilucci, M., Aringoli, D., & Pambianchi, G. (2022). Analysis of landslide susceptibility and tree felling due to an extreme event at mid-latitudes: Case study of storm Vaia, Italy. *Land*, 11(10), 1808, S.I. Analysis of slope stability in relation to climate change and multivariate models for landslide activation. https://doi.org/10.3390/land11101808.

Aringoli, D., Pambianchi, G. (2009). La conca intramontana di Montelago: un esempio di Geopark nell'Appennino Umbro-Marchigiano. In: Coratza, P., Panizza M. (eds) Il patrimonio geomorfologico come risorsa per un turismo sostenibile pp. 7–24, SAP Società Archeologica s.r.l. , Mantova.

Aringoli, D., Farabollini, P., Pambianchi, G., et al. (2021). Geomorphological hazard in active tectonics area: Study cases from Sibillini Mountains Thrust System (Central Apennines). Land, *10*(5), 1–21. (MDPI).

Aringoli, D., Buccolini, M., Coco, L., et al. (2015). The effects of in-stream gravel mining on river incision: An example from Central Adriatic Italy. *Zeitschrift Fur Geomorphologie, 59*(2), 095–107.

Aringoli, D., Calista, M., Gentili, B., Pambianchi, G., & Sciarra, N. (2008). Geomorphological features and 3D modelling of Montelparo mass movement (Central Italy). *Engineering Geology, 99*, 70–84.

Aringoli, D., Farabollini, P., Gentili, B., Materazzi, M., & Pambianchi, G. (2007). Climatic influence on slope dynamics and shoreline variations: Examples from Marche region (Central Italy). *Physio-Géo, 1*, 1–20.

Aringoli, D., Gentili, B., Materazzi, M., et al. (2010). Mass movements in Adriatic Central Italy: Activation and evolutive control factors. In E. D. Werner (Ed.), *Landslides: Causes, types and effects* (pp. 1–72). Nova Science Publishers Inc.

Aringoli, D., Gentili, B., Materazzi, M., Pambianchi, G., & Sciarra, N. (2013). DSGSDs induced by post glacial decompression in central Apennine (Italy). In C. Margottini, P. Canuti, & K. Sassa (Eds.), *Landslide Science and Practice* (Vol. 4, pp. 417–424). Springer.

D'Elia, B., & Tancredi, G. (1979). Colate permanenti e temporanee. *Geologia Applicata e Idrogeologia, 14*, 23–39.

Guzzetti, F., Peruccacci, S., Rossi, M., Stark, C. P. (2008). The rainfall intensity–duration control of shallow landslides and debris flows: An update. *Landslides*, 5, 3–17. https://doi.org/10.1007/s10346-007-0112-1.

Hutchinson, J. J. (1998). Morphological and geotechnical parameters of landslides in relation to geology and hydrogeology. In: *Proceedings of 5th International Symposium on Landslides* (pp. 3–35 (1)). Balkema, Lausanne.

Materazzi, M., Gentili, B., Aringoli, D., Farabollini, P., & Pambianchi, G. (2010). Elements of slope and fluvial dynamics as evidence of late Holocene climatic fluctuations in the central Adriatic sector, Italy. *Geogr. Fis. Dinam. Quat., 33*, 193–204.

Minu, T., Abraham, M. T., Satyam, N., et al. (2020). Using field-based monitoring to enhance the performance of rainfall thresholds for landslide warning. *Water*, 12(12), 3453. https://doi.org/10.3390/w12123453.

Pierson, T. C. (1980). Erosion and deposition by debris flow at Mt. Thomas, north Canterbury, New Zeland. *Earth Surface Processes*, 5(3), 227–247.

Takahashi, T. (1981). Debris flow. *Annual Review of Fluid Mechanics, 13*, 57–77.

Varnes, D. J. (1978). Slope movement. Types and processes. Transp. Res. Board rep. N.176: Landslides. Analysis and control, (pp. 11–33).

Whipple, K. X. (1993). Intepreting debris flow hazard from study of fan morphology. In: H. W. Shen, W. F. Wen (Eds.), *Hydraulic Engineering '93 Proceedings*, (Vol. 2, pp. 1302–1307).

Landslide Susceptibility Analysis with Artificial Neural Networks Used in a GIS Environment

Fabrizio Bendia, Guido Antonetti, Domenico Aringoli, Piero Farabollini, Matteo Gentilucci and Gilberto Pambianchi

Abstract

The identification of areas potentially subject to hydrogeological risk is an important issue, especially for Institutions that spend substantial sums each year on remediation. In 2018, north-eastern Italy was hit by an extreme event, Storm Vaia, which produced extensive damage, such as the collapse of trees and the reactivation of numerous landslide phenomena. In particular, the study area of this research, located in the municipality of Feltre, saw the reactivation of 9 landslide phenomena, which prompted the assessment of landslide susceptibility for the area in question. The objective of this research is to identify areas of high potential landslide hazard, using GIS-type software to prepare the spatial analysis and generating an algorithm inspired by the weight of evidence technique. For the calculation of susceptibility, eight environmental variables were considered, which can substantially influence the activation of landslide phenomena: permeability of the lithotype, type of vegetation cover, exposure, slope, ancient and stable deposits, recent slope deposits, bedrock, and existing landslides. The first level contains all the areas in which is present at least one of the conditions under our phenomena, the second at least two and so on. The algorithm works with three main processing tools, extraction by position, clip and merge, clipping the layers according to a number of simple combinations, without repetition, increasing for each susceptibility level: 8, 28, 56, and 70. The final map contains four levels of susceptibility in ascending order of hazard, and as further proof of the validity of the model it overlapped very well with already well-known and surveyed phenomena. The method used, unlike the weight of evidence, assigns the same weight to each underlying condition, so that the map can be more conservative and not underestimate risks. Potentially, the analysis could be further extended in the future by adding additional factors influencing slope stability, such as the amount of precipitation.

Keywords

Landslide susceptibility · GIS · Weight of evidence · Hydrogeological risk · Slope stability

1 Introduction

This research is an assessment of the effects on plants and landscape after the extreme climatic event Vaia that has hit the north-east of Italy on the last days of October 2018 focusing on the area of the municipality of Feltre. After extreme climatic event like the one examined, substantial modification of the landscape, changes in the vegetational cover and activation of landslide and other gravitative processes related mainly to the high amount of precipitation and the reduction of the stabilizing action of plant roots are to be expected. In this area, the need has emerged for research on risk assessment to enable the policy maker to be able to evaluate the most appropriate choices for safeguarding human life, territory, and economic resources. After Vaia has been censused by Comune di Feltre a set of 9 landslides activated by the storm, this led to the need to assess slope stability, creating a landslide susceptibility map. The research employed a simplified method of probabilistic analysis taking inspiration on the weight of evidence (WOE) technique (Gentilucci et al., 2021). WOE is a very efficient technique for mapping susceptibility or risk of landslides and other kind of gravitative phenomena (Sarkar & Kanungo, 2004).

F. Bendia (✉) · G. Antonetti · D. Aringoli · P. Farabollini · M. Gentilucci · G. Pambianchi
University of Camerino, 62032 Camerino, Italy
e-mail: fabrizio.bendia@unicam.it

H. Chenchouni et al. (eds.), *Recent Advancements from Aquifers to Skies in Hydrogeology, Geoecology, and Atmospheric Sciences*, Advances in Science, Technology & Innovation, https://doi.org/10.1007/978-3-031-47079-0_65

2 Materials and Methods

Using the collected data an algorithm was created to identify susceptible areas. This was performed using QGis and three main processing tools: extraction by location, clip, and merge. This analysis aims to plot the areas susceptible under specific stress conditions, not ordinary but limit conditions, where the cause-effect relationship occurs in a very short period of time, taking into account some environmental factors that favour them. The evaluation has been done using a statistical and probabilistic method; practically considering more than a fixed number of the same underlying condition in the site of the phenomena; this means that in every place in which are present that conditions is probable to have the insurgence of the same kind of phenomena.

For the evaluation have been taken eight categories of existing conditions, that can most be correlated with the activation mechanism: permeability of the lithotype, type of vegetational cover, exposition, slope, type of substrate, deposits, bedrock or soil, and the already existent phenomena.

3 Results

In the present research, each factor of probable instability was subdivided into classes.

The slope factor has been subdivided into six categories according to the degrees of inclination of the versant. The aspect, more properly exposition, has been subdivided into eight categories coincident with the portion of the azimuthal quadrant of the main cardinal point. The permeability of the lithotype is connected to the mechanical behaviour of the lithologies, so have been divided in bedrock and deposits and each one in four categories according to the porosity and fractures presence. For the substrate, there are three different datasets, bedrock, ancient-stable deposits, and recent versant deposits. For the vegetational cover has been used a dataset that comprehend only the high vegetated areas classified according to the wood type and dominant taxa present.

The first level of the map contains all the areas in which is present at least one of the conditions under our phenomena, the second at least two and so on. This system allows the production of a pyramidal system whit increasing number of superimposed layers till reaching the number of layers utilized for the evaluation.

The first operation made by the algorithm is the command extraction by location, to extract all the features for each layer that are located under every gravitative phenomena individuated by the census then it merges them. Due to the previous preparation when the algorithm extracts a feature, it will forcibly extract all features belonging to the same class. This allowed us not to extrapolate the single feature under the phenomenon, but all features in the same category even if there is no an activated landslide above. This first level so, individuates all the area of the municipality in which insist the presence of at least one of the conditions presents under again at least one of the phenomena censused. Under a gravitative phenomenon censused, we have eight layers of conditions, the first passage extracts all that features and merge them, so we have all those areas, belonging to different factors, plotted already in the first level allowing us to have already all the underling conditions taken into account. The second part of the algorithm is devoted to the production of the second level in which we have all the areas of the municipality that share at least two background categories with the phenomena. The first passage of the algorithm is to clip, with the function polygon clipping, all the feature selected by the first part of the algorithm among each other. So, it will perform a several number of clips according to the number of simple combinations (without repetition) with length two between the layers. This means that, the features extracted by the first layer of background conditions is clipped with the second, the third with the fourth and so on till every layer has been clipped with the others in all the possible combinations, precisely 28 clips and then merged together. The third and the fourth levels are created with the same method of the second but with increasing number of combinations, respectively 56 and 70.

Obviously, the algorithm should reach eight level, like the number of starting categories, since the combination add a new superimposed layer every time. After the algorithm has run, also the fifth level, it is clear that every superior level should have almost similar in terms of areal expansion than the previous. So, has been decided to reach at most the accuracy of the fourth level. Another reason why we stopped is the very specific situation of the areas evidenced, they are mainly exposed to south, are present only in a range of elevation and slope very specific and other unique factors.

The final map contains the four levels of susceptibility, and the probability of occurrence of a gravitative phenomenon is higher in the fourth level. The area is very complex and the susceptible portions are not confined but spread all over the territory (Fig. 1). The first level cover almost the entire area of the municipality. The second and the third level are smaller, but again very wide. They are instead confined in a range of elevation that varies in accordance with the exposition and is reduced from second to third level. The fourth, border instead a very confined parts of the municipality. Like was expectable we can see a concentration along the versant with high or moderate slope and

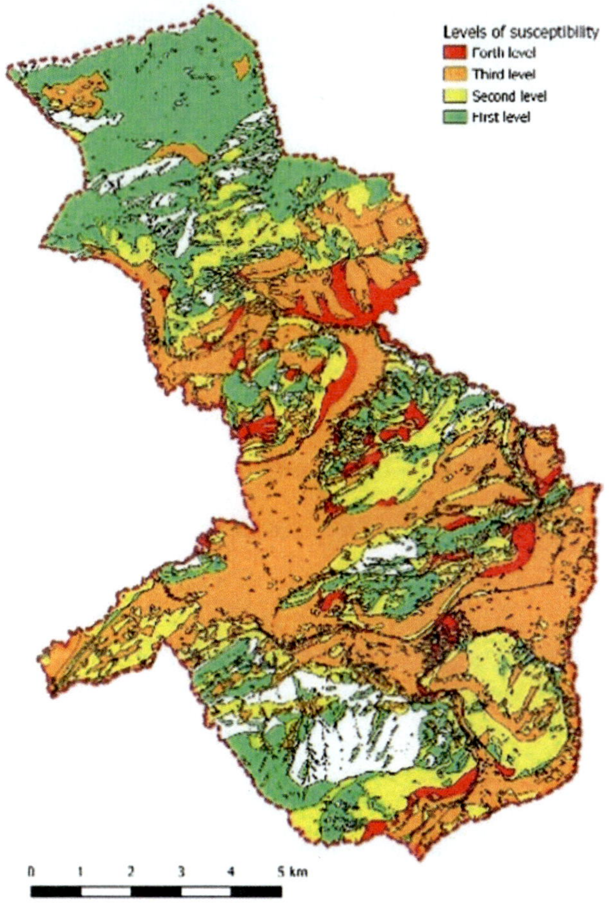

Fig. 1 Map of susceptibility of the municipality

where the incision of the hydrographic net is more marked. These areas share also the same main exposition, towards south, south-east, and an elevation that ranges from 350 and 600 m mainly with few pics till 900 m. If we search more in deep this fourth level areas have in common also the substrate typology, they are mainly over the Flysch Bellunese formation that is one of the formations with worst mechanical behaviour in the area. In the portion individuated there are also consolidated moraine deposits, that associated with the inclination favourable of the underling Flysch formation and the incised hydrographic net provide good ground for the insurgence of gravitative phenomena.

4 Discussion

We can see that more or less all the areas of the municipality share at least one of the underling conditions that are present under the phenomena censused and till the fourth level we cannot appreciate a confined area. This condition picture the complexity of the area, that in the last few years challenged the municipality with a request of measures of intervention to regime the waters. This also poses a question, if most parts of the municipality, where are present the housing settlements share three underling conditions with the areas in which we have had an activation of gravitative phenomena, a possible future event with same or higher intensity should be very challenging to manage. The algorithm could be improved; already the analysis have been taken only eight conditions and the algorithm is not considering a lot of other factors, like the amount of precipitations that can influence the activation of these phenomena.

In the WOE we have that every single feature has a weight according to the number of landslides above, so the evaluation is very precise on a statistical point of view. Instead in the analysis performed in this project the features have not a weighted count in the processing, every feature that has a phenomenon positioned above, is counted with weight one, so at the end the map could be statistically less accurate but more conservative, permitting to not underestimate the risk. An important aspect during the assessment of the gravitative phenomena, with methodologies using a GIS software, are the accuracy of the dataset used for the calculation (Park et al., 2013). In other words, a greater scale is more accurate and should be interesting use different coefficient according to precision of the maps source of the data analysed. Vaia storm has recharged the groundwater system for several month also delaying the spring hydrograph recession. This additional recharge has sustained also the baseflow of the following dry season (Lucianetti et al., 2019). This higher amount of water in the aquifers should be a trigger of gravitative phenomena also after a consistent period of time from the event and the fourth level of the map, product of this work, should border the areas in which we could have those manifestations.

5 Conclusion

Unfortunately, a lot of fallen trees are still stacked in the forests, this because the number of plants killed during this event is equal to five years of normal harvesting. This particular situation further exposes the slopes, which are already depleted by the stabilizing action of the roots, gaining more weight on their shoulders. For sure to have a better evaluation of the mechanism going on we should take into consideration also a lot of other factors. Extreme weather events are becoming more frequent, with rainfall becoming less frequent but more intense, this highlights the importance of research of this kind, where the cause-and-effect relationship is clear and instantaneous. The opinion is that especially in this last years a probabilistic and predictive approach is the right way to do prevention and focus the areas on which make the interventions saving public funds and solve a problem before could manifests its effects.

References

Gentilucci, M., Materazzi, M., & Pambianchi, G. (2021). Statistical analysis of landslide susceptibility, Macerata province (Central Italy). *Hydrology, 8*(1), 5.

Lucianetti, G., Mastrorillo, L., Mazza, R., & Partel, P. (2019). Groundwater response to precipitation extremes: the case of the "Vaia" storm (Eastern Italian Alps). *Acque Sotterranee-Italian Journal of Groundwater*, 8(4). https://doi.org/10.7343/as-2019-429

Park, S., Choi, C., Kim, B., & Kim, J. (2013). Landslide susceptibility mapping using frequency ratio, analytic hierarchy process, logistic regression, and artificial neural network methods at the Inje area. *Korea. Environmental Earth Sciences, 68*(5), 1443–1464.

Sarkar, S., & Kanungo, D. P. (2004). An integrated approach for landslide susceptibility mapping using remote sensing and GIS. *Photogrammetric Engineering & Remote Sensing, 70*(5), 617–625.

Mitigation of Climate Change Through Using Renewable Energy in Afghanistan

Noor Ahmad Akhundzadah

Abstract

Afghanistan's contribution to greenhouse gas emissions is negligible on a global scale however Afghanistan is adversely affected by the effects of climate change. Climate changes deeply affected Afghanistan's natural resources, including water, agriculture, forests, pastures, rangeland, and ecosystems. Livelihoods are dependent on agriculture and create the basis of the country's economy. Renewable energy sources, including hydropower, solar, wind power, and biomass is abundant, and their potential is much more than the national energy and electricity demands. The protracted war and political instability damaged the socio-economic and environmental infrastructure of the country. Therefore, Afghanistan mostly relies on energy transition from neighboring countries however the country possesses huge renewable energy resources. Also, renewable energy resource utilization greatly influences environmental and socio-economic development. In this study, it is investigated how renewable energy contributes to mitigating climate change in Afghanistan. Also, a rough estimation of renewable energy resources' potential is conducted. In addition, the role of renewable energy resources in the sustainable development of the country is reviewed.

Keywords

Renewable energy · Climate change · Sustainable development

N. A. Akhundzadah (✉)
College of Agriculture and Life Sciences, Cornell University, Ithaca, NY, USA
e-mail: naa54@cornell.edu

1 Introduction

The main cause of climate change and global warming is the greenhouse gas (GHG) emissions by continued use of fossil fuels (Arias et al., 2021). Renewable energy is an alternative to nonrenewable energy and is the basis for climate change mitigation and adaptation (Edenhofer et al., 2012). Afghanistan is one of the most vulnerable countries to climate change impacts however GHG emission is negligible on the global and regional scale (INDC, 2015; NEPA & UNEP Afghanistan, 2016). Also, vulnerability to climate change is aggregated by protracted conflicts, instability, and environmental changes in Afghanistan (Price, 2019).

Since 1960, the mean annual temperature over Afghanistan has increased by 0.6 °C and the mean annual rainfall has decreased at an average rate of 0.5 mm per month (McSweeney et al., 2007; Savage et al., 2009). The future climate projections of Afghanistan under the optimistic (RCP 4.5) scenario, show a trend of approximately 1.4 °C warming by 2050 and about 2.6 °C at the end of the century. The pessimistic (RCP 8.5) scenario is extreme; warming is increasing by approximately 2 by 2050 and 6 °C by 2100 across the country (NEPA & UNEP Afghanistan, 2016). Precipitation projections are more uncertain, and most models predict a distinct decrease in precipitation during the spring season mostly in rain-fed agriculture months (NEPA & UNEP Afghanistan, 2016). The main climate change-induced disasters are drought and flood that severely affected the major economic sectors of the country such as water resources and agriculture. Afghanistan has substantial renewable energy (RE) resources such as solar, wind, hydro, biomass, and geothermal (Ludin et al., 2017; World Bank Group, 2018). The use of domestic renewable energy resources will have the following benefits: (1) Will complement the electricity transition from the neighboring countries which costs tens of millions per year. (2) The internal RE resource utilization is cheap and affordable and bars the current massive

H. Chenchouni et al. (eds.), *Recent Advancements from Aquifers to Skies in Hydrogeology, Geoecology, and Atmospheric Sciences*, Advances in Science, Technology & Innovation, https://doi.org/10.1007/978-3-031-47079-0_66

nonrenewable energy resources use that produce tons of GHG. (3) The influencing CO_2-producing sectors as energy and its related sectors will reach the Net Zero Emission (NZE) target. In addition, there are no studies conducted about mitigating climate change using domestic RE, and the role of RE in the economic and environmental sustainability of the country. Moreover, Afghanistan needs to build its coping capacity against climate change risks through the development and utilization of its potential renewable energy resources. In conclusion, the main purpose of this paper includes but is not limited to a rough estimation of RE and CO_2 emissions in Afghanistan. Climate change mitigation through using RE resources of the country, and a general review of the role of RE in the sustainable development of the country.

2 Methods and Materials

In this study, mixed qualitative and quantitative research methods were used. The issue of climate change in Afghanistan was mostly drawn from published papers and documents. Current energy consumption, total GHG emission, and renewable energy resources potentials of Afghanistan were collected from several national resources and reliable international databases. For annual GHG emissions, Climate Watch online database was used (Afghanistan Available & online, 2022). For hydropower, the Afghanistan Power Sector Master Plan was used (Fichtner Power Sector Master Plan, 2013). Solar and wind energies were estimated using Global Solar and Wind Atlas (Global Wind Atlas, 2022). The data were analyzed using ArcGIS Pro and MS. Excel. In addition, some of this paper's raw materials are drawn from the author's own experience, supplemented by intensive desk-based literature on related topics and the life experiences of the author in the topics.

3 Results and Discussion

3.1 Current Energy System and Renewable Energy Resources Potential

Based on a survey conducted in 2017, about 97% of households have some kind of access to electricity but only 30.9% received their power from the grid (IRoA_CSO, 2016). A large portion of the grid is provided by imported electricity from neighboring countries and supplemented by domestic hydropower and thermal plants (IRoA_CSO, 2016). Afghanistan imported a total of 900 Mega Watt (MW) of electricity from Tajikistan, Uzbekistan, Turkmenistan, and Iran during the Government of the Islamic Republic of Afghanistan (GoIRA) before collapsed on August 15, 2021. The internal sources provided 604.6 MW, shared 263.4 MW by Hydropower, 93.3 MW by a diesel generator, and 247.9 MW by thermal power plants (Slimankhil et al., 2020). However, the total 1504.6 MW electricity was less than the national energy demands, and often the electricity transmission cables were disconnected by the insurgences operation which created more costs. The total hydroelectric capacity of Afghanistan was estimated at 23,000 MW, solar electricity was estimated at 222,000 MW and wind electric capacity was estimated at 150,000 MW (World Bank Group, 2018). The geothermal and biomass resources for electricity production are huge. The total electricity demand of the country is less than the existing renewable energy capacity. Access to electricity supplied by renewable resources is one of the higher development priorities for Afghanistan. The exploitation of the existing RE resources will stop the further use of most of the nonrenewable energy resources in the country. Especially the energy-related sectors will be reached zero emissions. Therefore, the utilization of existing renewable energy resources provides the increasing energy demand and has a strong influence on the economic and environmental sustainability of the country (Shoaib & Ariaratnam, 2016).

3.2 Annual CO_2 Emission

In 2019 the total GHG emission of Afghanistan was estimated at 28.79 million tons of CO_2 equivalent (Mt CO_2 eq.) representing 0.06% of global emissions, and the annual per capita emission was 0.76 (tCO_2 eq.) (Afghanistan Available & online, 2022). The annual national CO_2 emission was 12.16 (Mt CO_2 eq.), and the per capita CO_2 emission was 0.3 metric tons as shown in Fig. 1. The neighboring countries GHG emissions are Iran with 745 (Mt CO_2 eq.) at the top and then Pakistan with 235, Uzbekistan with 112, and Turkmenistan with 75, respectively (Ritchie et al., 2020). Therefore, Afghanistan's annual GHG emissions are negligible on a global and regional scale. Also, the small amount of CO_2 emission is possible to eliminate by utilizing the existing RE resources.

The steep increase after 2006 in CO_2 emission shows the rapid massive industrial and infrastructure development during GoIRA with international community support in the country fueled by fossil fuel energy.

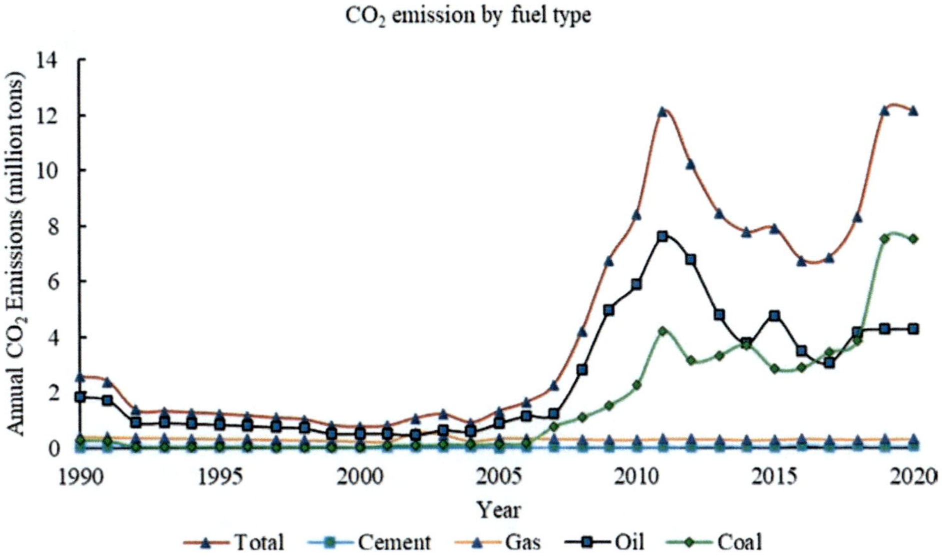

Fig. 1 Annuawl CO_2 emissions from different fuel types in Afghanistan

3.3 Renewable Energy, Climate Change Mitigation, and Sustainability

Climate change mitigation policies are set based on energy transition from nonrenewable to RE energy use (Edenhofer et al., 2012). As shown in Fig. 1, the main sources of CO_2 emissions in Afghanistan are oil and coal used in major industrial processes. Most of the air pollution in Afghanistan's big cities is due to the use of nonrenewable energy resources. Oil and electricity are imported from neighboring countries that bear huge costs yearly whereas Afghanistan has huge RE resources as alternatives for the imported electricity and oil and internal coal. RE production is affordable, economical, and sustainable for Afghanistan. For RE production and its infrastructure, the country has sufficient materials such as copper, iron, lithium, and a lot of other mining (Richard, 2014). Also, developers are interested in RE resources exploitation and its use will attract UNFCCC climate finance resources as well. With this potential, energy-related sectors will reduce their CO_2 emission nearly to zero. In addition, renewable energy is directly linked with SDGs 7, 11, and 13 and indirectly influences other SDGs (UNDESA_DSDG, 2018). Therefore, the development and utilization of the existing RE resources are vital for climate change mitigation, environmental restoration, and digital transformation which cumulatively ends up in the socio-economic development of Afghanistan.

4 Conclusion

Afghanistan has been adversely affected by climate change impacts where the GHG emission of the country is insignificant on regional and global scales. Afghanistan has rich renewable energy resources of hydro, solar, wind, and biomass. The renewable energy resources potentials are much more than the total national energy demand. With abundant water resources, wide arable land, and energy resources Afghanistan has a high capacity for climate change mitigation, adaptation, and reducing air pollution in largely populated cities. Water resources conservation and management for hydropower, irrigation, and use of the existing RE resources are the basis for the country's climate change mitigation and a strong foundation for the country's sustainable development. However, the protracted war and conflict, and political instability hampered sustainable development in the country. Establishing a responsible government, peacebuilding, and rule of law with international partnership are prerequisites for sustainability and prosperity in the country.

References

Arias, A. P., Bellouin, N., Trewin, B., Achuta Rao, K., den Hurk, B. van. (2021). Technical Summary. *In climate change 2021: The physical science basis. Contribution of working group I to the sixth assessment report of the intergovernmental panel on climate change*. Cambridge (pp. 33–144).

Climate Watch. (2022). *Afghanistan* Available online: https://www.climatewatchdata.org/countries/AFG?end_year=2019&start_year=1990. Accessed on October 29, 2022.

Edenhofer, O., Pichs-Madruga, R., Sokona, Y., Seyboth, K., Eickemeier, P., Matschoss, P., Hansen, G., Kadner, S., Schlömer, S., Zwickel, T. et al. (2012). *Renewable energy sources and climate Change mitigation: Special report of the IPCC* (Vol. 49)

Fichtner. (2013). *Power sector master plan: Ministry of energy and water, Islamic republic of Afghanistan*; Kabul, Afghanistan

Global Wind Atlas. (2022). https://Globalwindatlas.Info/En/Area/Afghanistan. Available online: https://globalwindatlas.info/en/area/Afghanistan?download=print. Accessed on October 29, 2022.

INDC. (2015). *Intended nationally determined contribution submission to the United Nations framework convention on climate change, islamic republic of Afghanistan*

IRoA_CSO. (2016). *Afghanistan living conditions Survey 2016–17*. Kabul: Afghanistan (Vol. 14).

Ludin, G. A., Amin, M. A., Aminzay, A., & Senjyu, T. (2017). Theoretical potential and utilization of renewable energy in Afghanistan. *AIMS Energy, 5*, 1–19. https://doi.org/10.3934/energy.2017.1.1

McSweeney, C., New, M., Lizcano, G. (2007). *UNDP climate change country profiles documentation*.

NEPA & UNEP. (2016). *Afghanistan: Climate Change Science Perspectives*. Kabul, Afghanistan.

Price, R. (2019). *Climate change as a driver of conflict in Afghanistan and other fragile and conflict affected States*. Brighton: UK

Richard, S. (2014). Mother of All Lodes. *Science, 345*, 725–727. https://doi.org/10.2307/24917140

Ritchie, H., Roser, M., Rosado, P. (2020). CO_2 and greenhouse gas emissions. Published online at OurWorldInData.Org. Available online: https://ourworldindata.org/co2/country/afghanistan#what-are-the-country-s-annual-co2-emissions. Accessed on September 10, 2020.

Savage, M., Dougherty, B., Hamza, M., Butterfield, R., Bharwani, S. (2009). *Socio-economic impacts of climate change in Afghanistan a report to the department for international development*; Stockholm Sweden.

Shoaib, A., & Ariaratnam, S. (2016). A study of socioeconomic impacts of renewable energy projects in Afghanistan. *Procedia Eng, 145*, 995–1003. https://doi.org/10.1016/j.proeng.2016.04.129

Slimankhil, A. K., Anwarzai, M. A., Sabory, N. R., Danish, M. S. S., Ahmadi, M., & Ahadi, M. H. (2020). Renewable energy potential for sustainable development in Afghanistan. *Journal of Sustainable Energy Revolution, 1*, 8–15. https://doi.org/10.37357/1068/jser.1.1.02

UNDESA_DSDG. (2018). *HLPF review of SDG implementation: SDG 7-ensure access to affordable, reliable, sustainable and modern energy for all*.

World Bank Group. (2018). *Afghanistan renewable energy development issues and options*; Energy Stu.; World Bank: Kabul, Afghanistan.

Assessment of the Potential of Renewable Energy with Bias Correction Due to Climate Change Over South Asia Using Global Atlas Dataset

Muhammad Abid Khan and Koji Dairaku

Abstract

In order to meet the rising energy needs in South Asia as a result of population growth and societal development, the development of renewable energy is of paramount importance. With the hindcast data from the global atlas (GWA3 and Solargis) dataset, the aim of this study is to determine how climate change impacts wind power density (WPD) at 50 m above the ground level (AGL) through the power law method and photovoltaic (PV) potential over the subcontinent by using the climate change historical evaluation. The developing countries require the use of renewable energy, because of this, knowing the ideal location is important, and implementing wind and solar energy initiatives is crucial in this regard. The validation of GWA3 is based on the wind atlas methodology and the wind atlas analysis and application program (WAsP). Specifically, when analyzing threshold wind speed factors that influence wind energy production, according to the novelty of this research that applies the bias correction techniques quantile mapping based on Weibull distribution. Plus, global solar atlas uses Solargis PV modeling software and a high-resolution database of solar resources. The duration of data for wind energy and long-term reference data (ERA 5) is 2 and 12 years, respectively. In addition, overall subcontinent the average daily/yearly totals of solar PV plant that produces 1 kW-peak electricity, calculated for the period of 20 recent years. For WAsP modeling, the mean wind speeds of the predicted wind climates have an estimated ~ 6.5% standard uncertainty in the region of Pakistan and Maldives, and Bangladesh, respectively. For GWA3 modeling, the mean wind speeds of the predicted wind climates have an estimated mean absolute bias of 17% ± 8% and ± 2% standard deviation (assuming no bias or uncertainty on the WAsP modeling) in the region of Pakistan and Maldives, respectively. Moreover, for Bangladesh 1% ± 0 percent standard deviation is. Furthermore, in the simulation, losses due to dirt and soiling were estimated at 3.5%, while the total loss from all conversion losses was estimated at 7.5%. Another point worth mentioning was to estimate the energy density based on the data results on wind speed (m/s) at the hub height of 50 m AGL over the subcontinent. For instance, maximum and minimum energy density regions are Afghanistan and Maldives, respectively. Meanwhile, the PV system consists of crystalline silicon PV modules mounted at a fixed position, tilted to maximize yearly energy production. Consequently, understanding long-term WPD and PV power potential for micro-scale energy yields is crucial for understanding subcontinental variations. Additionally, the RE system reduces GHGs and limits global climate change. The CO_2 can be switched to low with this method.

Keywords

Renewable energy · Subcontinent · Simulations · Bias correction · Weibull distribution

1 Introduction

A growing number of greenhouse gas (GHG) emissions from burning fossil fuels make climate change the biggest environmental issue today (Burton, 2014). As a result, scientists are trying to come up with a source of energy that doesn't harm the environment and generates enough energy (Khan, 2018). Renewable energy sources such as wind and solar can generate electricity cleanly and at a relatively low cost. Using a photovoltaic (PV) module and a wind

M. A. Khan (✉) · K. Dairaku
Department of Engineering Mechanics and Energy, Graduate School of Science and Technology, University of Tsukuba, Tsukuba 305-8577, Ibaraki, Japan
e-mail: s2130195@u.tsukuba.ac.jp

H. Chenchouni et al. (eds.), *Recent Advancements from Aquifers to Skies in Hydrogeology, Geoecology, and Atmospheric Sciences*, Advances in Science, Technology & Innovation, https://doi.org/10.1007/978-3-031-47079-0_67

turbine, wind and solar energy can be converted into electricity (Khan, 2020). Several countries have met their electricity needs with wind and solar power. Wind and solar atlases generated from weather forecast models could assist address "grand challenges" in the wind and photovoltaic energy science of them. To model the interaction between microscales and mesoscales of atmospheric motion, wind atlases have been used in the past.

There have been a lot of scientific studies on using wind and solar energy to generate electricity. Zhou et al. (2011) studied the potential for onshore wind energy in China using meteorological data from 1979 to 2008. About 146,336 GWh can be produced by onshore wind farms in the study region. Additionally, Rehman et al. (2017) examined whether a grid-connected photovoltaic (PV) system of 10 MW could be feasible in 44 locations in Saudi Arabia. A future PV plant should be installed at Bisha, they concluded. Based on the Paris Agreement and the International renewable energy agency (IRENA), it should limit global warming, decrease the temperature and explore switching from high carbon to low carbon energy by wind yield to limit climate change for the next generation, respectively. In addition, based on the IPCC fourth and fifth reports global warming is caused by human action and needs to decrease greenhouse gas (GHG) emissions to stop climate change, sequentially.

The wind characteristics at turbine hub height over the region under changing climate for micro-scale energy yield have been investigated in only a few studies. Renewable energy (RE) resources are abundant in South Asian countries, which can produce a great deal of electricity in a short period of time. The paper followed a method based on subcontinental results for fast-cycle correction of windspeed forecast biases. Hence, a novel perspective is presented on the new method for windspeed forecast corrections. In this regard, it is appropriate to determine how RE should be utilized as well as to implement appropriate technologies to maximize its potential.

2 Methods

The validation of GWA3 is based on the wind atlas methodology and the wind atlas analysis and application program (WAsP). Specifically, when analyzing threshold wind speed factors that influence wind energy production, according to the novelty of this research that applies the bias correction techniques quantile mapping based on Weibull distribution. For bias correction of variables are applied quantile mapping with Weibull distribution (QWD), expressed in the following Eqs. 1 and 2 (Haas, 2014). Further, the power law method was used for extrapolation. Moreover, global solar atlas uses Solargis PV modeling software and a high-resolution database of solar resources. The duration of data for wind energy and long-term reference data (ERA-5) is 2 and 12 years, respectively. In addition, overall subcontinent the average daily/yearly totals of solar PV plant that produces 1 kW-peak electricity, calculated for the period of 20 recent years.

$$W_{\text{cor. m/s}} = F_{\text{ref}}^{-1}(F_{\text{atlas}}(W_{\text{atlas}})) \tag{1}$$

$$W_{\text{cor. m/s}} = \alpha_{\text{ref}}\left[-\ln\left(1-\left(1-\exp^{-\left(\frac{W_{\text{atlas}}}{\alpha_{\text{atlas}}}\right)^{\beta_{\text{atlas}}}}\right)\right)\right]^{\frac{1}{\beta_{\text{ref}}}} \tag{2}$$

Weibull parameters scale (α) and shape (β) are calculated by the method of moments.

3 Results

3.1 Energy Potential Calculations

Afterward, the correction of the climate variable based on QWD, the author investigated the energy protentional based on the global atlas sources for the wind yield in the first part of the research. Meanwhile, the second part of the research is solar yield by global solar atlas through Solargis did the work.

3.2 Wind Energy Yield

Figure 1 illustrated the wind energy density at 50 m above the ground level for the subcontinent on average at 6.02 m/s velocity. In this research, the area shows the maximum energy level in Afghanistan, followed by Nepal, and the minimum comes out in the Maldives, followed by the second last minimum yield in Bangladesh. This data is for 10% of the windiest areas in this research. Therefore, on mean value is 370.5 w/m^2 for the energy density, based on the average wind speed is 6.02 m/s. In the case of validation sites for GWA3 modeling 12 meteorology stations were examined in Pakistan, 1 and 2 for Bangladesh and Maldives, respectively.

For WAsP modeling, the mean wind speeds of the predicted wind climates have an estimated ~ 6.5% standard uncertainty in the region of Pakistan and Maldives, and Bangladesh, respectively. For GWA3 modeling, the mean wind speeds of the predicted wind climates have an estimated mean absolute bias of 17% ± 8% and ± 2% standard deviation (assuming no bias or uncertainty on the WAsP modeling) in the region of Pakistan and Maldives, respectively. Moreover, for Bangladesh 1% ± 0% standard deviation is.

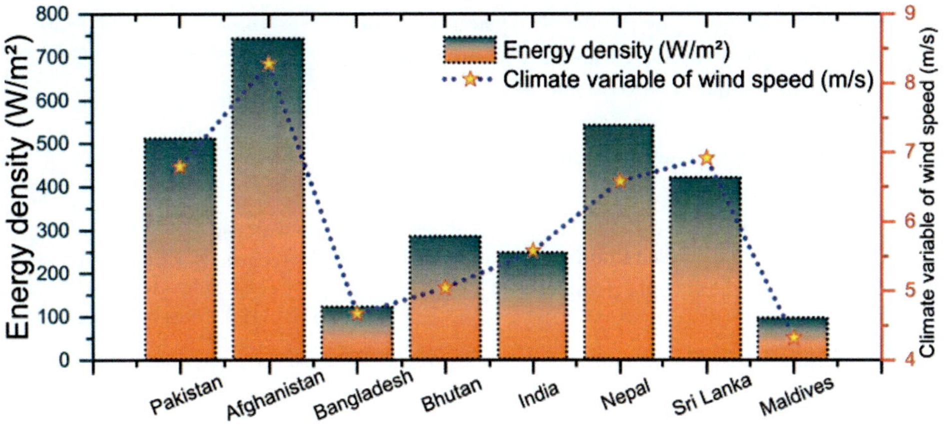

Fig. 1 Wind energy density at 50 m above the ground level (AGL) presenting for the subcontinent countries at an average of 6 m/s

3.3 Photovoltaic Power Potential Yield

In the second part of the research simulation for PV power potential, losses due to dirt and soiling were estimated at 3.5%, while the total loss from all conversion losses was estimated at 7.5%. There's no downtime on the power plant. Based upon a 30-min time step analyzed by satellite and atmospheric data, the solar resource database is determined with a spatial tenacity of 1000 m and is computed from meteorological and satellite data. PV system consists of crystalline silicon PV modules mounted at a fixed position, tilted to maximize yearly energy production presented in an overview in Fig. 2.

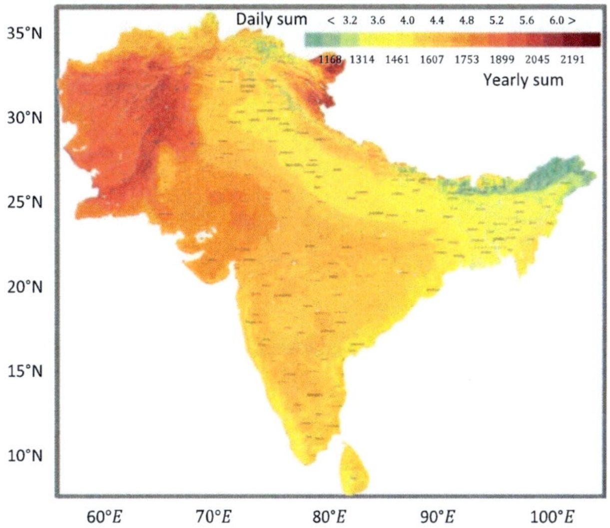

Fig. 2 Photovoltaic power yield consists of crystalline silicon PV modules mounted at a fixed position, tilted to maximize yearly energy production over the subcontinental countries

4 Discussion

An additional point worth mentioning was to estimate the energy density based on the data results on wind speed (m/s) at the hub height of 50 m above the ground level over the subcontinent. For instance, maximum and minimum energy density regions are Afghanistan and Maldives, respectively. Meanwhile, the PV system consists of crystalline silicon PV modules mounted at a fixed position, tilted to maximize yearly energy production. Tilting the telescope toward the equator at least 0.0° but not more than 39.0° is optimal. Inverters that are high-efficient are assumed to be used.

As in previous research such as (Haas, 2014), modeling uncertainties arising from model choice affect methodology when assessing climate change's impact on renewable energy yield. According to Khan (2020), Rehman et al. (2017), the studies applied RETScreen models for photovoltaic power yield which raise the uncertainties in the collection and analysis of the dataset without bias correction. Consequently, in the present research to decrease the uncertainties in the investigation of renewable energy such as wind and solar yield by the bias correction techniques quantile mapping based on Weibull distribution applied.

For the future, it should be calculated at 100 m above the ground level to know the performance of QWD, but the data availability level is limited. However, most of the studies applied the power law method to extrapolate to the desired height without validation. Undoubtedly PV systems should consider the dual position tilted to maximize yearly energy yield after applying the QWD method.

5 Conclusion

This research examines the wind power density, photovoltaic power, and static atmospheric stability of two state-of-the-art wind and solar atlases. Consequently, firstly, long-term WPD and PV power potential for micro-scale energy yield assist understand the variations in regions that has be highly susceptible in the subcontinent. Secondly in the additional research, applying the RE system can reduce GHG and limit global climate change and break the temperature. So, this method can switch to low CO_2. Estimated the energy density at the hub height of 50 m AGL for the subcontinental region renewable energy yield was maximum and minimum energy density regions are Afghanistan and Maldives, respectively. Besides, there is no downtime on the PV power plant. Based upon a 30-min time step analyzed by satellite and atmospheric data.

References

Burton, L. (2014). *Agriscience: Fundamentals and applications*. Cengage.

Haas, R., Pinto, J. G., & Born, K. (2014). Can dynamically downscaled windstorm footprints be improved by observations through a probabilistic approach? *Journal of Geophysical Research: Atmospheres*, 119, 713–725.

Khan, M. A., Çamur, H., & Kassem, Y. (2018). Modeling predictive assessment of wind energy potential as a power generation sources at some selected locations in Pakistan. *Modeling Earth Systems and Environment*, *5*, 555–569. https://doi.org/10.1007/s40808-018-0546-6

Khan, M. N. K. A. R. (2020). Power generation development based on solar potential at urban and rural regions selected locations in Pakistan. *International Journal of Mechanical and Production Engineering Research and Development (IJMPERD)*, 2601–2618.

Rehman, S., Ahmed, M., Mohamed, M., & Al-Sulaiman, F. (2017). Feasibility study of the grid connected 10 MW installed capacity PV power plants in Saudi Arabia. *Renewable and Sustainable Energy Reviews*, *80*, 319–329. https://doi.org/10.1016/j.rser.2017.05.218

Zhou, Y., Wu, W., & Liu, G. (2011). Assessment of onshore wind energy resource and wind—Generated electricity potential in Jiangsu, China. *Energy Procedia*, *5*, 418–422. https://doi.org/10.1016/j.egypro.2011.03.072

Mapping the Shoreline Evolution in Response to Sea Level Rise Along Agadir Bay, Morocco: Geospatial and Empirical Approach

Abdelhaq Aangri, Mounir Hakkou, Yann Krien, Toufik Chtioui, Zakaria Elmostafa and Aicha Ben Mohammadi

Abstract

Global warming resulting from meteorological and oceanic perturbations manifests itself, among other things, in the rise of sea level and the increase in the energy of marine storms, as well as their frequency. The IPCC predictions put forward the probability of sea level rise (SLR), which could even exceed 1m by 2100 for the most pessimistic scenarios (i.e., RCP8.5). As a result, coastal erosion events will increase. Sandy coasts are generally considered to be sensitive to erosion and SLR. They become more vulnerable when they are heavily occupied by socio-economic activities and marine structures. To predict Agadir's bay shoreline movement due to erosion and sea level rise in the long term (2100), we used multi-date aerial photographs to initially deduct the historical rates of shoreline movement (1969–2016) and spatially delimit eroding areas and accreting areas, using Digital Shoreline Analysis System (DSAS). The calculated and geospatialised rates were then used to predict the displacement of the shoreline by 2100, using the empirical models of (Bruun in J Waterways Harbors Div 88:117–130, 1962) adjusted by Ferreira (2006) as well as the IPCC projections. The results show a considerable impact of marine structures on the erosion, balance and accretion processes along Agadir Bay (7 km), which reach 38%, 20%, and 42%, respectively. The shoreline prograde in the northern zone (1.35 m/year), and retreated in the south (− 1.54 m/year). Predictions of shoreline displacement by 2100 under RCP8.5 reveal a potential risk for tourist infrastructures bordering the sea (hotels, cornices, etc.) and call on coastal managers to develop protection and adaptation strategies. The predicted average setback rates of the shoreline are between − 16.8 m and − 33.7 m for 2050 and 2100, respectively, exceeding the width of the beach in several localities.

Keywords

Erosion area · Shoreline change · Sea level rise · Prediction rule · DSAS

1 Introduction

Shoreline position changes are considered one of the most dynamic processes in coastal areas (Boak & Turner, 2005; Mills et al., 2005). The position of the interface between land and sea changes in time and space as a response to variations in a certain number of influencing factors (Forbes et al., 2004; Cooper et al., 2004; Jana et al., 2013). Shoreline change is controlled by a complex interplay of natural and anthropogenic processes and influences (Burningham & French, 2017).

Shoreline management is the long-term adaptation of the shoreline to the impacts of climate change and its consequences (Burningham & French, 2017). For many applications, such as the standard design for buildings and coastal infrastructures, construction projects, tourism and industrial activities, and emergency management, the long-term shoreline change must be quantified. Observed shorelines indicate either coastal erosion or accretion (Aangri et al., 2022; Hakkou et al., 2018).

A. Aangri (✉) · T. Chtioui · A. B. Mohammadi
Natural Resources and Sustainable Development Laboratory, Geosciences and Environment Team, Faculty of Science, Ibn Tofail University, Kenitra, Morocco
e-mail: aangri.abdelhaq@gmail.com

M. Hakkou
Department of Earth Sciences, Scientific Institute, Mohamed-V University, Rabat, Morocco

A. Aangri · Y. Krien
Université Paul Sabatier Toulouse III, Toulouse, France

Z. Elmostafa
Royal Naval School, Casablanca, Morocco

H. Chenchouni et al. (eds.), *Recent Advancements from Aquifers to Skies in Hydrogeology, Geoecology, and Atmospheric Sciences*, Advances in Science, Technology & Innovation, https://doi.org/10.1007/978-3-031-47079-0_68

The coastline of the study area is 7 km and located in south of the Moroccan Atlantic coast. This meso-tidal coast is aligned at about 10° in the north–south direction, and with a gentle foreshore slope between 0.6° and 1.2°. The backshore of Agadir's beach is equipped with tourist and commercial infrastructures (e.g., hotels, cornices, restaurants, harbor), implanted on the dune which was decapitated in the sixties to diversify and motivate the region's economy. Concerning the forcing factors, the tide is semidiurnal with about 2.9 m of mean range, induced currents on the continental shelf on the order of 0.2–0.3 m/s, while the most frequent swells are those coming from the north-west (Aangri et al., 2022). The main objective of this study is to understand and assess historical and potential future changes in the Agadir Bay shoreline position.

2 Materials and Methods

The methodological approach used in this work is based on the analysis of historical shoreline movement and the prediction of the SLR impact by 2100. The details of this approach are presented (Fig. 1).

Historical shoreline behavior is evaluated here by delineating shoreline positions and analysis of rate change over the last 47 years. Mean High Water (MHW) shorelines (considered as a reference line) are digitized from aerial photographs covering the period between 1969 and 2016.

When to shoreline prediction, the method known as the Bruun (1962). Rule is the most commonly used to predict shoreline recession due to SLR. However, this model has been criticized by numerous researchers (Rosati et al., 2013; Ranasinghe et al., 2012; Aangri et al., 2022) who tried to develop new formulations that give predictions close to reality and propose adjusting Bruun's model (2006) by adding a term that considers the average annual rate of historical retreat of the shoreline. This adjusted model (Eq. 1) is used in this work to attempt to predict the future evolution of the shoreline.

$$R_f = R_h + \Delta s * L/(B + h) \quad (1)$$

whereas h is the maximum depth of exchange of material between nearshore and offshore, L is the horizontal distance from the shoreline to closer depth h, B is the berm or dune elevation estimate for the eroded area, Δs is SLR scenarios, the severe pathway RCP 8.5 scenario of the recent global projection model of SLR (IPCC, 2013) is used, R_h is the average annual rate of historical retreat of the shoreline and R_f is the horizontal extent of coastal recession (Ranasinghe et al., 2012).

Uncertainties in the computed rates' accuracy were quantified and accounted for in subsequent predictive modeling. Several studies (Moore, 2000; Morton et al.,

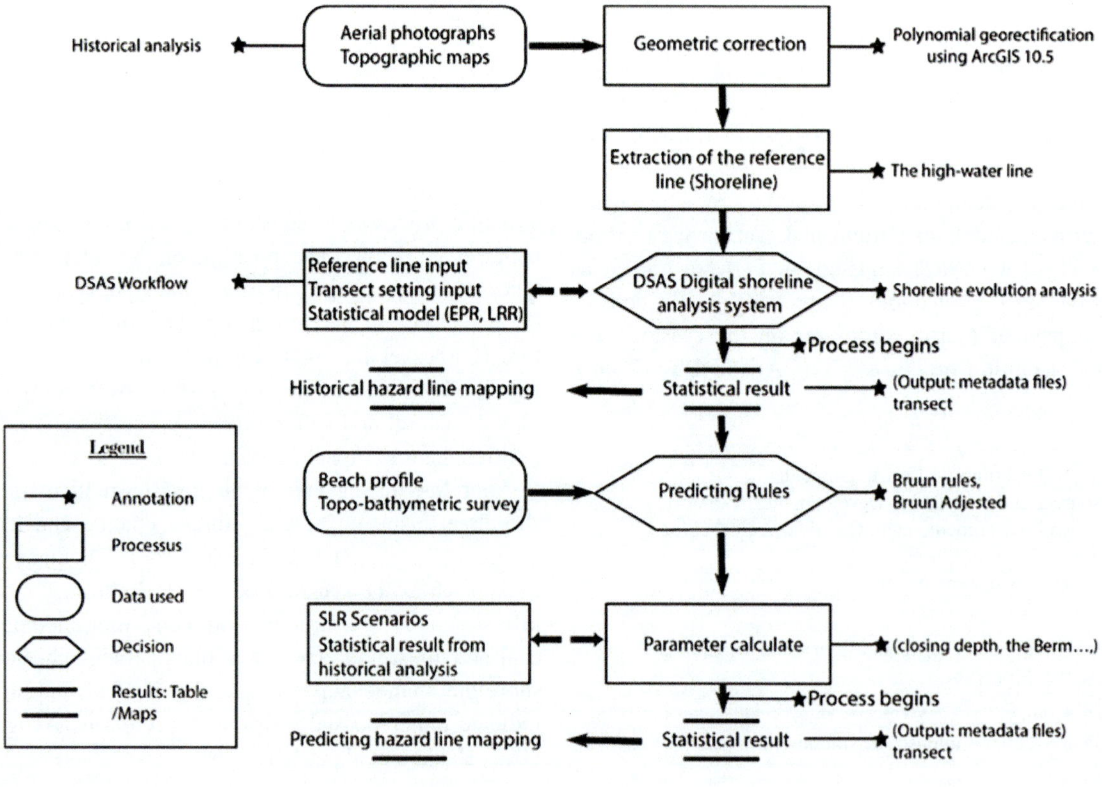

Fig. 1 Methodology flowchart

2004; Thieler & Danforth, 1994) estimated typical measurement errors associated with mapping methods and shoreline digitizing. The total shoreline position erroris is the square root of the sum of the squares of the digitizing error (E_d), Root-Mean Square Error RMSE (E_r), pixel error (E_p), and the variations in position of the high water line of the oscillations induced by the high tide level (E_v). The maximum error of uncertainties was the sum of all errors associated with this data processing, estimated on average between ± 0.449 m/year.

3 Results

Results of shoreline historical analysis between 1968 and 2016 show an overall negative sediment balance, with an average annual retreat rate between − 0.2 and − 0.3 m/year according to the statistical method used (EPR, LRR). The northern sector from the Agadir port complex to the Hotel Robinson boundary shows a tendency toward accretion with an average annual rate of 0.7 m/year (Fig. 2). On the other hand, the central and southern sector, which extends to the mouth of Souss River, show tendency toward equilibrium in central sector and retreat in southern sector with an average annual rate of 0.26 m/year and − 1.5 m/year, respectively.

The shoreline's prediction model for 2050 and 2100 shows critical retreat motions with an average of − 16.8 m and − 33.7 m, respectively, affecting the quasi-totality of Agadir Bay.

The shoreline prediction for 2100 for the northern part of Agadir Bay results (for transects between t1 and T29) in an average evolution of about of + 64 m. In contrast for all the central and southern sectors, between transects T29 and T285, the results obtained show an evolution of shoreline toward the continent, which increases toward the southern end of the study area, with an average evolution of about of − 97 m. The resilience seen in the north of Agadir Bay is due to the protection offered by the harbor dike (Fig. 3).

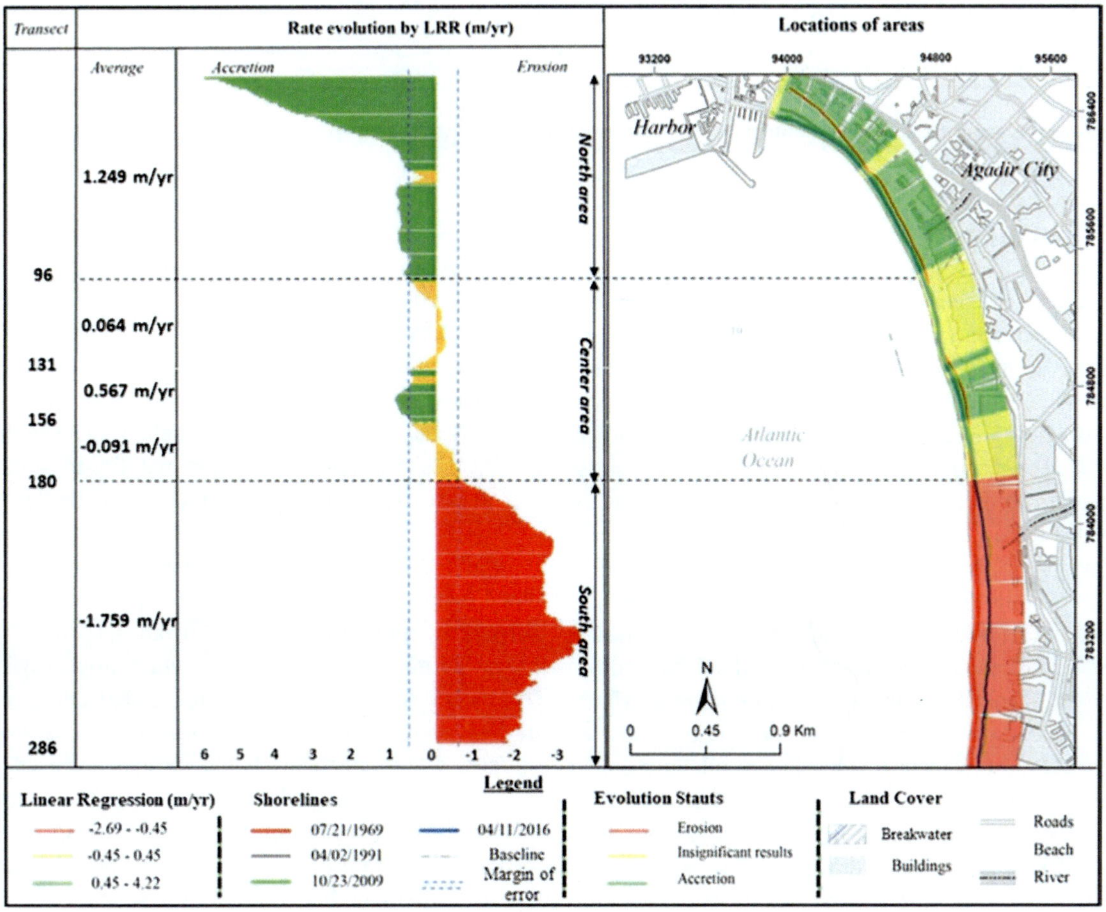

Fig. 2 Sectorial evolution of the shoreline position between 1969 and 2016 (erosion and accretion), calculated by the Linear Regression Rate (LRR)

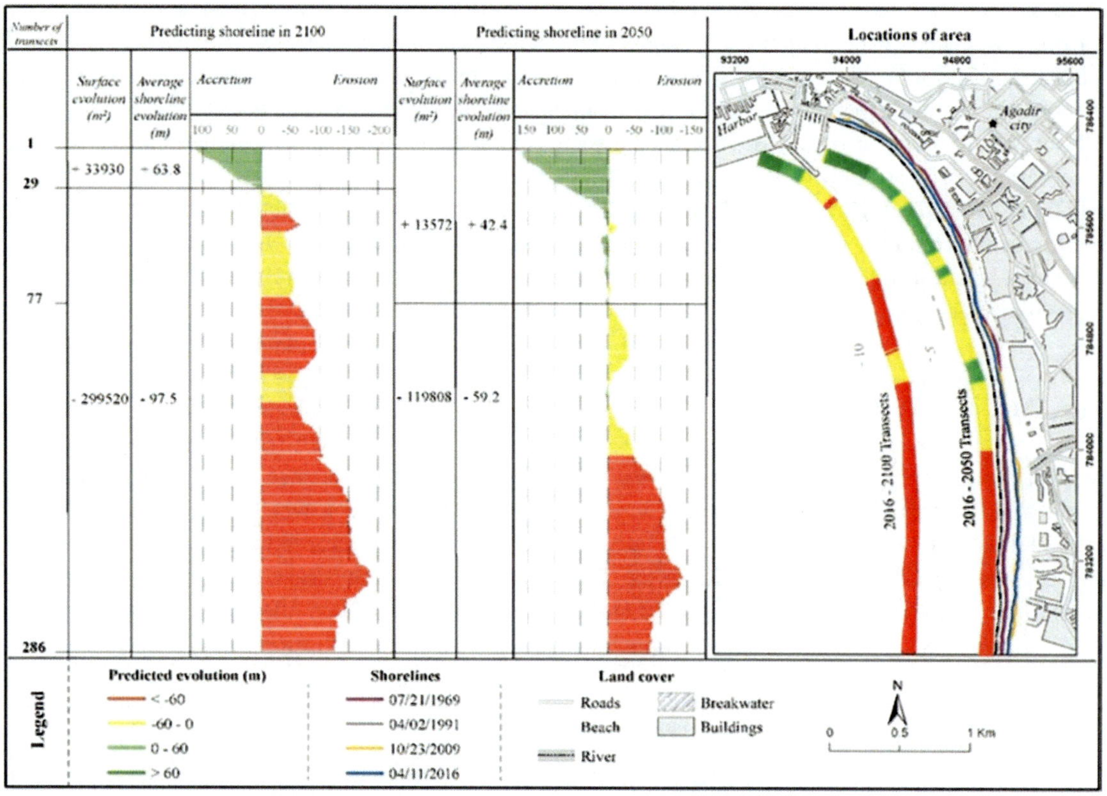

Fig. 3 Sectorial evolution of the shoreline position in 2050, 2100 horizons (erosion and accretion)

4 Discussion

Historically, the accretion of the northern sector, the stability of the central sector and the retreat of the southern sector can be explained by hydrodynamic and anthropogenic factors. Hydrodynamically, the northern sector is sheltered from the offshore swells due to the protection offered by the harbor dike. On the other hand, the central and southern sectors are openly exposed to the swells. The more sloping isobaths to the north of the bay compared to the south induce greater wave refraction in the north than in the south. The dominant offshore swell comes from the NNW sector arriving north of the bay in a W direction and generates a northward expansion current. On another side, the swells diffracted by the harbor dike generate locally a Northward current, which favors the deposition of sands in this sector. Beyond the breakwater toward the south (central and Southern sector), the offshore swell is less refracted at the coast compared to the northern sector. It generates a Southward longshore drift current and grows more in this direction. This explains the retreat of the sector, which is accentuated toward the south. The breakwater position is the boundary between the two areas, each appearing to be a sedimentary cell.

On the other hand, the retreat results obtained using the prediction approach must be interpreted carefully because of the uncertainty due first to the nature of the model used and secondly to the sensitivity of the input parameters, in terms of morphology and in terms of estimating SLR. Concerning the model used, its main limitation is that it considers the evolution of the shoreline linear, although it is subject to irregularities due to variability in forcing from year to year. The multitude of input parameters, in particular the slope, the depth of closure theoretically deduced and the width between it and the shoreline (even if it is estimated from the topo-bathymetric profile because the topo-bathymetric profile can change from season to season), makes the model sensitive to the morphological variations. The model is also sensitive to the global connotations of the SLR inputs. Therefore, the results obtained should be interpreted carefully and should be updated every five to ten years depending on the availability and accuracy of the data and on the progress of science in this field. The results of this work can now be considered and used by the public authorities for such intervention by soft actions to maintain the beach structure.

5 Conclusion

The present study predicts the shoreline evolution of Agadir Bay in response to the SLR, using the results of its historical evolution, morphological measurements and IPCC scenarios forcing the adjusted Bruun model. All the results obtained on the historical kinematics and the future prediction of the shoreline on Agadir Bay verified that the anthropic modifications generated new hydrodynamics totally different from the natural hydrodynamic regimes of the bay. While in terms of climate change, specifically SLR, these short- and long-term impacts on the bay are critical, while simultaneously accelerating coastal erosion and thus the retreat of the shoreline.

References

Aangri, A., Hakkou, M., Krien, Y., & Benmohammadi, A. (2022). Predicting shoreline change for the Agadir and Taghazout coasts (Morocco). *Journal of Coastal Research, 38*(5), 937–950. ISSN 0749-0208.

Boak, E. H., & Turner, I. L. (2005). Shoreline definition and detection: A review. *Journal of Coastal Research*, 688–703.

Bruun, P. (1962). Sea-level rise as a cause of shore erosion. *Journal of the Waterways and Harbors Division, 88*, 117–130.

Burningham, H., & French, J. (2017). Understanding coastal change using shoreline trend analysis supported by cluster-based segmentation. *Geomorphology, 282*, 131–149. https://doi.org/10.1016/j.geomorph.2016.12.029

Cooper, J. A., Jackson, D., Nava, F., Mckenna, J., & Malvarez, G. (2004). Storm impacts on an embayed high energy coastline, Western Ireland. *Marine Geology, 210*, 261–280.

Ferreira, O., Garcia, T., Matias, A., Taborda, R., & Dias, J. (2006). An integrated method for the determination of set-back lines for coastal erosion hazards on sandy shores. *Continental Shelf Research*. https://doi.org/10.1016/j.csr.2005.12.016.

Forbes, D., Parkers, G., Manson, G., & Ketch, K. (2004). Storms and shoreline retreat in the Southern Gulf of St. Lawrence. *Marine Geology, 210*(1–4), 169–204.

Hakkou, M., Maanan, M., Belrhaba, T., El khalidi, K., El Ouai, D., & Benmohammadi, A. (2018). Multi-decadal assessment of shoreline changes using geospatial tools and automatic computation in Kenitra coast, Morocco. *Ocean and Coastal Management, 163*, 232–239.https://doi.org/10.1016/j.ocecoaman.2018.07.003

IPCC, Climate Change. (2013). *The physical science basis, contribution of working group 1 to the fifth assessment report of the intergovernmental panel on climate change* (p. 1523). Cambridge University Press.

Jana, A., Biswas, A., Maiti, S., & Bhattacharya, A. K. (2014). Shoreline changes in response to sea level rise along Digha Coast, Eastern India: An analytical approach of remote sensing, GIS and statistical techniques. *Journal of Coastal Conservation, 18*(3), 145–155. https://doi.org/10.1007/s11852-013-0297-5

Mills, J. P., Buckley, S. J., Mitchell, H. L., Clarke, P. J., & Edwards, S. J. (2005). A geomatics data integration technique for coastal change monitoring. *Earth Surface Processes and Landforms, 30*(6), 651–664.

Moore, L. (2000). Shoreline mapping techniques. *Journal of Coastal Research, 16*, 111–124.

Morton, R. A., Miller, T. L., & Moore, L. J. (2004). *National assessment of shoreline change. Part1: Historical shoreline change sand associated coastal land loss along the US Gulf of Mexico*. U.S. Geological Survey Open File Report 2004-1043.

Ranasinghe, R., Callaghan, D., & Stive, M. J. F. (2012). Estimating coastal recession due to sea level rise: Beyond the Bruun rule. *Climatic Change, 110*, 561–574. https://doi.org/10.1007/s10584-011-0107-8

Rosati, J. D., Dean, R. G., & Walton, T. L. (2013). The modified Bruun rule extended for landward transport. *Marine Geology, 340*, 71–81. https://doi.org/10.1016/j.margeo.2013.04.018

Thieler, E. R., & Danforth, W. W. (1994). Historical shoreline mapping I: Improving techniques and reducing positioning errors. *Journal of Coastal Research, 10*(3), 549–563.